The Cost of Development in China

Guangyu Hu

The Cost of Development in China

Guangyu Hu
Beijing
China

ISBN 978-981-10-4174-7   ISBN 978-981-10-4175-4 (eBook)
DOI 10.1007/978-981-10-4175-4

Jointly published with People's Publishing House

The print edition is not for sale in China Mainland. Customers from China Mainland please order the print book from: People's Publishing House.

Library of Congress Control Number: 2017936645

© Springer Nature Singapore Pte Ltd. and People's Publishing House 2017
This work is subject to copyright. All rights are reserved by the Publishers, whether the whole or part of the material is concerned, specifically the rights of translation, reprinting, reuse of illustrations, recitation, broadcasting, reproduction on microfilms or in any other physical way, and transmission or information storage and retrieval, electronic adaptation, computer software, or by similar or dissimilar methodology now known or hereafter developed.
The use of general descriptive names, registered names, trademarks, service marks, etc. in this publication does not imply, even in the absence of a specific statement, that such names are exempt from the relevant protective laws and regulations and therefore free for general use.
The Publishers, the authors and the editors are safe to assume that the advice and information in this book are believed to be true and accurate at the date of publication. Neither the Publishers nor the authors or the editors give a warranty, express or implied, with respect to the material contained herein or for any errors or omissions that may have been made. The Publishers remains neutral with regard to jurisdictional claims in published maps and institutional affiliations.

Printed on acid-free paper

This Springer imprint is published by Springer Nature
The registered company is Springer Nature Singapore Pte Ltd.
The registered company address is: 152 Beach Road, #21-01/04 Gateway East, Singapore 189721, Singapore

Contents

Part I Proposition of the Cost of Development in China

1 Learning and Practice of Scientific Outlook on Development: Considering Maximization of Net Welfare as the Scientific Political Achievement View................................. 3
   1.1 Scientific Outlook on Development: Reviewing the Cost of Development in China 3
   1.2 The Study on Development Cost Would Build the Political Achievements View with Maximization of Net Welfare 5
   1.3 Development Cost Evaluation Is Alternative Judgment of Comprehensive National Power 9

2 Scientific Development of China in the Future: Inspiration of Chinese Dream to Development Cost........................ 15
   2.1 China in the Future: High Income, High Technology, High Welfare and High Human Development 15
   2.2 Inspiration of Chinese Dream to Development Cost 17

3 Innovation Powerhouse: Build a Society of Joint Development and Common Prosperity................................. 19
   3.1 A Society of Joint Development and Common Prosperity 19
   3.2 Innovation Powerhouse: Control the Development Cost of Economy and Society................................ 21
   Appendix: Definitions 22

Part II Economic Cost of Development in China

4 Economic Growth Cost in China 27
   4.1 Connotation of Economic Growth in China 27
   4.2 Expression Forms of Economic Growth Cost in China 29

		4.2.1	Cost from Expansion of Investments	30
		4.2.2	Cost from Expansion of Domestic Demands	31
		4.2.3	Cost from Expansion of Exports	34
	4.3	Generating Causes of Economic Growth Cost in China		37
		4.3.1	Increasingly Worsening of Structural Conflicts in China	37
		4.3.2	Systematic Causes Affecting Economic Growth	37
		4.3.3	Decision-Making Mistakes in Economic Construction Process	37
		4.3.4	Marketization Degree of Production Factors Are Still not High	38
	4.4	Countermeasures to Economic Growth Cost in China		39
		4.4.1	Adjust and Optimize Industrial Structure to Promote Coordinated Economic Growth	39
		4.4.2	Continuously Deepen the Economic System Reform	39
		4.4.3	Prevent Blind Decision-Making During Economic Construction Process	41
		4.4.4	Cultivate Paid Using Mechanism of Resource-Type Products	41
		4.4.5	Transformation from "Investment Attraction" to "Investment Selection"	42
	Annexed Table			42
5	**Cost of Economic Transformation in China**			**45**
	5.1	Connotation of Economic Transformation in China and Its Cost		45
	5.2	Expression Forms of the Cost of Economic Transformation in China		47
		5.2.1	Cost of System Transformation	50
		5.2.2	Cost of Structural Transformation	51
		5.2.3	Cost of Development Model Transformation	55
		5.2.4	Cost of Globalization Transformation	57
	5.3	Forming Causes of the Cost of Economic Transformation in China		60
	5.4	Countermeasures for the Cost of Economic Transformation in China		63
	Annexed Tables			63
6	**Cost of Economic Disturbance in China**			**69**
	6.1	Overview of Economic Disturbance		69
	6.2	Expression Forms of the Costs of Economic Disturbance in China		70

		6.2.1	Inflation Cost	70
		6.2.2	Cost of Exchange Rate Fluctuation	71
		6.2.3	Risk of Economic Crisis	73
	6.3	Forming Causes of the Cost of Economic Disturbance in China		75
	6.4	Countermeasures to the Cost of Economic Disturbance in China		78
	Annexed Tables			80
7	**Cost of Economic Regulation and Control in China**			**83**
	7.1	Connotation of Economic Regulation and Control in China		83
	7.2	Expression Forms of the Costs of Economic Regulation and Control in China		83
		7.2.1	Cost of Regulation and Control Through Economic Means	85
		7.2.2	Cost of Regulation and Control Through Legal Means	89
		7.2.3	Cost of Regulation and Control Through Administrative Means	90
	7.3	Forming Causes of Chinese Economic Regulation Cost		90
	7.4	Countermeasures to Handle Chinese Economic Regulation Cost		93

Summary of Economic Cost of Development in China

Part III Political Cost of Development in China

8	**Political Reform Cost in China**			**101**
	8.1	Connotation of Political Reform in China		101
	8.2	Forms of Political Reform Cost in China		104
		8.2.1	Cost of Political Struggle	104
		8.2.2	Cost of Political Risk	111
	8.3	Forming Causes of Political Reform Cost in China		112
		8.3.1	Forming Causes of Political Struggle Cost	112
		8.3.2	Forming Causes of Political Risk Cost	115
	8.4	Countermeasures to Handle Chinese Political Reform Cost		116
		8.4.1	Prevent the Occurrence of Political Reform in Field of Ideology and Law	116
		8.4.2	Make Efforts to Reduce Risk of Turmoil and Control the Widening Gap Between Rich and Poor	118
9	**Chinese Political System Construction Cost**			**121**
	9.1	Connotation of Chinese Political System Construction		121
	9.2	Forms of Chinese Political System Construction Cost		122

		9.2.1	Corruption Cost．．．．．．．．．．．．．．．．．．．．．．．．．．．．．．．	123
		9.2.2	System Reform Cost．．．．．．．．．．．．．．．．．．．．．．．．．．	124
		9.2.3	Political System Setting Cost．．．．．．．．．．．．．．．．．．．．	129
		9.2.4	Political System Destruction Cost．．．．．．．．．．．．．．．．	130
	9.3	Forming Causes of Chinese Political System Construction Cost．．．		131
		9.3.1	Historical Causes of Chinese Corruption Problem．．．．．	131
		9.3.2	Flaws in Chinese Political System．．．．．．．．．．．．．．．．．	131
		9.3.3	Incomplete Specific System of Political Participation System Is the Most Direct Factor that Restrains Political Participation in Modern China．．．．．．．．．．．．．	132
		9.3.4	Chinese Citizens' Political Cultural Quality Is Still not High, and Their Maturity of Political Participation Is Relatively not Sufficient．．．．．．．．．．．．．．．．．．．．．．．．．	132
	9.4	Countermeasures for Chinese Political System Construction Cost．．．		134
		9.4.1	Establishment and Improvement of System to Punish and Prevent Corruption．．．．．．．．．．．．．．．．．．．．．．．．．．	134
		9.4.2	Gradually and Stably Promote Political System Reform．．．．．．．．．．．．．．．．．．．．．．．．．．．．．．．．．．．．．．．	134
		9.4.3	Make Efforts to Enhance the Public's Awareness and Ability of Political Participation．．．．．．．．．．．．．．．．．．．	135
10	Cost of Chinese Political Decision-Making System．．．．．．．．．．．．．．			137
	10.1	Connotation of Chinese Political Decision-Making System．．．．		137
	10.2	Forms of Chinese Political Decision-Making System Cost．．．．．		138
		10.2.1	Changing Cost of Political Decision-Making System．．．	139
		10.2.2	Cost of Political Decision-Making Mistake．．．．．．．．．．	140
		10.2.3	Improving Cost of Political Decision-Making System．．．	144
	10.3	Forming Causes of Chinese Political Decision-Making System Cost．．．		146
		10.3.1	Immaturity of Political Decision-Making System．．．．．	146
		10.3.2	Lack of Decision-Making Knowledge, Experience, Ability and Quality of Decision Makers (Group)．．．．．．	146
	10.4	Countermeasures for the Cost of Chinese Political Decision-Making System．．．．．．．．．．．．．．．．．．．．．．．．．．．．．．．．		148
		10.4.1	Establish and Improve Political Decision-Making Regulations and Systems, Promote the Theorization, Scientification, Democratization and Legalization of Decision Making According to Law．．．．．．．．．．．．．．．	148
		10.4.2	Well Selection, Use and Education of People to Establish Highly Efficient Decision-Making Team．．．．．	148

11	**Cost of Chinese Ruling Party Construction**........................		151
	11.1 Connotation of Chinese Ruling Party Construction............		151
		11.1.1 Forms of Chinese Ruling Party Construction Cost....	152
		11.1.2 Ideological and Cultural Construction Cost..........	152
		11.1.3 Organizational Construction Cost..................	153
		11.1.4 Working Style Construction Cost..................	154
		11.1.5 Cost of Democracy Construction Within the Party....	156
		11.1.6 Forming Cost of Political Civilization...............	156
	11.2 Forming Causes of the Cost of Chinese Ruling Party Construction..		157
		11.2.1 The Need for Chinese Communist Party to Continuously Develop and Strengthen, and to Consolidate the Ruling Foundation.................	157
		11.2.2 Input of Party Style Rectification and Loss of Expansion..	158
		11.2.3 Necessary Input to Prevent Occurrence of Crisis Within the Party	158
	11.3 Countermeasures for the Cost of Chinese Ruling Party Construction..		158
		11.3.1 Strengthening of the Ideological and Moral Construction	158
		11.3.2 Improve the Democracy Within the Party System and Reduce the Cost of Decision-Making Mistakes........	161
		11.3.3 Realize Scientific Planning and Optimized Development for the Establishment of Basic-Level Party Organization and Development of Party Member Team....................................	161
12	**Cost of Chinese Political Consultation and Crossing of Information Gap**..		163
	12.1 Connotation of Chinese Political Consultation		163
	12.2 Forms of the Cost of Chinese Political Consultation and Crossing of Information Gap		165
		12.2.1 Cost of Political Consultation Evolution	165
		12.2.2 Cost of Crossing Information Gap..................	167
	12.3 Forming Causes of the Cost of Chinese Political Consultation and Information Gap Crossing		170
		12.3.1 Historical Cause of the Evolving Cost of Chinese Political Consultation System	170
		12.3.2 Existence of Information Gap in the Process of Communication in Different Organizations or Systems...	170
	12.4 Countermeasures for the Cost of Chinese Political Consultation and Information Gap		172

		12.4.1	Improve Working Method of CPPCC and Fully Play the Role of Information Democracy	172
		12.4.2	Establish Information System and Implement Scientific Management to Reduce Barrier of Information Delivery	172

13 Cost of Chinese Democracy Construction 175
    13.1 Connotation of Chinese Democracy Construction 175
    13.2 Forms of Chinese Democracy Construction Cost 177
        13.2.1 Democracy Explaining Cost 178
        13.2.2 Democracy Development Cost 178
    13.3 Forming Causes of the Cost of Chinese Democracy Construction 180
        13.3.1 Necessary Requirement for China to Actively or Passively Make Democracy Explanation and Promote National Image in International Community 180
        13.3.2 Chinese Democracy Construction Development Faces Restriction of Historical Factors and Basic National Situation 180
    13.4 Countermeasures for the Cost of Chinese Democracy Construction 181
        13.4.1 Actively Expand Foreign Exchange and Cooperation and Demonstrate Chinese Image of Democracy to Reduce Passive Democracy Explaining Cost 181
        13.4.2 Continuously Improve China's Democracy Construction System and Enhance China's Democratic Level 183

Summary of Chinese Political Development Cost

Part IV Cost of Chinese Social Development

14 Chinese Social Livelihood Issues and Their Cost 189
    14.1 Definition and Connotation of Livelihood Issue 190
    14.2 Forms of Social Livelihood Cost 192
        14.2.1 Education Development Cost 192
        14.2.2 Employment Promotion Cost 195
        14.2.3 Social Security Lacking Cost 197
        14.2.4 Income Distribution Imbalance Cost 199
        14.2.5 Medical Service Development Cost 200
    14.3 Forming Causes of Social Livelihood Cost 202
    14.4 Countermeasures for Social Livelihood Cost 206
        14.4.1 Establish Public Fiscal System, Expand Social Livelihood Investments and Build Service-Type Government 206

		14.4.2	Improve Social Redistribution System Policies and Enhance Social Equity and Efficiency	206
		14.4.3	Enlarge Investments in Fiscal Education Expenditure and Guarantee the Balanced Distribution of Education Resources in Urban and Rural Areas.	207
		14.4.4	Improve Market Employment System and Create Fair Employment Environment	208
		14.4.5	Enlarge Government Investments in Public Health Undertakings and Establish a Civil and Integrated Medical Service System	208
15	Cost of Population Change in Chinese Society			209
	15.1	Definition and Connotation of Population Change		209
	15.2	Forms of the Cost of Population Change in Society		210
		15.2.1	Population Increase Cost	211
		15.2.2	Aging Cost	213
		15.2.3	Population Management Cost	216
	15.3	Forming Causes of the Cost of Population Change in Society		218
	15.4	Countermeasures for the Cost of Population Change in Society		218
		15.4.1	Implement Soft-Landing Policies to Undertake Birth Model Adjustment, Control the Growth of Total Population and Optimize Population Structure.	220
		15.4.2	Transform Economic Development Method, Actively Encourage Start-Ups and Employment, Expand Labor Supply, Reduce the Number of Advanced Retirees, and Relieve the Pressure of Social Security Expenditure	220
		15.4.3	Establish Social Security System and Face the Peak of Elderly Population.	221
		15.4.4	Establish the System to Protect the Related Rights and Interests of Floating Population and Accelerate the Citizenization of Migrant Workers	221
16	Chinese Social Management Cost.			223
	16.1	Definition and Connotation of Social Management.		223
	16.2	Forms of Social Management Cost		224
		16.2.1	Social Harmony Cost.	224
		16.2.2	Administrative Management Expenditure	227
	16.3	Causes of Social Management Cost		229
		16.3.1	Mistakes in the Positioning of Government Functions and Perspectives	229
		16.3.2	Backward Social Management System Causes Confusion and Failure in Social Management	229

		16.3.3	The Social Management and Operation System Lacks Complete Management Means and Organizational System.	230

16.4 Analysis of Countermeasures for Social Management Cost 231

17 Chinese Social Stability Cost 235
17.1 Definition and Connotation of Social Stability Cost 235
17.2 Forms of Social Stability Cost 236
   17.2.1 National Defense Expenditure 236
   17.2.2 Stabilization Cost 240
   17.2.3 Economic and Social Price of the Events that Destructs Social Stability....................... 240
17.3 Forming Causes of Social Stability Cost. 242
17.4 Analysis of Countermeasures for Social Stabilization Cost..... 244
   17.4.1 Increase Necessary National Defense Expenditure, Reduce Unnecessary Stabilization Expenditure and Optimize Social Stabilization Cost Structure 244
   17.4.2 Strengthen Long-Term Macroeconomic Planning, Determine Correct Stabilization Idea and Rights Idea............................... 244
   17.4.3 Emphasize Social Management Innovation, Establish Effective Expression Mechanism of People's Interest 246

18 Chinese Social Advancement Cost 247
18.1 Definition and Connotation of Social Advancement Cost...... 247
18.2 Forms of Social Advancement Cost 248
   18.2.1 Innovation Cost.............................. 248
   18.2.2 Full Information Symmetry Cost.................. 249
   18.2.3 Comprehensive Human Development Cost 255
18.3 Forming Causes of Social Advancement Cost 255
18.4 Countermeasures for Social Development Cost.............. 257

Summarization of Chinese Social Development Cost

Part V Chinese Cultural Development Cost

19 China's Cost of Civilization Inheriting and National Customs Protection.. 265
19.1 Definition of Civilization and Recall of Chinese History of Civilization....................................... 265
   19.1.1 Definition of Civilization and Connotation of Its Inheriting Cost............................... 265
   19.1.2 Recall of the Process of Chinese Civilization 266
19.2 Forms of the Cost of Civilization Inheriting................ 268

		19.2.1	Civilization Conflict and Peace Maintenance Cost.....	269
		19.2.2	Civilization Eliminating and Evolving Cost..........	270
		19.2.3	Cost of Civilization Inheriting and Protection	275
	19.3	Connotation, Definition of National Customs Protection and Forms of Cost..		279
		19.3.1	Ethnical Conflict Handling Cost	280
		19.3.2	National Fusion and National Customs Protection Cost..	283
		19.3.3	Religion and Faith Protection Cost	284
	19.4	Forming Causes of the Cost of Civilization Inheriting and National Customs Protection............................		286
	19.5	Countermeasures for the Cost of Civilization Inheriting and National Customs Protection............................		289
20	Cost of Chinese Ideological Evolution and Modern Media System Construction			291
	20.1	Connotation and Forms of the Cost of Chinese Modern Ideological Evolution		291
		20.1.1	Cost of Ideological Evolution Since the Establishment of New China	291
		20.1.2	Regular Cost of Current Chinese Ideological Construction	293
		20.1.3	Cost of Risks in Current Chinese Ideological Construction	298
		20.1.4	Cost of International Ideological Contradictions and Conflicts	298
	20.2	Connotation and Forms of the Construction Cost of Chinese Modern Media System		301
		20.2.1	General Management Cost of the Media System with Chinese Characteristics	303
		20.2.2	Reform Cost of Modern Media System..............	305
	20.3	Forming Causes of the Cost of Ideological Change and the Cost of Modern Media System Construction		310
	20.4	Countermeasures for the Cost of Ideological Change and the Cost of Modern Media System Construction		312
21	Cost of Chinese Soft Power Construction and Response to International Cultural Invasion			315
	21.1	Connotation of the Cost of Soft Power Construction		315
	21.2	Forms of the Cost of Soft Power Construction on Different Levels and the Cost of Response to Foreign Cultural Invasion		317
		21.2.1	National Level: Cost of Fighting for Speaking Right in International Community......................	318

		21.2.2	Regional Level: Cost of Regional Soft Power Construction	320
		21.2.3	Organizational (Enterprise) Level: Cost of Modern Organization (Enterprise) Establishment and Soft Power Enhancement	323
		21.2.4	Citizen Level: Cost of Enhancing Scientific Quality and Cultural Quality	324
	21.3	Connotation of Social Civilization Cost in Modernization Construction Process and Forms of Cost		326
		21.3.1	Cost of Sliding Collective Morality in Society	328
		21.3.2	Cost of Change in Social Relations	330
		21.3.3	Cost of Contradictions and Conflicts Caused from Formation of New Social Groups	331
	21.4	Forming Causes of the Cost of Soft Power Construction and the Cost of Response to Foreign Cultural Invasion		332
	21.5	Countermeasures for the Cost of Soft Power Construction and the Cost of Response to Foreign Cultural Invasion		333

Summary of the Cost of Chinese Cultural Development

Part VI Cost of Opening to the Outside World and Development

22	**China's Opening Development Thoughts and Their Cost**			341
	22.1	Overview of Opening Development Thoughts and Their Cost		341
	22.2	Forms of Opening Development Thoughts and Their Cost		342
		22.2.1	Price of No Opening-up and Cost of Thinking Game	342
		22.2.2	Forming Cost of Comprehensive Opening-up	348
	22.3	Forming Causes of Opening Development Thoughts and Their Cost		349
		22.3.1	Long-term No Opening-up Due to Decision-Making Mistakes of Leader	349
		22.3.2	Necessary Price for Exploration and Development of New System	349
		22.3.3	Associated Effect of Globalization Trend	350
	22.4	Countermeasures for Opening Development Thoughts and Their Cost		350
		22.4.1	Implement Democratic Centralism Through System Reform	350
		22.4.2	Correctly Treat and Handle the Necessary Cost of the Formation of Reform and Opening-up Idea	351
		22.4.3	Countermeasures for Associated Risks of Globalization	352

23	**China's Cost of International Exchange and Consensus**		355
	23.1	Overview and Forms of the Cost of International Exchange and Consensus	355
		23.1.1 Cost of International Exchange and Participation	356
		23.1.2 Cost of Organizing International Exchange	358
		23.1.3 Cost of Fulfilling International Consensus	360
	23.2	Forming Causes of the Cost of International Exchange and Consensus	363
	23.3	Countermeasures for the Cost of International Exchange and Consensus	364
24	**China's Cost of Investment and Construction of Investment Environment**		365
	24.1	Overview and Forms of the Cost of Investment and Construction of Investment Environment	365
		24.1.1 Cost of Providing Investment Environment	366
		24.1.2 Cost of China's Foreign Investment	368
	24.2	Forming Causes of the Cost of Investment and Construction of Investment Environment	372
		24.2.1 Forming Causes of the Cost of Providing Investment Environment	372
		24.2.2 Forming Causes of the Cost of Foreign Investment	373
	24.3	Countermeasures for the Cost of Investment and Construction of Investment Environment	374
		24.3.1 Countermeasures for the Cost of Providing Investment Environment	374
		24.3.2 Countermeasures for the Cost of Foreign Investment	376
25	**Cost of International Trade Development**		379
	25.1	Overview of the Cost of International Trade Development	379
	25.2	Forms of the Cost of International Trade Development	380
		25.2.1 Tax Rebate Burden	380
		25.2.2 Opportunity Cost and Risk of High Foreign Exchange Reserve	382
		25.2.3 Cost of Handling Trade Frictions	385
	25.3	Forming Causes of the Cost of International Trade Development	386
		25.3.1 Conflict Between "Rebating All Rebates" and Finance Support Capacity	387
		25.3.2 Necessity of Foreign Exchange Reserve for Economic Development	388
		25.3.3 Cause of Intensifying Trade Frictions	389
	25.4	Countermeasures for International Trade Cost	390

		25.4.1	Reform the Export Tax Rebate System	390
		25.4.2	Improve Foreign Exchange Reserve Management	390
		25.4.3	Actively and Strategically Handle Trade Frictions	390
26	**Construction Cost of China's Three-Dimensional Transportation**			393
	26.1	Overview of the Construction Cost of Three-Dimensional Transportation		394
	26.2	Forms of the Construction Cost of Three-Dimensional Transportation		394
		26.2.1	Input of Transportation Infrastructure Construction	395
		26.2.2	Opportunity Cost of Transportation Infrastructure Construction	396
		26.2.3	Loss of Management Efficiency	400
	26.3	Forming Causes of the Construction Cost of Three-Dimensional Transportation		401
		26.3.1	Promote the Demand of National Economic Development	401
		26.3.2	Lack of General Vision in Planning	402
		26.3.3	Incomplete Management System	403
	26.4	Countermeasures for the Construction Cost of Three-Dimensional Transportation		405
		26.4.1	Improve the Investment and Financing System of the Construction of Chinese Transportation Infrastructure	405
		26.4.2	Realize the Comprehensive Management of Traffic & Transportation Industry	405
		26.4.3	Strengthen the Coordinated Development Planning and Layout	406
27	**The Cost of China's Creditability Construction and Handling of Threats**			409
	27.1	Overview and Forms of the Cost of China's Creditability Construction and Handling of Threats		409
		27.1.1	Maintaining Cost of China's Territory and Sovereignty Integrity	409
		27.1.2	Cost of National Unity	412
		27.1.3	Cost of Foreign Aids	412
		27.1.4	Cost of Bearing International Responsibilities	415
		27.1.5	Communication Cost of Major Power Image	417
	27.2	Forming Causes of the Cost of China's Creditability Construction and Handling of Threats		419
		27.2.1	The Demand for China to Enhance Comprehensive National Strength and Build National Image	419
		27.2.2	The Ulterior "Chinese Threat Theory"	419

	27.3	Countermeasures for the Cost of China's Creditability Construction and Handling of Threats	420
		27.3.1 Enhance Communication Ability of National Image ...	420
		27.3.2 Enhance China's Comprehensive National Strength ...	421

Summary of the Cost of Foreign Opening-up Development

Part VII Cost of China's Natural Development

28 Cost of China's Resource and Energy 427
    28.1 Overview of the Cost of China's Resource and Energy 427
    28.2 Forms of the Cost of Resources and Energy................ 431
            28.2.1 Farmland Cost................................ 431
            28.2.2 Cost of Forest Land........................... 432
            28.2.3 Cost of Mineral Resource 433
            28.2.4 Cost of Coal 433
            28.2.5 Cost of Oil 434
            28.2.6 Cost of Natural Gas........................... 435
    28.3 Forming Causes of the Cost of Resource and Energy......... 436
    28.4 Countermeasures for the Cost of Resource and Energy 440
            28.4.1 Countermeasures for the Control of Resource Cost.... 440
            28.4.2 Countermeasures for the Control of Energy Cost 441
    Appendix 442

29 Cost of China's Ecological Environment 453
    29.1 Overview of the Cost of the China's Ecological Environment ... 453
    29.2 Forms of Environmental Cost........................... 454
            29.2.1 Cost of Water Pollution........................ 456
            29.2.2 Emitting Cost Sulfur Dioxide and Dust............. 457
            29.2.3 Discharging Cost of Industrial Dust 458
            29.2.4 Cost of Solid Waste 459
    29.3 Forming Causes of the Cost of Ecological Environment....... 460
    29.4 Countermeasures for the Cost of Ecological Environment 463
    Appendix 465

30 Cost of Disasters in China 473
    30.1 Overview of the Cost of Disasters in China 473
    30.2 Forms of the Cost of Disasters in China................... 476
            30.2.1 Cost of Geological Disaster..................... 476
            30.2.2 Cost of Forest Fire 477
            30.2.3 Cost of Oceanic Disaster....................... 478
    30.3 Forming Causes of the Cost of Disaster in China............ 479
    30.4 Countermeasures for the Control of the Cost of Disaster in China 482
    Appendix 486

31	China's Cost of Handling Climate Change	489
	31.1 Overview of China's Cost of Handling Climate Change	489
	31.2 Forms of China's Cost in Handling Climate Change	491
	31.2.1 Cost of Carbon Dioxide Emission	492
	31.2.2 Cost of Methane Emission	492
	31.2.3 Emission Cost of Nitrous Oxide	493
	31.3 Forming Causes of the Cost of Handling Climate Change	494
	31.4 Countermeasures for the Cost of Handling Climate Change	498
	31.4.1 Strengthen the Ecological Protection Countermeasure in Water Source Management	500
	31.4.2 Strengthen Environmental Construction in Coastal Regions	501
	31.4.3 Strengthen International Cooperation to Handle Global Climate Change	501
	31.4.4 Reduce the Deforestation	501
	31.4.5 Innovate the Domestic Carbon Trade Mechanism	502
	Appendix	503
32	Cost of Harmony Between Human and Nature	507
	32.1 Overview of the Cost of Harmony Between Human and Nature	507
	32.2 Forms of the Cost of Harmony Between Human and Nature	510
	32.2.1 Cost of Health Loss Resulted from Environmental Pollution	511
	32.2.2 Transformation from "Forest Deficit" to "Forest Surplus"	511
	32.3 Cost of Maintaining Harmony Between Human and Nature	513
	32.4 Countermeasures of the Cost of Harmony Between Human and Nature	514

Summary of China's Cost of Nature Development

Part VIII Inspiration of China's Development Cost Theory for China's Development

33	Indicator System Foundation of Development Cost	521
34	Promote the Enhancement of Total Factor Productivity and Reduce Development Cost	525
35	Benchmarking of Scientific Achievement View	527

References ... 529

About the Author

Dr. Guangyu Hu professor, researcher and doctoral supervisor, incumbently holds following positions: Deputy Headman for the leading group of Shanghai High People's Court's Thinking Tank, Deputy Director for the Development and Research Center of Shanghai High People's Court, Deputy Chairman and Secretary-general for Shanghai Judicial Thinking Tank Institute, Dean for Rule of Law National Situation Institute of East China University of Political Science and Law. The National Energy Development Institute of North China Electric Power University has been established in his charge as deputy chief dean. During his ten-year stay in Tsinghua University as a teacher, he participated in the preparation and establishment effort of National Situation Institute and was invited by Center for China Study of Chinese Academy of Sciences/Tsinghua University as deputy director. He put on field practices in Hengshui City of Hebei Province, the Enterprise Bureau of National Development Bank and Shanghai No. 1 Intermediate People's Court as deputy mayor, deputy chief and assistant to the dean, respectively.

Once a visiting scholar of Cambridge University, Prof. Hu has long been engaging in researches of national situation, policies and strategies. He has published more than 30 works that can be categorized into national situation series, theoretical research series and internationalization series.

The national situation series include: *The Cultural Construction of the Communist Party of China; China Maritime: New Positioning of Economic Development Mode; On Development Costs of China; Governance in China: Chinese Experiences; Governance in China: Rule of Law; The Relations between Development Finance and National Development; University Thinking Tank; The Revolution of Energy System; China: Gender and Development; China's Cause of Astronomy; Comparison of Corporate Governance between China and Foreign Countries; Going-out Strategy of Chinese Enterprises.*

The theoretical research series include: *The Political and Economic History of China (1946–1956)/(1957–1965)/(1966–1976); Foundation of Strategic Qualitative Research: Implementation and Control; Foundation of Strategic Quantitative Research: Prediction and Decision-making.*

Internationalization series include: *China in the World Economy: The Domestic Policy Challenges*, translation works, originally by OECD; *Governance in China*, translation works, originally by OECD; *Achievement Evaluation of IFI Assistance Loans to China; Aid and Development*; *World Development Report* from 2005 to 2017, translation works, originally by World Bank; *Cities in a Globalizing World: Governance, Performance and Sustainability*, translation work, originally by World Bank Institute.

Introduction

The Cost of Development in China focuses on the future of China and its sustainable development and summaries the connotations, forms, causes, countermeasures, and related rules of the main costs generated during a country's period of development, so as to provide theoretical reference and decision-making and consulting tools for decision-making institutions and decision makers in their scientific governance and management.

This book integrates the national situation and development characteristics of China; with country as unit, it uses case studies to propose the concept of cost theory and the theoretical system of national development cost. The main contents focus on the goals of innovation in nation building, common development, common prosperity, and enhancement of people's net welfare, make systematic conclusion and summarization of the fields in national development, including economic development cost, political development cost, social development cost, cultural development cost, foreign opening-up development cost, and nature development cost, preliminarily establish the indicator system foundation of national development cost for the promotion of the enhancement of full factor productivity and reduction of development cost, and provide theoretical basis for the implementation of the scientific political achievement view.

Part I
Proposition of the Cost of Development in China

Chapter 1
Learning and Practice of Scientific Outlook on Development: Considering Maximization of Net Welfare as the Scientific Political Achievement View

1.1 Scientific Outlook on Development: Reviewing the Cost of Development in China

Outlook on development means the basic view and fundamental standpoint that people have regarding the development of economy, society and human being as a whole. Outlook on development only appeared in modern world history, which reflects that development is already "Self-conscious" and has transcended "Spontaneity". Modern times was an era dominated by rationality, rationalism found its way across the world with optimistic mood, as a result, the measuring standard of social development was defined as—advancement of rationality. Rationality is generally classified into two types, subjective rationality (or value rationality) and objective rationality (or scientific rationality or instrumental rationality). What prevailed in modern times was instrumental rationality. Value rationality has been gradually declining, the development of goal of human survival, significance and value of life, social responsibility, human destiny and other ultimate faith has almost entered dormancy stage.

Such instrumental rationality even considers human as a means and instrument, only economic growth and accumulation of wealth are the final goals, such economic growth and social development are "Material-centered". People are considered as what Adam Smith described, "Economic Persons" who are selfish and having endless pursuit to material interests, human nature is dissimilated.

In modern history, several major historical events, such as Religious Reform, the Enlightenment Movement, and the French Revolution, enabled the awakening of people's subject consciousness and strengthening of subjective ability of practice, "Anthropocentrism" replaced "Nature Centrism", the "Disenchantment" process of the nature is also underway. Habermas believed that the determination of people's subjective position was a long process, the key step of which was the Religious

Reform, the Enlightenment Movement, and the French Revolution.[1] Dallmayr indicated that the purpose of Anthropocentrism is to "Seek supremacy of human being".[2] Lu Feng indicated that the value of Anthropocentrism believes that human rationality has infinite epistemological beliefs.[3] Such outlook on development would directly direct to the dark development road of large-scale consumption of natural resources, environmental pollution and ecological destruction, while neither developed nations nor developing nations could avoid walking such a development path, and then enter a stage of new outlook on development and development path of reflection and re-exploration. As for dark development path, to reflect it most directly, we call it having very high cost of development, such cost is not a micro cost, but a macro concept, to be more exactly, it's the price of development, including the price of environment, ecology, resource, etc. The traditional outlook on development is a outlook on development that ignores price and cost.

China is the biggest developing nation in the world, when choosing development model, China also interpreted development as economic growth for a long time. Economic growth and conflict accumulation proceed in parallel. Such conflict includes conflict between human and nature, conflict between human and human (hidden by the relation between object and object), conflict between human and society as well as structural conflict in the fields of economy, politics, society, culture, etc.

The proposition of Well-off Society was to build a society with high material level and comprehensive advancement in economy, politics, culture, etc., and was a strategy proposed in order to overcome the problems in development. Marxism theories regarding comprehensive development of human being indicated that the various abilities, social relationships, personalities and needs of human being need to develop comprehensively, so the society that bears such "Comprehensive Human" must be comprehensively developed in all aspects. The 3rd Plenary Session of the 16th Central Committee proposed the Scientific Outlook on Development, which inherited and developed this theory and emphasized the comprehensive development of economy, society and people, so such outlook on development must be people-based, and such outlook on development is comprehensive, coordinated and sustainable. The so-called "Coordination" is to coordinate the relationship between urban and rural areas, the relationship between different regions, the relationship between economy and society, the relationship between human and nature, and the relationship between domestic development and opening to outside world. Only the comprehensive and coordinated outlook on development is the sustainable outlook on development, meanwhile, sustainability is also

---

[1]Yang Liangcai: *From Traditional Development Outlook to Scientific Development Outlook: Modern Transition of Development Outlook,* Journal of Yan'an University (Social Science Edition), 5th Issue in 2004.
[2]Dallmayr, *Twilight of Subjectivity*, Shanghai People's Publishing House, 1992 Edition.
[3]Lu Feng: *After Enlightenment,* Hunan University Press, 2003 Edition.

reflected in the five subsystems, resource, environment, economy, human and society.

The Scientific Outlook on Development can be penetrated into each field of economy and society, but there is also a symbiosis, meaning the places with development must have development cost. Scientific and comprehensive cognition of development cost could enable us to scientifically conclude the net welfare of development, so as to benefit forming the scientific political achievements view.

The author believes that The Cost of Development in China would be another academic perspective to study the scientific outlook on development in China. The economics of sustainable development has a whole set of complete system, including evaluation and index of sustainable development, population growth and sustainable development, natural resources and sustainable development, ecological environment and sustainable development, sustainable development of agriculture, sustainable development of industries, regional sustainable development, technological advancement and sustainable development, human resources and sustainable development, price regulatory mechanism of sustainable development, property system of sustainable development, cultural construction of sustainable development, economic development cost and other related contents. Scientific Outlook on Development is pursuit on higher level based on sustainable development, seeking comprehensiveness and coordination, the requirements on Scientific Outlook on Development might be more refined, for more detailed analysis, please refer to Charts 1.1 and 1.2.

1.2 The Study on Development Cost Would Build the Political Achievements View with Maximization of Net Welfare

Net welfare means the residual welfare of the entire economic welfare brought by economic development deducting the part consumed by compensation of the negative effect generated from economic growth.

Currently, nations in the world usually use gross domestic product (GDP) and gross national product (GNP) as economic measurement scale. In 1992, Samuelsson proposed "Net Economic Welfare" (NEW) to correct GDP or GNP, because GDP or GNP would ignore the factors such as "Dirty air", "Crowded population", "Expenses paid to deal with pollutions", "Urban lives remain unchanged", which factors are reflection of severe environmental pollution and highly deteriorated ecology.[4] In western economics, economic net welfare is also referred to as "People's Net Welfare", which emphasized that economic income

---

[4]Zhang Lizhi: *Discussion on "Economic Net Welfare Growth": Redefining of "Fourth Industry" and Economic Growth,* Qinghai Social Science, 5th Issue in 2004.

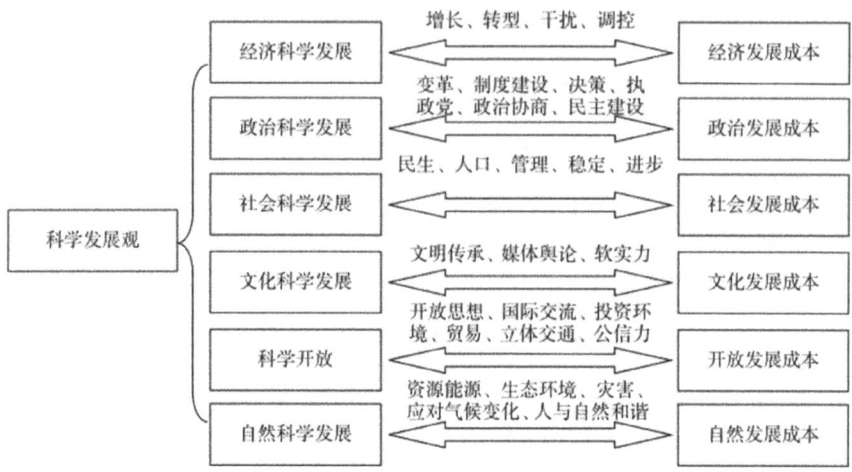

科学发展观	Scientific development outlook	经济科学发展	Scientific economic development
政治科学发展	Scientific political development	社会科学发展	Scientific social development
文化科学发展	Scientific cultural development	科学开放	Scientific opening-up
自然科学发展	Scientific natural development	增长、转型、干扰、调控	Growth, transformation, disturbance and regulation
变革、制度建设、决策、执政党、政治协商、民主建设	Reform, system construction, decision making, ruling party, political negotiation, democratic construction	民生、人口、管理、稳定、进步	Livelihood, population, management, stability, advancement
文明传承、媒体舆论、软实力	Transfer of civilization, public opinion on media, soft power	开放思想、国际交流、投资环境、贸易、立体交通、公信力	Open mind, international exchange, investment environment, trade, three-dimension transportation, creditability

Chart 1.1 Scientific development outlook and development cost

1.2 The Study on Development Cost Would Build the Political Achievements … 7

资源能源、生态环境、灾害、应对气候变化、人与自然和谐	Resource and energy, ecological environment, disaster, response to climate change, harmony between human and nature	经济发展成本	Economic development cost
政治发展成本	Political development cost	社会发展成本	Social development cost
文化发展成本	Cultural development cost	开放发展成本	Open development cost
自然发展成本	Nature development cost		

Chart 1.1 (continued)

received needs to deduct the cost of economic and social development, such cost may also include inflation and other factors.

The traditional political achievements view of China is GDP-oriented, the pursuit to high-speed GDP growth in short term would definitely cause the government to focus on short-term and immediate interests. The blind pursuit to high investment index, high export index and expanding domestic demands, meaning the traditional "Troika"—"Investment, Consumption, Export", played active role at certain stage. But the problems caused are also relatively severe, such as pursuit to quantity than quality and benefit, lack of motive force for technological innovations; the preference to big investments benefited state-owned enterprises and manufacturing industry, but the medium and small-sized private enterprises and service industry that are most capable of absorbing employments had poor development; the vicious competition of foreign trades, exports and investment attraction resulted in worsening export benefits and outflow of interests.

Political achievements, by definition, mean the achievements, performances and results realized in political office. Administering for people, the political achievements view of creating high welfare for people shall "always consider benefiting the country and people as the basis, strengthening the country and benefitting the people as the standard, and enriching the country and strengthening the people as the benchmark".[5] A scientific political achievements view should be having the ability and level in promoting scientific development, and promoting social harmony. Therefore, comprehensive and key political achievements evaluation indicators system is continuously developing.

---

[5]Li Jingjing: *Correct the Concept of Political Achievements to Accelerate Economic Development Ways,* 12th Issue of China's Foreign Trade in 2011.

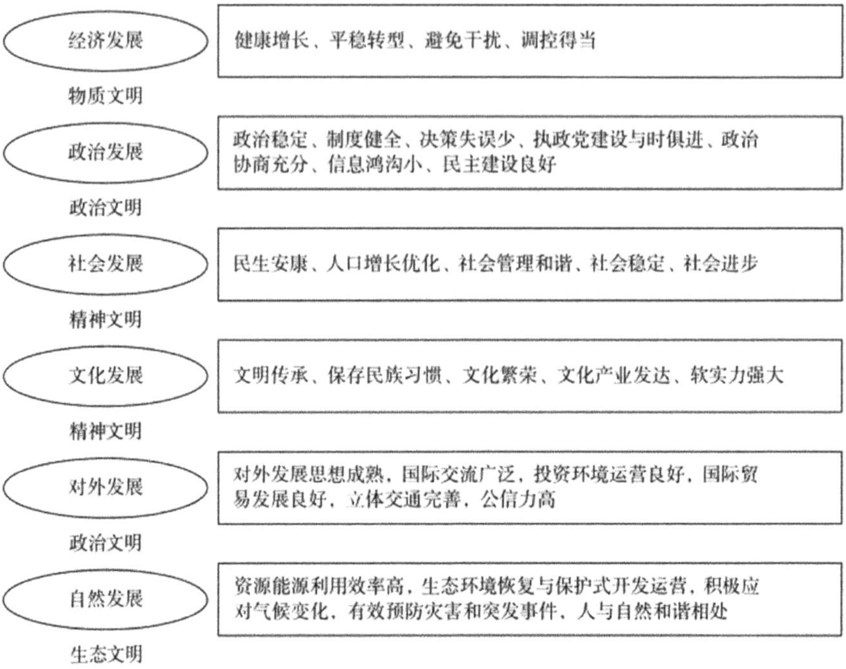

Chart 1.2 (diagram)

- 经济发展 — 物质文明 : 健康增长、平稳转型、避免干扰、调控得当
- 政治发展 — 政治文明 : 政治稳定、制度健全、决策失误少、执政党建设与时俱进、政治协商充分、信息鸿沟小、民主建设良好
- 社会发展 — 精神文明 : 民生安康、人口增长优化、社会管理和谐、社会稳定、社会进步
- 文化发展 — 精神文明 : 文明传承、保存民族习惯、文化繁荣、文化产业发达、软实力强大
- 对外发展 — 政治文明 : 对外发展思想成熟、国际交流广泛、投资环境运营良好、国际贸易发展良好、立体交通完善、公信力高
- 自然发展 — 生态文明 : 资源能源利用效率高、生态环境恢复与保护式开发运营、积极应对气候变化、有效预防灾害和突发事件、人与自然和谐相处

经济发展	Economic development	物质文明	Material civilization
政治发展	Political development	政治文明	Political civilization
社会发展	Social development	精神文明	Spiritual civilization
文化发展	Cultural development	精神文明	Spiritual civilization
对外发展	Foreign development	政治文明	Political civilization
自然发展	Nature development	生态文明	Ecological civilization
健康增长、平稳转型、避免干扰、调控得当	Healthy growth, stable transformation, avoiding disturbance, proper regulation	政治稳定、制度健全、决策失误少、执政党建设与时俱进、政治协商充分、信息鸿沟小、民主建设良好	Stable politics, complete system, construction of ruling party up to date, full political negotiation, small information gap, excellent democracy construction

Chart 1.2 Specific forms of scientific development outlook

The author believes that the transformation of political achievements view from GDP maximization to net welfare maximization is a trend. To study net welfare, firstly we need to study the cost of economic and social development, during the learning and practice process of Scientific Outlook on Development, we need to

民生安康、人口增长优化、社会管理和谐、社会稳定、社会进步	Peaceful and prosperous livelihood, optimized population growth, harmonious social management, social stability, social advancement	文明传承、保存民族习惯、文化繁荣、文化产业发达、软实力强大	Heritage of civilization, preservation of ethnic custom, cultural prosperity, developed cultural industry, strong soft power
对外发展思想成熟，国际交流广泛、投资环境运营良好、国际贸易发展良好、立体交通完善，公信力高	Mature foreign development thinking, broad international exchange, excellent operation of investment environment, complete three-dimensional transportation, high creditability	资源能源利用效率高，生态环境恢复与保护式开发运营，积极应对气候变化，有效预防灾害和突发事件，人与自然和谐相处	High utilizing efficiency of resource and energy, recovery of biological environment and protective development and operation, active response to climate change, effective prevention of disasters and emergency events, and harmonious coexistence between human and nature

Chart 1.2 (continued)

correctly recognize, study, manage, control and adjust cost, which is a necessary path of scientific development.

Firstly, development cost is a cost system, which integrates various types and forms of costs into a unified framework for discussion. It is not only comprehensive, but also having key characteristics, and focusing on those cost factors that substantially affect the development of economy, politics, society, culture, ecology, etc. (Chart 1.3).

1.3 Development Cost Evaluation Is Alternative Judgment of Comprehensive National Power

When we study The Cost of Development in China, domestically, we advocate the political achievements view with maximization of net welfare, internationally, we adopt alternative perspective and judgment to form the evaluation framework of comprehensive national power, which may lead to such a result that those superpowers in traditional sense may become hard to get rid of confirmed habits and having unsustainable development; while for those happy nations that still maintain

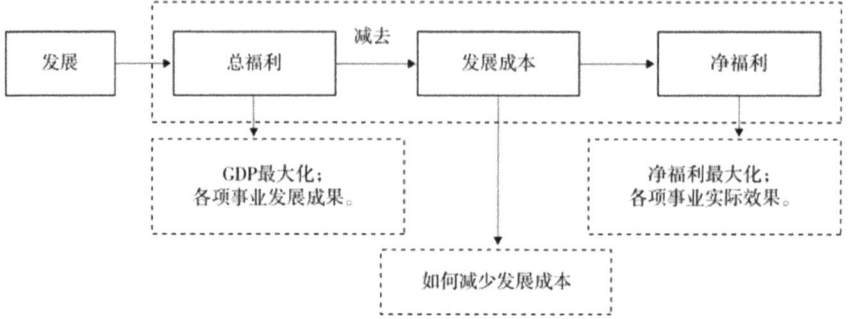

发展	Development	总福利	Total welfare
减去	Minus	发展成本	Development cost
净福利	Net welfare	GDP 最大化；各项事业发展成果	Optimized GDP; development results of various social courses
净福利最大化；各项事业实际效果。	Optimized net welfare; actual effect of various social causes.	如何减少发展成本	How to reduce development cost

Chart 1.3 Research logic

green mountains, clear waters and blue skies, although they are not strong in terms of GDP, technological strength and soft power, their developments are sustainable. Study of development cost would reveal an alternative perspective to judging comprehensive national power, which is what we are looking forward to.

The core of Deng Xiaoping's thinking of strengthening comprehensive national power was to strengthen economic construction and comprehensively enhance the development level of social productivity,[6] he used to say: "We are all discussing and handling issues with our own national interests as the highest principle."[7] Therefore, the foundation for competition of comprehensive national power is technological innovation and talents cultivation, China's fundamental interests lie in the enhancement of comprehensive national power, and building of a rich, strong, democratic, civilized, socialistic and modernized nation.

Comprehensive national power generally means the composite forces of physical force, spiritual force and synergy force that a sovereign nation needs for survival

---

[6]Gu Dong: *In-depth Historical Significance of Deng Xiaoping's Thinking of Strengthening Comprehensive National Power,* Journal of Nantong Textile Vocational Technology College (Comprehensive Edition), 3rd Issue in 2007.

[7]*Selected Works of Deng Xiaoping,* Volume 3, People's Publishing House, 1993 Edition, Page 330.

Table 1.1 Four major academic schools of research on comprehensive national strength in modern times

	School	Representative	Main ideas
1	Staatenkunde	Conring from Germany	Four factors determining national matters: land and population of a country (material factor), state system and form of government (form factor), financial and military force of a country (dynamic factor), purpose of country building (purpose factor)
2	Political Arithmatic School	William Petty	Comparison and analysis of realistic power and potential power, pioneering comparison research
3	Mercantilist Financial School	Adam Smith	There is critical correlation between fiscal and financial strength and power politics, fiscal and credit strength is scale of the relative durability of a country
4	Geopolitical School	Alfred Mahan; Mackinder	Mahan emphasized sea power national strength theory, while Mackinder emphasized on land power national strength theory

Source Shi Zuhui: *Research of Comprehensive National Strength Theories in Foreign Countries*, Foreign Economics and Management, 1st Issue in 2000. Concluded from summarization by the author

and development, it is a standard or scale to measure a nation.[8] The four academic schools in modern times to study comprehensive national power are as follows in Table 1.1. Modern times and contemporary era are the golden age of comprehensive national power studies, the representative figures and thinking of which are as follows in Table 1.2.

To summarize, the author believes that the influencing factors of comprehensive national power include economic factor (industry, agriculture, commerce, finance, technology, public finance, etc.); state political power factor (politics, policy, sovereignty, cohesive force, etc.); social factor (population, popular feelings, etc.); cultural factor (ethnicity, religion, thinking, morality, etc.); opening factor (infrastructure, internationalization, diplomacy, etc.); natural factor (national land, energy, resource, geography, etc.). Development cost will become an alternative perspective to study comprehensive national power. Any nation has historical cost that has already been generated in history or ongoing or happening development cost. When we undertake international comparison study, form consensus in global governance, refer development models, evaluate development results and development achievements, the study of development cost will not only open an alternative perspective of the Scientific Outlook on Development in China, but also integrate the framework of historical study and comparison study, provide the vein-like thinking of "Past—Present—Future" for the study of comprehensive national

---

[8]Shi Zuhui: *Research of Foreign Comprehensive National Strength Theory,* Foreign Economy and Management, 1st Issue in 2000.

Table 1.2 Summarization of modern contemporary researches on comprehensive national strength

	Period	Representative	Main ideas
1	Versailles—Washington system after WWI	[UK] Liddell Hart	The big strategy is to comprehensively use various powers of a country to realize the political goals regulated by national policies, including various powers such as national politics and policy, military, economic, diplomatic, national quality, ideological, moral and powers and various spiritual factors
2	Yalta system after WWII	[US] Hans Morgan	National strength include: the first is composition of physical factors (geographic and natural resources); the second is composition of human factors (population, national character, popular feelings, diplomatic quality and government quality); the third is composition of combining factors between human and materials (industrial strength and military reserve)
3	Deterrence strategy during the Cold War	[Germany] William Fox	Formula of strong power: $Mt = [(Ms)t + (Me)t] (1/2)$, $Ms = Pa* Sb$, $Me = Pa* Eb$; in which Mt means the comprehensive national strength indicator during period t, (Ms)t and (Me)t respectively mean the steel indicator and energy indicator during period t, Pa, Sb and Eb respectively mean the population, steel output and energy production
4	Deterrence strategy during the Cold War	[US] R.S. Klein	Power evaluation of "National Strength Formula": $PN = (C + E+M) * (S + W)$; PN means comprehensive national strength indicator; C means the basic entity composed by population and territory; E means economic strength (GDP, energy, key non-fuel minerals, industrial production capacity, food production and foreign trade); M means military strength; S means strategic goal; and W means will to pursue national strategy
5	Deterrence strategy during the Cold War	[Japan] Fukushima Kangren	Modified "National Strength Formula": $P = (C + E+M) * (G + D)$; C = Population + Territory + Natural Resource; E = (GNP + Per Capita GNP + Actual Growth Rate Of GNP) + (Sum Of Agricultural,

(continued)

1.3 Development Cost Evaluation Is Alternative ...

Table 1.2 (continued)

	Period	Representative	Main ideas
			Industrial And Commercial Strength); M = Military Strength; G = Political Capacity; D = National Diplomatic capacity
6	Comprehensive competitive strategy during diversification period	[Japan] Japan Research Institute	Japan's "comprehensive three factors" national strength measurement theory; the comprehensive national strength evaluation system composed of three layers, including international contribution capacity (basic, economic, financial, technological, policy, fiscal, foreign activity strength and other strengths), survival ability (population, geography, resource, economic strength, defense capacity, national will, friendly allies) and enforcement capacity (military strength, strategic material and technology, economic strength, diplomatic capacity)
7	Comprehensive competitive strategy during diversification period	[US] Josef S. Ney	National strength evaluation system of "soft power and hard power"; "hard power" includes basic materials, military power, economic power and technological power. "soft power" includes national coherence, degree of wide acceptance culture and degree of participation in international institutions
8	Comprehensive competitive strategy during diversification period		Lausanne "international competitiveness" research model and "four in one" national strength theory. "international competitiveness" research model: 8 modules including economic strength, international degree, role of government, financial environment, infrastructure, degree of management, science and technology, population structure and quality. Four in one means organic integration of economy, technology, military and cultural value

Source Shi Zuhui: *Research of Comprehensive National Strength Theories in Foreign Countries*, Foreign Economics and Management, 1st Issue in 2000. Concluded from summarization by the author

power. It can be said that whichever nation having the lowest development cost in the future would be more likely to become the most advanced developed nation. This is the domestic and foreign significance of development cost study.

Chapter 2
Scientific Development of China in the Future: Inspiration of Chinese Dream to Development Cost

2.1 China in the Future: High Income, High Technology, High Welfare and High Human Development[1]

As for realization of Scientific Outlook on Development, China preliminarily included Scientific Outlook on Development during the "Eleventh Five Year Plan" period and basically included Scientific Outlook on Development during the "Twelfth Five Year Plan" period, the "Thirteenth Five Year Plan" period is going to be the period when China is comprehensively included in the track of Scientific Outlook on Development, which means, by 2020, China will realize the targets of socialist harmonious society and comprehensive completion of well-off society, and complete the mission at the stage.

Based on the "Three-step" Strategy proposed by Comrade Deng Xiaoping, the third step would be that by the Mid 21st Century, China's per capita gross domestic product reaches the level of that in moderately developed countries, people are relatively well-off in their lives, and China basically realizes modernization. Compared with the "First Two Steps", the third step was only a general concept. Comrade Jiang Zemin further proposed the "New Three Steps", which indicated that by 2010, China shall double the gross domestic product of that in 2000, people are more well-off in their lives, and China has formed a relatively complete socialist market economy system; during the period from 2010 to 2020, China's gross domestic product in 2020 shall quadruple that in 2010; during the period from 2020 to 2050, after hard works during the gold 30 years, China shall basically realize modernization, such target of "Modernization" was positioned to be a rich, strong, democratic, civilized, socialist and modernized nation. The Chinese dream of 2050

---

[1]"High Human Development" means that the HDI ranges between 0.8 and 0.9. HDI (Human Development Index) is the standard that the UNDP began releasing since 1990 to measure the social and economic development degree of different countries, and based on which classify different countries into four groups, extremely high, high, medium and low.

is a dream about modernization and also a dream about supporting the world of universal harmony.

The general standard of developed country includes per capita GDP and social development level. By the standard around 1995, a country with per capital GDP of over USD 8000 (converted at nominal exchange rate) plus certain degree of social development level can be basically defined as developed country. By the standard of 2005, the figure should be increased to around USD 10,000. The developed countries in 2010 are as follows in Column 2-1. The standard that the UN currently uses to measure developed country is that the countries with human develop index no less than 0.9 shall be considered as developed countries. The author summarizes the descriptions of China in the future as a modernized country with high income, high technology, high welfare and high human development.

High income means reasonable economic development cost, high net welfare and high net income of economic development.

High technology means building a strong innovation country, enter the rank of "Strong Power in Science and Technology", and reach world advanced level in several important fields of science and technology.

High welfare means high education level and high health level in a country.

High human development means human development index of higher than 0.8 and lower than 0.9.

Column 2.1 Developed Countries in 2010

The United Nations Development Program (UNDP) revised the classification of countries in the world in the *2010 Human Development Report* released on Nov 4, 2010. After the revision, the number of developed countries or regions increased by 6 from 38 in 2009 to 44 in 2010. Including the developed economies in the Organization for Economic Cooperation and Development (28 countries) and developed economies that are not members of the Organization for Economic Cooperation and Development (16 countries or regions).

Developed economies in the Organization for Economic Cooperation and Development (28 countries).

Australia, Austria, Belgium, Canada, Czech, Denmark, Finland, France, Germany, Greek, Hungary, Iceland, Ireland, Italy, Japan, Luxemburg, Holland, New Zealand, Norway, Poland, Portugal, Slovakia, Spain, Sweden, the US, the UK.

Developed economies that are not the members of the Organization for Economic Cooperation and Development (16 countries or regions).

Andorra, Bahrain, Barbados, Brunei, Cyprus, Estonia, Hong Kong China, Israel, Liechtenstein, Malta, Monaco, Qatar, San Marino, Singapore, Slovenia, United Arab Emirates.

2.2 Inspiration of Chinese Dream to Development Cost

Chinese dream is first of all a "Dream of Rejuvenation"—meaning the great rejuvenation of Chinese people. China before modern times was in leading position of the world in terms of economic, political, social and cultural development. So "Rejuvenation" is a comprehensive Rejuvenation, besides the aspects of economy, politics, society and culture, China also needs to have the rejuvenation with solutions for ecological problems brought by industrial civilization in modern times. The Party's Eighteenth Conference Report stressed that the Scientific Outlook on Development should be "Integration of Five Parts" and supported by five constructions, economic construction, political construction, social construction, cultural construction and ecological construction, the construction of a "Beautiful China" was included in the concept. Moreover, this round of rejuvenation should be a rejuvenation that is scientific, open and confident among the peoples in the world. Such confidence was summarized by Comrade Xi Jinping as the confidence in theories, confidence in path and confidence in system of the socialism with Chinese characteristics. Therefore, we need to conclude the development experiences in development cost on the path to "Rejuvenation", the main dimensions of which include the five internal items, meaning economy, politics, society, culture and ecology, as well as one external item, meaning opening to outside world. Because opening to outside world is also a systematic project that involves various aspects including economy, politics, culture, etc.

Chinese dream is also a "Dream of Exploration"—meaning the dream that Chinese Communist Party constantly leads the people to create common development in the future. How China develops is always a hot topic. Currently, the most time-characteristic "Chinese Dream" stresses letting people share the opportunities of having brighter lives, stresses patriotism, reform, innovation and team work. Such "Dream of Exploration" has actively transferred the reign of exploration from few great statesmen and leaders to the people—the wisdom and power of 1.4 billion people. Kang Youwei, Sun Yat-sen and Mao Zedong all explored the realizing form of the "World of Universal Harmony" in China. As a comparison, the current Chinese dream is more realistic and practical. Its inspiration to the study of development cost lies in using the innovation ability of each person to deal with the development cost of every aspect, this is a kind of participation-type study of development cost. It reflects that the realization of our Chinese dream bears all aspects of cost and needs people from all walks of life to continuously use our strength, find, reduce and even eliminate those unnecessary development cost.

Chapter 3
Innovation Powerhouse: Build a Society of Joint Development and Common Prosperity

3.1 A Society of Joint Development and Common Prosperity

"Joint Development" was reinforced in the Party's Fifteenth and Sixteenth General Conference Reports, which stressed that China shall adhere to and improve the socialist market economy system with public ownership playing the dominant role and allowing diverse forms of ownership to develop side by side, promote various forms of ownership to develop their advantages, promote each other and jointly develop in market competitions. This term is also commonly heard in international relations and external exchanges, and requiring mutual tolerance, seeking common points while reserving difference, and joint development.

Comrade Mao Zedong used to stress that different ethnic peoples shall have joint development. Mao Zedong said that we must help the minorities in "Seeking their liberation and development in politics, economy and culture".[1] It can be seen that the connotation of joint development has been constantly developing, including joint development of different ethnic peoples, joint development of different regions, joint development of urban and rural areas, joint development of nature and society, joint development of different nations, etc.

Marxism established the revolutionary, scientific and prudent theories about human development, what they emphasized was joint development of majority of human race.

Joint development has its historical origin and sedimentation, this book believes that the *Common Program* integrated the thinking of joint development, and it expressed the consensus of Chinese Communist Party, different democratic parties and people of all ethnic groups to support the foundation of the New China.

---

[1]*Selected Works of Mao Zedong*, Volume 3, People's Publishing House, 1991 Edition, Page 1084.

The development goal of *Common Program* was to develop production, prosper the economy; the means adopted were integration of both public and private ownerships, benefiting both labor and owner, mutual help between urban and rural areas, domestic and foreign exchanges, national unity, political consultation, etc. which reflected "Commonality" and "Development Nature".

The common prosperity thinking of Marx and Engels was established on the basis of the full development of laissez-faire capitalism. Including three aspects: firstly, the social development in the future is on the purpose of prosperity for all people; secondly, common prosperity will not be realized without highly developed productivity; thirdly, the means of common prosperity is that means of production shall be owned by the society.[2]

In July, 1955, Comrade Mao Zedong clearly proposed the concept of "Common Prosperity" for the first time in the *Report Regarding Problems in Agricultural Cooperation*, he indicated that "While gradually realizing socialist industrialization and gradually realizing socialist transformation on handcraft industry and capitalist industry and commerce, we shall gradually realize the socialist transformation on the entire agriculture, which means realizing cooperation, eliminating the rich peasant economic system and individual economic system, enabling the entire rural people to prosper.[3]" He believed that the basic means of realizing common prosperity is to develop productivity, common prosperity couldn't be realized overnight, but a rather long historical process. However, Comrade Mao Zedong's thinking of common prosperity was a framework in nature, which didn't include specific measures.

Comrade Deng Xiaoping believed that poverty is not socialism, the superiority of socialism is common prosperity, he hoped to eventually eliminate gap by firstly admitting gap, he surpassed the "Simultaneous Prosperity" proposed by Comrade Mao Zedong and designed the means of firstly having those getting rich first and those getting rich later, and eventually realize common prosperity. He proposed the energetic development of productivity and shortening of the gap between rich and poor and the gap between different regions, etc. When meeting the former Japanese Prime Minister Masayoshi Ohira on Dec 6, 1979, he proposed the concept of "Well-off" for the first time based on China's circumstance, and proposed the conception of realizing the "Well-off Family"-type "Four Modernizations" by the end of 20th Century and reach the "Well-off Society".

From the formal citation of the "Well-off Society" concept in the Party's "Twelfth General Conference" to the comprehensive construction of well-off society in the first twenty years of the 21st Century proposed in the Party's "Sixteenth General Conference", Comrade Jiang Zemin further enriched the connotation of "Common Prosperity".

---

[2]Ju Wei: *Historical Observation and Inspiration of Socialist Common Prosperity*, Journal of Ningxia Party School, 1st Issue in 2011.

[3]*Selected Works of Mao Zedong,* Volume 6, People's Publishing House, 1999 Edition, Page 437, 496 and 329.

Comrade Hu Jintao further developed the thinking of "Common Prosperity" of Comrade Deng Xiaoping by integrating the construction of Common Prosperity-type society as one of the goal to realize socialist harmonious society—the trend of expanding development gaps between urban and rural areas and between different regions are gradually turned around, the reasonable and orderly income distribution pattern basically forms, the family assets commonly increase, and people have richer lives.

3.2 Innovation Powerhouse: Control the Development Cost of Economy and Society

By 2050, China shall enter the rank of "Great Powers of Science and Technology", reach world advanced level in several important fields of science and technology, and enable China to become a country with major contributions to the world in science and technology. China shall occupy major commanding heights in front-line sciences and high-tech fields, energetically guarantee the national economic security and national defense security, as well as play important role in protecting the world peace and development; science and technology shall become an major force to promote the advancement of Chinese economic development and social advancement; scientific and technological innovations shall fully meet the demands for sustainable development of economy and society, as well as continuous improvement of people's living quality.[4] Innovation is the necessary stop of human development.

Innovation is a major dynamic force mechanism to reduce the economic and social development cost. Innovation could not only reduce the micro cost of enterprises and individuals, but also reduce the macro cost of economy and society. Innovation doesn't just mean scientific and technological innovations, it also include idea innovation, management innovation, etc.

The innovation concept of Marxism is reflected in labor concept, innovative labor is a staged development of labor and a form of labor that transcends qualitatively identical labors. The self-creation behavior of human is the qualitative change from discovery and innovation to quantitative change of repeating and accumulation; the fundamental issue of innovative labor lies in the enhancement and innovation of labors' own quality; social innovation is innovative development of social human on social relations, as well as other thinking.

Schumpeter published the *Introduction to Economics of Development* in 1912, which explained the concept of innovation, he believed that the basic concept of innovation is to integrate new production factors and production conditions before bringing them into production system, including bringing in new products, bringing

---

[4]Wang Yong: *2020: A Big Technological Power; 2050: A Strong Technological Power,* Wen Hui Bao, Jan 17, 2003.

in new production methods, opening new market, obtain raw materials or unfinished products or new supply sources, and realizing new industrial organization. Innovation includes technical innovation and non-technical innovation.

Comrade Jiang Zemin expressed in the 1995 National Science and Technology Conference that innovation is the soul of advancement for a nation and an inexhaustible motive force for a country to prosper and develop. A nation without innovation ability would be difficult to stand among the advanced nations in the world. Over the decade, innovation has become a consensus of this nation. Economic growth can be interpreted from another level as innovation of new combinations that are continuously realized by the entire society.[5] Innovation is correlated with many aspects, such as economic development, political development, social development, cultural development, opening to the outside world, natural development, etc. Innovations can penetrate into each aspect of society, including technological innovation, managerial innovation, cultural innovation, system and mechanism innovation, development method innovation, conceptual and thinking innovation, etc. Innovation would not only benefit key factors in natural development, such as reduction of resource and energy consumption, reduction of environmental pollution and ecological destruction, reduction of climate change risks, reduction of natural disaster losses, etc., but also benefit the reduction of various costs during economic development, political development, social development, cultural development, development of external affairs, etc., such as construction cost, transaction cost, changing cost, stabilization cost, sustainable development cost, etc. Innovation powerhouse does not only use the means of innovation to realize development and prosperity; but also to realize the strengthening of total factor productivity (TFP), reduction of development cost, elimination of rich-poor gap and polarization, and enhancement of the connotation and significance of human welfare.

Appendix: Definitions

(1) "Economic Development Cost" in a broad sense means the entire costs and expenses that a country (or region) pays in order to obtain development, meaning the costs and expenditures of development and advancement of human society, including the part already paid and the part not yet paid. In a narrow sense, it means the costs and expenses that a country (or region) pays in order to obtain economic development, in this article, it means the sum of costs paid in an economic system in order to promote growth, promote transformation, prevent disturbance, regulate, manage and maintain stability.

---

[5]Pei Sensen: *Schumpeter's Innovation Theory and New Development Outlook,* Journal of Yancheng Industrial College (Social Science Edition), 3rd Issue in 2005.

Appendix: Definitions

(2) "Political Development Cost" means the expenses and costs that a country (or region) pays in order to obtain political advancement and development, including construction cost (system construction, ruling party construction, democracy construction), reform cost, decision-making cost, information cost (negotiation cost), etc.
(3) "Social Development Cost" means the costs and expenses that a country (or region) pays in order to obtain social advancement and development. Including the cost to solve existing problems, the cost to handle population migration, management cost, stabilization cost, and other cost to promote advancement and development, etc.
(4) "Cultural Development Cost" means the costs and expenses that a country (or region) pays in order to obtain cultural development and civilization heritage. Including the cost to inherit civilization and preserve folk custom, the cost to build and manage public opinion of modern media, ideological evolution cost, soft power construction cost, etc.
(5) "Opening Development Cost" means the costs and expenses that a country (or region) pays in order to expend opening. Including thinking reform cost, the cost to form and maintain international exchanges and consensus, the cost to build and maintain investment environment, the cost to promote and develop trades, the cost to build and maintain three-dimensional transportation, the cost to establish and maintain creditability, etc.
(6) "Nature Development Cost" means the entire costs and expenses that a country (or region) pays in order to compensate the damages of natural resources and environment caused from economic and social development, and in order to promote harmonious coexistence of human and nature.

Part II
Economic Cost of Development in China

Chapter 4
Economic Growth Cost in China

4.1 Connotation of Economic Growth in China

Economic growth usually means the continuous growth of per capita output (or per capita income) level of a country over a relatively long span of time. Economic growth rate reflects the growing speed of economic aggregate of a country or region during certain period, it is also a symbol to measure the growing rate of overall economic strength of a country or region. The direct factors that determine economic growth include investment volume, work quantity, and productivity level. GDP calculated at current prices could reflect the economic development scale of a country or region, while GDP calculated at unchanged prices could be used to calculate the speed of economic growth.

The stable and high-speed growth of Chinese economy since the reform and opening-up have been universally acknowledged by the world, and referred to by domestic and foreign scholars as "Chinese Miracle". Economic growth means the expansion of commodity producing and labor providing ability of a country, growth of gross domestic product (GDP) can be used as a basic index to measure economic growth. Along with the stable and high-speed growth of economy, China's GDP has been continuously growing, over thirty years from 1978 to 2009, the GDP grew from RMB 364.52 billion to RMB 34.05069 trillion, that's over 93 times of growth, the average annual GDP growth was 9.83%, during the period, although economic growth had temporary fluctuation, it maintained overall trend of stable and high-speed growth.

The progressive-type path of reform well coordinated the relationship between growth and stability. Studies indicated the average annual growth rate and time sequence of the Chinese GDP and CPI in each year of the four stages from 1978 to 2008. Since the reform and opening-up, the average annual GDP growth rate of China was 9.8% with average fluctuation rate of 0.07%. the two peaks of economic growth were 1984 and 1992, the GDP growth rate reached 15.2 and 14.2% respectively, while the three valleys were in 1981, 1989 and 1990, during which the GDP growth rates were only 5.2, 4.1 and 3.8% respectively, but there was no

Table 4.1 China's annual GDP data since reform and opening-up

年份	1979	1980	1981	1982	1983	1984	1985	1986	1987	1988
GDP 增长率	7.6%	7.8%	5.2%	9.1%	10.9%	15.2%	13.5%	8.8%	11.6%	11.3%
年份	1989	1990	1991	1992	1993	1994	1995	1996	1997	1998
GDP 增长率	4.1%	3.8%	9.2%	14.2%	14.0%	13.1%	10.9%	10.0%	9.3%	7.8%
年份	1999	2000	2001	2002	2003	2004	2005	2006	2007	2008
GDP 增长率	7.6%	8.4%	8.3%	9.1%	10.0%	10.1%	10.4%	11.6%	13.0%	9.0%

Source: announced through modification and adjustment of the data issued by State Statistics Bureau, concluded from the sorting and calculation of annual statistic bulletins issued by State Statistics Bureau in various years.

年份	Year	GDP 增长率	GDP Growth Rate

negative growth. At the four stages of reform and opening-up, as shown in Table 4.1, economic growth reflected the characteristic of acceleration stage by stage, which were 9.6, 9.6, 9.9 and 10.3% respectively.[1] The economy has always been on the track of high-speed growth, it was a typical exponential growth.

During the period from the Liberation to 1970s, China had experienced a long and winding history of development.[2] However, during the 30 years from 1978 to 2007, Chinese economy maintained continuously high-speed growth at the average annual rate of 9.85%, and became one of the main economies in the world as a developing country, which was extremely rare in the world economic history.[3] We can totally say that Chinese economy gradually stepped into the track of high-speed growth after the reform and opening-up, and maintained continuous growth. It means the Chinese economy after reform and opening-up entered what Kuznets described as the stage of modern economic growth, or entered what Rostow described as the stage of economic take-off.[4]

---

[1]Zhang Ping: *China's Economic Growth and Structural Reform for 30 Years of Reform and Opening-up*, Modern Economic Discussion, 7th Issue in 2008.

[2]In 1949, China ended a century of internal conflicts, foreign invasions and chaos since Modern Time, China won the victory of revolution and started a new history. Since them, China has gradually established the social, political and systematic frameworks that are needed for economic take-off. It can be recognized that China started economic take-off during the first half of 1950 s, however, due to various reasons, during the following decades, Chinese economy experienced turbulent and complex development path. After the decade-long "Cultural Revolution", by the mid of 1970 s, Chinese economy was on the brink of collapse. At the end of 1970 s, China began to implement the reform and opening-up policy, ended the decades-long era of political dominance and entered a new era of economic development.

[3]Zhang Jiangang: *Causes and Prospect of China's High Economic Growth Rate since the Reform and Opening-up*, Economic Horizon, 3rd Issue in 2009.

[4]Ying Hao: *Changes of China's Economic Growth and Labor Market since Reform and Opening-up and Prospect of the Future*, Population Journal, 5th Issue in 2001.

Not long after enforcing the reform and opening-up policy, China formed the "Three-step" development target of modernization in early 1980s.[5] In 1987, China realized the target of doubling gross national product 3 years in advance. After realizing the first step of development target, it was determined in the *10-year Planning for Development of National Economy and Society* formed in 1990 that the growth rate of gross national product shall maintain at 6% on average during the ten years to 2000, so as to realize the second step of development target. In fact, in 1995, the original second-step development target of quadrupling the gross national product in 2000 was realized in advance. As indicated by the data of the State Statistics Bureau, in 2000, China's gross domestic product reached RMB 9.9215 trillion, and the per capita gross domestic product reached RMB 7858. Based on the materials released by World Bank, in 1998, China's gross national product ranked No. 7 in the world, and ranked No. 1 among developing nations. Based on the three-step development strategy, China further proposed the target of doubling the gross domestic product in 2010 on the basis of that in 2000 (see Footnote 4).

We should notice that during the period of reform, opening-up and transformation, there are also costs of development within economic system, including the cost of economic growth, the cost of economic transformation, the cost of economic regulation and control and other initiative costs, as well as the disturbance and risks within economic system, meaning economic disturbance cost, which is a passive cost.

4.2 Expression Forms of Economic Growth Cost in China

"Troika" in complete sense means final consumption expenditure, gross fixed capital formation and export of products and service in output-expenditure approach, final consumption expenditure reflects demands for consumption; gross fixed capital formation reflects demands for investment; net outflow equals the net amount of goods and service outflow deducting of inflow, and reflects external demands.[6] The "Three Major Demands" are what we usually described as "Troika" to drive economic growth, expansion of investments, expansion of domestic demands and expansion of exports.

Along with the high-speed growth of Chinese economy, the corresponding development cost come with it, this Article mainly discusses the costs generated

---

[5]The first development goal is to use about 10 years of time to double the GDP compared with that in 1980, and basically solve the basic livelihood issue of the people. The second step of development goal is to double the GDP again by the end of 20th Century, and further improve people's living standard. The third development goal is to enable the per capita GDP to reach the level of medium developed nations by 2030 s to 3050 s, and basically realize the modernization of national economy.

[6]Cao Meifang: *What is the "Troika" Driving Economic Growth,* Statistical Science and Practice, 4th Issue in 2011.

Table 4.2 Forms of the cost of China's economic growth

Classification of cost	Forms	Specific contents
Cost of economic growth	Cost of investment expansion	A series of costs brought by the expansion of investment, expansion of domestic demands and expansion of export proposed in order to drive economic growth
	Cost of domestic demand expansion	
	Cost of export expansion	

from the process of driving the three major demands of economic growth, meaning the cost from expansion of investments, the cost from expansion of domestic demands and the cost from expansion of exports. Please refer to Table 4.2.

4.2.1 Cost from Expansion of Investments

Cost from expansion of investments means the related monetary expenditure generated from investments, or the related problems brought by active expansion of investments, and the price paid for them. Since the reform and opening-up, the total investment volume of fixed assets in China has been growing rapidly. The investment rate is relatively high and the growth of total investments is relatively fast, which make the operation of Chinese total investments always face the pressure of scale expanding, which has to certain extent exceeded the ranged allowed by China's national power. Under such circumstance, the room for using large-scale investments to drive economic growth is severely restrained, which is specifically reflected in Chart 4.2.

From the history of Chinese economic growth, expansion of investments would bring the problem of economic overheating, which would definitely cause radical change of economy, so as to cause tremendously negative influence on national economy. It can be seen from the survey of Chinese economic operation states since 1949 that China's economy generally experienced 9 development cycles, each cycle provided the lesson of overheated investments. Each cycle began with large volume of infrastructure investment, with the soaring-type expansion of annual investments and total investments of projects under construction, large batch of new projects commenced, the supply-demand conflict of investment objects became prominent, which caused rapidly rising prices of investment objects, which in turn stimulated the expansion of production department and construction department of investment. Compared with foreign countries at the same development stages, China had relatively more times of investment fluctuations and relatively larger ranges of fluctuation.[7] During decades of investment and construction process, China was always haunted by such unwelcome cycle, and the swing of such economic fluctuation becomes increasingly bigger, and the cycle becomes increasingly shorter, which is the opposite to the objective requirements of normal economic operation.

Besides, for a long period of time, China's extensive operation of investments mainly dominated by extension resulted in the continuous declining of marginal benefits of investments themselves and the extensive declining of their contribution to economic growth. Under the conditions of market economy, infrastructure investment, like all other investments, is a way to obtain present or future incomes. If an investment is unable to obtain expected effective income, such investment would lose its meaning economically.[8] The reason is that when there is no major change in investment and financing mechanism, large-scale infrastructure construction would result in repeated construction and generate economic bubble.[9] This book will summarize the afore-mentioned circumstances with Chart 4.1.

4.2.2 Cost from Expansion of Domestic Demands

Cost from expansion of domestic demands means the fiscal expenditures, related social problems and prices paid from active driving of domestic demands while bringing in certain economic benefits. Domestic demands include two aspects, one is demand incurred from investments, and the other one is demand incurred from consumptions. In order to cope with the global economic crisis, Chinese government released the program of expanding domestic demands and investments worth over RMB 4 trillion, including government direct investments of RMB 1.2 trillion and bank loans, enterprise investments, etc. of RMB 2.8 trillion. But the expansion of domestic demands would also cause certain social costs.

---

[7]Taking the "Eighth Five Year Plan" for an example, the fluctuation of annual investment scale was very radical, compared with growth rate, the highest year and the lowest year could vary by nearly 40%. During the Sixth, Seventh and Eighth Five Year Plan periods, due to the unstable investment growth, it all caused different degrees of fluctuation of national economy, and China was forced to make renovation.

[8]During the "Eighth Five Year Plan Period", the additional GDP per one hundred Yuan of total investment was only around 50 Yuan. From the perspective of total investment, for the GDP created in current year from every additional RMB 100 million of investment: the number was RMB 320 million during the "Sixth Five Year Plan Period" and RMB 220 million during the "Seventh Five Year Plan Period", but decreased to less than RMB 200 million during the "Eighth Five Year Plan Period", we assessed the investment benefit through the internationally recognized investment coefficient. Since 1994, this ratio has always been increasing, meaning the rate of return has been declining; the ratio was 1.04 in 1994, meaning that every RMB 1 of investment could increase the GDP growth by about RMB 1. In 1996, this ratio has become 1.92, meaning that about RMB 2 of investment could obtain only RMB 1 of GDP.

[9]Li Yubing: *Several Thoughts Regarding Expanding Investment to Drive Economic Growth*, Shandong Social Science, 1st Issue in 2000.

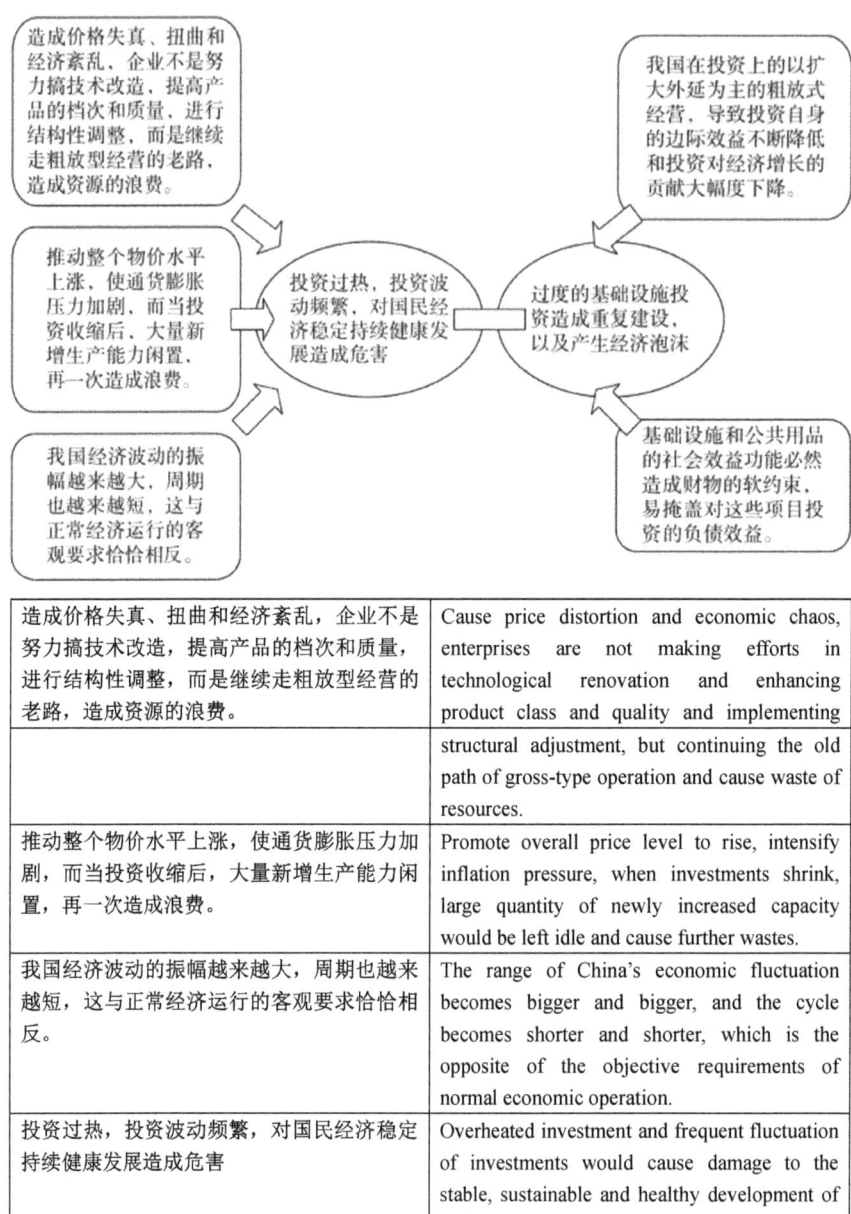

造成价格失真、扭曲和经济紊乱，企业不是努力搞技术改造，提高产品的档次和质量，进行结构性调整，而是继续走粗放型经营的老路，造成资源的浪费。	Cause price distortion and economic chaos, enterprises are not making efforts in technological renovation and enhancing product class and quality and implementing structural adjustment, but continuing the old path of gross-type operation and cause waste of resources.
推动整个物价水平上涨，使通货膨胀压力加剧，而当投资收缩后，大量新增生产能力闲置，再一次造成浪费。	Promote overall price level to rise, intensify inflation pressure, when investments shrink, large quantity of newly increased capacity would be left idle and cause further wastes.
我国经济波动的振幅越来越大，周期也越来越短，这与正常经济运行的客观要求恰恰相反。	The range of China's economic fluctuation becomes bigger and bigger, and the cycle becomes shorter and shorter, which is the opposite of the objective requirements of normal economic operation.
投资过热，投资波动频繁，对国民经济稳定持续健康发展造成危害	Overheated investment and frequent fluctuation of investments would cause damage to the stable, sustainable and healthy development of national economy

Chart 4.1 Cost generated from expansion of investment in China

我国在投资上的以扩大外延为主的粗放式经营，导致投资自身的边际效益不断降低和投资对经济增长的贡献大幅度下降。	China's gross-type operation of investments that are dominated by expansion has caused continuous decrease of marginal benefits of investments themselves and large-scale decrease of the investments' contribution to economic growth.
过度的基础设施投资造成重复建设，以及产生经济泡沫	Excessive infrastructure investments have caused repeated construction and generated economic bubble
基础设施和公共用品的社会效益功能必然造成财物的软约束，易掩盖对这些项目投资的负债效益。	The social benefit of infrastructure and public goods would inevitably cause soft binding of properties, and may hind the liability benefit of these project investments

Chart 4.1 (continued)

Firstly, domestic demands expanded through investments would need increase of fiscal expenditures, which would definitely increase pressure on public finance, the fixed assets investment scale of the entire society reached RMB 2.8852 trillion in 1998, while in 2008, China invested RMB 4 trillion to promote domestic demands, which enlarged the fiscal deficits in current year. The enlargement of fiscal deficits would not only bring inflation pressure, but also make national creditability rating decline.[10]

Secondly, expansion of domestic demands through investments would also increase monetary supply and increase M2, so as to bring extra-money-issue-type inflation. Demand-driven inflation means the situation when social total demands are higher than total supply during economic development process, so as to cause continuous rising of price level. When an economy hasn't reached full employment, there are usable resources in society, under such circumstance, "Extra Demands", while causing rising prices, would also promote growth of output; when an economy doesn't have idle resources, expansion of total demands would not promote increase of output, but only cause rising prices, such inflation is considered as "Too much money chasing too few goods".[11] After the financial tsunami erupted in 2008, China adopted large-scale economic stimulation policies and issued large quantity of money, too much currency was issued, which in theory was twice the GDP. So there were relatively more money flowing in the market, when money supply was too much, it caused the later inflation. Meanwhile, due to imbalance of exchange rates, large volume of hot money flew into China for speculations, which caused too much money supply and too high liquidity, so as to cause the inflation in China.

---

[10]The European Debt Crisis was to certain extent because the government increased investments after the financial crisis and made the deficit to exceed the healthy standard of 5%.

[11]Qin Na: *Several Thoughts about the Causes of China's Inflation,* Science and Management, 3rd Issue in 2011.

Besides, expansion of domestic demands through investments could easily cause look-ahead constructions, repeated constructions and blind constructions, especially in the aspect of infrastructure constructions and investments. Moreover, expansion of domestic demands would also bring trade protectionism, mercantilism begin to rise, which is not good for the globalization process; in comparison, demands of consumption is much "Healthier", but premature consumption, excessive consumption and large consumption on credit would also cause adverse consequences.

4.2.3 Cost from Expansion of Exports

Cost from expansion of exports means the sum of monetary price and non-monetary price that China paid for taking the related measures in order to expand exports.

Economic globalization provides opportunities for each country to expand exports. Every country doesn't want to miss this opportunity of development and actively participate in various international organizations and various regional alliances in order to expand exports and win the opportunities for development. However, using export expansion to boost economy could encounter serious social problems.

Some scholars indicated that high foreign reliance would slow economic growth down. Because the higher the foreign reliance, it means export/import have higher influence on the country's economic development, meaning higher reliance of the underdeveloped country to developed countries, only relying on continuous export of primary commodities could maintain economic growth of this country, then it will definitely lead to immiserizing growth.[12] Therefore, expansion of exports caused too high reliance of China's economy to foreign countries, the development room for national economy is narrow, and it is difficult to adjust the export structure.

Besides, expansion of exports would cause trade surplus, which enlarges the pressure of RMB appreciation and increases international trade frictions. Trade surplus causes the foreign currency supplies in domestic foreign exchange market to be higher than the demands for foreign currency, which would definitely cause the expectation of foreign currency devaluation and expectation of RMB appreciation, while the expectation of RMB appreciation would in turn enlarge foreign investment inflow and expansion of trade surplus, and further strengthen the pressure for RMB appreciation. Besides, trade surplus would also intensify

---

[12]Xu Qian: *Export Expansion and Economic Growth,* Northern Economics and Trades, 7th Issue in 2005.

4.2 Expression Forms of Economic Growth Cost in China

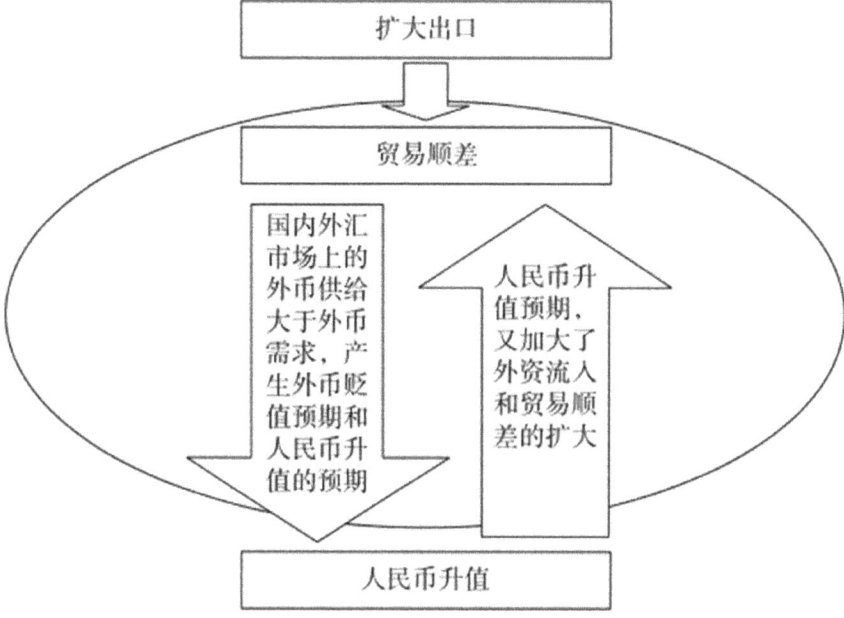

扩大出口	Expansion of export
贸易顺差	Trade surplus
国内外汇市场上的外币供给大于外币需求，产生外币贬值预期和人民币升值的预期	The supply of foreign currency in domestic foreign exchange market is more than the demands for foreign currency, which caused expectancy of devaluation of foreign currency and expectancy of RMB appreciation
人民币升值预期，又加大了外资流入和贸易顺差的扩大	The expectancy of RMB appreciation further increased foreign capital inflow and expansion of trade surplus
人民币升值	RMB appreciation

Chart 4.2 Cost from trade surplus caused from expansion of export

international trade frictions,[13] the trade frictions of China caused from trade surplus have been increasing in these years (Chart 4.2).

---

[13] Increase of trade surplus would also mean that the countries trading with China have deficits and would use WTO rules to restrain import of Chinese products in order to protect their own industries. After joining the WTO, due to the relatively large deficit of trade between China and the US, the country that used the most anti-dumping measures was the US, and China was the country that has received the most anti-dumping measures and ranked number one among the WTO members, and the number of anti-dumping cases against China has increased year by year.

Besides, trade surplus would weaken the effects of monetary policies, and reduce the using efficiency of social resources.[14] The combination of trade surplus and Chinese residents savings caused the using efficiency of social resources to be low.[15] Secondly, the rapid increase of foreign reserve means increase of domestic currency supply, and brought prominent problem of extra liquidity to domestic economy, so as to cause pressure of inflation.[16]

Expansion of exports would increase the cost of foreign reserve and increase fund outflow. There are always risks in foreign exchange operations in international financial market, the most prominent one is exchange rate risk. China adopts fixed exchange rate pegging to US dollar, whenever USD devalues or the US has inflation, China's foreign reserve would devalue accordingly, so as to cause loss of foreign reserve.[17]

Finally, trade surplus affects the process of interest rate liberalization of Chinese financial industry. Under the conditions of interest rate liberalization, trade surplus means increase of domestic monetary supply, since the second half of 2003, Chinese economy has been facing the pressure of inflation, in order to prevent negative influence of inflation on the economy, the Central Bank had to maintain the currently regulated interest rate for deposits and floating interest rate for loans, so as to delay the process of market rate liberalization.

---

[14]Inflow of foreign exchange would increase along with the increase of trade surplus, and the input of currency would also increase along with the increase of foreign exchange inflow. With large amount of RMB passively flew into the circulation, the fundamental currency account of the Central Bank would be further restrained by foreign exchange inflow, so as to weaken the effect of Central Bank's monetary policy.

[15]The large amount of foreign exchange reserve caused from trade surplus was as high as USD 400 billion, meanwhile, the residents' domestic deposit was as high as RMB 11 trillion, the sum of the two combined was over RMB 14 trillion. No effective investments were formed from this RMB 14 trillion and became idle funds in the economy, corresponding to these idle funds, it was the idling of production materials and human resources, and the low using efficiency of social resources.

[16]Wang Xiaoxiao: *Discussion on the Current Situation of China's Foreign Exchange Reserve*, Chinese Business Circle, 4th Issue in 2008.

[17]The more the trade surplus becomes, the more the foreign exchange to be operated in international financial market, and the higher the cost of national foreign exchange reserve. Trade surplus increases the outflow of funds. Under the foreign exchange sale and purchase system, because all foreign capital inflow need to be converted into foreign exchange reserve, which are mostly bonds in USD or Euro. The more the trade surplus, the more the foreign exchange reserve, the more the foreign bounds and the more the capital outflow. These two aspects have formed internal paradox: trade surplus caused foreign capital inflow, the more the export and the more the foreign direct investments, the more the capital inflow; trade surplus caused increase of foreign exchange reserve, the more the foreign exchange reserve, the more the capital outflow.

4.3 Generating Causes of Economic Growth Cost in China

4.3.1 Increasingly Worsening of Structural Conflicts in China

Chinese economy has experienced over thirty years of reform and opening-up, the structural conflicts are becoming increasingly prominent, which are mainly reflected in the following aspects: irrational supply structure[18]; irrational consumption structure[19] and irrational industrial structure.[20]

4.3.2 Systematic Causes Affecting Economic Growth[21]

With the deepening of reform, some systematic conflicts have been continuously appearing, and have become the pressing problems in order to have further economic growth. Currently the systematic causes that affect China's economic growth mainly include the following aspects: government institutional reform is not in place[22]; social security system is incomplete.[23]

4.3.3 Decision-Making Mistakes in Economic Construction Process

In China, a considerable part of economic decision-making powers are still held by government on different levels, while some leaders don't know about economy and

---

[18] On one hand, the production of daily goods and other light industrial products in China has severe overcapacity; on the other hand, China has insufficient capacity in precision machine tool, high-tech products and some other industries, China has insufficient production capacity, depends on import over long period, which not only increased China's burden of foreign exchange expenditure, but also made China subject to restriction by developed countries.

[19] China's policy of expanding domestic demands and promoting economic growth hasn't truly taken effect, the contribution ratio of consumption in economic growth has always been not high.

[20] The proportion of secondary industry in Chinese economy is still too high, meaning that we need to consume more energy to develop economy, while such growth method of depending on energy consumption is apparently not sustainable.

[21] Mi Jianhui: *Causes of China's Slowing Economic Growth Rate and Discussion of Countermeasures,* Business Era, 20th Issue in 2011.

[22] Firstly, government regulation has been too wide and too specific; secondly, the government is insufficient in serving economic growth.

[23] This is one of the many reasons of China's excessively low domestic consumption, from long-term perspective, economic growth should focus on releasing domestic consumption and solving the problem of too low contribution rate of domestic consumption in general economic growth, which is the core issue urgently to be addressed in China's economic growth.

make decisions only based on their experiences and tastes. Besides, they don't need to bear the losses caused from mistakes in investment decisions, so bad debts generated hereof would have nothing to do with them, so they could establish many projects that haven't been carefully evaluated. What's after decision-making is usually huge amount of investments, whether a decision-making is successful depends to most extent on the economic benefits and returns received, therefore, mistakes in project decision-making must cause loss of large amounts of investments. As estimated by World Bank,[24] during the period from the "Seventh Five Year Plan" period to "Ninth Five Year Plan" period, China's mistake rate of investment decisions was around 30%, funding waste and economic loss were about RMB 400 billion—RMB 500 billion.

4.3.4 Marketization Degree of Production Factors Are Still not High

After long-term market-oriented reform, the current micro foundation of Chinese economic operation already had fundamental changes, the basic framework of market economy has been preliminarily established. However, it shouldn't be neglected that in the field of factor market, the marketization degree is still relatively low. Government has very strong control on lands, capital and various mineral resources, and is able to decide the direction and use of the distribution of such resources, so the pricing mechanism of lands, funds and natural resources is severely distorted,[25] which were reflected as incomplete land market,[26] incomplete

---

[24]Huang Tiemiao and Cao Zheng: *Wastes Caused from Decision-making Mistakes are the Worst,* Nanfang Daily, Nov 4, 2011.

[25]Sun Yongzheng: *Brief Analysis of Risks in Urban Operation,* Jiangsu Construction, 1st Issue in 2003.

[26]The land resources are basically controlled by the government, its distribution is not realized through market transactions by micro individuals, but through administrative means by government. In order to attract investments, many local governments transfer lands to land developers at cheap price or even for free. Besides, some local governments acquire lands from farmers at low price, and change collectively owned lands in rural area into state owned, and then sell them to developers. In here, governments were just compensating farmers based on related regulations instead of based on market prices.

capital market,[27] lack of paid use mechanism and pricing mechanism of natural resources (Chart 4.3).[28]

4.4 Countermeasures to Economic Growth Cost in China

4.4.1 Adjust and Optimize Industrial Structure to Promote Coordinated Economic Growth

Targeting the unbalance between the primary, secondary and tertiary industry in Chinese economy, the development of three industrial sectors shall be coordinated through optimization and adjustment of industrial structure.

4.4.2 Continuously Deepen the Economic System Reform

Economic system reform has always been implemented, while the slowing-down of economic growth in China makes us have to continuously deepen the economic system reform, and solve some in-depth problems, so as to create an excellent environment for the long-term development of Chinese economy.

---

[27] The credit granting behaviors of state-owned commercial banks are to considerable extent subject to influences of different levels of government, which could easily borrow money from banks, therefore, the total credit of China's economy obviously favors the government. Considering the power and creditability of government, few banks would restrain loans to government, which is extremely easy to promote the in-debt operation of urban construction to exceed proper scope and cause the scale of urban construction to get out of control and cause the risk of excessively high debts. Currently the state-owned banks still have RMB 1.8 trillion of nonperforming assets, accounting for 25% of their loans. From the surface, poor operation of banks was the reason of increased nonperforming assets, but in fact it was caused from credit operation dominated by the government and blind expansion of urban construction scale.

[28] Currently, in many places across China, mining enterprises still use the "Approving System" method from the planning economy to obtain mining rights of various resources. After obtaining operating rights, except payment of few mineral resource taxes, enterprises could basically use the mineral resources for free. Besides, China's water price is 1/3–1/10 of that in foreign countries, the gasoline price is less than half of some European countries, and the price of coal is 50% cheaper than that of international market. The free or low-price use of resources is the fundamental cause of heavy ecological price of economic growth, it would not only cause severe resource waste, and low resource using efficiency, but would also cause severe mining without license or against regulation, chaotic order of mining industry and frequent occurrence of safety accidents.

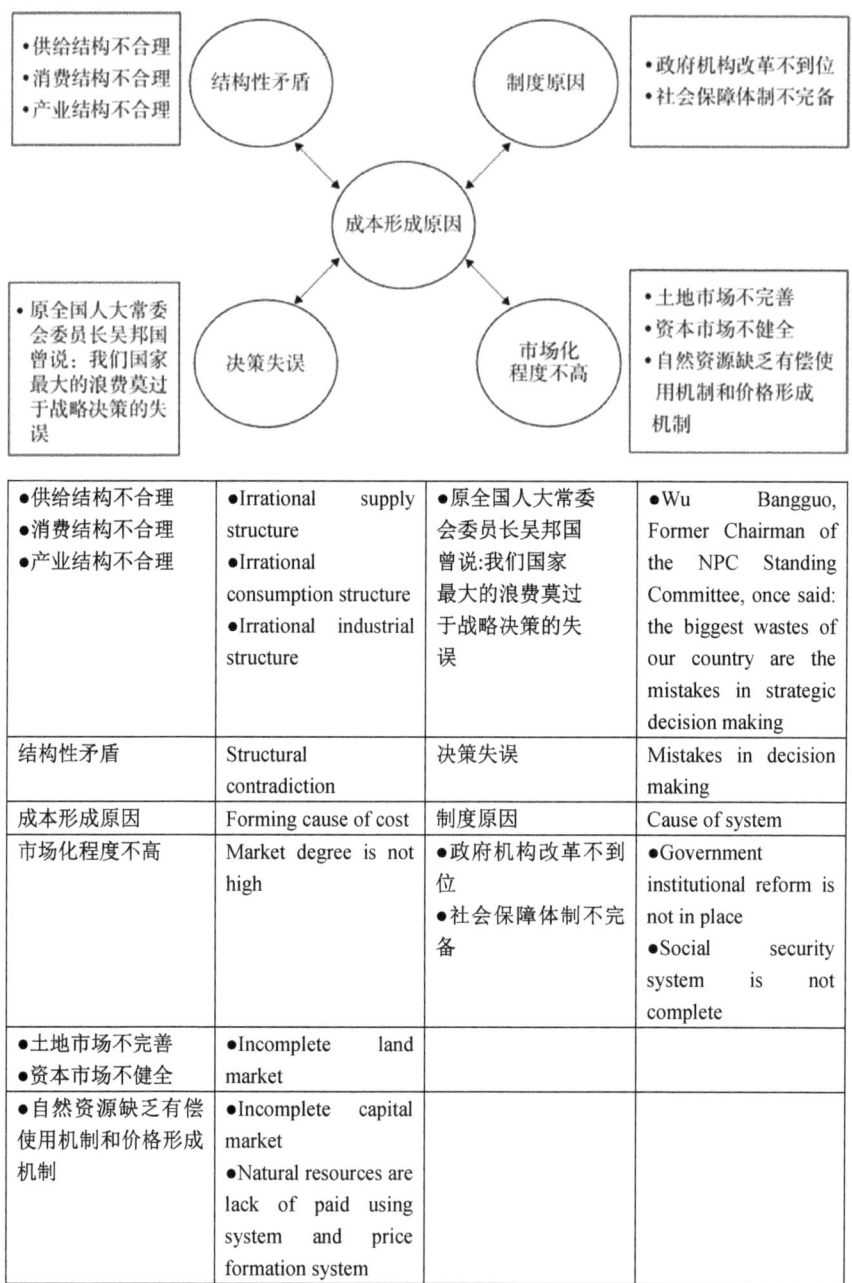

●供给结构不合理 ●消费结构不合理 ●产业结构不合理	●Irrational supply structure ●Irrational consumption structure ●Irrational industrial structure	●原全国人大常委会委员长吴邦国曾说:我们国家最大的浪费莫过于战略决策的失误	●Wu Bangguo, Former Chairman of the NPC Standing Committee, once said: the biggest wastes of our country are the mistakes in strategic decision making
结构性矛盾	Structural contradiction	决策失误	Mistakes in decision making
成本形成原因	Forming cause of cost	制度原因	Cause of system
市场化程度不高	Market degree is not high	●政府机构改革不到位 ●社会保障体制不完备	●Government institutional reform is not in place ●Social security system is not complete
●土地市场不完善 ●资本市场不健全 ●自然资源缺乏有偿使用机制和价格形成机制	●Incomplete land market ●Incomplete capital market ●Natural resources are lack of paid using system and price formation system		

Chart 4.3 Cause of the cost of China's economic growth

4.4.3 Prevent Blind Decision-Making During Economic Construction Process

Say-so decision-making and blind decision-making often result in vanity projects as well as jerry-built projects. Therefore, we must establish project decision-making mistake resignation system to guarantee democratic decision-making and prevent decision-making mistakes. Any leader or cadre who causes major economic loss due to individual decision-making mistake or work fault shall actively resign.[29]

4.4.4 Cultivate Paid Using Mechanism of Resource-Type Products

Only the implementation of paid using of resources and enhancement of resource using costs could fundamentally change the extensive economic growth method of "High Investment, High Consumption, High Emission and Low Efficiency", so as to reduce the ecological price of economic growth.[30] In other words, enterprises' mining rights and resource using rights after mining shall be obtained at certain prices, and rational tax and fee standards shall be determined.[31] A complete system in economic field is a necessary condition for transformation of development model.

---

[29]Some places have already issued such system, such as Hebei Province, which in May, 2005 issued the *Implementing Method (Trial) of Party and Political Leaders of Hebei Province to Resign out of Fault*. Considering that the major economic decision makings of party committees and especially governments and their working units are mainly focused on the two aspects of "Economic Construction Projects" and "Fiscal Funds and National Assets". The *Implementing Method (Trial) of Party and Political Leaders of Hebei Province to Resign out of Fault* clearly stipulated that in the process of approval, design, construction and production of economic construction projects, any and all leaders who have made serious mistakes in decision making, caused huge economic loss or bad consequences, or operate against regulation or make mistakes in decision making when arranging and using fiscal funds and handling state-owned assets, and bear main leadership responsibility, shall resign according to procedure.

[30]On Oct 29, 2005, the National Reform and Development Commission held the "Resource Price Reform Seminar" in Beijing, which was designated to promote the price reform of resource products, such as water, electricity, oil, natural gas, coal, land, etc. At this meeting, Zhu Zhigang, Deputy Minister of Finance, disclosed the basket of thoughts of raising resource using fees. Later on Oct 12 at the "2006 China Industry Report Conference", Lou Jiwei, Deputy Minister of Finance proposed again when mentioning China's tax reform in 2006 that the mineral resource tax reform is one of the important contents of China's tax reform in 2006. The two senior officials of the Ministry of Finance mentioned above disclosed the preliminary plan by the Ministry of Finance on mineral resource tax reform. The core content of such plan is the word "Paid".

[31]Lin Yu and Yan Pei: *Resource Reform to End the Era of Cheap Energy*, Economic Information Daily, Dec 14, 2005.

4.4.5 Transformation from "Investment Attraction" to "Investment Selection"

Since the reform and opening-up, China has always been considering attraction of foreign investments as an important driving force of economic development, the quantity and growth of foreign investments attracted in different regions have become an important indicator to assess political achievements of local governments.[32] Investment attraction should be applicable to both domestic investment and foreign investment.

Annexed Table

See Table 4.3.

---

[32]We must clearly realize that the reason why foreign capitals are willing to enter China in large scale is because, firstly, the vast market capacity of China; secondly, China's cheap factor price and various favorable policies granted by local governments; thirdly, they try to make China become the transferee of their pollutions. Therefore, we can't make investment attraction as a hard indicator to assess the political achievements of local governments, departments and enterprises, and we can't blindly introduce foreign investments at all costs, instead, we should change "Investment Attraction" into "Investment Selection", put project quality on top priority, actively introduce the intensive and environment-friendly industries with low resource consumption, less environmental pollution and high input-output ratio of resources, and resolutely reject the projects with high energy consumption, high material consumption and high pollution.

4.4 Countermeasures to Economic Growth Cost in China

Table 4.3 China's economic development and GDP in various years since reform and opening-up (*Unit* RMB 100 million)

年份	国民总收入	国内生产总值	第一产业	第二产业	第三产业	人均国内生产总值（元）
1978	3645.217	3645.217	1027.535	1745.2	872.4829	381.23
1979	4062.579	4062.579	1270.192	1913.5	878.8875	419.25
1980	4545.624	4545.624	1371.593	2192	982.0308	463.25
1981	4889.461	4891.561	1559.463	2255.5	1076.598	492.16
1982	5330.451	5323.351	1777.401	2383	1162.95	527.78
1983	5985.552	5962.652	1978.387	2646.2	1338.064	582.68
1984	7243.752	7208.052	2316.089	3105.7	1786.262	695.20
1985	9040.737	9016.037	2564.397	3866.6	2585.04	857.82
1986	10274.38	10275.18	2788.691	4492.7	2993.788	963.19
1987	12050.62	12058.62	3233.041	5251.6	3573.974	1112.38
1988	15036.82	15042.82	3865.362	6587.2	4590.261	1365.51
1989	17000.92	16992.32	4265.923	7278	5448.396	1519.00
1990	18718.32	18667.82	5062	7717.4	5888.422	1644.00
1991	21826.2	21781.5	5342.2	9102.2	7337.099	1892.76
1992	26937.28	26923.48	5866.6	11699.5	9357.376	2311.09
1993	35260.02	35333.92	6963.763	16454.43	11915.73	2998.36
1994	48108.46	48197.86	9572.695	22445.4	16179.76	4044.00
1995	59810.53	60793.73	12135.81	28679.46	19978.46	5045.73
1996	70142.49	71176.59	14015.39	33834.96	23326.24	5845.89
1997	78060.83	78973.03	14441.89	37543	26988.15	6420.18
1998	83024.28	84402.28	14817.63	39004.19	30580.47	6796.03
1999	88479.15	89677.05	14770.03	41033.58	33873.44	7158.50
2000	98000.45	99214.55	14944.72	45555.88	38713.95	7857.68
2001	108068.2	109655.2	15781.27	49512.29	44361.61	8621.71
2002	119095.7	120332.7	16537.02	53896.77	49898.9	9398.05
2003	135174	135822.8	17381.72	62436.31	56004.73	10541.97
2004	159586.7	159878.3	21412.73	73904.31	64561.29	12335.58
2005	185808.6	184937.4	22420	87598.09	74919.28	14185.36
2006	217522.7	216314.4	24040	103719.5	88554.88	16499.70
2007	267763.7	265810.3	28627	125831.4	111351.9	20169.46
2008	316228.8	314045.4	33702	149003.4	131340	23707.71
2009	343464.7	340506.9	35226	157638.8	147642.1	25575.48

Resource: sorted and concluded from *China Statistic Yearbook 2010*.

Table 4.3 (continued)

年份	Year	国民总收入	Gross national income
国内生产总值	Gross domestic product (GDP)	第一产业	Primary industry
第二产业	Secondary industry	第三产业	Tertiary industry
人均国内生产总值（元）	Per capita GDP (RMB)		

Chapter 5
Cost of Economic Transformation in China

5.1 Connotation of Economic Transformation in China and Its Cost

The continuous high-speed growth maintained during China's economic transformation process has been described as "Chinese Miracle" or "Chinese Mystery". The forming and development of planned economic system as well as the transformation from planned economic system to market economic system was one of the most significant political events that affected the human development in the 21st Century.[1] Transformation and development is currently one of the main themes of development in most countries.

Economic transformation, by definition, means the process of transformation of economic development stage or economic system, such as transformation from agricultural society to industrial society, or transformation from planed economic system to market economic system.

China's economic transformation has made tremendous achievements over the last thirty years, China realized average annual economic growth of 9.79% during 1978–2007 and became the country with the fastest economic growth rate during the same period in the world. Meanwhile, the overall citizen residential conditions, social transportation infrastructure, telecommunication facilities, urban public facilities popularization rate, etc. had significant improvements, the living quality of

---

[1]Bao Jianyun: *Natural Process and Political Control of Economic Transformation—Theoretical Assumption and A New Economic Explanation on Its Economic Transformation Method and Performance Difference*, Research of System Economics, 1st Issue in 2006.

residents continuously improved, which symbolized that China's economic transformation since 1978 obtained staged benefits.[2] Behind such benefits, we can't ignore the problem of costs. Because the costs are being "Delayed, Ignored and Hidden",[3] more terribly, it is continuously reflected in subsequent transformation stages in form of "Accumulation and Transfer", and increase difficulty for the subsequent transformation, the so-called statement that "Transformation has entered into the stage of overcoming difficulties" was well-founded (Table 5.1).

This book believes that the cost of economic transformation means the costs and prices generated in the process of transformation from planned economic system to market economic system, transformation from traditional economic structure to modern economic structure, transformation from extensive development model to intensive development model, and transformation from closed-type economy to open-type economy.

Economic transformation is not an unique phenomenon of socialist society, any country would face the issue of economic transformation in the process of modernization. Even for the western country with complete market economic system and highly developed economy, their economic system and economic structure are not exactly perfect, and have the process of transformation from current economic system to a more rational and more improved economic system, and also have the process of transition from certain economic structure to another economic structure. China began to propose the issue of economic transformation since the ninth five year plan.

Gradual-type transformation model can be interpreted as the main characteristic of the economic transformation in China: the selection of economic transformation model by China was determined by the fundamental realities of the country. Usually, the gradual-type economic transformation of China can be summarized in this way: in strategic deployment, economic transformation came first, political transformation came later; starting from the rural economic system, which was the weakest in planned economic system, rural area came first, urban area came later; in the periphery of original system, transformation was actively induced through the cultivation and development of incremental factors, increment was used to transform stock; China made first breakthrough in distribution field, gave "Emergency Oxygen Therapy" to the depressing economy, then progressed from surface to internal, and gradually went deep in the system and establishment (Table 5.2).[4]

From 1978 to 2007, the basic steps of Chinese economic transformation were as follows:

---

[2]Lv Wei: *Chinese-style Transition: Inherent Characteristics, Evolving Logics and Prospect—In Memorial of the 30th Anniversary of China's Reform and Opening-up,* Research of Financial and Economic Issues, 3rd Issue in 2009.

[3]Chen Dandan: *Measurement of the Cost of Chinese Economic Transformation: 1978–2008,* Research of Quantitative Economy and Technical Economy, 2nd Issue in 2010.

[4]Xu Bing: *Comparison Research of the Economic Transformation and Its Cost in China and Russia,* Siberia Research, 2nd Issue, Volume 38, April, 2011.

Table 5.1 Review of literatures about the cost of economic transformation

Scholar	Viewpoint
Khan (1995)	Cost of transformation means the political cost to be borne when creator of new system tries during forming policies in order to balance interests of social classes
Luis (1997)	When a transformation-type country faces economic reform, there will be a short term of reform cost, when the agent of economy reflects a stronger preference over the current system, they will delay the reform, and the cost of such delay will increase and cross the cost curve of adjustment and reform, so as to make transformation and reform reach more rational balance
Shen (2002)	Cost of economic transformation is an important content for the objective existence of system change, and it is believed that radical-type system change would probably cause internal conflict of economic system, so as to cause relatively high cost of economic transformation. Only when the cost of economic transformation is far less than the benefit generated from compensation of cost, economic transformation is an effective and reasonable system change
Gang (1993)	A new system change is likely to bring instability of social members or intensive expansion of income gap, as well as dissatisfaction of consumers and social disturbance, etc. but only if such dissatisfaction, complaint or disturbance don't cause actual loss of national income, they may not be counted as cost of reform, only when national income is damaged due to the reason, such loss would be counted as cost of reform
Hong (1994)	Cost of economic transformation is the cost of system transition, this includes giving up the maximum benefit of reform, meaning the opportunity cost of transformation and the failure of adjustment of flawed old system, so as to delay the optimal opportunity of transformation and therefore cause the cost of delay, in actual economic life, it is reflected as loss of national income and social members' complaints, slow-down, social disturbance, war and other losses of interests and welfare of social members
Xu (2003)	Cost of economic transformation means all the prices occurred during the reform and operating process of economic system, and reflected as increase of production expense, enhancement of transaction cost and increase of external expenses
Chen (2010)	Cost of economic transformation means the expense payment or efficiency loss generated from the non-equilibrium caused from the transition from the equilibrium state of planning economy to the equilibrium state of market economy system

Source Chen Dandan: *Measurement of the Cost of Chinese Economic Transformation: 1978–2008*, Research of Quantitative Economy and Technical Economy, 2nd Issue in 2010

5.2 Expression Forms of the Cost of Economic Transformation in China

The cost of economic transformation in China has the following characteristics. Firstly, locality. During the thirty years of transformation process, China didn't have the phenomenon of major political turbulence, social crisis or economic downturn, localized and short-term cost issues didn't jeopardize the process of

Table 5.2 Selection process of system goals of China's economic transformation

Time	Goal of system	Basic characteristics/goal	Remarks
Prior to 1978	Planning economy	Highly concentrated decision making, enterprises become subsidiaries of administrative organs, even distribution	
1978, the Third Plenary Session of the 11th Central Committee of CCP	Planning economy that develop the rule of value	Tried to use "Invisible Hands" to stimulate economic development and support resource distribution	Beginning of the theoretical and practical exploration of socialist market economy
1982, the 12th National Congress of CCP	Dominated by planning economy and supported by market regulation	Make efforts to quadruple the annual total output of Chinese industrial and agricultural industry	Core words: planning economy
October, 1984, the Third Plenary Session of the 12th Central Committee of CCP	China's socialist economy is a planned commodity economy on the basis of public ownership	Broke through the tradition viewpoint that equals planning economy with socialism and contradicts commodity economy with planning economy	Core words: market economy
1987, the 13th National Congress of CCP	The state adjusts market, and the market guides enterprise	Combine the direct nature of market adjustment and indirect nature of macroeconomic regulation. By the mid of next century, China's per capita GDP will reach the level of medium developed countries and China will basically realize modernization	Simultaneously use the two means of planning and market
1989	Organic integration of planning economy and market adjustment	Organic integration of planning economy and market adjustment	Both planning economy and market adjustment are means of resource distribution
October, 1992, the 14th National Congress of CCP	Socialist market economy system	Having general characteristics of market economy, as well as having the special characteristics of socialist basic economy system; adhere to the leadership of the party, accelerate reform and opening-up, and concentrate the efforts to improve economic construction	Core word: market economy
1993, the Third Plenary Session of the 14th Central Committee of CCP	Socialist market economy system	Adopted the reform strategy of "General Promotion and Key Breakthroughs", began the hard work of establishing socialist market system	

(continued)

5.2 Expression Forms of the Cost of Economic Transformation in China

Table 5.2 (continued)

Time	Goal of system	Basic characteristics/goal	Remarks
September, 1997, the 15th National Congress of CCP	Socialist market economy system	Dominated by public ownership, common development of multiple ownership economic systems, implemented the strategy of developing the country through science and technology and the strategy of sustainable development	Non-public ownership economy is an important part of China's socialist market economy
November, 2002, the 16th National Congress of CCP	Socialist market economy system	Walked the path of new industrialization, energetically implemented the strategy of developing the country through science and technology and the strategy of sustainable development; accelerated the process of urbanization; deepened the reform of state-owned assets management system; improved modern market system, improved social security system; comprehensively enhanced the level of foreign opening-up; and make all efforts to expand employment	Build well-off society in an all-around way
October, 2003, the Third Plenary Session of the 16th Central Committee of CCP	Socialist market economy system	Deepening period of China's economic transformation, the two aspects of China's economic growth and economic system conversion both achieved fundamental development	
October, 2007, the 17th National Congress of CCP	Socialist market economy system	Unswervingly consolidate the develop the public ownership economy, unswervingly encourage, support and guide the development of non-public ownership economy, adhere to equal protection of property, form the new situation of equal competition and mutual promotion among various ownership economies. Strengthen the coordination of development, expand socialist democracy, strengthen cultural construction, accelerate the development of social courses; build ecological civilization, basically form the industrial structure, growth mode and consumption model of conservation of energy and resources and protection of ecological environment	

Source Xu Bing: *Analysis on the Staged Characteristics of the Cost of China's Economic Transformation*, Value Project, 5th Issue in 2011

Table 5.3 Forms of the cost of China's economic transformation

Classification of cost	Forms	Specific contents
Cost of economic transformation	Cost of system transformation	The logic of system transformation includes two aspects, introducing competitive system to develop non-public ownership economy, reforming the internal structure of original property, and promote the reform of state-owned enterprises. The cost of system transformation includes three parts, the cost of system supply, the cost of system reform and the loss of fairness
	Cost of structural transformation	Phenomenon of uncoordinated development in the aspects of urban-rural structure, industrial structure, regional development structure, etc.
	Cost of development mode transformation	Due to the incompletion of market economic system, the price of factors is relatively distorted, enterprises don't have sufficient incentives in environmental protection, so as to cause the phenomenon of efficiency loss and environmental deterioration
	Cost of globalization transformation	Globalization has on one hand brought opportunities to China's economy and enabled economic system to be flexible, but also caused market loss and risks, such as the occupation of Chinese market by multinationals, etc.

transformation. Secondly, China basically adopted the cost apportionment method of "Step-by-step, Postponement and Average", the effect of cost apportionment was good. Secondly, at the latter stage of transformation, there were more cost factors accumulated, the domestic and foreign environment to solve cost factors became more severe, the great costs of transformation are testing the government, using the transformation of government administrative system as a breaking point to launch new round of efforts for transformation would benefit the solving of cost factors and pushing for the final completion of transformation (Table 5.3).[5]

5.2.1 Cost of System Transformation

Gang (1993) believed that the transformation of economic system in China is the process of system substitution and switch, he called such process as "Double-track Transition and Strengthening Reform". It can be seen from practices that such diversified property system was a relatively good selection, but analyzing from the

---

[5]Xu Bing: *Comparison Research of the Economic Transformation and Its Cost in China and Russia,* Siberia Research, 2nd Issue.

perspective of system evolution, it is still in a state of unbalanced system and has great uncertainty, so it must generate large quantity of friction cost, cognition cost, implementation cost, compensation cost, etc.[6]

By qualitative study of the cost of system transformation, this book believes that from the perspective of stage theory, cost of system transformation has rheology rule. The changing characteristics of cost from system transformation in China are as shown in Chart 5.1: which are basically divided into 4 stages.

This Article believes that, by quantitatively analyzing the cost of system transformation, cost of system transformation includes three parts, system supply cost, system reform cost and equality loss. System supply cost includes governmental subsidy for loss of state-owned enterprises, other policy subsidies, etc. system reform cost includes non-performing loans of state-owned commercial banks, cost of unemployed workers, etc. for economic transformation as system adjustment, different entities bear and receive different incomes, so it generates large quantity of equality losses, usually Gini coefficient[7] and Theil Index[8] could be used to measure the degree of inequality.

Chart 5.2 describes the policy subsidies during 1998–2006, it can be seen from the chart that policy subsidies have always been maintain at a relatively stable amount, only some years had extensive increases, so its regulatory effects should be determined based on the actual circumstances.

Chart 5.3 reflects the state of non-performing loans of Chinese commercial banks during 2009–2009, non-performing loans gradually declined year by year, which reflected that the governance of non-performing loans tends to be better.

5.2.2 Cost of Structural Transformation

The original state before China's economic take-off was backward development stage, low development level and lag-behind industrial structure. During the process of China's economic transformation and development, structural adjustment and upgrading were important themes. For the national economic operation of any underdeveloped country, the transformation from traditional sectors to modern

---

[6]Chen Dandan: *Measurement of the Cost of Chinese Economic Transformation: 1978–2008*, Research of Quantitative Economy and Technical Economy, 2nd Issue in 2010.

[7]GINI Coefficient is normally used to reflect a country or region's situation of wealth distribution, according to the regulations of related organization of United Nations, if the GINI Coefficient is lower than 0.2, the income gap is relatively small, if it is higher than 0.5, the income gap is high.

[8]Theil Index, or Theil's entropy measure, is an indicated use to measure the income gap between individuals or between regions. One of the biggest advantages of using Theil Index to measure inequality is that it could measure the contribution of the gap within group and gap between groups to general gap. Theil Index and Gini Coefficient have certain complementary nature between each other. GINI Coefficient is particularly sensitive to changes in middle income, while Theil Index is relatively sensitive to changes of upper income level.

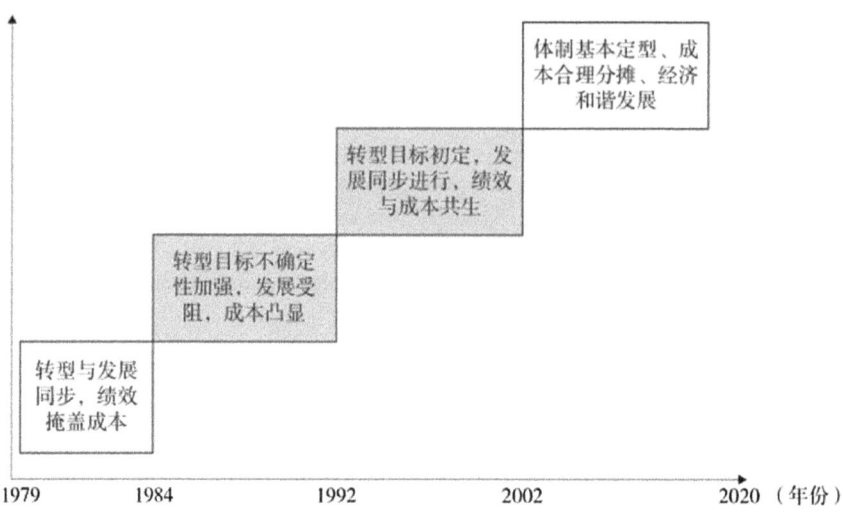

转型与发展同步，绩效掩盖成本	Transformation and development goes simultaneously, performance covers cost
转型目标不确定性加强，发展受阻，成本凸显	Uncertainty of transformation goal increases, development encounters obstacles, costs appear
转型目标初定，发展同步进行，绩效与成本共生	Transformation goal is preliminarily determined, development goes simultaneously, performance and cost coexist
体制基本定型、成本合理分摊、经济和谐发展	The system is basically fixed, costs are rationally distributed, the economy develops in harmony
（年份）	(Year)

Chart 5.1 Staged characteristics of the cost of system transformation

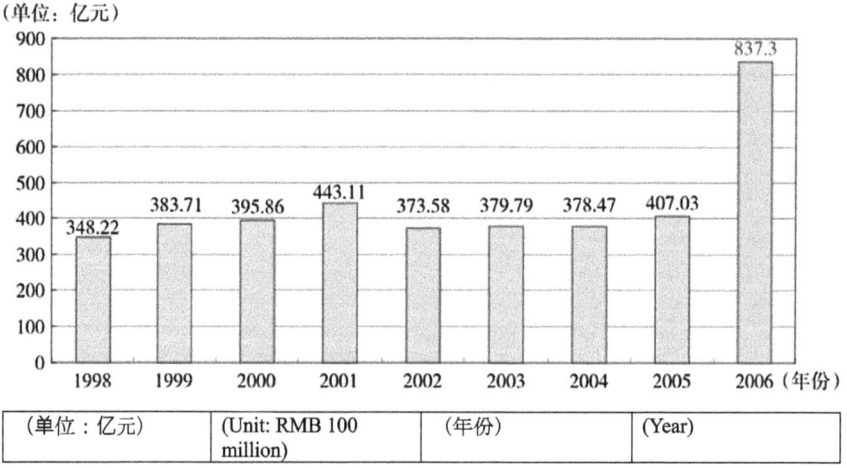

（单位：亿元）	(Unit: RMB 100 million)	（年份）	(Year)

Chart 5.2 Policy subsidies (1998–2006). *Source* Sorted and calculated from the data from *China Finance Yearbook 2007*

5.2 Expression Forms of the Cost of Economic Transformation in China 53

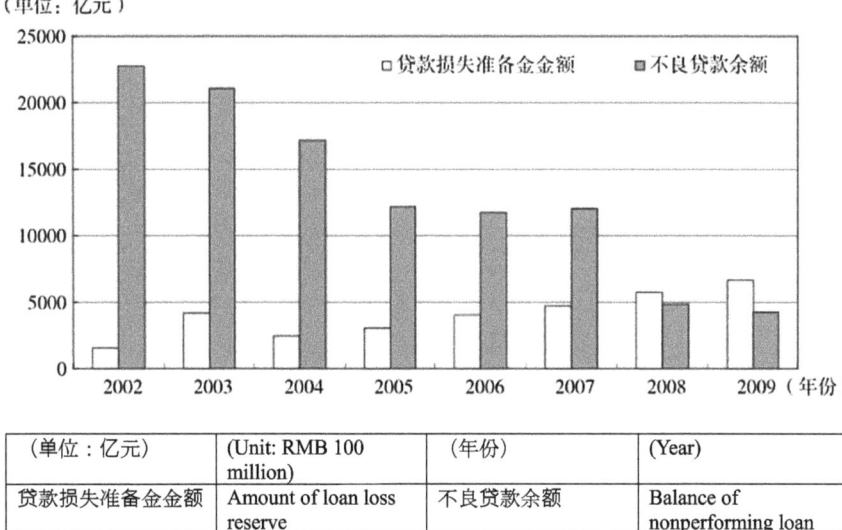

(单位：亿元)	(Unit: RMB 100 million)	(年份)	(Year)
贷款损失准备金金额	Amount of loan loss reserve	不良贷款余额	Balance of nonperforming loan

Chart 5.3 Situation of nonperforming loans of Chinese main commercial BANKS in various years (2002–2009). *Source* Sorted and calculated from the data from *China Finance Yearbook 2010*

sectors in economic structure would be very difficult, economic system and development stage are usually out of sync, which would cause various structure problems, the development of urban-rural structure, industrial structure and regional structure would become uncoordinated, and resource allocation would be irrational.

This book believes that as for corresponding problems of urban-rural structure, industrial structure and regional structure, the cost of structural transformation includes three parts: transforming cost of urban-rural dual structure, adjusting cost of industrial structure and changing cost of regional structure.

Transforming cost of urban-rural dual structure has two representative indicators, such as urban-rural income gap ratio and urban-rural Engle coefficient, although they couldn't directly calculate the specific cost measured with currency value, they could reflect the depth of cost.

The urban-rural income gap is as follows in Table 5.4:

The urban-rural Engle coefficient is as follows in Chart 5.4:

Adjusting cost of industrial structure could have two representative indicators, including the average wage ratio of different industries, especially the average wage ratio of financial industry and manufacturing industry; the other one is the proportion of national fiscal revenue used in agriculture expenditures (Chart 5.5 and Table 5.5).

The average wage ratio of financial industry and manufacturing industry increased from 1.64 times in 2003 to 2.25 times in 2009.

Besides, the other indicator representing industrial structure is the proportion of national fiscal revenue used in agriculture expenditures, which is shown in

Table 5.4 Urban-rural income gap

年份	1978	1990	2000	2008	2009
城镇居民人均可支配收入（元）	343	1510	6280	15781	17175
农村居民人均可支配收入（元）	134	686	2253	4761	5133
城乡收入差距比	2.56	2.20	2.79	3.31	3.35

Source Sorted and calculated from the data from *China Investment Yearbook 2010*.

年份	Year	城镇居民人均可支配收入(元)	Average disposable income of urban residents (RMB)
农村居民人均可支配收入(元)	Average disposable income of rural residents (RMB)	城乡收入差距比	Urban-rural income gap ratio

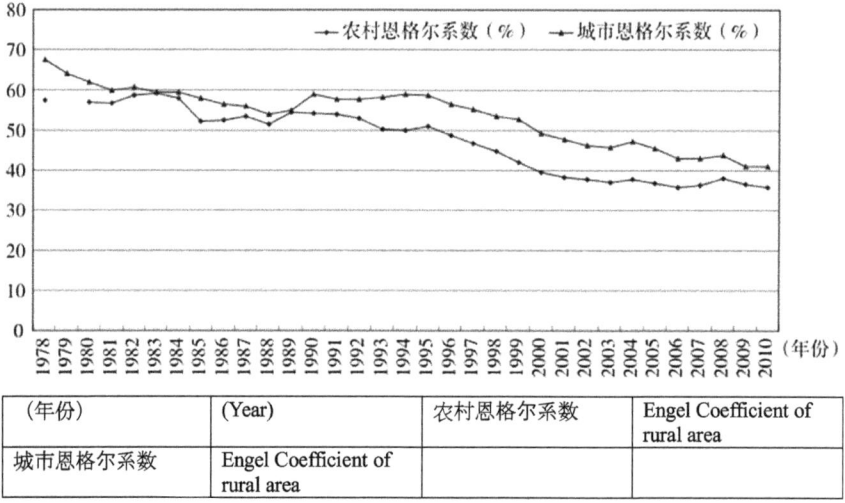

(年份)	(Year)	农村恩格尔系数	Engel Coefficient of rural area
城市恩格尔系数	Engel Coefficient of rural area		

Chart 5.4 Engel coefficient of urban and rural areas (1978–2010). *Source* Sorted and drawn from the data from *China Civil Affairs Statistic Yearbook 2011*

Charts 5.6 and 5.7. The national financial expenditures in agriculture had relatively extensive changes from 1952 to 1985. The state from 1985 to 2010 was relatively stable, the national financial expenditures on agriculture basically maintained at around 8%.

Changing cost of regional structure could usually measured with the ratios of eastern, middle and western regions in national GDP or other similar indicators. The GDP of eastern China has always been accounting for over 60% of national GDP. The specific circumstances are as follows in Table 5.6.

5.2 Expression Forms of the Cost of Economic Transformation in China

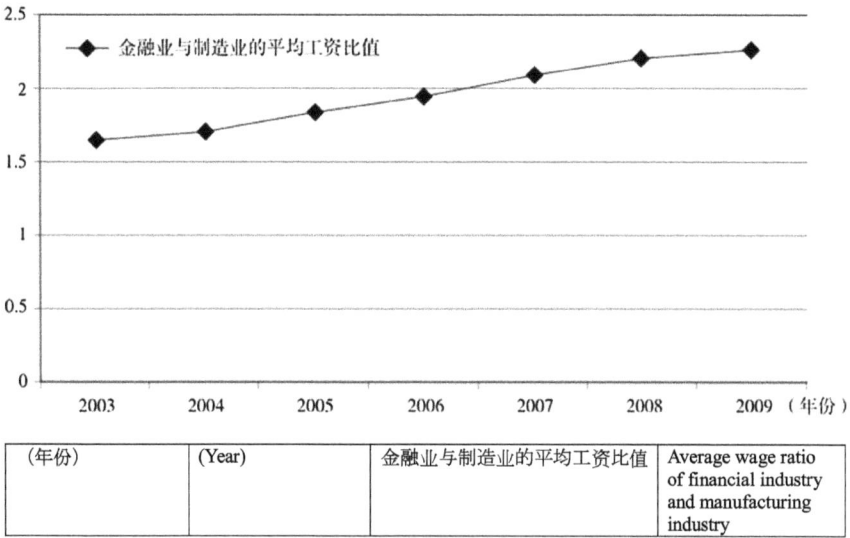

(年份)	(Year)	金融业与制造业的平均工资比值	Average wage ratio of financial industry and manufacturing industry

Chart 5.5 Average wage ratio of financial industry and manufacturing industry (2003–2009). *Source* Sorted and drawn from the data from *China Statistic Yearbook 2011*

5.2.3 Cost of Development Model Transformation

Cost of development model transformation means the efficiency loss, environmental deterioration, development unbalance or other prices caused under the circumstances that the market economic system is incomplete, the factor values are relatively distorted, and enterprises don't have sufficient incentives in environmental protection or other aspects of transformation development. It usually could be measured with the ratio of urbanization lagging behind industrialization,[9] or the ratio of consumption rate lagging behind investment rate.

The comparison of Chinese urbanization ratio and industrialization ratio is as follows in Chart 5.8. After 2003, the urbanization ratio approached and surpassed industrialization ratio, it can be seen that urbanization lagged long behind industrialization, this stage lasted as long as 25 years.

Besides, from the comparison of investment rate and consumption rate, China's investment rate was always lower than consumption rate until 2010 when it approached and surpassed consumption rate (refer to Chart 5.9).

---

[9]Industrialization ratio means the ratio of added industrial value in total production output.

Table 5.5 Comparison of average wage of 19 industries (2003–2009)

年份	2003	2004	2005	2006	2007	2008	2009
合计	13969	15920	18200	20856	24721	28898	32244
1. 农、林、牧、渔业	6884	7497	8207	9269	10847	12560	14356
2. 采矿业	13627	16774	20449	24125	28185	34233	38038
3. 制造业	12671	14251	15934	18225	21144	24404	26810
4. 电力、燃气及水的生产和供应业	18574	21543	24750	28424	33470	38515	41869
5. 建筑业	11328	12578	14112	16164	18482	21223	24161
6. 交通运输、仓储和邮政业	15753	18071	20911	24111	27903	32041	35315
7. 信息传输、计算机服务和软件业	30897	33449	38799	43435	47700	54906	58154
8. 批发和零售业	10894	13012	15256	17796	21074	25818	29139
9. 住宿和餐饮业	11198	12618	13876	15236	17046	19321	20860
10. 金融业	20780	24299	29229	35495	44011	53897	60398
11. 房地产业	17085	18467	20253	22238	26085	30118	32242
12. 租赁和商务服务业	17020	18723	21233	24510	27807	32915	35494
13. 科学研究、技术服务和地质勘查业	20442	23351	27155	31644	38432	45512	50143
14. 水利、环境和公共设施管理业	11774	12884	14322	15630	18383	21103	23159
15. 居民服务和其他服务业	12665	13680	15747	18030	20370	22858	25172
16. 教育	14189	16085	18259	20918	25908	29831	34543
17. 卫生、社会保障和社会福利业	16185	18386	20808	23590	27892	32185	35662
18. 文化、体育和娱乐业	17098	20522	22670	25847	30430	34158	37755
19. 公共管理和社会组织	15355	17372	20234	22546	27731	32296	35326

Source Sorted and calculated from the data from *China Statistics Yearbook 2011*.

年份	Year
合计	Total
1.农、林、牧、渔业	1. Agriculture, forestry, husbandry and fishery
2.采矿业	2. Mining industry
3.制造业	3. Manufacturing industry
4.电力、燃气及水的生产和供应业	4. Production and supply industry of electricity, gas and water
5.建筑业	5. Construction industry
6.交通运输、仓储和邮政业	6. Traffic, transportation, warehousing and postal industry
7.信息传输、计算机服务和软件业	7. Information transmission, computer service and software industry
8.批发和零售业	8. Wholesale and retail industry
9.住宿和餐饮业	9. Hotel and restaurant industry
10.金融业	10. Financial industry
11.房地产业	11. Real estate industry
12.租赁和商务服务业	12. Lease and commercial service industry

5.2 Expression Forms of the Cost of Economic Transformation in China

Table 5.5 (continued)

13.科学研究、技术服务和地质勘查业	13. Scientific research, technological service and geological prospecting industry
14.水利、环境和公共设施管理业	14. Water conservancy, environment and public facility management industry
15. 居民服务和其他服务业	15. Residents service and other service industry
16.教育	16. Education
17.卫生、社会保障和社会福利业	17. Health, social security and social welfare industry
18.文化、体育和娱乐业	18. Culture, sports and entertainment industry
19.公共管理和社会组织	19. Public administration and social organization

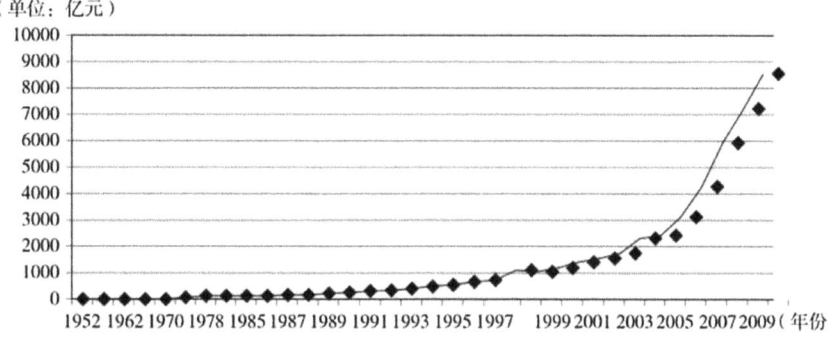

(单位：亿元)	(Unit: RMB 100 million)	(年份)	(Year)
国家财政农业支出	National fiscal expenditure on agriculture		

Chart 5.6 National fiscal expenditure on agriculture (1952–2010). *Source* Sorted from *China Agricultural Statistic Yearbook 2011*

5.2.4 Cost of Globalization Transformation

Globalization is a double-edged sword, on one hand, it enabled Chinese economic system to be more complete and more flexible; and on the other hand, it brought certain losses and risks, for example, multinationals occupied Chinese market and threatened domestic enterprises, etc., besides, during the process of integrating into globalization, the fluctuation of global economy would directly affect China, for example, the financial crisis caused by subprime mortgage crisis in the US also led to economic fluctuation and losses in China. The usual indicators to measure the

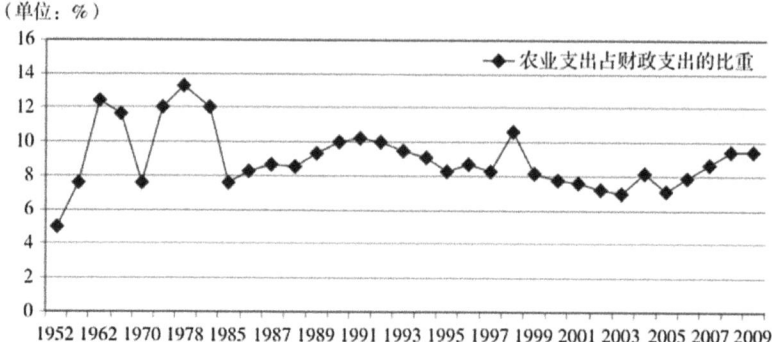

Chart 5.7 Ratio of agricultural expenditure in fiscal expenditure (1952–2009). *Source* Sorted from *China Agricultural Statistic Yearbook 2011*

Table 5.6 Ratio of the GDP of Eastern, central and western regions in national GDP (*Unit %*)

	2006	2007	2008
Eastern Region	61.8	61.4	60.6
Central Region	25.3	25.6	26.2
Western Region	12.9	13	13.2

Source Sorted from *China Statistic Yearbook 2009*

cost of globalization transformation include ratio of dependence on foreign trade,[10] capital flight, etc.

The ratio of dependence on foreign trade had been rising since the beginning of reform and opening-up to 2006 when it peaked at 65.17%, the period of 2006–2010 was the adjustment period, ratio of dependence on foreign trade being too high is not good for healthy and rapid economic growth, the details of which are as follows in Chart 5.10.

Capital flight means the situation that the domestic and foreign investors in a country or economy massively unload financial assets in such country out of worry that this country could have economic downturn or other economic or political uncertainty, and transfer funds to overseas.

The causes of capital flight mainly include transfer of illegal gains, realization of privatization, risk-avoiding regulation, seeking profit and avoiding risks, as well as transfer of individual assets.

---

[10] Foreign trade dependence is also referred to as foreign trade coefficient (traditional foreign trade coefficient), meaning the ratio of a country's total import/export amount in such country's GNP or GDP.

5.2 Expression Forms of the Cost of Economic Transformation in China 59

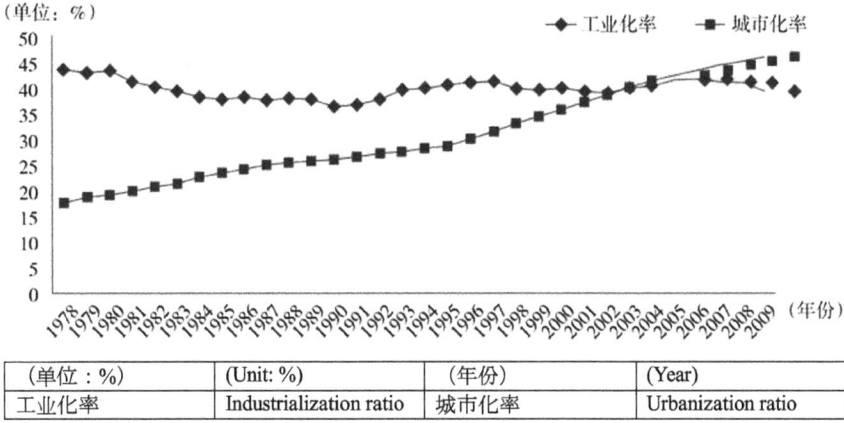

(单位：%)	(Unit: %)	(年份)	(Year)
工业化率	Industrialization ratio	城市化率	Urbanization ratio

Chart 5.8 China's urbanization ratio and industrialization ratio in various years (1978–2009). *Source* Sorted from *China Statistic Yearbook 2010*

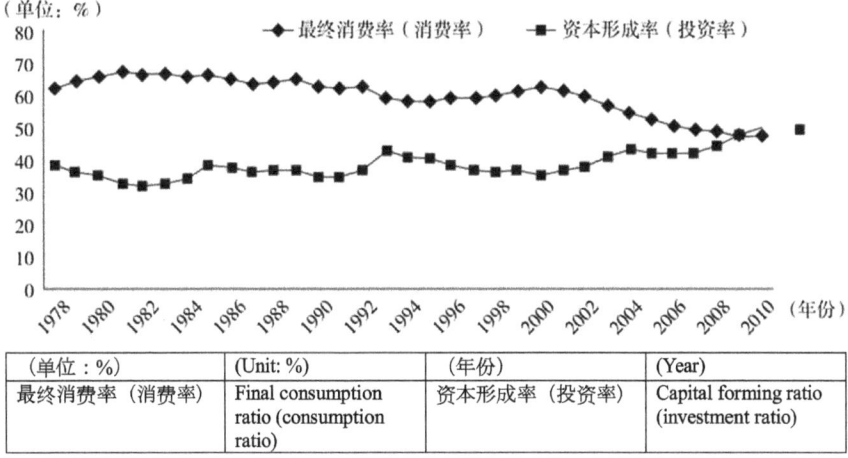

(单位：%)	(Unit: %)	(年份)	(Year)
最终消费率（消费率）	Final consumption ratio (consumption ratio)	资本形成率（投资率）	Capital forming ratio (investment ratio)

Chart 5.9 China's consumption ratio and investment ratio in various years (1978–2009). *Note* Capital formation ratio means the ratio of total capital formed in total GDP; final consumption ratio means the ratio of final consumption expenditure in total GDP. Sorted from *China Statistic Yearbook 2010*

The total volume of capital flight is relatively high and reflects obvious rising trend, and also reflects the coexistence of capital flight and capital inflow, increase rate of capital flight reflects obvious stage characteristics, and has strong sensitivity to the macroeconomic environment, besides, the scale of capital flight couldn't be reduced or restrained in short term. For the detailed data, please refer to Table 5.7.

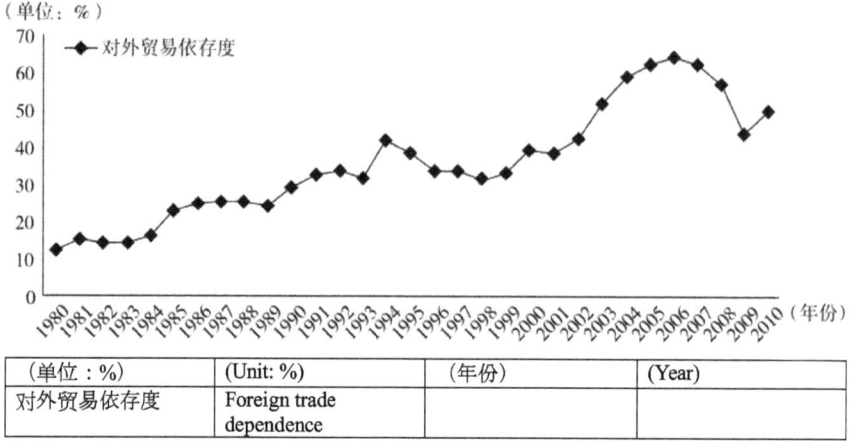

Chart 5.10 Foreign trade dependence (1980–2010). *Source* Calculated and sorted from the data from *China Commercial Yearbook 2011* and *China Foreign Trade Statistic Yearbook 2011*

5.3 Forming Causes of the Cost of Economic Transformation in China

The economic transformation in China once enabled China to become one of the countries in the world with the fastest growth rate of per capita gross domestic product, it can be said that almost every person, sector and region was benefited from the reform. China is also one of the countries in the world with the fastest and most extensive growth of Gini coefficient that reflects inequality.[11] Economic transformation is about transforming structure and adjusting model, as well as a tremendous task facing the domestic system and facing the international market. The forming causes are diversified, transformation is a long-term process, guaranteeing the continuous economic growth is very important for government and the people to reach the targets of increasing income and reducing poverty.[12]

Representation of transformation cost is exactly the representation of problems occurred during transformation, the causes of cost of economic transformation occurred during economic transformation are as follows in the following chart (Chart 5.11):

---

[11]Wang Shaoguang: *China: Facing the Challenge of Inequality and Response,* National Situation Report, 34th Issue in 1998.

[12]Lee Kuan Yew well interpreted the development of Singapore in his book *From Third World to First World: Singapore's Story during 1965–2000,* "Just 41 years ago, independent Singapore started the journey unknown. With high unemployment rate and lack of its own industry, the future of Singapore was bleak". Former Premier Lee Kuan Yew wrote, "I nervously started off along the small path nobody knows where it leads".

5.3 Forming Causes of the Cost of Economic Transformation in China 61

Table 5.7 Calculation results and related data of China's escaped capitals during 1985–2008 (*Unit* USD 100 million)

年份	直接法		间接法	综合法	资本流入	实际利用外国直接投资净流入	外债增量
	游资 2	游资 3					
1985	−1.40	−31.67	−23.55	−18.87	16.59	10.30	37.4
1986	1.65	−14.03	9.03	−1.12	18.75	14.25	57.50
1987	8.24	−2.27	15.11	7.03	23.14	16.69	87.20
1988	22.76	14.00	13.51	16.76	31.94	23.44	98.00
1989	5.38	7.18	2.30	4.95	33.93	26.13	13.00
1990	34.32	36.73	33.75	34.93	34.87	26.57	112.50
1991	67.24	64.89	70.78	67.64	43.66	34.53	80.10
1992	149.50	150.07	87.02	128.86	111.56	71.56	87.60
1993	149.42	118.92	104.01	124.12	275.15	231.15	142.50
1994	145.32	109.89	101.55	118.92	337.87	317.87	92.40
1995	176.08	168.18	177.33	173.86	358.49	338.49	137.80
1996	172.15	154.71	161.94	162.93	401.80	380.66	96.90
1997	189.82	120.39	627.83	312.68	442.36	416.74	146.80
1998	309.64	346.97	576.41	411.01	437.52	411.18	150.80
1999	320.52	432.85	527.44	426.94	387.52	369.78	57.90
2000	217.77	257.68	670.99	382.15	383.99	374.83	−61.00
2001	−26.61	167.45	47.51	62.78	442.41	373.56	390.70
2002	−114.70	−11.28	74.27	−17.24	493.08	467.90	15.30
2003	54.47	−59.80	−34.36	−13.23	470.77	472.29	224.30
2004	−363.92	−560.82	−354.42	−426.39	549.36	531.31	387.30
2005	162.78	212.10	877.69	417.52	791.27	678.21	335.60
2006	35.34	710.91	1512.01	752.75	780.95	569.35	419.38
2007	870.04	683.32	1343.09	965.48	1384.13	1214.18	506.30
2008	882.72	456.12	963.67	767.51	1477.91	943.20	10.82

Source Ying Weihua and Zhang Huanming: *Calculation of the Scale of China's Escaped Capitals 1985-2008,* Financial Development Research, 8th Issue in 2009.

年份	Year	直接法	Direct method
游资2	Hot money 2	游资3	Hot money 3
间接法	Indirect method	综合法	Comprehensive method
资本流入	Capital inflow	实际利用外国直接投资净流入	Net inflow of actually used foreign direct investment
外债增量	Increment of foreign debts		

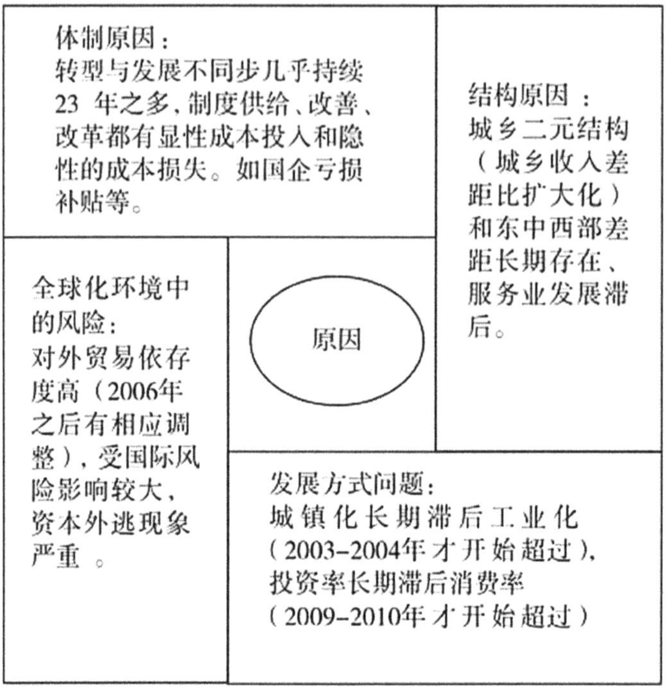

体制原因: 转型与发展不同步几乎持续23年之多，制度供给、改善、改革都有显性成本投入和隐性的成本损失。如国企亏损补贴等。	System cause: The out-sync situation between transformation and development lasted for almost 23 years, system supply, improvement and reform all had obvious cost input and potential cost loss, such as loss subsidy of state-owned enterprises, etc.
全球化环境中的风险: 对外贸易依存度高(2006年之后有相应调整)，受国际风险影响较大，资本外逃现象严重。	Risks in the environment of globalization: High foreign trade dependence (adjusted accordingly since 2006), subject to high influence from international risks, and severe phenomenon of capital escape.
结构原因: 城乡二元结构(城乡收入差距比扩大化)和东中西部差距长期存在、服务业发展滞后	Structural cause: Urban-rural dual structure (widening urban-rural income gap) and long-term existence of gap among eastern, central and western regions, and service industry lagging behind
发展方式问题: 城镇化长期滞后工业化(2003-2004年才开始超过)，投资率长期滞后消费率(2009-2010年才开始超过)	Development mode issue: For a long time, urbanization lags behind industrialization (only caught up in 2003-2004), and investment ratio lags behind consumption ratio (only caught up in 2009-2010)
原因	Cause

Chart 5.11 Causes of the cost of China's economic transformation

5.4 Countermeasures for the Cost of Economic Transformation in China

Michael Spence (2008) believed that in fact, high-speed growth would generate bottlenecks that couldn't be fully predicted. The success of China's growth strategy was partially reflected as the quick response to the bottlenecks or restriction of blocks.[13] Zou (2002) evaluated that the economic transformation occurred in China was to most extent a kind of government action designed to realize economic modernization of China, although private motive for profit also played certain role.[14] So the government is obliged to handle the phenomenon of the excessively high cost generated during the control of economic transformation. In fact, the government is handling these cost problems in four aspects. The regional policies, industrial policies, external opening policies, gradual reform policies, etc. are continuously solving new problems and settling leftover problems. Controlling the cost of economic transformation is a systematic project. Because reform and transformation themselves need many preparations, including political preparation, cultural preparation, etc. Its benefit is significant, so lower cost is not necessarily better. We should control costs within rational extent, especially the control of expansion of invisible costs, reduction of equality loss is an aspect where more efforts need to be made (Chart 5.12).

Annexed Tables

See Tables 5.8, 5.9 and 5.10.

---

[13]Michael Spence: *Successful Experiences and New Challenges of China's Reform and Opening-up*, Overseas China Research, 8th Issue in 2008.

[14][US] Zou Zhizhuang: *China Economic Transformation, Chinese Edition*, China Renmin University Press, 2005 Edition, Page 2.

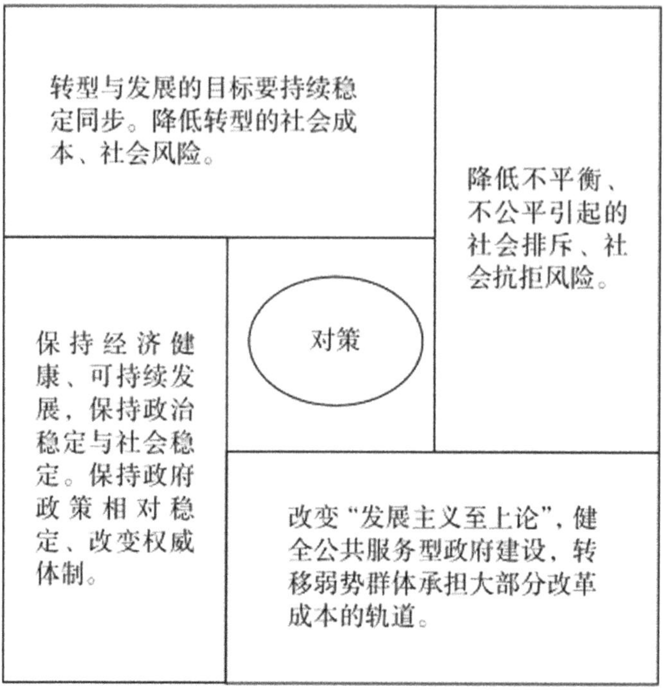

转型与发展的目标要持续稳定同步。降低转型的社会成本、社会风险。	The purpose of transformation and development should be continuously and stably in-synchronization. Reduce the social cost and social risk of transformation.
降低不平衡、不公平引起的社会排斥、社会抗拒风险。	Reduce the risk of social rejection and social resistance caused from unbalance and inequality.
保持经济健康、可持续发展，保持政治稳定与社会稳定。保持政府政策相对稳定、改变权威体制。	Keep economic health, sustainable development, and maintain political stability and social stability. Keep the relative stability of government policies, and change the authoritative system.
改变"发展主义至上论"，健全公共服务型政府建设，转移弱势群体承担大部分改革成本的轨道。	Change the "Development Upmost Theory", improve the construction of public service-type government, and change the situation that the disadvantageous groups of society bear the most reform cost.
对策	Countermeasure

Chart 5.12 Countermeasures for the cost of China's economic transformation

Table 5.8 Statistics of China's national fiscal expenditures on agriculture in various years (1952–2010)

年份	农业支出（亿元）	农业支出占财政支出的比重（%）
1952	9	5.1
1957	23.5	7.7
1962	38.2	12.5
1965	55	11.8
1970	49.4	7.6
1975	99	12.1
1978	150.7	13.4
1980	150	12.2
1985	153.6	7.7
1986	184.2	8.4
1987	195.7	8.7
1988	214.1	8.6
1989	265.9	9.4
1990	307.8	10
1991	347.6	10.3
1992	376	10
1993	440.5	9.5
1994	533	9.2
1995	574.9	8.3
1996	700.4	8.8
1997	766.4	8.3
1998	1154.8	10.7
1999	1085.8	8.2
2000	1231.5	7.8
2001	1456.7	7.7
2002	1580.8	7.2
2003	1754.5	7.1
2004	2337.6	8.2
2005	2450.3	7.2
2006	3173	7.9
2007	4318.3	8.7
2008	5955.5	9.5
2009	7253.1	9.5
2010	8579.7	9.5

Table 5.8 (continued)

Note: 1. Since 1998, the "Expenditure for basic agricultural construction" included expenditure from additional issuance of national bonds. 2. Since 2007, the caliber of national fiscal expenditure on supporting farmers had been different from that of previous years due to adjustment of financial statement adjustment. The expenditure on supporting farmers in this table only refers to the expenditure from Central Financial Authority on "Agriculture, Rural Area and Farmers".

Source Sorted from the data from *China Rural Statistic Yearbook 2011*.

年份	Year	农业支出（亿元）	Agricultural expenditure (RMB 100 million)
农业支出占财政支出的比重（%）		Ratio of agricultural expenditure in financial expenditure (%)	

Table 5.9 China's industrialization ratio, urbanization ratio, consumption ratio and investment ratio in various years (1978–2010)

	工业化率 (%)	城市化率 (%)	消费率 (%)	投资率 (%)
1978	44.1	17.9	62.1	38.2
1979	43.6	19	64.4	36.1
1980	43.9	19.4	65.5	34.8
1981	41.9	20.2	67.1	32.5
1982	40.6	21.1	66.5	31.9
1983	39.9	21.6	66.4	32.8
1984	38.7	23	65.8	34.2
1985	38.3	23.7	66	38.1
1986	38.6	24.5	64.9	37.5
1987	38.0	25.3	63.6	36.3
1988	38.4	25.8	63.9	37
1989	38.2	26.2	64.5	36.6
1990	36.7	26.4	62.5	34.9
1991	37.1	26.9	62.4	34.8
1992	38.2	27.5	62.4	36.6
1993	40.2	28	59.3	42.6
1994	40.4	28.5	58.2	40.5
1995	41.0	29	58.1	40.3
1996	41.4	30.5	59.2	38.8
1997	41.7	31.9	59	36.7
1998	40.3	33.4	59.6	36.2
1999	40.0	34.8	61.1	36.2
2000	40.4	36.2	62.3	35.3
2001	39.7	37.7	61.4	36.5
2002	39.4	39.1	59.6	37.8
2003	40.5	40.5	56.9	40.9
2004	40.8	41.8	54.4	43
2005	41.8	43	52.9	41.6
2006	42.2	43.9	50.7	41.8
2007	41.6	44.9	49.5	41.7
2008	41.5	45.7	48.4	43.9
2009	39.7	46.6	48.2	47.5
2010		49.6	47.4	48.6

Source Sorted from *China Statistic Yearbook 2010*.

工业化率 (%)	Industrialization Ratio (%)	城市化率 (%)	Urbanization Ratio (%)
消费率 (%)	Consumption Ratio (%)	投资率 (%)	Investment Ratio (%)

Table 5.10 Calculation of China's GDP, total import/export amount and foreign trade dependence in various years (1980–2010)

	GDP (亿元)	进出口总额 (亿元)	对外贸易依存度 (%)
1980	4545.6	570	12.54
1981	4891.6	735.3	15.03
1982	5323.4	771.3	14.49
1983	5962.7	860.1	14.42
1984	7208.1	1201	16.66
1985	9016	2066.7	22.92
1986	10275.2	2580.4	25.11
1987	12058.6	3084.2	25.58
1988	15042.8	3821.8	25.41
1989	16992.3	4155.9	24.46
1990	18667.8	5560.1	29.78
1991	21781.5	7225.8	33.17
1992	26923.5	9119.6	33.87
1993	35333.9	11271	31.90
1994	48197.9	20381.9	42.29
1995	60793.7	23499.9	38.66
1996	71176.6	24133.8	33.91
1997	78973	26967.2	34.15
1998	84402.3	26849.7	31.81
1999	89677.1	29896.3	33.34
2000	99214.6	39273.2	39.58
2001	109655.2	42183.6	38.47
2002	120332.7	51378.2	42.70
2003	135822.8	70483.5	51.89
2004	159878.3	95539.1	59.76
2005	184937.4	116921.8	63.22
2006	216314.4	140974	65.17
2007	265810.3	166863.7	62.78
2008	314045.4	179921.5	57.29
2009	340902.8	150648.1	44.19
2010	401202	201722.1	50.28

Source Calculated and sorted from the data from *China Business Yearbook 2011* and *China's Foreign Trade Statistic Yearbook 2011*.

GDP (亿元)	GDP (RMB 100 million)	进出口总额 (亿元)	Total Import/Export Amount (RMB 100 million)
对外贸易依存度 (%)	Foreign trade dependence		

Chapter 6
Cost of Economic Disturbance in China

6.1 Overview of Economic Disturbance

Disturbance means the behavior that intervenes or obstructs an action or procedure. In Systematic, the redundant, unwanted and mandatory "input" is referred to as disturbance to system. The reduction of input, restriction on output and excessive output are referred to as disturbance. The so-called disturbance includes political disturbance and economic disturbance. Economic disturbance means the phenomenon that an external economic activity has harmful influence on an economic entity, or impedes the input or output of economic system. For example, excessive exchange rate fluctuation would usually cause disequilibrium of balance of international payments, excessive market speculations, excessive impacts on the real economy or other similar phenomenon. The problems such as forced appreciation of RMB or low transparency of exchange rate policies would often affect the normal economic operation.

When a country or region is connected to world economy, such country or region would be inevitably putting itself in the position of being vulnerable to external impacts, meaning it may be affected by economic disturbance caused by foreign events.[1]

The main causes of economic disturbance generated from external impact are as follows in Chart 6.1.

This book believes that both exchange rate fluctuation and economic crisis could generate external impact on a country or district, so as to form economic disturbance. Besides, inflation could be imported inflation, which is also an important factor of the output of economic system and economic result. Therefore, this book will analyze the influential factors of inflation, exchange rate fluctuation, economic crisis as the key points in the following chapters and sections.

---

[1]Wang Liling: *Discussion on the Defense Measures for Developing Countries Facing Foreign Impacts*, Journal of Qinghai Normal University (Philosophy Social Science Edition), 4th Issue in 1999.

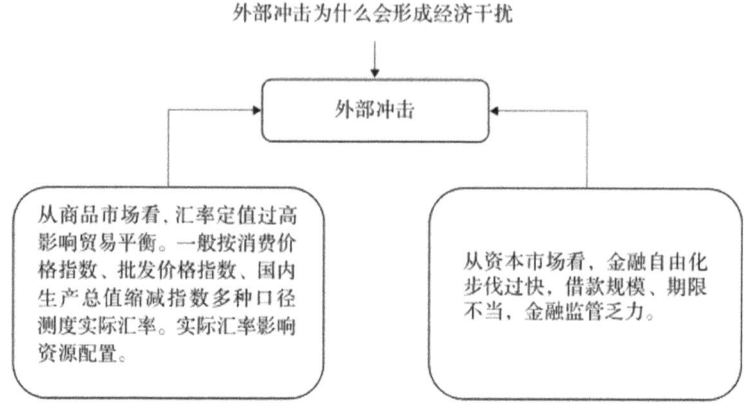

外部冲击为什么会形成经济干扰	Why foreign impacts could form economic disturbance
外部冲击	Foreign impacts
从商品市场看，汇率定值过高影响贸易平衡。一般按消费价格指数、批发价格指数、国内生产总值缩减指数多种口径测度实际汇率。实际汇率影响资源配置。	From the perspective of commodity market, excessively high exchange rate would affect trade balance. The actual exchange rate is normally measured through consumer price index, wholesale price index, GDP shrinking index and various other calibers. Actual exchange rate would affect resource distribution.
从资本市场看，金融自由化步伐过快，借款规模、期限不当，金融监管乏力。	From the perspective of capital market, too fast steps of financial liberalization, improper loan scale and period and insufficient financial supervision.

Chart 6.1 Formation of economic disturbance from foreign impacts

6.2 Expression Forms of the Costs of Economic Disturbance in China

This book believes that cost of economic disturbance includes three kinds of expression forms, for the specific contents, please refer to Table 6.1.

6.2.1 Inflation Cost

Inflation is generally defined as the phenomenon that under the conditions of fiduciary circulation, due to more currency supplied than actual demands of currency, meaning more actual purchasing power than production supply, currency would devaluate, as a result, prices continuously and widely rise during certain period. The reason why we consider inflation as a cost is because when inflation is

6.2 Expression Forms of the Costs of Economic Disturbance in China

Table 6.1 Forms of the cost of China's economic disturbance

Classification of cost	Forms	Specific contents
Cost of economic disturbance	Cost of inflation	Cost from currency devaluation and continuous rising prices generated from excessive currency supply
	Cost exchange rate change	Cost brought to China's economic development from the upwards or downwards fluctuation of foreign value of currency, including currency devaluation and currency appreciation
	Risk of economic crisis	Risk brought to China's economic growth from foreign economic crisis during certain period

too high or too low, inflation would have negative influence on economic growth (by affecting production factors).[2]

After the financial tsunami in 2009, in order to cope the negative influences, China adopted a series of economic stimulation programs and measures, which pulled up the economic growth speed within short period of time, but the inflation expectation of citizens also continuously increased. The academic circle has three viewpoints on the relationship between inflation and economic growth (Tables 6.2 and 6.3).

Take a look at the inflation of China from the details of consumer price index, as shown in Chart 6.2, the peak of consumer price index occurred in 1994, and the valley occurred in 1999.

It can be seen from Chart 6.2 that during the thirty one years from 1978 to 2009, based on the study conclusion of Hou (2010), there were 16 years when the index was higher than 5.1% or lower than 1.4%, during over half of the years, the index was not within the optimal range.

The damage of inflation is reflected in the following aspects, which are as follows in Chart 6.3.

6.2.2 Cost of Exchange Rate Fluctuation

First of all, let's study the fluctuation of exchange rate, the following data analysis is based on the statistics of RMB market exchange rates during 1981–2010.

It can be seen from Chart 6.4 that RMB relatively appreciated from 1994 to 2000, but it maintained at relatively stable level during the 11 years from 1994 to 2004, after 2005, the pressure of RMB appreciation increased.

---

[2]Hou Qing: *Whether China has Optimal Span of Inflation—Empirical Research of the Relationship between Inflation and Economic Growth,* Value Project, 2nd Issue in 2010.

Table 6.2 Three viewpoints regarding the relations between inflation and economic growth

Viewpoint	Detailed explanation
No correlation	There is no significant correlation between economic growth and inflation
Tobin effect	Inflation has positive influence on economic growth
Reverse Tobin effect	Inflation has negative influence on economic growth

Table 6.3 Viewpoints of nonlinear relations between inflation and economic growth

Viewpoints	Detailed explanation
Sarel (1996)	When inflation is higher than 8%, inflation has negative influence on economic growth, when inflation is lower than 8%, inflation has positive influence on economic growth
Christoffersen and Doyle (1997)	In the research of transforming countries, it was found that the threshold of different roles that inflation plays in economic growth is 13%
(Korea) Jaerany Lee and (US) Kar-yin Wong (2005)	In a research of Japan, it was found that the optimal inflation range of Japan is 2.5–9.7%, inflations higher than 9.7% or lower than 2.5% would both have negative influence on the economic growth
Liu and Xie (2003)	There is threshold effect in the mutual role between inflation and growth rate, when inflation rate is higher or lower than certain level, it would impede economic growth, meaning there is a optimal inflation level, which could not only benefit proper growth of economy, but also prevent virulent inflation from happening
Hou (2010)	The optimal inflation range of Chinese economy is 1.4–5.1%, when the inflation rate is higher than 5.1% or lower than 1.4%, inflation would have negative influence on economic growth (through influence on production factors)

Source Concluded and sorted by the author

With the launch of RMB exchange rate formation mechanism reform on Jul 21, 2005, under the managed floating exchange rate system, fluctuation range of RMB exchange rate obviously increased. Under the major background of strengthening flexibility of RMB exchange rate, how to handle risks of exchange rate fluctuation became an important content of risk management for domestic enterprises, especially the foreign trade enterprises. Exchange rate is a double-edged sword. Exchange rate risk is derived from the uncertainty of global exchange rate fluctuations, it could bring both damage and benefit to enterprises. Therefore, as an active management strategy, enterprises could, based on correct expectation of exchange rate, profit from fluctuation of exchange rate. Currently China's goal of monetary policy is positioned as maintaining the stability of RMB currency value, the harms of RMB appreciation include: (Chart 6.5)

6.2 Expression Forms of the Costs of Economic Disturbance in China

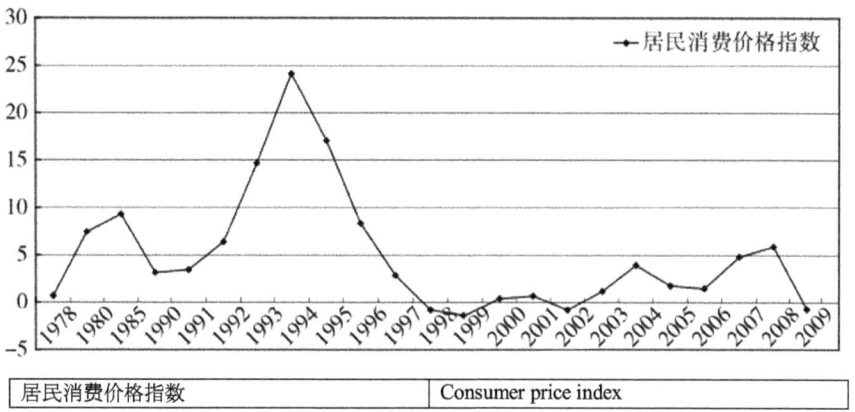

Chart 6.2 Consumer price index (1978–2009). *Source* Sorted and calculated from the data from *China Statistics Yearbook 2011*

6.2.3 Risk of Economic Crisis

In 1840s, Marx studied the periodic economic crisis of capitalism with universal overproduction, the publication of *Capital* symbolized the formation of Marx's economic crisis theories. Marx believes that overproduction of capitalism only means the overproduction of the means of production—means of labor and means of livelihood—that can be used to execute functions for capital, meaning the exploitation of labor based on certain degree of exploitation; when such degree of exploitation drops below certain point, it would cause chaos, stagnation, crisis and destruction of capital during capitalism production process.[3] Marx believed that economic cycle is composed of four stages, crisis, recession, recovery and prosperity, economic crisis is the inevitable result of capitalism economic development. In the book *Wage, Price and Profit*, Marx wrote "Productions of capitalism always have to go through certain periodic circulation. It will go through stages of recession, gradual recovery, prosperity, overproduction, crisis, stagnation, etc.[4]

Economic Crisis means that one or multiple national economies or the entire world economy continuously shrinks during a relatively long period of time (negative economic growth rate).

Since the reform and opening-up, the financial crisis that affected China mainly include the Asian Financial Crisis in 1998 and the financial crisis on global scale caused from the US subprime mortgage crisis in 2008 (Chart 6.6).

---

[3] *Karl Marx and Frederick Engels,* Volume 25, People's Publishing House, 1974 Edition, Page 285.
[4] *Karl Marx and Frederick Engels,* Volume 3, People's Publishing House, 2009 Edition, Page 71.

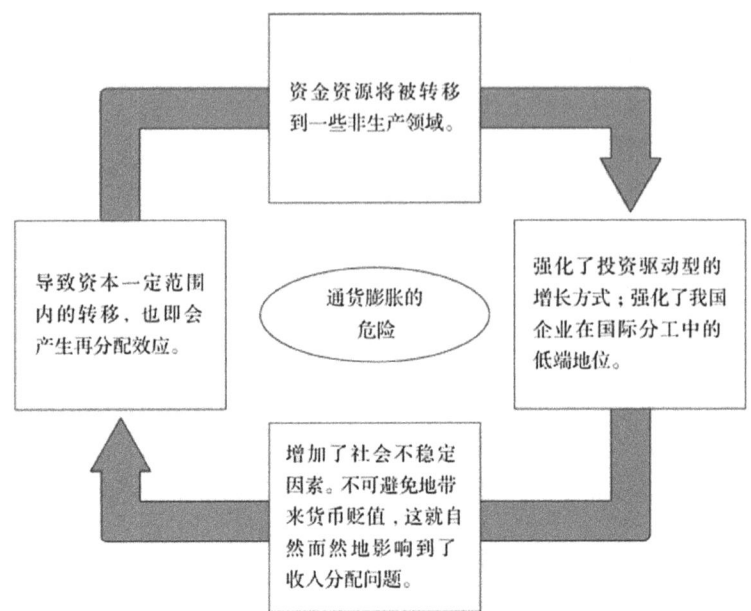

资金资源将被转移到一些非生产领域。	Resource of funds will be transferred to certain nonproduction fields
导致资本一定范围内的转移，也即会产生再分配效应。	Cause transfer of capital in certain scope, and cause redistribution effect
强化了投资驱动型的增长方式;强化了我国企业在国际分工中的低端地位。	Strengthen the investment-driven growth mode; strengthen the low-end status of Chinese enterprises in international division of work
增加了社会不稳定因素。不可避免地带来货币贬值，这就自然而然地影响到了收入分配问题。	Increase the unstable factors of society. Inevitably cause currency devaluation, which would naturally affect income distribution
通货膨胀的危险	Danger of inflation

Chart 6.3 Risk of inflation

Since the fourth quarter of 2008, the international financial crisis quickly spread, the world economy sunk into deep recession, the external environment rapidly deteriorated, China's economy was subject to severe impact, the foreign trades drastically declined, enterprises encountered difficulties in production and operation, the number of unemployed workers massively increased, fiscal income decreased, economic growth rate declined extensively, the confidence in society was subject to serious influence. Therefore, the Central Party Committee and the State Council quickly made decisions of expanding domestic demands, and released "New China Policy" first in the world.

As a response to the crisis, China issued a series of economic stimulation programs, including four keys and seven aspects, which are as follows in Table 6.4.

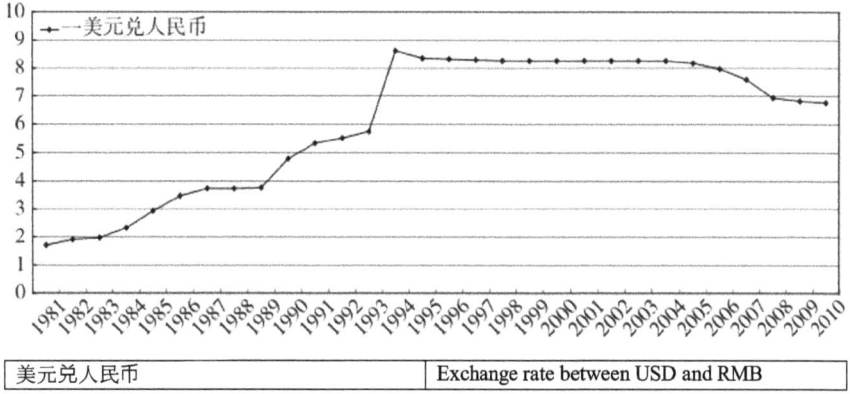

Chart 6.4 Exchange rate between USD and RMB (1981–2010). *Source* Sorted and calculated from the data from *China Foreign Trade Statistic Yearbook 2011*

6.3 Forming Causes of the Cost of Economic Disturbance in China

The cost of China's inflation can be classified into two levels:

The relatively superficial cause: the pressure of inflation was directly reflected as excessive currency issue. Due to the excessively fast economic growth and excessively fast currency and credit issue, the actual interest rate was negative, the cost of investment was relatively low, which further intensified excessive investments. Excessive investment resulted in rising prices of raw materials and other upstream products, there are two possibilities for the influence of upstream product price on downstream consumer goods price: if price transmission is smooth, high growth of product price would finally transmit to consumer goods prices, which would lead to comprehensive inflation; if the rising prices of preliminary and intermediate products are unable to transmit to final products due to restriction of final consumption demands or other reasons, there would be overcapacity.

In-depth cause: low investment efficiency is a main problem in China's economic field and also the fundamental cause of inflation. China's economic growth has always been depending on high investment and high consumption. According to the latest statistics released by World Resources Institute in the US, among the top 10 economies in the world, in terms of energy consumption for producing every USD 1 of gross national product, China ranks No. 1, which is 7.78 times of that in Japan, 4.98 times of that in France, 4.71 times of that in Germany, 3.1 times of that in the US, and 1.65 times of that in India. It can be seen that the energy

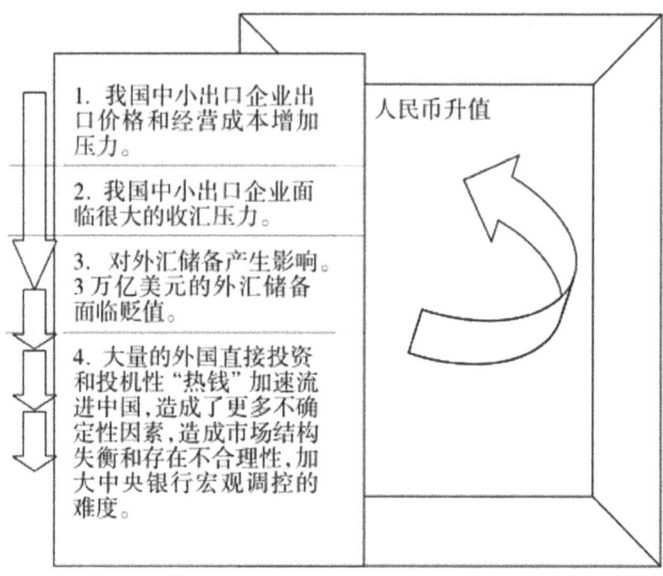

1.我国中小出口企业出口价格和经营成本增加压力。	1. Increase the pressure of the export price and operating cost of Chinese medium and small export enterprises.
人民币升值	RMB appreciation
2.我国中小出口企业面临很大的收汇压力。	2. Chinese medium and small enterprises face a lot of pressure of collective foreign exchanges
3.对外汇储备产生影响3万亿美元的外汇储备面临贬值。	3. Cause influence on foreign exchange reserve, the USD 3 trillion of foreign exchange reserve faces devaluation.
4.大量的外国直接投资和投机性"热钱"加速流进中国，造成了更多不确定性因素，造成市场结构失衡和存在不合理性，加大中央银行宏观调控的难度。	4. Large amount of foreign direct investments and speculative "Hot Money" accelerate in flowing into China, causing more uncertain factors, leading to market structural imbalance and more irrationality, and increasing the difficulty in Central Bank's macroeconomic regulation.

Chart 6.5 Negative effect of RMB appreciation

consumption per unit of output value was not only several times of that in developed countries, but also much higher than that of developing countries.[5]

There are several aspects of factors that affect the change of RMB exchange rate: the enhancement of China's economy aggregate and status as well as the expectation of continuous growth are the fundamental cause that promotes the slow appreciation of RMB; because RMB exchange rate is determined in reference to "A Basket of Currencies", so its exchange rate to US Dollar is to great extent subject to

---

[5]Tang Jing: *Causes of China's Current Inflation Pressure and Countermeasures of Relief*, Young Thinkers, 3rd Issue in 2004.

6.3 Forming Causes of the Cost of Economic Disturbance in China

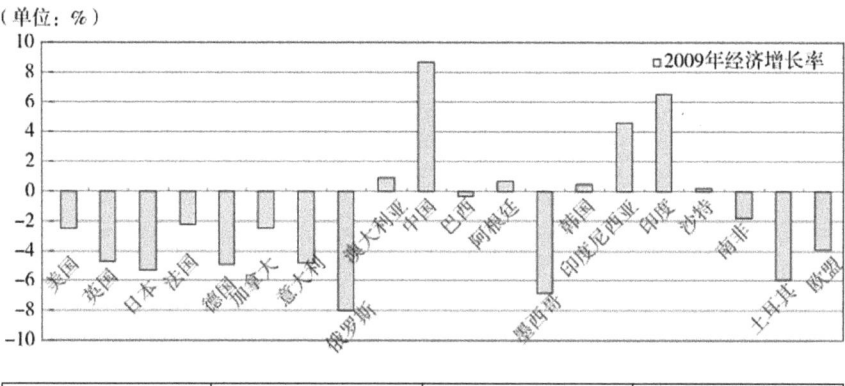

(单位：%)	(Unit: %)	美国	United States
英国	United Kingdom	日本	Japan
法国	France	德国	Germany
加拿大	Canada	意大利	Italy
俄罗斯	Russia	澳大利亚	Australia
中国	China	巴西	Brazil
阿根廷	Argentina	墨西哥	Mexico
韩国	Republic of Korea	印度尼西亚	Indonesia
印度	India	沙特	Saudi Arabia
南非	South Africa	土耳其	Turkey
欧盟	EU	2009年经济增长率	Economic growth rate in 2009

Chart 6.6 Economic growth rate of G20 countries in 1009. *Source* Concluded and drawn by the Author based on the data released by International Monetary Fund (IMF)

the influence of the overall performance of US Dollar in international exchange market; besides, RMB exchange rate is facing severe challenges at current stage, in recent years, there have been constant calls in domestic and foreign markets for RMB appreciation, such irrational expectations of appreciation were mainly based on the realistic foundation such as the studies of RMB Equilibrium exchange rate and unbalance of China's international payments. Meanwhile, because the change of foreign reserve scale is closely correlated with the monetary supply in China and trends of foreign exchange market, the growth of foreign reserve and growth of corresponding position for foreign exchange purchase by the central bank have become the main channel of release of monetary base in China currently, and also becomes a key factor that affects the regulatory effect of monetary policy.

There are several aspects for the causes of current world economic crisis. Firstly, the direct causes on operating level of economic system include excessive expansion of virtual economy compared with real economy; excessive development of financial derivatives, excessive times of leverage operation; financial deregulation, lack of transparency of transaction process information; deepened vulnerability of

Table 6.4 A basket of economic stimulus plan (four keys and seven aspects)

Four keys	Increase government investments in large scale, implement the two-year investment plan with total amount of RMB 4 trillion, in which the Central Government plans to increase additional RMB 1.18 trillion, implement structural tax cut and expand domestic demands
	Implement the plan in large scale to adjust the booth industries, and enhance the overall competitiveness of national economy
	Energetically promote independent innovation, strengthen scientific and technological support and strengthen economic potential
	Enhance social security level extensively, expand urban and rural employment, promote the development of various social courses
Seven aspects	Strengthen and improve macroeconomic regulation, adhere to flexible and diligent regulatory guidelines, implement active fiscal policy and moderately loose monetary policy
	Actively expand domestic demands, especially consumption demands, strengthen the driving role of domestic demands on economic growth
	Consolidate and strengthen the basic status of agriculture, promote the stable development of agriculture and stable income increase of farmers
	Accelerate the change of development mode, energetically promote the strategic adjustment of economic structure
	Continue to deepen reform and opening-up, further improve the systems and mechanisms that benefit scientific development
	Energetically develop social courses, and focus on security and improvement of livelihood
	Promote the construction of government itself, enhance the ability to handle the overall development of economy and society

financial system in the era of globalization; and severe setbacks of the world economic growth model with US financial industry as the center. Secondly, the basic economic system of capitalism is the source of crisis, the production purpose of capitalism for profit maximization results in the conflict between infinite production expansion and the relative shrinkage of demands by the wide labors with purchasing power; the free market system intensifies the conflict between the organized production of individual enterprises and the anarchy state of social productions (Chart 6.7).

6.4 Countermeasures to the Cost of Economic Disturbance in China

Targeting the inflation in economic expansion caused from backward economic structure and low efficiency, we must pay full attention of the negative role of increasingly expanding structural gap in it, we should solve the problem of resource distribution in market through the means of market economy, so as to cool the overheated economy and eliminate the investment behaviors without restrictions or

6.4 Countermeasures to the Cost of Economic Disturbance in China

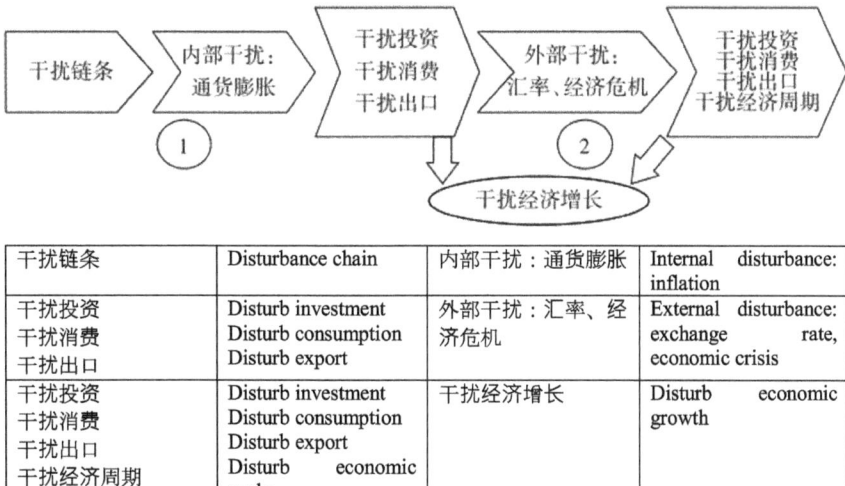

Chart 6.7 Economic disturbance affects economic growth

with soft restrictions: firstly, inflation pressure shall be eased in advance, as for each field, when consumption demands are insufficient, we need to pay attention to the rising pressure of investments and production prices, in the field of investment, we should pay attention to the pressure of repeated construction and blind investments on rising prices, in the field of production circulation, we should pay attention to the pressure of price fluctuation of enterprise commodities; enlarge guidance to investments, since currently it is mainly about high pressure of production-type inflation, we should control the overinvestment in certain industries, so as to avoid it from gradually transmitting to consumption prices and promoting the comprehensive and continuous rising prices; enlarge the guidance of inflation expectation; the most basic and most important issue is to comprehensively promote reforms and enhance investment efficiency.

In order to enable the economic effect of RMB exchange rate fluctuation to promote advantages and prevent disadvantages, the following countermeasures could be adopted: adhere to implementation of moderately tight fiscal and monetary policies, control the money supply and demand expansion, curb inflation, continuously create a good macroeconomic environment for RMB exchange rate to maintain a basic stability, further consolidate and strengthen the positive effect of the stably rising trend of RMB exchange rate; further reform the foreign trade administrative system, enhance the operating level and change response ability of foreign trade enterprises; further strengthen the foreign exchange management of capital account, properly control the scale of foreign investments, reduce the pressure of rising RMB exchange rate; further improve the RMB exchange rate formation mechanism and regulatory method, better develop the regulatory role of foreign exchange market on foreign exchange rate; properly increase the import of technical equipments and raw materials needed in China, and reduce the pressure of foreign exchange oversupply on rising RMB exchange rate.

Facing the leftover problems from economic crisis, China needs to drastically change its strategies in economic reform and development, the most crucial two are as follows: firstly, on the level of basic economic system, timely adjust the ownership structure, recover and adhere to the dominant position of public ownership economy and leading status of state-owned economy. Meanwhile, adjust the distribution of national income, adhere to the policy of distribution according to work, and gradually eliminate polarization. The first thing is to massively increase the proportion of labor remuneration in the primary distribution, so as to help workers to get rid of low-wage dilemma; the second thing is to massively increase of the proportion of citizen income in the redistribution of national income, and reduce the proportion of government income and capital gains, so as to change the abnormal current state that the poorest 20% of people in China receive 4.7% of national income, while the richest 20% people receive over 50% of national income. Secondly, on the level of economic system and economic operation, we must change over to new ways by completely abandoning the Keynesianism and Neo-liberalism policies, so as to be highly alert to possible phenomenon of "Stagnation". With socialist sharing economy theories with Chinese characteristics as the guidance, China shall independently build the economic development mechanism that opposes "Stagnation".

Annexed Tables

See Tables 6.5 and 6.6.

Table 6.5 Various price indexes (previous year = 100)

年份	居民消费价格指数	城市居民消费价格指数	农村居民消费价格指数	商品零售价格指数	工业品出厂价格指数	原材料、燃料、动力购进价格指数	固定资产投资价格指数
1978	100.7	100.7		100.7	100.1		
1980	107.5	107.5		106.0	100.5		
1985	109.3	111.9	107.6	108.8	108.7		
1990	103.1	101.3	104.5	102.1	104.1	105.6	
1991	103.4	105.1	102.3	102.9	106.2	109.1	109.5
1992	106.4	108.6	104.7	105.4	106.8	111.0	115.3
1993	114.7	116.1	113.7	113.2	124.0	135.1	126.6
1994	124.1	125.0	123.4	121.7	119.5	118.2	110.4
1995	117.1	116.8	117.5	114.8	114.9	115.3	105.9
1996	108.3	108.8	107.9	106.1	102.9	103.9	104.0
1997	102.8	103.1	102.5	100.8	99.7	101.3	101.7
1998	99.2	99.4	99.0	97.4	95.9	95.8	99.8
1999	98.6	98.7	98.5	97.0	97.6	96.7	99.6
2000	100.4	100.8	99.9	98.5	102.8	105.1	101.1
2001	100.7	100.7	100.8	99.2	98.7	99.8	100.4
2002	99.2	99.0	99.6	98.7	97.8	97.7	100.2
2003	101.2	100.9	101.6	99.9	102.3	104.8	102.2
2004	103.9	103.3	104.8	102.8	106.1	111.4	105.6
2005	101.8	101.6	102.2	100.8	104.9	108.3	101.6
2006	101.5	101.5	101.5	101.0	103.0	106.0	101.5
2007	104.8	104.5	105.4	103.8	103.1	104.4	103.9
2008	105.9	105.6	106.5	105.9	106.9	110.5	108.9
2009	99.3	99.1	99.7	98.8	94.6	92.1	97.6

Source Sorted from the data from *China Statistic Yearbook 2010*

年份	Year	居民消费价格指数	Consumer price index
城市居民消费价格指数	Urban consumer price index	农村居民消费价格指数	Rural consumer price index
商品零售价格指数	Commodity retail price index	工业品出厂价格指数	Ex-factory price indices of industrial product
原材料、燃料、动力购进价格指数	Raw material, fuel and power purchase price index	固定资产投资价格指数	Fixed assets investment price index

Table 6.6 Exchange rate of 100 units of USD, Euro, JPY, HKD and GBP to RMB (1981–2010)

年份	美元	欧元	日元	港币	英镑
1981	170.5		0.77		30.41
1982	189.25		0.76		31.15
1983	197.57		0.83		27.36
1984	232.7		0.98		29.71
1985	293.67		1.25		37.57
1986	345.28		2.07		44.22
1987	372.21		2.58		47.74
1988	372.21		2.91		47.7
1989	376.51		2.74		48.28
1990	478.32		3.32		61.39
1991	532.33		3.96		68.45
1992	551.46		4.36		71.24
1993	576.2		5.2		74.41
1994	861.87		8.44		111.53
1995	835.1		8.92		107.96
1996	831.42		7.64		107.51
1997	828.98		6.86		107.09
1998	827.91		6.35		106.88
1999	827.83		7.29		106.66
2000	827.84		7.69		106.18
2001	827.7		6.81		106.08
2002	827.7	800.58	6.62		106.07
2003	827.7	936.13	7.15		106.24
2004	827.68	1029	7.66		106.23
2005	819.17	1019.53	7.45		105.3
2006	797.18	1001.9	6.86		102.62
2007	760.4	1041.75	6.46	97.46	1522.13
2008	694.51	1022.27	6.74	89.19	1286.64
2009	683.1	952.7	7.3	88.12	1071.97
2010	676.95	897.25	7.73	87.13	1045.72

Source Sorted from the data from *China Foreign Trade Statistic Yearbook 2011*

年份	Year	美元	USD
欧元	Euro	日元	JPY
港币	HKD	英镑	GBP

Chapter 7
Cost of Economic Regulation and Control in China

7.1 Connotation of Economic Regulation and Control in China

Macroeconomic regulation and control means that the government implements various policies and measures to adjust the operation of market economy. In a market economy, supply and demand of commodities and services are affected by the price rules and free market mechanism. Market economy would bring economic growth, but would also cause inflation, the recession following prosperity would make economy stagnant or even recede, such periodic fluctuation would constitute severe influence on social resources and productivity. So macroeconomic regulation and control emphasizes on the economic operation of the society as a whole, and adjusts supply and demand manually in order to reach the goals of economic plans (Table 7.1).

7.2 Expression Forms of the Costs of Economic Regulation and Control in China

The costs of economic regulation and control in China can be classified, based on types of regulation and control, into regulation and control through economic means, regulation and control through legal means, regulation and control through administrative means. For more details, please refer to Table 7.2.

Table 7.1 Three stages of economic regulation

Stage	Main regulation means	Remarks
Prior to 1978	China implements unitary planning economy system, the government mainly relies on administrative and planning means to realize macroeconomic management and regulation, although fiscal and credit means are important tools of the government's macroeconomic regulation, but it hasn't formed actual practice of fiscal and monetary policy	
1978–1991	Reform stage of traditional planning economy and old system, China was still in the state of shortage economy, the main tasks of macroeconomic management is to control inflation, and mainly adopted administrative and planning methods, and began to introduce the concepts and practice of fiscal and monetary policies	This stage experiences three economic fluctuations during 1978–1981, 1982–1986 and 1987–1991, adopted the adjustment of 1980–1981, the "Soft Landing" in 1986, and the "Governance Renovation" and "Double Tightening" during 1989–1990
1992–present	Stage of preliminary establishment of market economy system, shortage economy is gradually ended, some aspects have oversupply, there are both pressure of inflation and possibility of deflation, the government's macroeconomic regulation has developed from the original direct administrative and planning means to indirect means in economy and law, supported with necessary administrative, government investment and other direct means, the role of fiscal and monetary policy has become bigger and bigger	Experienced the inflation during 1993–1996, the Asian financial crisis during 1997–1998, the deflation during 1999–2002, the partial overheating in 2003–2004, the running of economy is slightly overheated, the government adopted the tightening fiscal policy and moderately tightening monetary policy during 1993–1997, active fiscal policy and robust monetary policy during 1998–2003, robust fiscal policy and robust monetary policy during 2004–2007, and the robust fiscal policy and tightening monetary policy proposed in December, 2007

Table 7.2 Forms of the cost of China's economic regulation

Classification of cost	Forms	Specific contents
Cost of economic regulation	Cost of regulation by economic means	Cost of regulation by the state through fiscal policy and monetary policy
	Cost of regulation by legal means	Cost of regulation by the state through economic legislation and judicial enforcement
	Cost of regulation by administrative means	Cost of regulation by the state depending on administrative institutions through enforcement orders, index, regulation, instructive tasks and other administrative methods

7.2 Expression Forms of the Costs of Economic Regulation … 85

Table 7.3 Change of the effective room of China's fiscal policy

Time	Period of planning economy (1956–1977)	First stage of reform and opening-up (1978–1997)	Second stage of reform and opening-up (1997–present)
Economic operation environment	The national economy ran tightly, the main restraining factor of economic development was system restraint	The national economy ran tightly, the main restraining factor of economic development was energy restriction supply restriction	The economy has been growing rapidly; but structural conflicts become increasingly sharp
Fiscal policy	The fiscal authority was the main body of national income distribution, the fiscal policy was the sole support	Fiscal became poor; the fiscal policy went silently	Concentration degree of the central fiscal authority increased, fiscal policy emerged again

Source Cui Jianjun: *Historical Evolution of the Effective Room of Fiscal and Monetary Policies and Its Inspiration—Based on the Practices of Chinese Fiscal and Monetary Policies,* 3rd Issue in 2008

7.2.1 Cost of Regulation and Control Through Economic Means

Economic means is the main means of macroeconomic regulation and control under market economy. It is the method that government adopts economic leverages (such as price, taxation, credit, interest rate, exchange rate, etc.) to restrain or guide economic activities of enterprises and operation of national economy. Regulation and control are implemented mainly through the two methods of fiscal policy and monetary policy (Table 7.3).

With the high-speed economic growth, the room for regulation and control through fiscal policies tends to expand. Fiscal revenue has also been growing at high-speed rate, and its proportion in GDP is increasing. Please refer to Table 7.4. Besides, the enforceability of taxation is good for the implementation of regulation and control effect of fiscal policy[1] (Table 7.5).

Excessive Liquidity, by definition, means excessive money supply, which is also referred to as "Excessive Capital Turnover". US economist Goldsmith believed that the higher a country's monetization rate (M2/GDP), the higher degree of financial development of the country. Related statistic data indicated that China's

---

[1]On May 8, 2007, in order to restrain excessive loan growth and stock market speculation, the People's Bank of China adopted three measures at the same time: raising the deposit reserve ratio, hiking interest rate and expanding the fluctuation range of exchange rate between RMB and USD. But the stock market went against the trend, rise instead of fall; On May 30, 2007, the Ministry of Finance announced to adjust the stamp duty of stock transactions, increasing it from 1 to 3‰, the stock market immediately fell, over 900 stocks fell to limit down.

Table 7.4 Change of the ratio of China's fiscal income in GDP

	1978 年	1990 年	2000 年	2009 年	2010 年
GDP（亿元）	3645.2	18667.8	99214.6	340902.8	397983.3
财政收入（亿元）	1132.3	2937.1	13395.2	68518.3	83080.3
财政收入占GDP的比重	31.06%	15.73%	13.50%	20.10%	20.88%

Source Calculated from the related data from *2011 China Business Yearbook*.

GDP（亿元）	GDP (RMB 100 million)	财政收入（亿元）	Fiscal income (RMB 100 million)
财政收入占GDP的比重	Ratio of fiscal income in GDP		

Table 7.5 Change of the effective room of China's monetary policy

Time	Period of planning economy (1956–1977)	First stage of reform and opening-up (1978–1997)	Second stage of reform and opening-up (1997–present)
Economic operation environment	The national economy ran tightly, the main restraining factor of economic development was system restraint	The national economy ran tightly, the main restraining factor of economic development was energy restriction supply restriction	The economy has been growing rapidly; but structural conflicts become increasingly sharp
Monetary policy	Banks were the subsidiary of fiscal authority and didn't have relatively independent monetary policy	Banks' rights expanded; monetary policy had solo performance	Deflation and excessive liquidity, the regulatory function of monetary policy was weakened

Source Cui Jianjun: *Historical Evolution of the Effective Room of Fiscal and Monetary Policies and Its Inspiration—Based on the Practices of Chinese Fiscal and Monetary Policies,* 3rd Issue in 2008

monetization rate is already as high as 165.03%, which is already the highest in the world.[2] It means China has excessive liquidity, and the distribution efficiency of financial resources is low. Moreover, since 1995, China's financial institutions began to have constant loan-deposit gap, in 2006, the loan-deposit ratio was already as high as 148.89%, financial institutions had very abundant funds. In the same year, China's foreign exchange reserve broke the record of USD 1 trillion and reached USD 1.066344 trillion, becoming the country with the largest foreign

---

[2]Cui Jianjun: *Historical Evolution of the Effective Room of Fiscal and Monetary Policies and Its Inspiration—Based on the Practices of Chinese Fiscal and Monetary Policies,* 3rd Issue in 2008.

Table 7.6 Two stages of China's financial macroeconomic regulation

Stage I (1984–1997)	Stage II (1998–present)
The excess reserve of commercial bank system was negative, and heavily relied on central banks in funds, and the central bank had strong ability in financial macroeconomic regulation	The excess reserve of commercial bank system was positive and the excess reserve became more and more, the liquidity was abundant. The Central Bank's tool of legal deposit reserve has lost its strong ability in financial and macroeconomic regulation

Source Sorted by the author

exchange reserve in the world. Under the circumstance of excessive liquidity, the effect of regulation and control through monetary policy would be weakened (Table 7.6).

New China's fiscal policy has experienced three stages of being "Strong, Weak and Strong", correspondingly, the monetary policy has experienced the three states of being "Weak, Strong and Weak". The rooms for the effects of fiscal policy and monetary policy have the conflict of trading off and taking turns.

Column 7-1 After Effects of Fiscal and Monetary Policies

1. After effect of fiscal policy

Whenever the general economy slides into recession, the basic method for the government to increase total demands and stimulate economy is to use fiscal policies to mainly concentrate government investments in the construction of social infrastructure, government investments would reflect the characteristics of simplicity. Although such method of economic stimulation would get effect instantly, but if the investments in social infrastructure couldn't realize proper scale and rational purpose, such government investments are usually lack of efficiency. In the process of increasing government expenditure and expansion of total demands, the risk of government debts is an important expression form of "After Effect of Policies". The harm of such After Effect of Policies" is that government debts would often result in severe economic and social problems. As indicated by the data of the National Audit Office of the People's Republic of China, as of the end of 2009, the total balance of local government debts was RMB 2.79 trillion. As for the ratio of debt balance and available fiscal strength in current year, the debt risks of provisional and municipal level and western regions were relatively concentrated. Among the surveyed regions, the ratio of 7 provinces, 10 cities an d14 countries exceeded 100%, the highest one reached 364.77%. Therefore, governments on all levels should pay high attention to the government debt risk and control it within rational level, so as to prevent "After Effect of Policies".

2. After effect of monetary policy

In the aspect of monetary policy, the "Dilemma" of interest rate adjustment causes the regulating and controlling orientation of policies to be distorted, which is sourced from the "After Effect of Policies" of monetary policies. The so-called

"Dilemma" of interest adjustment means that facing economic recession or expectation of economic recession, the usual method adopted by central bank is usually interest rate cut (interest rate reduction); while facing inflation or expectation of inflation, the usual method adopted by central bank is to increase interest rate. But the facts have proved that both interest rate cut and interest rate increase would generate "After Effect of Policies".

Firstly, "After Effect of Policies" brought by interest rate cut, it is more obvious reflected in stock market or real estate market. In order to cope with international financial crisis, China implemented moderately loose monetary policy. Such policy on one hand stimulated economic growth, and on the other hand caused credit funds to flow into the real estate market, which promoted the continuous rising housing prices. The related data by State Statistics Bureau indicated that in 2009, China completed real estate development investment of RMB 3.6232 trillion, which grew by 16.1% compared with the same period in previous year.

Secondly, the implementation of cheap monetary policy with low interest rate as the goal would also often cause the occurrence of "After Effect of Policies". Some economists advocated that the Central Bank should implement the cheap monetary policy with relatively low real interest rate as the goal, which is to stabilize real interest rate at certain very low level. The Central Bank should follow the "Rules of Real Interest Rate" rather than the rules of monetary growth or rules of inflation rate. This is the task with minimum damage that a central bank could set for itself. The low-interest-rate policy of China before 1984, although not a monetary policy in real sense, had the purpose of using a long-term low-interest operation process to enable investments to expand at low costs, so as to speed up the process of industrialization.[3] Since 2003, China's interest rate level has been generally maintained at low level.

Thirdly, highly interest rate policy would also cause "After Effect of Policies". China's CPI in 1988 and 1989 was 18.8 and 18.0% respectively. The academic circle commonly believed that the inflation was demand-driven type, the economic policy to cope with it should be restraining aggregate social demands. Whoever, the tough dilemma the central bank faced back then was that if they restrain the aggregate demand through reducing investment scale, the production would decline, so as to cause supply shortage of consumable goods, they worried about repetition of the bank run and panic buying that they just experienced, so they chose the policy of hiking interest rate, they raised the deposit and loan interest rate twice, adopted inflation proof savings deposits and raised legal reserve ratio for the purpose of reducing consumption tendency by raising savings.

Source: Liu Xiuguang: "After Effect of Policies" in process of macroeconomic control—an interpretation based on social cost of macroeconomic control, Journal of Wuyi University (Social Science Edition), 2nd Issue of Volume 13, May, 2011.

---

[3]Liu Xiuguang: *Operating Mechanism Theory of Interest Rates,* Fujian People's Publishing House, 2006 Edition, Page 163–165.

7.2.2 Cost of Regulation and Control Through Legal Means

Legal means of regulation and control means that a state adopts legal means to adjust various economic relations and economic activities, it has the characteristics of authoritativeness, normalization, enforceability, relative stability, etc. Content of legal means of regulation and control include two aspects, legal protection and punishment. Cost of economic legislation means the consumption of labor or materialized labor in the process of economic legislation, meaning the consumed resources in economic legislative activities, such as labor, money, material, time, information, etc.[4] The cost is composed of:

Cost of economic legislation = Office and living expenses of economic legislation institutions and their staff + Expenses incurred from collection of economic legislation information and formation of economic legislation draft + Expenses incurred from review of economic legislation drafts and revision of economic legislation texts + Expenses incurred from production and release of economic laws and regulations texts + Expenses incurred from promotion of economic laws and regulations information.

As of the end of June, 1994, NPC and its standing committee as well as the State Council and its subsidiary ministries, committees and bureaus formed and issued 1581 currently effective economic laws, regulations and rules in total, in which only 12 were formed prior to December, 1978.[5] As of March, 2009, China had 231 effective laws, 223 of which were issued after the Third Plenary Session of the 11th Central Committee. The issuance of these economic laws basically ended the situation that Chinese social and economic lives had no law to abide by, and promoted the establishment of the socialist market economic system. However, while Chinese economic legislation has been rapidly developing, there are also some problems to be noticed. Such as the supply-demand imbalance of economic laws, deviation of enforcement effective of economic laws with legislative expectation, prominent conflict from lack of financial support for economic laws to operate in society, etc.

China usually has relatively low economic decision-making cost, therefore has relatively high systematic benefit of legislative decision-making. Legislative benefits are mainly reflected as follows; legislative review procedures can be launched only after legislative suggestions are determined as formal legal proposals; the unilateral review procedure enables Chinese legislative drafts to only need to pass the "First Review" by NPC Standing Committee and the review and modification procedure by its subsequent special committees. Besides, the independence of legislative interest usually determines conflict of interest and its coordination cost. But relatively low legislative decision-making cost would, while bringing procedural efficiency, generate relatively high institutional cost. Besides, the conflict

---

[4]Zhou Xianzhi: *Thoughts Regarding Reducing Economic Legislation Cost and Enhancing Economic Legislation Quality,* Journal of Jinan (Philosophy Social Science), 5th Issue in 1999.

[5]Li Shenglan, Zhou Linbing and Qiu Haiyang: *Legal Cost and the Construction of China's Economic Rule of Law,* Chinese Social Science, 4th Issue in 1997.

between economic law operation and relative lacking of financial support from the society is very prominent. Since the reform and opening-up, the teams for economic legislation, justice and law enforcement have been continuously expanding, the related fiscal expenditures and financial expenditure of the society have been constantly increasing, but it is still insufficient to support the normal operation of economic laws. According to incomplete statistics, from 1987 to 1995, the annual growth rate of judicial and administrative case-handling expenses within and out of fiscal budgets on national and local levels was higher than 25%, but the phenomenon of incomplete law enforcement due to lack of law enforcement force and shortage of case-handling funds in practices are still very common.[6] We need to pay special attention to the issue of legal means of regulation and control as well as the issue of hidden cost.

7.2.3 Cost of Regulation and Control Through Administrative Means

Administrative means are a kind of means that rely on administrative institutions to adopt mandatory administrative orders, administrative instructions and administrative regulations and other administrative methods to regulate and control and enable macroeconomic control goals to be realized (Charts 7.1 and 7.2).

Regulation and control of economy through administrative means mainly have the following problems:

Monopoly often causes loss of social benefits, in the industries of natural monopoly regulated and controlled by administrative means, the phenomenon of inaction even caused huge loss. The investments by regulation and control authorities are still stably growing (Refer to Table 7.7).

7.3 Forming Causes of Chinese Economic Regulation Cost

Economic regulation is a means to promote economic growth, increase employment, stabilize general price level and balance international payments. The Chinese Government made 4 times of relatively large macroeconomic regulations in total. The first one was promoted after the South Talks by Comrade Deng Xiaoping in 1992; the second one was promoted after the Asian Financial Crisis in 1996; the third one began in early 2004; and the fourth one was promoted after the occurrence of global financial crisis in 2008. Economic regulation usually has "Three Combinations", meaning using economic means, legal means and administrative means.

---

[6]Li Shenglan, Zhou Linbing and Qiu Haiyang: *Legal Cost and the Construction of China's Economic Rule of Law,* Chinese Social Science, 4th Issue in 1997.

7.3 Forming Causes of Chinese Economic Regulation Cost

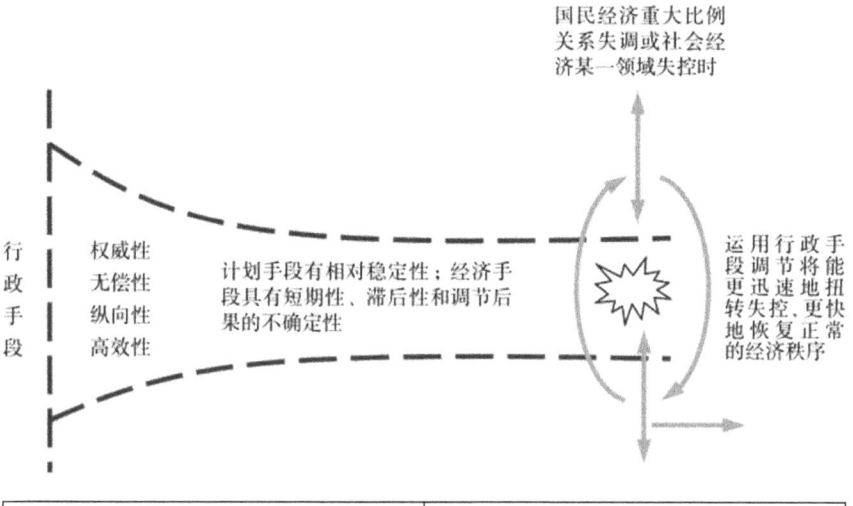

行政手段	Administrative means
权威性	Authority
无偿性	No compensation
纵向性	Vertical
高效性	Highly effective
计划手段有相对稳定性;经济手段具有短期性、滞后性和调节后果的不确定性	Planning means has relative stability; economic means are short-term, lagging and uncertain in adjustment results
国民经济重大比例关系失调或社会经济某一领域失控时	When major ratio relation of national economy is out of balance or certain aspect of social economy becomes out of control
运用行政手段调节将能更迅速地扭转失控,更快地恢复正常的经济秩序	Using administrative means to regulate could more quickly turn out-of-control situation around, and more quickly recover normal economic order

Chart 7.1 Effect of administrative means on recovery of economic order

The traditional government debt issue, the problem of repeated construction of government investments, blind expansion of investments at low currency cost and other old problems, as well as the excess liquidity in market; growth of cost and demands of agricultural products; high dependence on iron ores, crude oil and other bulk international commodities, facing the higher risk of imported-type inflation from international market and other constantly occurring new factors, promoted Chinese economic regulation to expand in scope and deepen in degree. The construction of legal system for market economy was mainly completed after the reform and opening-up, the laws, regulations and systems issued during this period need large quantity of revision works in order to adapt to the new requirements for the development of socialist market economy.

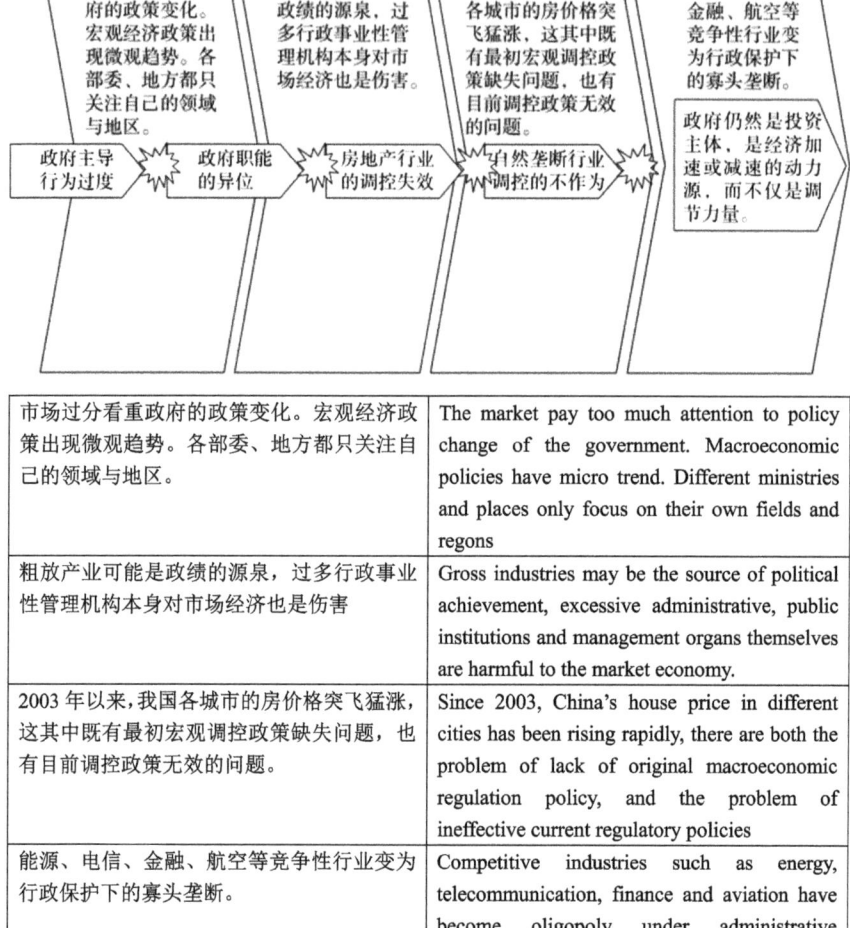

市场过分看重政府的政策变化。宏观经济政策出现微观趋势。各部委、地方都只关注自己的领域与地区。	The market pay too much attention to policy change of the government. Macroeconomic policies have micro trend. Different ministries and places only focus on their own fields and regons
粗放产业可能是政绩的源泉，过多行政事业性管理机构本身对市场经济也是伤害	Gross industries may be the source of political achievement, excessive administrative, public institutions and management organs themselves are harmful to the market economy.
2003年以来,我国各城市的房价格突飞猛涨,这其中既有最初宏观调控政策缺失问题，也有目前调控政策无效的问题。	Since 2003, China's house price in different cities has been rising rapidly, there are both the problem of lack of original macroeconomic regulation policy, and the problem of ineffective current regulatory policies
能源、电信、金融、航空等竞争性行业变为行政保护下的寡头垄断。	Competitive industries such as energy, telecommunication, finance and aviation have become oligopoly under administrative protection
政府主导行为过度	Excessive government guiding behavior
政府职能的异位	Dislocation of government functions
房地产行业的调控失效	Ineffective regulation in real estate industry
自然垄断行业调控的不作为	Inaction of regulation in natural monopoly industries
政府仍然是投资主体，是经济加速或减速的动力源，而不仅是调节力量	The government is still the main investment body, and the dynamic source of economic acceleration or slowdown rather than just a regulatory power.

Chart 7.2 Problems and performances of administrative regulatory means

7.3 Forming Causes of Chinese Economic Regulation Cost

Table 7.7 Investment growth in China's several "regulated" industries (2003–2005)

	2003 年		2004 年		2005 年	
	同比增长	比重	同比增长	比重	同比增长	比重
全部投资			27.6	100	27.2	100
房地产业	33.1	33.8	29.1	24.7	20.5	23.3
非金属矿物制品业	61.3	3.1	43.6	1.9	26.6	1.9
黑色金属冶炼及压延加工业	87.2	3.1	26.9	3	27.5	3
有色金属冶炼及压延加工业	68.7	1.1	23.4	1	32.4	1

Source Sorted and calculated from *2006 China Economic Yearbook*.

同比增长	Increase compared with the same period in previous year	比重	Ratio
全部投资	All investments	房地产业	Real estate industry
非金属矿物制品业	Nonmetallic mineral product industry	黑色金属冶炼及压延加工业	Ferrous metal smelting and rolling industry
有色金属冶炼及压延加工业	Non-ferrous metal smelting and rolling industry		

Under the socialist market economic system of China, cost of economic regulation is necessary, the two hands of "Market" and "Government" would coexist for long term. Economic regulation cost is an active cost, what we need to avoid is the immeasurable loss caused from improper means, improper degree, improper decision-making or other aspects (Chart 7.3).

7.4 Countermeasures to Handle Chinese Economic Regulation Cost

In order to control economic regulation cost, we need to correspond the power and responsibility of macroeconomic regulation, such responsibility not only includes the moral and political responsibility to the people, but also include legal responsibility. Economic regulation is certain power equipped for government to exercise economic management function, and also a kind of responsibility. However, due to the relatively many entities of economic regulation, such as the Ministry of Finance and Central Bank for economic means; the NPC and its Standing Committee for legal means; and various departments and local governments for administrative means, when exercising management powers, there are no management goals and

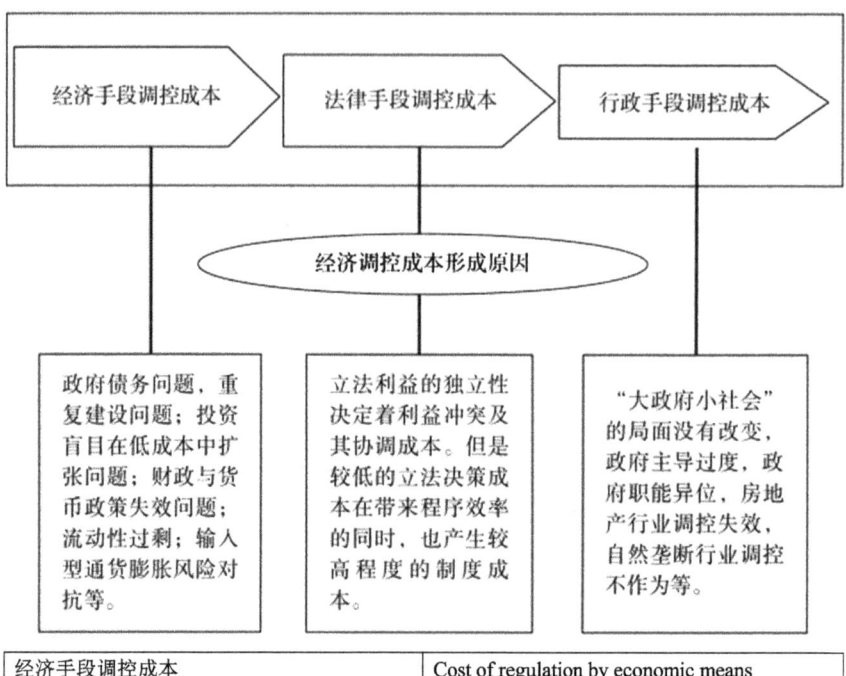

经济手段调控成本	Cost of regulation by economic means
法律手段调控成本	Cost of regulation by legal means
行政手段调控成本	Cost of regulation by administrative means
经济调控成本形成原因	Forming causes of cost of economic regulation
政府债务问题,重复建设问题;投资盲目在低成本中扩张问题;财政与货币政策失效问题;流动性过剩;输入型通货膨胀风险对抗等。	Government debt issue, repeated construction issue; blind investment and expansion at low cost issue; ineffective fiscal and monetary policy issue; excessive liquidity, imported inflation risk and resistance, etc.
立法利益的独立性决定着利益冲突及其协调成本。但是较低的立法决策成本在带来程序效率的同时,也产生较高程度的制度成本。	Independence of legislative interest determines the conflict of interest and its coordination cost. But relatively low legislative decision-making cost, while bringing procedural efficiency, would also generate relatively high system cost.
"大政府小社会"的局面没有改变,政府主导过度,政府职能异位,房地产行业调控失效,自然垄断行业调控不作为等。	The situation of "Big Government and Small Society" is not changed yet, excessive government dominance, dislocation of government functions, ineffective regulation of real estate industry, inaction of regulation in natural monopoly industry, etc.

Chart 7.3 Forming causes of economic regulation cost

7.4 Countermeasures to Handle Chinese Economic Regulation Cost

management responsibilities with specific operability that are correspondingly formed for functional departments. Macroeconomic regulation is a matter of general situation, mistake in decision-making would cause imbalance of overall economy, so as to result in general economic disturbance.

The economic regulation with Chinese characteristics, while obtained actual effects, also delayed the improvement of micro market foundation, and impeded the development of the transmission mechanism of monetary policy. Macroeconomic regulation should gradually change to market means. Economic regulation cost is a kind of necessary economic cost, being too high or too low isn't a good sign. Surrounding the goal of "promoting economic growth, increasing employment, stabilizing general price level and maintaining balance of international payments", we should undertake necessary management by objectives and performance management, and evaluate the effectiveness of economic regulation from the perspective of cost-performance.

Summary of Economic Cost of Development in China

According to estimation, Chinese gross economy output is going to surpass that of the US in 2018 and become the biggest economy in the world. Discussions about Chinese economic growth and cost are generally made around two indicators that reflect economic growth, which are economic growth rate per capita GDP. It is generally believed that during certain period of regional development, without consideration of other factors, per capital GDP and various development costs are in relations shaped like U upside down . In order to maintain high-speed and stable growth, China maintained or needs to still maintain certain input in the aspects of investment expansion, consumption expansion and export expansion, which are necessary. But China also faces problems such as overcoming economic transformation, reducing economic disturbance, proper and suitable regulation, etc.

Over the thirty years of reform and opening-up, China didn't have major political turbulence, social crisis or economic slide in its transformation process, partial and short-term cost problems didn't jeopardize the transformation process, the system problem, structural problem and development method problem gradually draw attention, which all exist in form of hidden and long existing cost, when crossing the middle-income trap, we need to pay special attention to such costs, which means we still have 7–8 years of time, around 2012–2018 when the per capita GDP reaches USD 10,000–12,000, such key issues of transformation need to be solved, the historical debts need to be solved well step by step. In the aspects of internal and external balance of economy, prevention of disturbance from internal and external factors, we need to focus on cultivation of crisis-defense economic governance ability, use improved match of economic, legal and administrative means in order to maintain rapid, healthy and sustainable development of economy.

This part mainly stresses the development cost in economic aspect, meaning economic system cost, in Chinese economy of reform and opening-up, the external opening cost should also be included (refer to the statement in Part VI).

Part III
Political Cost of Development in China

Chapter 8
Political Reform Cost in China

8.1 Connotation of Political Reform in China

Political reform means change and reform occurred in political field, usually it would be reflected in form of political struggle, and bringing certain negative influence on the development of economy and society. At the beginning of the establishment of New China, China still had class antagonism, the capitalist class and its exploitation system still existed. With the increasing reflection of conflict between worker class and capitalist class, in order to consolidate the political power of people's democratic dictatorship, China launched a series of top-down political campaigns. Those major political transformations included the Counter-revolutionary Suppression Campaign,[1] the Land Reform,[2] the Struggle against the "Three Evils" and the "Five

---

[1]After the establishment of New China, targeting the large batch of spies, bandits, local tyrants, counter-revolutionary organization leaders, counter-revolutionary gangsters and other counter-revolutionaries that KMT left in Mainland China before escaping to Taiwan, and their various sabotage activities, in March, 1950, CCP Central Committee issued the instruction of oppressing the counter-revolutionary activities. Since the winter of 1950, the Counter-revolutionary Oppression Campaign was developed across China, the key objects oppressed in this campaign includes bandits, spies, local tyrants, counter-revolutionary gangsters and counter-revolutionary organization leaders. The Counter-revolutionary Oppression Campaign, along with the War to resist US aggression and aid Korea and Land Reform, were referred to as the Three Major Campaigns, which was a destroying crackdown upon the hidden forces and dispatched spies of KMT, and consolidated the new-born political system. While the campaign was no doubt expanded, and the standard of criminal punishment was far beyond regulations.

[2]After the establishment of New China, according to the Common Program of Chinese People's Political consultation Conference, the State shall "step-by-step change the feudal and semi-feudal land ownership system into the farmers" land ownership system. From the winter of 1950 to the spring of 1953, in the rural areas of new liberated areas which accounted for over half of national population, the party led farmers to complete the reform of land system.

Evils" Campaign,[3] the "Socialist Industrialization and Three Reforms",[4] the "Party-wide Cadre Reform Campaign",[5] the Rectification Movement, the Struggle

---

[3]After the land reform, the income gap in rural areas began to widen, the leaders decided to implement large-scale and radical socialist revolution, so the Struggle against the "Three Evils" and the "Five Evils" Campaign was developed. In December, 1951, CCP Central Committee made the decision of implementing better troops and simpler administration, increasing production, saving consumption, opposing corruption, opposing wastes and opposing bureaucracy. A total of 3.86 million people participated in the campaign (government agencies above county level), over 105 thousand of corrupt officers were punished, 9942 people received term sentence, 67 people received life imprisonment, and 51 people received immediate death penalty or suspended death penalty. In January, 1952, the CCP Central Committee issued the *Instruction of Developing the Struggle against Five Forces* in large and medium cities, meaning the struggle against bribery, tax evasion, theft of national assets, cheating on works and materials, and theft of national economic intelligence. Large batch of private industrial and commercial owners stopped business. Over 1 million business owners participated in the campaign, 1059 received criminal punishment, 5% of which were severe law violation and complete law violation. In April, 1952, the CCP Central Committee released the *Instruction on Several Issues in the Completion of the "Five Opposing" Campaign*. In October, the Party Central Committee approved the two reports by An Ziwen and Liao Luyan regarding the completion of the "Three Opposing" and "Five Opposing" campaigns, which marked the successful completion of the two campaigns. The development of the "Three Opposing" and "Five Opposing" campaigns consolidated the leadership status of the worker's class and the leadership status of socialist state-owned economy in national economy, and created advantageous conditions for the further socialist reform of the capitalism industrial and commercial industry and capitalist class, but due to the influence of extreme leftist thought, there were also many unjust, false and erroneous cases created. Refer to Hu Angang: *History of China's Politics and Economy (1949–1976)*, Tsinghua University Press, 2008 Edition, p. 133.

[4]At the end of 1952, CCP Central Committee proposed the general guideline of the party during the transition period: which is to, during considerable long period, gradually realize the country's socialist industrialization, and gradually realize the socialism reform the state on agriculture, handicraft industry and capitalism industrial and commercial industry. The general guideline of this transition period was also referred to as the "Socialist Industrialization and Three Reforms". In March, 1955, Comrade Mao Zedong proposed, during the address at the opening ceremony of the CCP National Congress, to complete the general task proposed during the transition period within the third Five Year Plan period (meaning 1968). After that, China accelerated the process of the "Socialist Industrialization and Three Reforms", and basically completed it in 1956, which only took 3 years. Although the "Voluntary" principle was repeatedly stressed during the "Three Major Reforms", but actually they were almost all realized by mobilizing the mass, political mobilization and the means of "Forced Order" and by means of class struggle. Even Comrade Mao Zedong himself referred to the "Three Major Reforms" as a radical mass class struggle. Therefore, the "Socialist Industrialization and Three Reforms" was a social revolution with forced nature, and a mandatory system change from top down. Refer to Mao Zedong: *Speech at the CCP National Congress Representative Meeting*, in March, 1955, refer to *Selected Works of Mao Zedong*, Volume 6, People's Publishing House, 1999 Edition, p. 392.

[5]"Gao Rao Event" was the first serious event of senior leadership group conflict and splitting since the establishment of New China. The event of Gao Gang and Rao Shushi directed led to the cadre review campaign throughout the party since 1953. In October, 1953, CCP Central Committee issued the *Decision Regarding Cadre Investigation*, and made comprehensive political investigation on cadres, and eliminated the counter-revolutionaries mixed into the party and political organs, alien class elements, the degenerated and corrupted. The campaign to eliminate internal counterrevolutionaries began since July, 1955. According to the figure released in 1957, over 81 thousand of counterrevolutionaries were eliminated, including over 3800 active

8.1 Connotation of Political Reform in China

against Rightists,[6] the "Four Clean-ups" Movement[7] and the Great Proletarian Cultural Revolution (hereunder referred to as the "Cultural Revolution").[8]

(Footnote 5 continued)
counterrevolutionaries, the political and historic issues of over 1.3 million people were cleared. At the later stage of the campaign, some unjust, false and erroneous cases were gradually corrected. Refer to Pang Song: *China in the Era of Mao Zedong (1949–1976) (I)*, CCP Party History Press, 2003 Edition, pp. 437–443.

[6]The Rectification Movement and the Struggle against Rightists were launched by Comrade Mao Zedong personally, later under the instructive thought of "Class Struggle as Guideline", China's class struggle became bigger and bigger, from top down, and from within the party to outside the party, and lasted for nearly 20 years. Starting from correctly handling the contradictions among the people, on Apr 27, 1957, the CCP Central Committee formally issued the *Instruction Regarding Rectification Movement*, which indicated that the party is already in ruling status, targeting the new breeding of bureaucracy out of touch and out of reality, sectarianism and subjectivism, it is necessary to have a rectification movement throughout the party that is against bureaucracy, sectarianism and subjectivism. With the deepening of the Rectification Movement, the open and large-scale criticism against the ruling by CCP made Comrade Mao Zedong decide to turn the Rectification Movement to the struggle against "Rightist". The struggle against Rightist quickly expanded, a large of intellectuals, patriots and cadres within the party were wrongly classified as "Rightists", and caused unfortunate consequences. When the whole movement ended in 1958, over 550 thousand people were classified as "Rightists" throughout China. Refer to Bo Yibo: *Review of Several Major Decisions and Events*, Volume B, CCP History School Press, 1993 Edition, pp. 619–620.

[7]From 1963 to 1966, CCP Central Committee developed socialism education campaign in urban and rural areas of China, which is al referred to as "Four Clean-up Movement", meaning cleaning up politics, cleaning up thoughts, cleaning up organization and cleaning up the economy. At the beginning, the Four Clean-up Movement mainly depended on basic-level organizations and basic-level cadres, and the struggle was against the corrupted in cities and rural areas, later the leaders of working teams began to lead large movement, and the struggle was turned against the "Five Black Categories", and the phenomenon of undisciplined struggle, beating, searching, collective lecturing, classification and undisciplined punishment, etc. were gradually occurred in the struggle. The Four Clean-up Movement gradually changed from education in nature to class struggle.

[8]At later stage of the Four Clean-up Movement, Comrade Mao Zedong believed that the Four Clean-up Movement and cultural criticism are not sufficient to solve the fundamental problems, so he prepared and launched the "Cultural Revolution". In May, 1966, the decade-long "Cultural Revolution" formally erupted. In June, 1981, the Sixth Plenary Session of the 11th Central Committee of CCP passed the *Resolution Regarding Several Historical Issues of the Party since the Establishment of New China*, and made historical conclusion and comment on the ten years of "Cultural Revolution. The resolution believed that the ten years of "Cultural Revolution" was the ten years when the party, country and people received the most severe setbacks and losses since the establishment of New China. The "Cultural Revolution" was not and can't be any meaningful revolution or social advancement. The history has proved that the "Cultural Revolution" was a political turmoil that was wrongfully launched by leader, taken advantage of by counterrevolutionary group and brought severe disaster to the party, country and people of all nationalities. Refer to the *Resolution Regarding Several Historical Issues of the Party since the Establishment of New China*, which was unanimously passed by the Sixth Plenary Session of the 11th Central Committee of CCP on Jun 27, 1981, refer to, edited by CCP Central Committee Literature Research Office: *Compilation of Important Literature since the Third Plenary Session* (B), People's Publishing House, 1982 Edition, pp. 808 and 811.

Certain political reform, despite its good intention, but due to its instructive thinking of "class conflict as the guiding principle", would usually evolve into fierce political campaign and political struggle, and bring different degree of influence and harm to China's economic construction and social development. Especially the "Cultural Revolution" and "Ten Chaotic Years", which made China lost the golden era of development. A scholar estimated that in the aspect of loss in human capital accumulation, the "Cultural Revolution" made the potential human capital stock decrease by 14.3%.[9] From certain point of view, any political reform would need a country to pay certain price of labor, material and fund, and cause different degree of loss to the country in specific practical process, it is the cost and price to pay for the country in its development process.

8.2 Forms of Political Reform Cost in China

It is believed in this book that political reform cost is the cost generated from substitution of political system or adjustment of political relations in the scope of society. Political reform cost has two forms, political struggle cost and political risk cost. Refer to Table 8.2. In which political struggle cost mainly means the cost incurred from a series of political struggle events listed in Sect. 8.1, and is a type of already incurred cost; political risk cost means the risk of having social disturbance after a country's GINI Coefficient[10] reaches certain level, and is a type of expected cost. These two types of cost will be further explained in this section.

8.2.1 Cost of Political Struggle

Political struggle is a major change from top down, and would bring influence to every aspect of political, economic, cultural and social life, and even cause losses, therefore, measurement and calculation of political struggle cost would involve comprehensive management and operation of the entire country, and involve the input and consumption of human, fund and material in each aspect, these input and

---

[9]The influence of "Cultural Revolution" on the accumulation of human capital is long-term and far-reaching, currently, a large proportion of laid-off workers in urban areas were deeply harmed by these political movements, when these people enter the retired and old population, they will become low-income group or poor population. Hu Angang: *History of China's Politics and Economy (1949–1976)*, Tsinghua University Press, 2008 Edition, p. 541.

[10]At the beginning of the 20th Century, Italian Economist GINI found the indicator determining distribution equality degree based on the Lorenz Curve, which is referred to as GINI Coefficient or Lorenz Coefficient, and it is an effective method to measure wealth gap. Such coefficient could be any value between 0 and 1. Related organization of UN regulated that GINI Coefficient lower than 0.2 means absolute equality; 0.2–0.3 means relatively equal; 0.3–0.4 means relatively reasonable; 0.4–0.5 means relatively big income gap; GINI Coefficient above 0.6 means wide income gap.

consumption are all cost of political struggle. The main political reform events since the establishment of New China as listed in Table 8.1 mostly evolved into political struggle, some of them indeed promoted the advancement of productive relations, and made contributions to the elimination of feudal exploitation system and consolidation of the people's democratic dictatorship. However, they all impeded development of productivity to various degrees, and made the huge amount of labors, materials and funds that were urgently needed for the development of New China consumed in blind political struggle, and was a retrograde of comprehensive national power (see Table 8.2).

Analyzing the influence of political struggle on China from economic perspective, it can be seen that Chinese economy had obvious fluctuation during the period of 1965–1977. The four economic peaks occurred in 1965, 1970, 1973 and 1975 respectively, while the four bottoms occurred in 1967, 1972, 1974 and 1976 respectively. Refer to Chart 8.1. Obviously, the occurring time of each economic bottom basically coincided with the occurring times of political struggle events, which fully reflects that political struggle is a major influence factor of economic fluctuation.

From specific analysis, the fluctuation coefficient of GDP growth during 1966–1977 was 126%, in which the highest value was 19.4% (1970) and the lowest value was −5.7% (1967), the two varied by 25.1%; the fluctuation coefficient of fixed asset investment growth was 240%, in which the highest value was 62.9% (1969), and the lowest value was −26.3% (1967), the two varied by 25.1%. This was the period in the economic history of the People's Republic of China with the second biggest economic fluctuation, only next to the economic fluctuation of the period during 1958–1966 (the fluctuation coefficient of GDP growth was 245%, and the fluctuation coefficient of fixed assets investment growth rate was 242%).

Cost is usually measured in form of currency, for political struggle cost, this book tries to use occurring frequency of events as the indicator to measure its degree of influence. From 1949 to 1976, China had as many as 67 times of big or small political campaigns, which were 2.5 times per year on average. In which the 1950s was the peak period of various political campaigns, for a period of only ten years, there were 31 political campaigns launched, which were more than 3 times per year.[11] Because it is difficult to comprehensively and precisely measure and calculate the input and loss of political struggle, this book will use the indicator of the occurring times of political struggle (frequency) to measure the degree of influence of the political struggles in China on the national development (refer to Chart 8.2).

Chart 8.2 indicated the times of political struggles of China in each year during the period from 1949 to 1976. The chart and data reflected that firstly, China had frequent political struggle events during this period, and the country was in the

---

[11]Hu Angang: *History of China's Politics and Economy (1949–1976)*, Tsinghua University Press, 2008 Edition, p. 546.

Table 8.1 Political reforms and contents during 1949–1976

Time	Political reform	Contents
1949–1953	Wars against bandits	Eliminated over 2.6 million of bandits and armed spies
1950	Crackdown of speculative capitalists	Bankrupted the hoarded and speculative capitalists
1950–1953	Rectification of party	Overcame mistakes committed in works, overcame the cocky and complacent mood of considering themselves as meritorious statesmen, overcame bureaucracy and authoritarianism, over 328 thousand people left party organization, in which 238 thousands of bad elements and degenerated and corrupted people were eliminated from the party, and 1.07 million advanced people joined CCP
1950–1953	Land reform	Canceled the feudal and exploitative land ownership by land owner class, implemented the farmers' land ownership, so as to liberate rural productivity, develop agricultural production and open the path for the industrialization of New China. However, the land reform movement implemented in form of mass campaign made many people wrongfully classified as land owners, caused a batch of unjust, false and erroneous cases, and objectively and to certain extent weakened the productivity development of New China
1950–automn of 1953	Oppression to counterrevolutionaries	Primarily crack down bandits, spies, local tyrants, counterrevolutionary gangsters and counterrevolutionary organization leaders. The campaign expanded and had severe phenomenon of abused arrest and killing
1951–1952	"Three opposing" and "five opposing" campaign	The struggle against private capitalism, which consolidated the leadership status of workers' class and the leadership of socialist state-owned economy in national economy, but due to influence of extreme Leftist thoughts, a lot of unjust, false and erroneous cases were created

(continued)

Table 8.1 (continued)

Time	Political reform	Contents
1953–1954	Opposition against "Gao and Rao anti-party group"	Criticized Gao Gang and Rao Shushi
1953–1956	Socialist industrialization and three reforms	General path of the party during transition period: in a considerably long period, gradually realize the state's socialism industrialization, and gradually realize the state's socialist reform on agriculture, handicraft industry and capitalist industrial and commercial industry. It was completed within only 3 years, and was a radical mass class struggle, and a top-down mandatory system change
1953–1954	Party-wide cadre investigation campaign	Made comprehensive political investigation on cadres, eliminated the counterrevolutionaries, alien class elements and the degenerated and corrupted officials
1955–1957	Campaign to eliminate internal counterrevolutionaries	Struggled against and eliminated the hidden counterrevolutionaries
1955	Criticize the "Hu Feng counterrevolutionary group"	An event evolved from artistic dispute to political investigation, and became a large-scale political elimination and cleanses movement of the literature and art circle since the establishment of the People's Republic of China. In 1980, the CCP central committee decided to rehabilitate the "Hu Feng counterrevolutionary group" case
1956–1957	Oppose to the rightist conservative thought	Criticized Deng Zihui and accelerated the movement of agricultural cooperative
1957–1958	Criticized "opposition to rash advance"	Criticized Zhou Enlai, Chen Yun, Li Fuchun, Li Xiannian, Bo Yibo and other comrades
1957–1959	Rectification movement and anti-rightist struggle	Had the rectification movement throughout the party to oppose bureaucracy, sectarianism, and subjectivism. The anti-rightist struggle expanded, and a large batch of intellectuals, patriots and cadres within the party were wrongfully classified as "rightists" and caused unfortunate consequences

(continued)

Table 8.1 (continued)

Time	Political reform	Contents
1959–1961	Criticize the "anti-party group of Peng, Huang, Zhang and Zhou"	Criticized Peng Dehuai, Huang Kecheng, Zhang Wentian and Zhou Xiaozhou
From August, 1959 to the first half of 1960	Struggle against right deviations	Over three million of cadres and party members were criticized and classified into "right deviation opportunists"
1960–1961	Rectification of people's commune	The "three opposing" movement and rectification movement were developed in rural area
1962	Criticize the "individual farming trend"	In September, 1962, the party's tenth plenary session of the 8th central committee of CCP developed the criticism against the "individual farming trend" (meaning Deng Zihui) and "tendency to reverse past verdicts" (meaning Peng Dehuai and Xi Zhongxun) and Xi Zhongxun
1963–1965	Four clean-up movement	For the socialism education movement developed by CCP central committee in Chinese urban and rural areas, the struggling targets turned from the corrupt people in urban and rural areas to the "five black categories", and gradually changed from education in nature to class struggle
November, 1965	Criticize the *Hai Rui Dismissed from Office*	Planned by Jiang Qing, wrote by Yao Wenyuan, supported by Mao Zedong, criticized Wu Han, and targeted the CCP municipal committee of Beijing city
February–April, 1966	Counterattack against the black line of literature and art circle	Political alliance between Lin Biao and Jiang Qing
April–May, 1966	Criticize the "anti-party group of Peng, Luo, Lu and Yang"	Criticized Peng Zhen, Luo Ruiqing, Lu Dingyi and Yang Shangkun, and criticized the *february program*
1966–1967	Eruption of "cultural revolution"	
August–December, 1966	Red guard movement and sweeping of the "four old"	
August–December, 1966	Criticize the bourgeoisie counterrevolutionary path	Criticized the path of Liu Shaoqi and Deng Xiaoping
January, 1967	Criticized Tao Zhu	Criticized Tao Zhu, Wang Renzhong, Xiao Hua, etc.
January, 1967	Comprehensive power takeover	Overthrew local leaders

(continued)

8.2 Forms of Political Reform Cost in China

Table 8.1 (continued)

Time	Political reform	Contents
February, 1967	Counterattacked the "February countercurrent"	Criticized Tan Zhenglin, Chen Yi, etc., the political bureau was forced to stop working
March, 1967	Arrest traitors	Arrested Bo Yibo and others
July–August, 1967	Eliminate capitalist roaders within the military	Criticized Chen Zaidao and others; on July 22, 1967, Jiang Qing approved the slogan of "cultural attack and armed safeguarding", on July 25, 1967, Lin Biao proposed to completely crack down the "hell leadership of the general political department" and proposed the "elimination of the capitalist roaders within the military"
Spring, 1968	Struggle against rightist tendency of reversing past verdicts	Criticized Yang Chengwu, Yu Li Jin and Fu Chongbi
1968	Clean-up of class team	The objects of clean-up include "capitalist roaders", traitors, spies, landlords, counterrevolutionaries, bad elements, rightists and bad leaders
1968–1971	Struggle, criticizing and rectification	Established revolution committee, grand revolutionary criticizing, struggle against privatization and revisionism
1968–1971	Party reorganization movement	
1970	One strike—three anti campaign	Strike counterrevolutionaries, anti speculation, and anti wasting
1970–1976	Investigation of the "516 group"	
1970–1971	"Criticizing Lin and rectification" movement	Struggle that criticized Lin Biao and crashed the counterrevolutionary coup by the Lin Biao and Chen Boda anti-party group
1973–1974	"Criticize Lin and criticize Confucius" movement	Criticized Zhou Enlai and Guo Moruo
1975	Comment on Song Jiang in *water margin*	Criticized Deng Xiaoping
1976	Criticizing Deng and counterattack on the rightist tendency of reversing past verdicts	Criticized Deng Xiaoping
October, 1976	Crashed the "gang of four"	Arrested Jiang Qing, Zhang Chunqiao, Yao Wenyuan and Wang Hongwen

Source Hu Angang: *Chinese Political and Economic History* (1949–1976), Tsinghua University Press 2008 Edition, pp. 560–567

Table 8.2 Forms of political reform cost in China

Classification of cost	Forms	Specific contents
Cost of political reform	Cost of political struggle	The labors, materials and funds input by China in a series of political struggles, events or campaigns since the establishment of New China, and the loss caused to the nation from expansion of political movements
	Cost of political risk	When a country's GINI Coefficient reaches certain level, there is risk of social turmoil, which is an expected cost

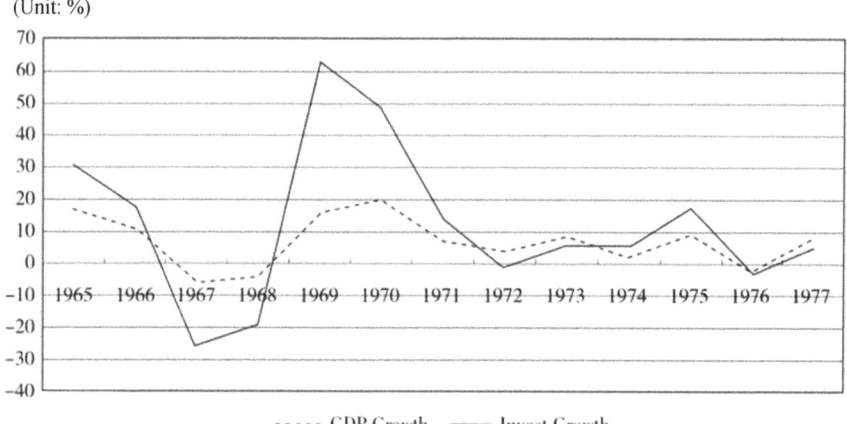

Chart 8.1 Economic fluctuation during 1965–1977 (*Source* State Statistics Bureau: *Compilation of Statistic Materials of New China in Fifty Years*, China Statistics Press, 1999 Edition, pp. 5, 7)

environment of continuous political struggle; secondly, in 1950 when New China was just established and 1966–1967 when the "Cultural Revolution" began, the occurring times of political struggle events reached two peaks respectively, especially during the preliminary stage of "Cultural Revolution". The excessively frequent and continuous political struggle generated meaningless waste of China's originally weak economic capital and human resources, and impeded the productivity development of China during this period.

"Class Conflict as the Guiding Principle" was the fundamental instructive thinking of every work of the entire party and country during this period, and even evolved into the core content of later theory of the "Continuous revolution under the dictatorship of the proletariat". Until the Third Plenary Session of the 11th Central Committee of the Chinese Communist Party, during which the proposition was made to make economic development for our central task, the situation was changed, since then, the cost of political struggle continuously decreased.

8.2 Forms of Political Reform Cost in China

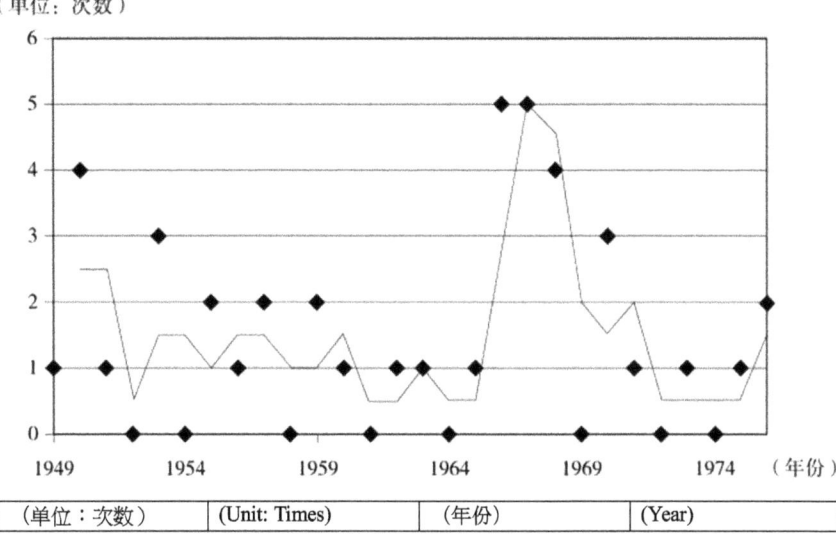

Chart 8.2 Change of frequency of political struggles during 1949–1976 (*Source* The author-concluded this chart from the situation of political struggle stated in the book of Political andEconomic History of China (1949–1976) wrote by Hu Angang [Tsinghua University Press, 2007edition])

8.2.2 Cost of Political Risk

Political risk defined in this book is a type of expected cost, and the cost of risk of having political turmoil existing in national development process. Turmoil cost could be measured based on the expected value of turmoil risk. Based on international consensus, it is extremely easy to have social turmoil when GINI Coefficient is higher than 0.4 (the probability is recognized as 100% in this book), in China's general environment of stable political and economic situation, it is assumed in this book that China would begin to have risk of social turmoil when the GINI Coefficient reaches 0.4 (the probability of having turmoil when GINI coefficient is lower than 0.4 is zero), and the probability of having social turmoil is 1 when the coefficient reaches 0.7%. Besides, it is assumed that the probability distribution function is uniform distribution. Then the calculation formula of Chinese turmoil risk cost is:

$$\text{Turmoil risk cost in current year} = (\text{Gini coefficient in current year} - 0.4)/(0.7 - 0.4) \\ \times \text{Turmoil loss in current year} \tag{8.1}$$

"Turmoil loss in current year" adopted the research theory of Dani Rodrik, Professor of International Economics from Kennedy School of Government of

Harvard University, when he researched the Eastern Asia Financial Crisis.[12] The specific calculation method was that: this book assumed that the influence of turmoil on economic growth is about 1%, therefore, "Turmoil loss in current year" equals GDP value of current year times 1%. As for GINI Coefficient, the result calculated by Yu et al. (2010)[13] was adopted, as for the lack of data in some years, the fitting compensation was made in grey system method. Please refer to Table 8.3 for the turmoil risk cost during the period from 1981 to 2009.

8.3 Forming Causes of Political Reform Cost in China

8.3.1 Forming Causes of Political Struggle Cost

The forming of Chinese political struggle cost has its in-depth historical causes and background. In summarization, there are two aspects of cause: firstly, at the beginning of New China, in order to consolidate the regime of people's democratic dictatorship, China needed to fight against the remaining feudal force and capitalist class force back then, and establish the socialist regime led by the proletariats as well as the forms of ownership of socialist public ownership; secondly, in the process of transforming from revolutionary political party to ruling party, the first generation of leadership had the guideline of socialist country construction in form of revolution, so the democratic centralism within the party was undermined. The specific causes are analyzed as follows:

1. At the beginning of New China after establishment, the regime was not stable, various forces existed, including the capitalist class and survivals of feudalism, various economic sectors coexisted. Although the Common Program regulated that the five economic sectors[14] shall, under the leadership of state-owned economy, share out the work and cooperate with each other, take care of their respective duties, cover both public and private interests, and benefit both labors and owners, however, not long after this guideline, it was replaced by large-scale socialization and collectivization of production materials. Those

---

[12]Dani Rodrik believed that the formula "Economic growth = External Impact × (Potential social conflict/Conflict management organ)" is tenable.

[13]Yu Haiqing and others: *Empirical Analysis of the Regional Economic Difference of Shandong Province based on GINI Coefficient Decomposition,* Journal of Ludong University, January, 2010.

[14]Five economic sectors means state-owned economy, cooperative economy, individual economy, private capitalism economy and state capitalism economy.

8.3 Forming Causes of Political Reform Cost in China

Table 8.3 Turmoil risk cost during 1981–2009

年份	GDP（亿元）	基尼系数	动乱损失（亿元）	动乱风险成本（亿元）
1981	4891.6	0.288	48.916	−18.262
1982	5323.4	0.2494	53.234	−26.7235
1983	5962.7	0.2641	59.627	−27.011
1984	7208.1	0.297	72.081	−24.7478
1985	9016	0.2656	90.16	−40.3917
1986	10275.2	0.2968	102.752	−35.3467
1987	12058.6	0.3052	120.586	−38.1052
1988	15042.8	0.382	150.428	−9.02568
1989	16992.3	0.349	169.923	−28.8869
1990	18667.8	0.343	186.678	−35.4688
1991	21781.5	0.324	217.815	−55.1798
1992	26923.5	0.376	269.235	−21.5388
1993	35333.9	0.3592	353.339	−48.0541
1994	48197.9	0.436	481.979	57.83748
1995	60793.7	0.445	607.937	91.19055
1996	71176.6	0.458	711.766	137.6081
1997	78973	0.403	789.73	7.8973
1998	84402.3	0.403	844.023	8.44023
1999	89677.1	0.397	896.771	−8.96771
2000	99214.6	0.417	992.146	56.22161
2001	109655.2	0.49	1096.552	328.9656
2002	120332.7	0.454	1203.327	216.5989
2003	135822.8	0.53	1358.228	588.5655
2004	159878.3	0.46	1598.783	319.7566
2005	184937.4	0.45	1849.374	308.229
2006	216314.4	0.496	2163.144	692.2061
2007	265810.3	0.48	2658.103	708.8275
2008	314045.4	0.49	3140.454	942.1362
2009	340902.8	0.47	3409.028	795.4399

Source: GDP data in history was sourced from the official website of the State Statistics Bureau. Source of GINI Coefficient: the data of 1981, 1984, 1988, 1989, 1992 and 1998 was cited from Zeng Guo'an: *Discussion about the Characteristics, Causes and Countermeasures of the Income Gap of Chinese Citizens*, Journal of China University of Geo-science (Social Science Edition), the 4th Issue in 2001; the data of 1990, 1999 and 2000 was the calculation result of the World Bank and Chinese Academy of Social Science, the data of 1996 was the calculation result by Li Qiang, and all cited from Gan Wenxiu and Yin Yanlin's *Summarization of Research on the Income Gap of Chinese Residents*, Economic Research Reference, the 83rd Issue in 2003; the data of 1982, 1983, 1985, 1986, 1987, 1991 and 1993 was cited from Xiang Shusheng's *Calculation and Regression Analysis of Income Distribution Gini Coefficient of Chinese Residents*, Financial and Economic Theories and Practice, the 1st Issue in 1998; the data of 2001 was the result of sampling survey and data calculation by the studying team of "Studies of Chinese Social Structural Change" in 2001 from the Chinese Academy of Science, cited from Li Chunling: *Chinese Social Stratification and New Trend of Life Style*, Scientific Socialism, the 1st Issue in 2004; the data of 2002 was cited from Liang Meina: *Will the Chinese Income Gap Become Wider?* Chinese Operation News, Jan 15, 2005; the data of 2003 was a result of cooperated investigation between China Renmin University and Hong Kong University of Science and Technology, cited from Liang Liping: *A Modernized Harmonious Society --- Interview with Professor Wu Zhongmin from CPC Central Party School*. Forum of Chinese Party and Political Cadres, the 11th Issue in 2004; the data of 2005 was cited from the *2005 Human Development Report* of the UN; the data of 2007 was from Business Education and Economic Research, the 23rd Issue in 2008.

年份	Year	GDP（亿元）	GDP (RMB 100 million)
基尼系数	GINI Coefficient	动乱损失（亿元）	Loss from turmoil (RMB 100 million)
动乱风险成本（亿元）	Cost of turmoil risks (RMB 100 million)		

speculative capitalists were cracked down, the struggles against the "Three Evils"[15] and the "Five Evils"[16] were the most prominent representatives of such political struggle events.

2. Failure of democratic decision-making system, the disagreement of different opinions within the party was raised to class struggle. As early as the Seventh National Congress of CPC in 1945, the party constitution was deemed as a system, everyone is equal before this system, there was no special leader.[17] During 1949–1956, the democratic atmosphere was high within the party, consensus was easy to make for major issues, during this period, the socialist cause in China was developing smoothly. After 1957, the party central committee's ruling method was transformed, collective leadership decision-making

---

[15]Opposing the "Three Evils" means opposing corruption, wasting and bureaucracy.

[16]Opposing the "Five Evils" means opposing bribery, tax evasion, theft of state assets, cheating on works and materials, and theft of national economic intelligence.

[17]Liu Shaoqi: *Reports Regarding Amendment of Party Constitution*, in May, 1945, refer to *Selected Works of Liu Shaoqi*, Volume A, People's Publishing House, 1981 Edition, pp. 316 and 360.

was changed into "Supreme Instruction" by individual.[18] The wrong guiding ideology of "Class Struggle as Guideline" made the country trapped in endless political campaigns and struggles, as well as caused large amount of unjust, false and erroneous cases, and severely affected the development of productivity.

8.3.2 Forming Causes of Political Risk Cost

The fundamental cause of the forming of political risk cost is the widening gap of income caused by unequal distribution of income in the country's development process. Since the reform and opening-up, under the call of "Encouraging and mobilizing some people to get rich first" and "the rich first pushing those being rich later", the national government formed different policies for the economic development of different regions, which enabled China to achieve extensive growth of gross economy within short 30 years, China currently has become the second biggest economy in the world.[19] However, due to China's vast territory, huge population, and based on different natural resources, cultural and historical conditions as well as the related policies, while China has achieved development results that amazed the world, China also has various problems, such as imbalance of development in different regions, imbalance of development in urban and rural areas, continuously widening gap of income among residents, etc. To be specific in the issue of distribution, China firstly adopted combination of distribution according to work and distribution according to productive factors, and adhered to the principle of efficiency first and taking fairness into account, but in actual enforcement, income from productive factors outweighed labor income, so "Efficiency First" was emphasized, but the principle of taking fairness into account was not realized, which caused continuously widening gap of incomes among residents. The widening gap between rich and poor was directly reflected as increase of GINI Coefficient, it caused increase of social unrest factors, as a result, the risk of turmoil was increased.

For the summarization of forming causes of Chinese political reform cost, please refer to Chart 8.3.

---

[18]On Jun 27, 1981, the Sixth Plenary Session of the 11th Central Committee of CCP unanimously passed the *Resolution Regarding Several Historical Issues of the Party since the Establishment of New China*, refer to, edited by CCP Central Committee Literature Research Office: *Compilation of Important Literature since the Third Plenary Session* (B), People's Publishing House, 1982 Edition, pp. 819.

[19]As indicated by the data released by the Cabinet of Japan, calculated by comparable price, the nominal GDP of Japan in 2010 was USD 5.4742 trillion, which was USD 400 billion less than China's GDP released in 2010. This is the first time since 1968 when Japanese economy receded to the fifth place in the world.

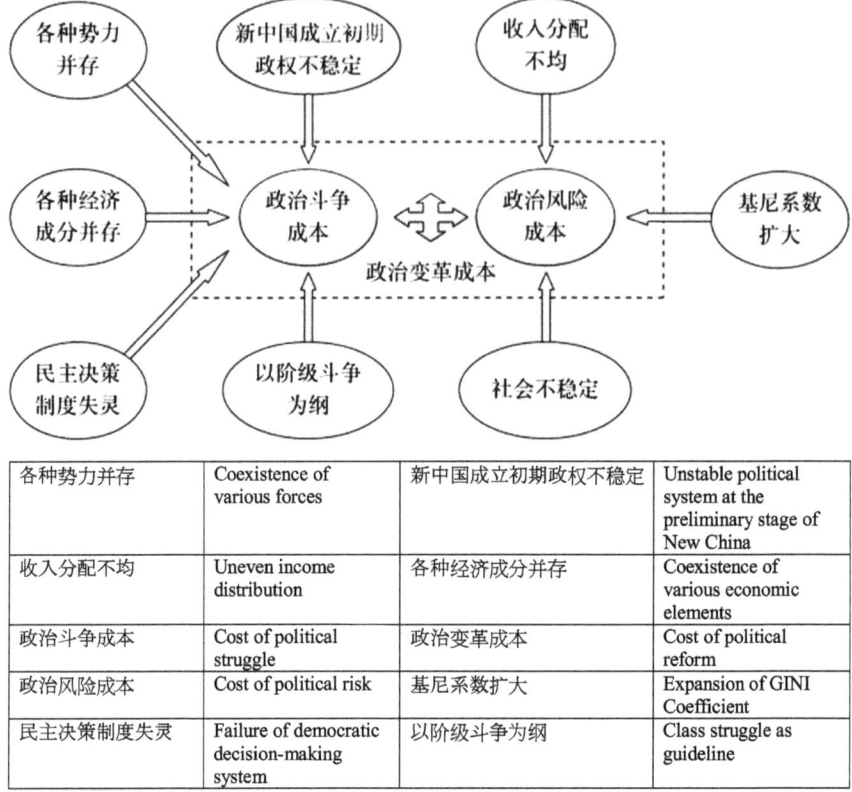

Chart 8.3 Forming causes of Chinese political reform cost

8.4 Countermeasures to Handle Chinese Political Reform Cost

8.4.1 Prevent the Occurrence of Political Reform in Field of Ideology and Law

In order to avoid repeating of history, firstly, we need to make necessary investment in ideological and political field, build good party conduct and government style, strengthen the political and cultural quality of leaders and cadres on different levels, establish the public servant awareness of wholeheartedly serving the people, and maintain a stable political environment. Secondly, we need to eliminate the authoritarian style within the party, realize democratic decision-making, scientific decision-making, and legal decision-making, and realize equal communication of

8.4 Countermeasures to Handle Chinese Political Reform Cost

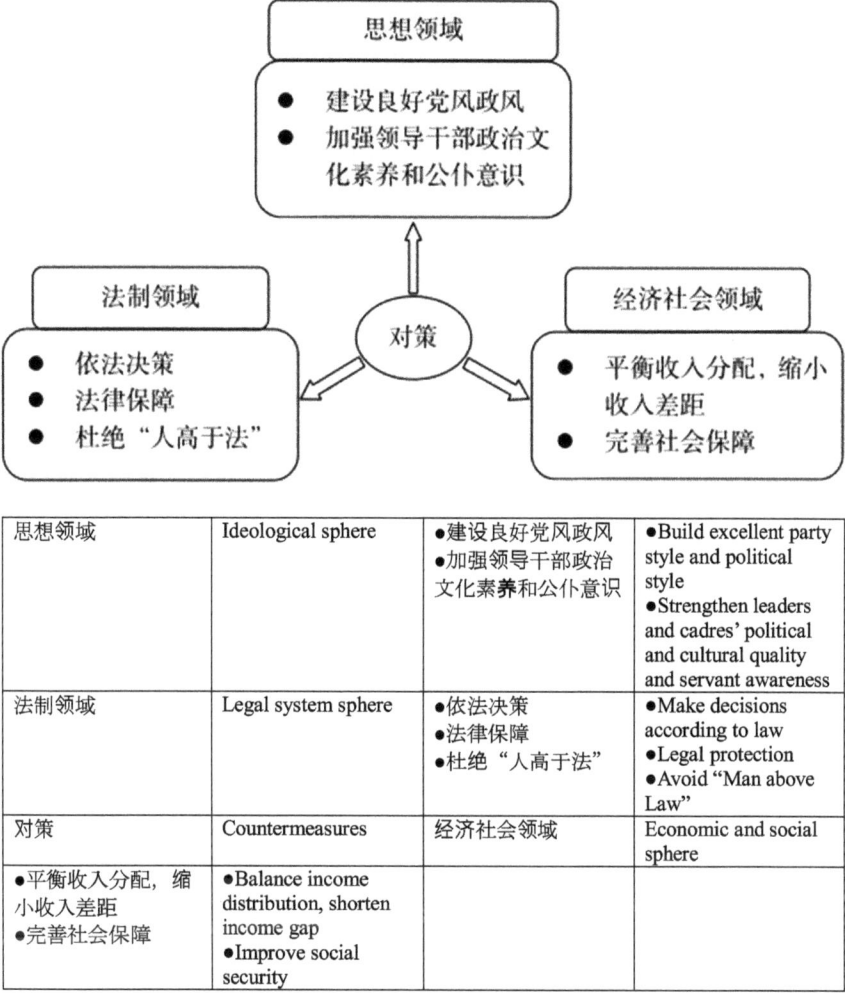

Chart 8.4 Countermeasures for Chinese political reform cost

different opinions. The measures mentioned above all need protection through policy, laws and systems, and shall be strictly enforced, the occurrence of "People above Law" phenomenon is absolutely not allowed. For the specific measures, please refer to Chart 8.4.

8.4.2 Make Efforts to Reduce Risk of Turmoil and Control the Widening Gap Between Rich and Poor

In order to maintain stability, the national government needs to balance income distribution, and try the best to realize the balancing role of primary distribution,[20] secondary distribution[21] and tertiary distribution[22] to income, this is mainly a countermeasure for political reform cost in economic and social field. We shall consider distribution according to work as the main body, limit the distribution quota of excessively high income in monopoly industries; use taxation policies to regulate and control income gap among different classes; and use multiple forms of social security measures to ensure the basic living income level of low-income groups. For the specific situation, forming causes and handling strategy of income distribution gap, please refer to Chart 8.5.

---

[20] Primary distribution means the distribution of total national income (meaning GDP) directly correlated to production factors.

[21] Secondary distribution, or referred to as redistribution of national income, means the income of economic entities that can be obtained without directly through labor, such as pension, (official's) retirement payment, unemployment subsidy, etc.; or voluntary subsidy by the state or other economic entities.

[22] Tertiary distribution system means proper exemption of income tax for donation that is designed to encourage wealthy class to actively devote to public welfare and charity, and is to rationally allocate social wealth through development of charity courses, so as to ease the problem of widening income gap.

8.4 Countermeasures to Handle Chinese Political Reform Cost

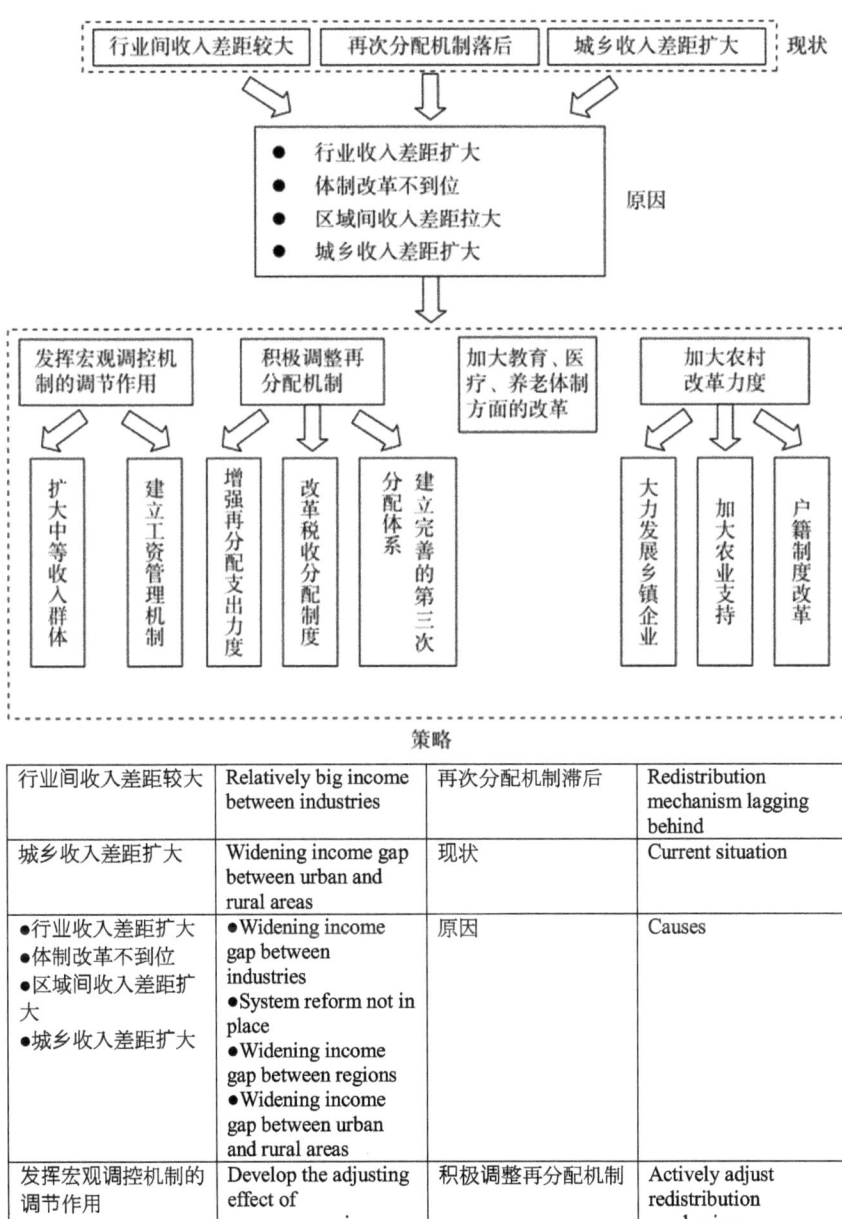

行业间收入差距较大	Relatively big income between industries	再次分配机制滞后	Redistribution mechanism lagging behind
城乡收入差距扩大	Widening income gap between urban and rural areas	现状	Current situation
●行业收入差距扩大 ●体制改革不到位 ●区域间收入差距扩大 ●城乡收入差距扩大	●Widening income gap between industries ●System reform not in place ●Widening income gap between regions ●Widening income gap between urban and rural areas	原因	Causes
发挥宏观调控机制的调节作用	Develop the adjusting effect of macroeconomic regulatory mechanism	积极调整再分配机制	Actively adjust redistribution mechanism

Chart 8.5 Current situation, forming causes and countermeasures of income distribution gap

加大教育、医疗、养老体制方面的改革	Enlarge the reforms in the aspects of education, medical and endowment systems	加大农村改革力度	Enlarge the degree of rural reforms
扩大中等收入群体	Enlarge middle-income group	建立工资管理机制	Establish salary management mechanism
增强再分配支出力度	Strengthen expenditure degree of redistribution	改革税收分配制度	Reform the tax distribution system
建立完善的第三次分配体系	Establish a complete tertiary distribution system	大力发展乡镇企业	Energetically develop rural enterprises
加大农业支持	Increase agricultural support	户籍制度改革	Household system reform
策略	Strategies		

Chart 8.5 (continued)

Chapter 9
Chinese Political System Construction Cost

9.1 Connotation of Chinese Political System Construction

System construction of a country is a process, meaning the long historical process of system innovation, system construction and system implementation based on Chinese situation after fully learning, absorbing and referring to the experiences and knowledge of modernized nations in human history.[1] Political system usually means organizational form of national political power and the related systems, its determination and change both need to adapt to the fundamental nature of the country. Marx called it "special object that has regulations to manage all special and common senses".[2]

The result of political system construction is reflected as change of political system. Change of political system is a complex and systematic change, such change is sometimes partial change[3] and sometimes general change.[4] From macro perspective, the logical cause of political system change is fundamentally basic social conflict movement between productivity and productive relations, between economic foundation and superstructure, it is the fundamental cause of political system change, and is the basic power of historical advancement. From micro

---

[1] State system construction will be good for reducing administrative cost of a country, regulate the conflict of interests among different social groups, maintain the necessary social order and transaction rules for economic activities, it provides wide income distribution and public service in the entire society, and improve the human capital conditions for all people of the country. Therefore, system construction itself is not economic construction, but it is a basic condition to promote economic construction and continuous development.
[2] *Discussion of Politics and Political System by Marx, Engels, Lenin and Stalin (A)*, Achieve Publishing House, 1998 Edition, p. 15.
[3] Such partial change is also normally peaceful and gradual.
[4] Such general change is also usually radical.

Table 9.1 Five forms of political participation

Forms	Contents	Effect
Voting	In the election of NPC members of various levels and rural committee directors of various villages, citizens express their political will through voting and election of candidates	Promote political system construction
Electoral activities	Election is composed of a series of political activities, such as organization, coordination, promotion, public mobilization, and voting is just the last part of electoral activities	Promote political system construction
Regional activity	Regional activity includes participating in citizen activities and residents activities	Promote political system construction
Individual contact	Citizens could participate in politics by individually contacting NPC members	Promote political system construction
Violence	Physically damage other individuals or their properties, so as to affect the decision making by government	Damage political system construction

perspective, at certain historical point, what decides the direction of political system change is comparison of powers among different parties, and depends on the political attitude and political participation degree[5] of majority people. There are mainly five forms of political participation: voting, elective activity, regional activity, individual contact and violence. Refer to Table 9.1.

9.2 Forms of Chinese Political System Construction Cost

The political system construction cost studied in this book, just as its name implies, means the cost of expenditure and loss in the process of continuous improvement of Chinese political system, including corruption cost, system reform cost, political system setting cost and political system destruction cost (Refer to Table 9.2). In which corruption cost and system reform cost is the input and loss of political system construction considered from the perspective of national rulers; while political system setting cost and political system destruction cost are the cost of political system promoting and destructing activities involved from the perspective of political participation by citizens.

---

[5]Political participation means common citizens' activities that would affect decision making by government. Quoted from [Japan] Pudao Yufu; *Political Participation*, Translated by Jie Lili, Economic Daily Publishing House, 1989 Edition, p. 4.

9.2 Forms of Chinese Political System Construction Cost

Table 9.2 Forms of Chinese political system construction cost

Classification of cost	Forms	Specific contents
Cost of political system construction	Cost of corruption	Loss from corruption, input to prevent corruption
	Cost of system reform	Cost of institution setting, merging and cancellation
	Cost of setting of political system	Cost paid to promote political system setting in political participation process
	Cost of damage of political system	Loss from illegal damage of political system in political participation process

9.2.1 Corruption Cost

Since the reform and opening-up in 1978, the problem of corruption has become a very prominent political and social problem faced by China at transformational stage. With the continuous increase of anticorruption efforts, the growing trend of corruption cases is weakened. According to related data, first of all, the total number of corruption cases gradually decreased from the peak during 1992–1997. According to the work reports of the Supreme People's Procuratorate and related statistical materials in previous years, the prosecuting authorities on each level across China investigated 98 thousand corruption cases in total during 1979–1982, the number increased to 155 thousand cases during 1983–1987, 21 thousand cases during 1988–1992, and even increased to 387 thousand cases during 1993–1997 (and the registration criterion for embezzlement, bribery and appropriation of public funds and other corruption cases had been raised), then the number decreased to 207 thousand cases during 1998–2002, 180 thousand cases during 2003–2007, which decreased by 13.2%[6] compared with that in five years before. Secondly, the corruption-involved amount was also peaked during 1993–1997. As for the amount of direct economic losses that the prosecuting authorities at each level across China recovered for the country and collective group through investigating economic cases, it was RMB 1.63 billion during 1983–1987, RMB 2.58 billion during 1988–1992, and as high as RMB 22.92 billion during 1993–1997, then the number was RMB 22 and 22.48 billion respectively during the two subsequent five-year periods.[7]

The corruption cost studied in this book includes not only criminal income by officials through taking advantage of their privilege, but also the part of grey income caused from systematic cause, which is "Corruption Cost" in broad sense. Because the data of corruption cost couldn't be directly obtained, this book will use the

---

[6]Source: Sorted from the Work Reports of the Supreme People's Procuratorate of the People's Republic of China in various years.

[7]Source: Sorted from the Work Reports of the Supreme People's Procuratorate of the People's Republic of China in various years.

existing data to estimate the corruption cost in China in different years. The source of data is the economic losses recovered by prosecuting authorities for the country through investigating corruption cases in each year, the estimation and calculation method is the adoption of the research on Chinese corruption by Wang Xiaolu[8] and the conclusion of high correlation between corruption income and GDP proposed by Sun Fenghua[9] and others in *Geographic Research of Chinese Crimes*. The calculation method of corruption cost studied in this book is as follows:

$$\text{Corruption cost} = \text{Economic loss recovered for country by investigating corruption cases} \times \text{Loss and investigation ratio} \quad (9.1)$$

$$\text{Loss and investigation ratio} = (\text{GDP in 2008} \times 13\%/\text{Loss recovered in 2008} \\ + \text{GDP in 2007} \times 13\%/\text{Loss recovered in 2007} \\ + \text{GDP in 2006} \times 13\%/\text{Loss recovered in 2006} \\ + \text{GDP in 2005} \times \text{Loss recovered in 2005})1/4 \quad (9.2)$$

This book adopted the research result by Wang Xiaolu on ratio of corruption income in GDP, which believed that the amount of economic loss recovered for the country through investigating corruption cases is highly correlated with actual amount involved in corruption cases.[10] It needs to be pointed out that such calculation method is very limited in measurement of corruption cost, because the economic loss recovered by prosecuting authorities is only one aspect to measure corruption. For Chinese corruption cost during 1978–2008 calculated by the formula, please refer to Table 9.3 (in which the data during 1978–1982 is not available).

9.2.2 System Reform Cost

Political system reform is self-improvement of existing political system promoted by the leadership of Chinese Communist Party, and the basic channel to liberate and develop productivity, develop socialist democratic politics, and construct socialism political civilization, and the dominant factor to firmly adopt the socialism political

---

[8]Wang Xiaolu: *Grey Income and Distribution of National Income,* Comparison, 48th Issue, CITIC Press, 2010 Edition.

[9]Sun Fenghua and Wei Xiao: *Geographic Research of Chinese Crimes,* Journal of Liaoning Normal University (Natural Science Edition), April, 2006.

[10]Wang Xiaolu's research result showed that corruption income accounts for 10–15% of GDP, this book adopts 13%.

9.2 Forms of Chinese Political System Construction Cost

Table 9.3 Estimation of Chinese corruption cost during 1978–2008

年份	腐败涉案金额（亿元）	GDP（亿元）	腐败成本（亿元）	腐败成本占GDP比重
1978	—	3645.20	—	—
1979	—	4062.60	—	—
1980	—	4545.60	—	—
1981	—	4891.60	—	—
1982	—	5323.40	—	—
1983	0.60	5962.70	4.73	0.08%
1984	0.90	7208.10	7.10	0.10%
1985	2.68	9016.00	21.13	0.23%
1986	8.00	10275.20	63.08	0.61%
1987	4.12	12058.60	32.49	0.27%
1988	4.23	15042.80	33.35	0.22%
1989	4.83	16992.30	38.07	0.22%
1990	8.10	18667.80	63.87	0.34%
1991	5.00	21781.50	39.42	0.18%
1992	3.64	26923.50	28.71	0.11%
1993	22.00	35333.90	173.47	0.49%
1994	34.00	48197.90	268.08	0.56%
1995	49.00	60793.70	386.36	0.64%
1996	67.80	71176.60	534.59	0.75%
1997	56.40	78973.00	444.70	0.56%
1998	43.80	84402.30	345.35	0.41%
1999	40.90	89677.10	322.49	0.36%
2000	47.00	99214.60	370.59	0.37%
2001	6.80	109655.20	53.62	0.05%
2002	81.50	120332.70	642.61	0.53%
2003	43.00	135822.80	339.05	0.25%
2004	45.60	159878.30	359.55	0.22%
2005	74.00	184937.40	583.48	0.32%
2006	48.00	216314.40	378.47	0.17%
2007	34.20	265810.30	269.66	0.10%
2008	21.00	314045.40	165.58	0.05%

Source sorted out based on the work reports of the Supreme People's Procuratorate of the People's Republic of China in previous years, the rest of it was calculated based on the formula (9-1) and formula (9-2) explained above.

年份	Year	腐败涉案金额（亿元）	Involved amount of corruption (RMB 100 million)
GDP（亿元）	GDP (RMB 100 million)	腐败成本（亿元）	Cost of corruption (RMB 100 million)
腐败成本占GDP比重	Ratio of corruption cost in GDP		

Table 9.4 Development stages of Chinese political system reform

Time	Stage	Main contents
December, 1978–December, 1989	Preliminary stage	Separation of party and politics, unveiled the political system reform
1990–October, 2003	Comprehensive development stage	Law-based governance, stable promotion of political system reform
October, 2003–December, 2010	Continuous promotion stage	Democratic ruling, in-depth development of political system reform

path with Chinese characteristics, it involves the future and destiny of Chinese Communist Party and the socialism cause. Deng Xiaoping believed that the main contents of Chinese political system reform should include: "firstly the separation of party and politics, solving the problem of how the party be good at leadership. This is the key and should put in priority. The second content is to delegate power to lower levels, solving the relationship between the center and local regions, meanwhile, different local levels also have the issue of delegating power to lower levels. The third content is to streamline organizations, which is correlated with delegation of power to lower levels.[11]" For the 30 years since reform and opening-up, Chinese political system reform could be mainly divided into three development stages, refer to Table 9.4.

The system reform at different stages included separation between party and politics, institutional streamlining, setting and merge of big ministries and commissions, as well as the various obstacles and setbacks faced in system reform process, which all need to pay certain cost.

(I) Preliminary Stage

The establishment of socialist country was a unprecedented enterprise, there was no existing experiences to follow for the establishment of its political system, so it can't be very complete at once. China's current political system was derived from the age of revolution and war, and affected by the model of Soviet Union, plus the long-term restriction of China's feudal autocracy in history, the "Leftist" thinking continued to develop and even to overflow, which caused the party and country's leadership system and other political systems to have problems such as undivided party and politics, replacing politics with party, excessive power centralization, bureaucracy, etc. which was the general root of problems in Chinese political system.[12]

---

[11] *Selected Works of Deng Xiaoping,* Volume 3, People's Publishing House, 1993 Edition, p. 177.

[12] Deng Xiaoping pointed out that "Excessive concentration of power to individual or few people, most handling people don't have the right to make decisions, few powerful people have too much burden, which would inevitably cause bureaucracy, and make various mistakes, and will harm the democratic life, group leadership, democratic concentration system, and individual division of work responsibility system of the party and government on different levels". Refer to *Selected Works of Deng Xiaoping,* Volume 2, People's Publishing House, 1994 Edition, p. 329.

9.2 Forms of Chinese Political System Construction Cost

Table 9.5 The first 4 government institutional reforms after reform and opening-up

No.	Time	Contents of reform
1	1982	The number of departments of the state council was reduced from 100 to 61, and the size of personnel force was reduced from 51 thousand to 30 thousand
2	1988	The number of ministries and commissions of the state council was reduced from 45 to 41, after the reform, the size of personnel force was reduced by over 9700
3	1993	The composed departments and directly-governed agencies of the state council was reduced from 86 to 59, and the number of personnel was reduced by 20%
4	1998	15 ministries and commissions were no longer preserved, 4 ministries and commissions were newly established, 3 ministries and commissions were renamed. After the reform, except the general office of the state council, the composing departments of the state council were reduced from the original 40 to 29

(II) Comprehensive Development Stage

The domestic political disturbance in 1989 and the occurrence of the collapse of the former Soviet Union and tremendous change to East Europe generated major influence and impact to Chinese political stability, Chinese political system reform was once trapped in a state of stagnation. In September, 1997, at the party's Fifteenth National Congress, Jiang Zemin explained in details the meaning of the Rule of Law, since then, "Rule of Law"[13] was wrote into the basic guideline of the party. This indicated that our party's ruling method fundamentally abandoned the tradition of "Rule of Man", realized transformation to rule of law, it is a revolutionary change to Chinese Communist Party's governance in idea and method, it was an important step of Chinese political system reform, so it has important epoch-making significance. Table 9.5 listed the contents of the first 4 times of government institutional reforms since the reform and opening-up.

(III) Continuous Promotion Stage

This stage mainly included the institutional reforms in 2003, 2008 and 2013. Refer to Table 9.6. The institutional reform in 2003 was made under the major back ground of China joining the WTO, and based on the requirements of coordinated three powers, "Decision-making, Enforcement and Supervision", it focused on the adjustment of the systems in the aspects of state-owned assets management, macroeconomic regulation and control, financial supervision, circulation management, food safety and safety production supervision, population and family

---

[13]"Rule of Law" is the basic strategy for CCP leaders to govern the country and a major development of the rule of law construction of China's political system reform.

Table 9.6 Government institutional reform at continuous promotion stage

No.	Time	Contents of reform
1	2003	Established the state-owned assets management commission and banking regulatory commission, set up the ministry of commerce, China food and drug administration, administration of work safety, reorganized the national development planning commission into national development and reform commission, and changed the composing departments of the state council into 28
2	2008	"Large ministries system" reform. Implemented merging, adding and other adjustments on the institution, function and affiliation of the original 14 ministries and commissions

planning, etc.[14] As for the institutional reform of the State Council in 2008 (or "Large Ministries System" reform,[15] refer to Table 9.7), the main purpose was to promote good and rapid development of the economy and society. This reform made important steps in some key fields: focused on the promotion of transformation of government functions, transferred over 60 functions of the departments under the State Council; further cleared the division of works and duties of different departments, and collectively solved the problem of crossing and unsmooth relations among macroeconomic regulation, environmental resources and 70 other duties. Later, CPC Central Committee passed the *Opinion Regarding Institutional Reform of Local Governments*, so the institutional reforms of local governments are also actively and orderly promoted. Besides, the Large Ministries Reform in 2013 was even bigger in reform efforts, which was reflected as ① reflected more the efforts in socialism market economic reform, such as the realization of separation between politics and enterprise for railway; ② focused more on the planning of roads with socialism characteristics, such as the establishment of the National Health and Family Planning Committee, and included population development strategy and other planning into the State Development and Reform Commission; ③ emphasized more on the correlated combination of functions, such as the establishment of the State Food and Drug Administration, which involved combination of related functions in four departments, as well as the establishment of the State Administration of Radio, Film and Television; ④ focused more on ocean development space, such as the reestablishment of the State Oceanic

---

[14]Except the General Office of the State Council, in the reform, the 29 composing departments of the State Council were adjusted to 28, the State Economic and Trade Commission and the Ministry of Foreign Economics and Trade were no longer preserved, and their functions were taken by the newly established Ministry of Commerce.

[15]Large Ministries System, based on the proposition by related expert, is to promote the comprehensive management and coordination of government affairs, merge government agencies based on the comprehensive management functions of government, and form large-ministries government organization system. The characteristic is to expand the business scope managed by certain one ministry, and let one ministry handle various affairs with internal connections, so as to avoid government function crossing, each department action on its own, multiple management to most extent, and to enhance administrative efficiency, and reduce administrative cost.

9.2 Forms of Chinese Political System Construction Cost

Table 9.7 List of institutional reforms of the state council in 2008

Name of ministries and commissions after adjustment	Original ministries and commissions merged or included	Added subsidiary institutions	Adjustment of affiliation
National development and reform commission	NDRC (ministry system remain unchanged)	National energy administration	
Ministry of industry and information	State commission of science and technology for national defense industry, ministry of information industry, information work office of the state council, state tobacco monopoly bureau	SASTIND	The state tobacco monopoly bureau was changed to be under the management by the ministry of industry and information
Ministry of transport	Ministry of transport, civil aviation administration, state post bureau	State civil aviation bureau	The state post bureau was changed to be under the management by the ministry of transport
Ministry of labor and social security	Ministry of personnel, ministry of labor and social security	State bureau of civil servants	
Ministry of environmental protection	State administration of environmental protection		
Ministry of housing and urban-rural construction	Ministry of construction		
Ministry of health	Ministry of health, administration of drug		The state administration of food and drug was changed to be managed by the ministry of health

Administration; ⑤ focused more on the strategic combination of electric power and energy, such as the reorganization of National Energy Administration, and canceled the independent Electricity Supervision Commission.

9.2.3 Political System Setting Cost

As indicated in Table 9.8, Chinese political system setting cost means the cost paid to promote political system setting in political participation process. By integrating the effectiveness of the five political participation forms, it can be concluded that the political system setting cost specifically includes the cost of voting, election, regional activity and individual contact.

Table 9.8 Political system setting cost

Name of cost	Forms	Scale of cost
Setting cost of political system	Voting	Relatively low
	Electoral activity	Relatively high
	Regional activity	Relatively low
	Individual contact	Relatively low

Voting is the political activity participated by most citizens. Chinese citizens have the chances of voting in elections of NPC members on different levels, Village Committee Directors, etc. Citizens could express their political will to decision makers through voting and election of candidates. However, under most circumstances, elections through voting are not often held, but held on regular basis according to certain procedure. Compared with other political activities, the time spent for voting is very short, and the participation price paid is relatively low.

Election is composed of a series of political activities, such as organizational coordination, promotion, mobilization, voting is just the last stage of elective activity. These activities require input of large volume of labors, materials and funds, the cost of entire election process is relatively high.

Regional activity includes participation in citizen activities and resident activities.[16] In China, the actual occurring frequency of such activities is not high, therefore the cost is not high.

In China, citizens can participate in politics by individual contact with NPC members. The influence brought by individual contact is mostly and depending on individuals, its scope of influence is narrow, and cost is relatively low.

9.2.4 Political System Destruction Cost

Since the establishment of New China, the most severe political system destruction period was the "Cultural Revolution". This is a top-down civil political movement, which caused severe destruction to the political system that was built not long after the establishment of New China and the whole country's democratic and legal system. Besides, in the process of Chinese citizens participating in politics, some people, in order to realize their own interests, adopted illegal forms and channels of political participation, impacted the operation of legal political participation system. Some citizens participated in politics and even did it to vent their discontent, they were unable to rationally adopt normalized and standard participation form, but participated in politics with enthusiasm, or even adopted violence to destruct political system, therefore caused certain loss.

---

[16]Such as environmental protection movement, anti-pollution movement, peace movement, consumer movement, etc.

9.3 Forming Causes of Chinese Political System Construction Cost

9.3.1 Historical Causes of Chinese Corruption Problem

In order to tackle down corruption, Deng Xiaoping proposed anticorruption issue three times in 1982, 1986 and 1989 respectively,[17] and made in-depth conclusion on the forming causes of corruption: firstly, the "Decade-long Cultural Revolution" severely damaged our democracy within party and promotion of ethical party and government, enabled anarchism and extreme individualism to be continuously abused, people acted in thinking and behaviors as hedonism and seeking quick success and instant benefits, they let their alert down in thinking; secondly, some people within the cadres team with weak willpower couldn't resist the test of reform and opening-up in the new era, under the temptation of material interest, power and money, they began to lose alert in mind, put profit first, be blinded by lust for money, and walked on the path of law violation and crime; thirdly, the influence of the party's ideological and political work on the promotion of ethical party and government was not well developed, the advantage and dominating role of the party's ideological and political work was not developed in the society and politics, and we were not aware of the severity and particularity of corruption problem in the party's political life. Besides, the degree of crackdown and punishment towards corruption problem was not sufficient, we were lack of basic political mechanism and legal measures to restrain and prevent corruption phenomenon, which made various corruption problems[18] relating to economic activities to be more prominent.

9.3.2 Flaws in Chinese Political System

China's traditional political system has severe flaws of excessive centralization of power, lack of democracy, and incomplete rule of law, which caused the situation of undivided party and politics, and replacing politics with party. Excessive

---

[17]Deng Xiaoping pointed out that "Since we implemented the two aspects of policy of foreign opening-up and domestic activation of economy, just one or two years, considerable number of cadres was corrupted." Refer to *Selected Works of Deng Xiaoping,* Volume 2, People's Publishing House, 1994 Edition, p. 402.

[18]Various renamed and disguised forms of corruption continued to appear and covered with various theories and legal coating, such as wasting was called "Public Entertainment", "Attracting Investments", giving and accepting bribery was named "Sales" and "Fair Transaction". Some leaders and cadres went "Rent Seeking", "Money Seeking" and "Profit Seeking" using their powers, and considered their powers as a kind of potential intangible assets or the capital to seek personal wealth through their power operation. Including approving and seeking lands, arranging relatives and friends to secure position and serve senior posts in highly profitable units, or taking advantage of the various opportunities during assets reorganization to embezzle state assets.

centralization of power is the hotbed to breed corruption, meanwhile, the lack of supervising and punishment measures made us unable to effectively find and restrain corruption, so as to cause the spreading of corruption phenomenon.

9.3.3 Incomplete Specific System of Political Participation System Is the Most Direct Factor that Restrains Political Participation in Modern China

In a modern society, the operation of political participation is realized through a whole set of complete system. Therefore, China has established the People's Congress System, the CPC-led Multi-party Cooperation System, Village Self-government System and System of Regional National Autonomy, these systems provide fundamental guarantee for the operation of political participation. However, the specific measures and implementing details of these systems are still not complete and improved.[19] This is bound to restrain the operation of political participation and undermine the public's activeness in political participation.

9.3.4 Chinese Citizens' Political Cultural Quality Is Still not High,[20] and Their Maturity of Political Participation Is Relatively not Sufficient

This is highly correlated to the development of Chinese economic level and degree of education of the public. The remaining of traditional and backward feudal political culture still exists in objective sense, which still has certain negative influence on the operation of political participation. The traditional political culture enabled traditional political ideas to pass on generations, such as feudal patriarchal ideas, imperial power ideas, hierarchy ideas, attachment ideas, grass root ideas, rule of man ideas, etc., while the public's independence awareness, equality awareness, participation awareness, responsibility awareness, etc. are always difficult to form into shape in the long evolving process of Chinese traditional politics.

---

[19] For example, the existence of the rubber stamp phenomenon in the election of NPC representatives is to great extent because the nomination system of the NPC and the candidate determining system are not complete enough.

[20] A survey that analyzed Chinese citizens' political quality from the three aspects of political participation attitude, political knowledge and technology, and political participation indicated that the average score of Chinese citizens' political quality is only 3.3 points, which still has very big gap with the full score of 10 points (the standard of full score includes the political awareness to actively participate in socialist democratic politics, sufficient political knowledge, skilled participation technique, and abundant participating experiences). Refer to Zhang Mingpeng, *China's "Political Person"*, China's Social Science Press, 1994 Edition, p. 190.

9.3 Forming Causes of Chinese Political System Construction Cost

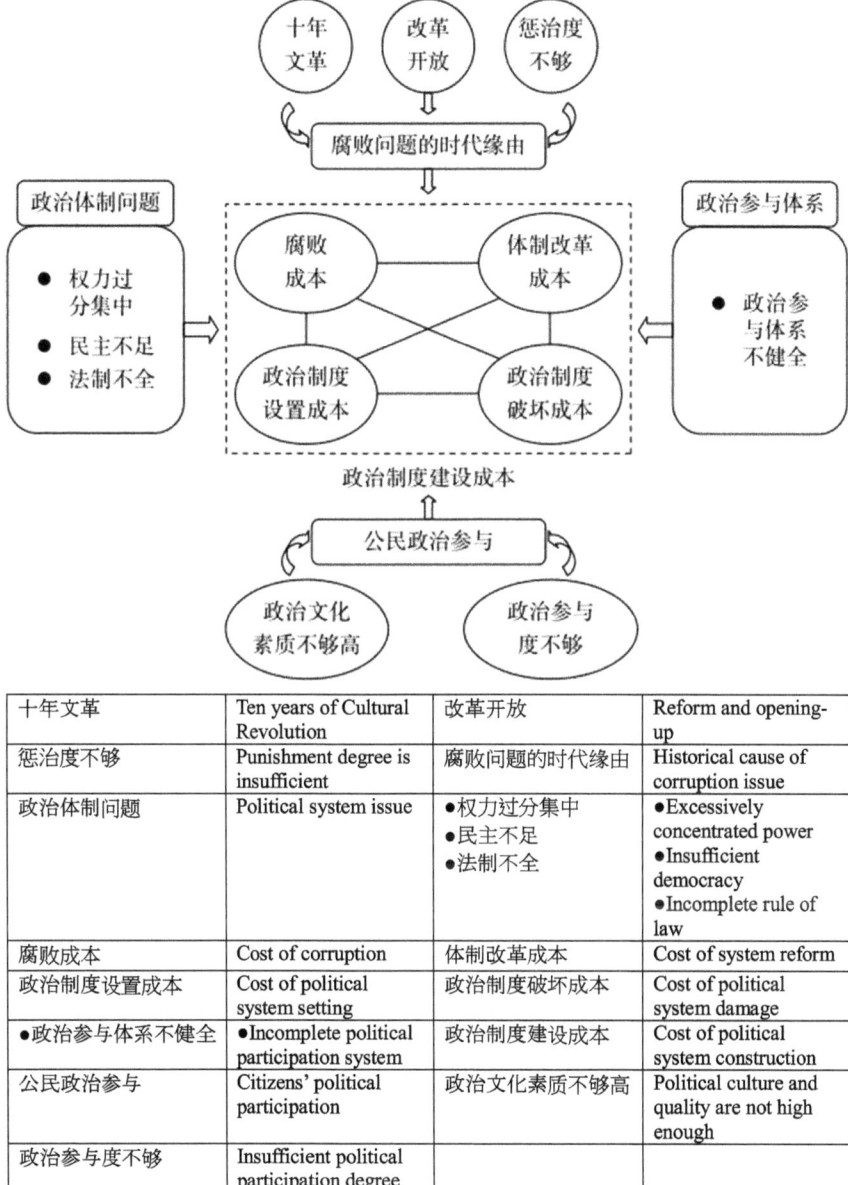

十年文革	Ten years of Cultural Revolution	改革开放	Reform and opening-up
惩治度不够	Punishment degree is insufficient	腐败问题的时代缘由	Historical cause of corruption issue
政治体制问题	Political system issue	●权力过分集中 ●民主不足 ●法制不全	●Excessively concentrated power ●Insufficient democracy ●Incomplete rule of law
腐败成本	Cost of corruption	体制改革成本	Cost of system reform
政治制度设置成本	Cost of political system setting	政治制度破坏成本	Cost of political system damage
●政治参与体系不健全	●Incomplete political participation system	政治制度建设成本	Cost of political system construction
公民政治参与	Citizens' political participation	政治文化素质不够高	Political culture and quality are not high enough
政治参与度不够	Insufficient political participation degree		

Chart 9.1 Forming causes of Chinese political system construction cost

For the conclusion of the forming causes of Chinese political system construction cost, please refer to Chart 9.1.

9.4 Countermeasures for Chinese Political System Construction Cost

9.4.1 Establishment and Improvement of System to Punish and Prevent Corruption

Resolutely oppose and prevent corruption is a major political task of the party and nation. In order to improve the system to punish and prevent corruption, we need to strengthen the implementation of combating corruption and upholding integrity, adhere to in-depth reform, innovate the system and mechanism, prevent and punish corruption from the origin. In order to punish and prevent corruption, we shall not only make symptomatic relieves, but also make radical treatment; we need both punishment and prevention. We shall implant anticorruption in various important policies and measures by deepening reform, innovating system, strengthening education, developing democracy, improving rule of law, and strengthening supervision. We shall prevent and treat corruption in developing thinking and reform, meanwhile, we shall use in-depth reform and system innovation to further expand the degree to prevent and treat corruption from the origin, gradually form a set of evolving corruption prevention and control mechanism, gradually widen the field of using reform system and mechanism to restrain corruption. For the key links and fields that generate corruption, we shall use the reform and innovation of system and mechanism to change the restriction of corruption to prevention of corruption, and focus on the breeding of potential corruption cost, so as to reach the purpose and effect of preventing corruption from the origin.

9.4.2 Gradually and Stably Promote Political System Reform

China's political system reform must have a stable political environment.[21] System reform can't be realized at once, this requires that we have to be brave to explore in the process of political system reform, meanwhile, we also need to be careful and prudent, we must focus on maintaining certain balance, without such balance, it is

---

[21]Deng Xiaoping pointed out that "For China's issues, the most important of all is stability. Without a stable environment, nothing can be done, and the achieved results would be lost". *Selected Works of Deng Xiaoping,* Volume 3, People's Publishing House, 1993 Edition, p. 284.

easy to cause political instability,[22] so as to cause high reform cost and waste. Therefore, in the process of Chinese political system reform, we shall eliminate possible deviation and tendency of instability, it must be promoted with organization, steps and order under the leadership of the party.

9.4.3 Make Efforts to Enhance the Public's Awareness and Ability of Political Participation

The public of society is the principal of political participation, the public's awareness of political participation and ability of political participation play critical role in political participation, strengthening the public's participation awareness and enhancing the public's ability of political participation are the main tasks faced by China. In the aspect of political culture, we need to prevent the feudal political mentality with the characteristics of respecting the superior and bullying the subordinate, inactiveness, personal attachment and the fanatical and extreme capitalistic political awareness, determine the "Participation-type" political culture, establish the common equality awareness, broad independence awareness, active participation awareness, strong sense of social responsibility, as well as establish the rule of law awareness, emphasize the role of education, enable the people to learn the operation method of modern democracy in practice, so as to actively and correctly participate in various political democratic activities.

For the countermeasures for Chinese political system construction cost, please refer to Chart 9.2.

---

[22]"This may have two circumstances: one is the political instability caused from completely defying the country's basic political system; the other one is the political instability caused from 'Explosion' of people's political participation due to negligence of priority and strategy during the process of promoting reforms". Zhou Tianyong, Wang Changjiang and Wang Anling as Chief Editors: *Difficulty Solving: Research Report of China's Political System Reform*, Xinjiang Production and Construction Corps Press, 2008 Edition, p. 9.

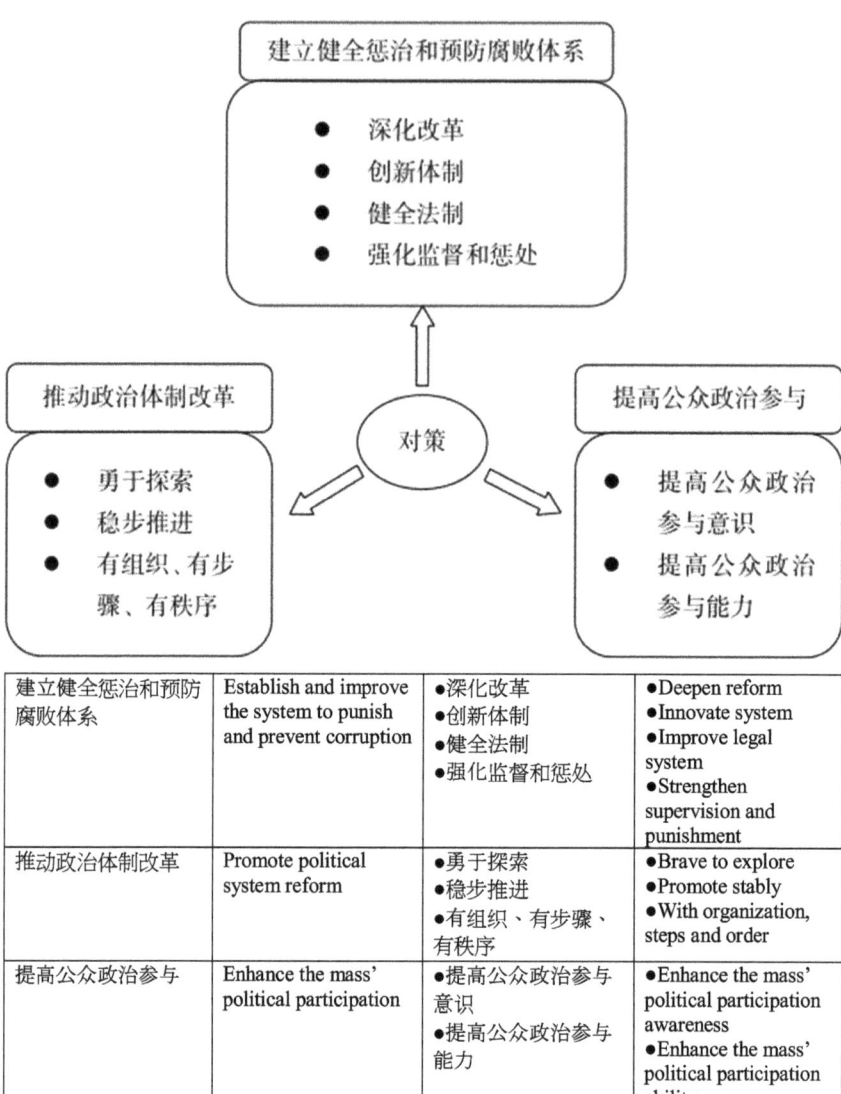

建立健全惩治和预防腐败体系	Establish and improve the system to punish and prevent corruption	●深化改革 ●创新体制 ●健全法制 ●强化监督和惩处	●Deepen reform ●Innovate system ●Improve legal system ●Strengthen supervision and punishment
推动政治体制改革	Promote political system reform	●勇于探索 ●稳步推进 ●有组织、有步骤、有秩序	●Brave to explore ●Promote stably ●With organization, steps and order
提高公众政治参与	Enhance the mass' political participation	●提高公众政治参与意识 ●提高公众政治参与能力	●Enhance the mass' political participation awareness ●Enhance the mass' political participation ability
对策	Counter measures		

Chart 9.2 Countermeasures for the cost of China's political system construction

Chapter 10
Cost of Chinese Political Decision-Making System

10.1 Connotation of Chinese Political Decision-Making System

Simon, a well-known scholar in the US proposed that "Management is decision-making",[1] referring to this view point, as for a government that fulfills public management functions, political decision making is the core and foundation of government's administrative function.

Political decision making is the process that a government decision-making subject uses its decision-making power to solve public issues of society and handle social and public affairs. This is a open political process that includes the participation of government and various social political forces. This process indicates that subject of government decision making is not sole but multiple. To understand from the level of internal structure of government decision-making system, political decision-making system specifically involves decision-making subject and its allocation of power, and involves which subjects to participate in decision making, in the process of decision making, different subjects have what kind of different decision-making power, and play which kind of different role; to understand from the internal operation of government decision-making system, political decision-making system involves the occurring and developing situation and trend of the mutual effect of different subjects in the process of decision making and implementation, as well as the process and method of the various supervision and control bodies in order to guarantee the correctness of decision making and guarantee the successful realization of decision-making goals in its implementation process. Political decision-making system have different structures in different social, political, economic and cultural conditions, and it will develop and change along with the development and change of social, political, economic and cultural conditions.

---

[1] Herbert Simon; *New Science of Management Decision Making,* Translated by Li Zhuliu and others, China Social Science Press, 1982 Edition, Page 168.

Traditional political decision-making system concept sees political decision-making more of decision-making activity of government body itself, it uses system to fixedly assume the duties, structures and mutual relationships of different government institutions and personnel in government decision-making tasks, it is mainly to solve the problem of who owns government decision-making power. Individual dictatorship decision-making or collective democratic decision-making involves the issue of decision-making system. The concept of modern political decision-making system[2] considers political decision-making as the decision-making process that government serves as main role, and different subjects in or out of government effect on each other.

10.2 Forms of Chinese Political Decision-Making System Cost

The waste caused by decision-making mistake is the biggest mistake.[3] Political decision-making, as the core process of the entire national management work, no doubt needs certain labor, material and fund, which are referred to as "Political Decision-making Cost". Political decision-making cost can be interpreted in broad and narrow sense. Political decision-making cost in broad sense means the sum of all labor, material, fund and other resources to be input in order to realize its certain political goal by political decision-making subject. Political decision-making cost in narrow sense means the input of political decision-making activity itself.[4]

---

[2] Shen Rong from China Society of Administrative Management believed that "Traditional studies of government decision making focused on decision-making activities of government agencies themselves, and ignored the participation and influence of the political parties, non-government organizations and mass outside the government on government decision making, which can be referred to as Administrative Decision Making; while the concept of modern government decision making extends the vision of government decision-making activities beyond government, and consider government decision making a decision-making process with government as the leading role and different entities inside and outside government interacting with each other..." Based on Shen Ronghua's explanation, government decision making can also be referred to as public decision making, which generates the necessity of analyzing government decision-making mechanism. Refer to Shen Ronghua: *Government Mechanism*, State Administration College Press, 2003 Edition, Page 210.

[3] When many local governments implement production projects, due to rash decision making, after completion of factory and production, products couldn't be sold, having severe loss or even bankruptcy. There are usually RMB millions or even hundreds of millions of investment costs that couldn't be recovered and caused huge wastes.

[4] Narrow sense of political decision-making cost include the sum of expenses, including the labors mobilized to development investigation and research, the collection of information and materials, the meetings opened, the number of experts consulted, related material expense, consultation expense, investigation expense, and meeting office expense.

10.2 Forms of Chinese Political Decision-Making System Cost

Table 10.1 Forms of Chinese political decision-making system cost

Classification of cost	Forms	Specific contents
Cost of political decision-making system	Cost of the change of political decision-making system	Price paid in the changing process of decision-making system, "personal decision making—collective decision making—common decision making"
	Cost of mistake in political decision making	Loss caused from events of major political decision-making mistakes
	Cost of the improvement of political decision-making system	Input and price paid to promote the establishment of a scientific, democratic and legal decision-making system

Compared with political decision-making cost in broad sense, political decision-making cost in narrow sense is naturally low, but it should also be controlled, or else it would cause waste.

This book divides political decision-making cost into 3 forms, please refer to Table 10.1.

10.2.1 Changing Cost of Political Decision-Making System

Based on different decision-making subjects, political decision-making process can be classified into individual decision making,[5] collective decision making[6] and common decision-making.[7] As for public policy, introduction of public participation is the inherent requirement of decision-making democratization and necessary requirement for economic and social development.

Since the establishment of New China, China has experienced the changing process of "individual decision making—collective decision making—common decision making". In the era of Mao Zedong, the individual decision making reached its peak, after that, the second generation of central leadership group with Deng Xiaoping as the core broke the model of individual decision making, and

---

[5]Individual decision making means the decision made by main leader of decision-making institution through the method of personal deciding, and based on his/her own judgment, knowledge, experience and will.

[6]The promotion of collective decision making is sourced from such a recognition that a person's ability, knowledge and experience are limited, if concentrating different people's power and wisdom, it could make the best of everyone, draw the wisdom of the masses, and thus make correct decisions.

[7]Common decision making has the effect of deeply promoting mass participation. Mass participation is a process that under certain social environment, the different parties subject to public decision making participate in the related decision-making process, imposes influence on decision making or even change the direction of decision making.

changed China's political decision-making system to the path of collective decision making again. With the advancement of society and enhancement of democratic degree, the public participation advocated in modern society enabled common decision making to become the goal and method for governments at different levels to make political decision making. Each change of decision-making system would pay certain cost and price.

10.2.2 Cost of Political Decision-Making Mistake

Since the establishment of New China, political decision-making mistakes caused huge loss to national economic construction, which is cost of political decision-making mistake. The "Great Leap Forward"[8] during 1958–1961 and the "Cultural Revolution" during 1966–1967 were two most typical examples. The fact of "Great Leap Forward" and the "Cultural Revolution" told us that the biggest mistake is strategic decision-making mistake. Final decision making decided by supreme leader is the root of strategic decision-making mistake.[9]

This book made quantitative estimation on the economic loss caused by the two most important political decision-making mistakes.[10] First of all, this book estimated that the long-term potential growth rate of China is about 9%. Since the establishment of New China, China's economic growth record is as follows: it was high during the first and last periods, and relatively low in the middle period, during the period of 1952–1957, the actual GDP growth rate was 9.2%, the actual GDP growth rate during 1978–2003 was 9.3%, and was 5.4% during the period of 1957–1978, which was about 4% lower than that of the first and last period.

Based on the economic growth rates in the first and last two periods, assuming the economic growth trend during the period of 1957–1978 to be 7.5–9.0% (refer to Table 10.2), in which 7.5% was the lower limit and 9.0% was the higher limit.

---

[8]The "Great Leap-forward" movement means the extreme Leftist path developed by Chinese Communist Party across China during 1958–1961, which determined the general guideline of "Be fully brave, make the best, build the socialism society in fast, quick, good and saving manner". It pursued high speed in production development, and considered the realization of high industrial and agricultural production as the target, and required doubled, several times and even dozens of times of growth of main industrial and agricultural products. The three years of Leftist" tendency and rash advancement during the three years since the "Great Leap Forward" caused major unbalance of the national economy and resulted in severe economic difficulty.

[9]Hu Angang, *History of China's Politics and Economy (1949–1976)*, Tsinghua University Press, 2008 Edition, Page 259.

[10]The main method adopted is to compare the long-term potential output growth rate and actual growth rate of Chinese economy, the former reflected the combined effect of the various factors that affect and determine China's economic growth, while the latter reflects the record of actual growth rate, the difference between the two reflects the degree of trend of actual growth rate deviated from long-term growth rate, and could be deemed as the economic loss caused from mistakes in political decision making.

During the impact and influence of the "Great Leap Forward" during 1958–1961 and the "Cultural Revolution", the actual GDP during 1957–1978 was 5.4%, which is lower than that simulated result of 7.5–9.0%, during the period of 1952–1978, the actual GDP growth rate was 6.1%,[11] which is 2–3% lower than the long-term growth trend (8.9%), therefore, it can be estimated that the economic loss of decision-making mistake accounts for about 1/4–1/3 of the economic growth rate; since 1978, during to the decrease of economic decision-making mistakes, China's actual economic growth rate approached or reached or even surpassed the long-term growth trend or potential.

It can be seen from the accumulated effect of economic growth that the actual GDP in 1978 was about 4.7 times of that in 1952, while the GDP of simulated result was equivalent to 7.2–9.5 times of that in 1952, because the economic loss of decision-making mistake accounted for about 1/2–1/3 of the simulated gross GDP in 1978.[12] Based on the same method, with other conditions (such as total population growth rate, employment growth rate) remaining unchanged, the actual per capita GDP growth rate and labor productivity growth rate were both lower than the simulated value (refer to Table 10.2).

The simulated result above indicates that without the "Great Leap Forward" during 1958–1961 and the "Cultural Revolution", China's economic growth would be even higher, or at least reach the long-term growth potential; judging from the record of Chinese actual economic growth since 1978, China indeed has a very high growth potential, which reaches 9.3–9.5%.

Besides, the "Movement of People's Commune"[13] which occurred at the same time of the "Great Leap Forward" was also a major mistake. The major campaign of "Movement of People's Commune" and energetic endeavor of "Communism Movement" was the first attempt and also the biggest attempt made by Comrade Mao Zedong in order to accelerate the Communism Construction and realize

---

[11]Hu Angang, *History of China's Politics and Economy (1949–1976)*, Tsinghua University Press, 2008 Edition, Page 536.

[12]This has similarities with the research results of Y Kwan and C Chows. Their main conclusion was that if there was no major decision-making mistake, the "Big Leap Forward" and "Cultural Revolution" would have never happened, then the historic record of Chinese economic growth performance since 1957 would be much different, the labor productivity assumed in 1993 was 2.7 times of the actual labor productivity. Y Kwan and C Chow, Estimating Economic Effects of Political Movements in China, *Journal of Comparative Economics*, Vol. 23, 1996, pp. 192–208; Refer to Cai Fang and Lin Yifu: *Chinese Economy*, China Fiscal Economics Press, 2003 Edition, Page 9.

[13]This was another "Great Leap Forward" launched and led by Comrade Mao Zedong to reform the production relations. This campaign completed the transformation from advanced cooperative to people's commune within very short period. However, the "Supply System" with communism factor implemented in People's Commune only barely maintained for couple of months before quickly disappearing, the mandatory "Communist Campaign" greatly frustrated the incentive of the broad farmers. Refer to Hu Angang, *History of China's Politics and Economy (1949–1976)*, Tsinghua University Press, 2008 Edition, Page 306.

Table 10.2 Comparison of Chinese economic growth performance (unit: %)

	GDP 增长率	人口增长率	人均 GDP 增长率	就业增长率	劳动生产率 增长率
实际情况					
1952—1957 年	9.2	2.4	6.8	2.8	6.4
1957—1978 年	5.4	1.9	3.5	2.5	2.9
1952—1978 年	6.1	2.0	4.0	2.6	3.5
模拟情况					
1957—1978 年 A	9.0	1.9	7.1	2.5	6.5
B	7.5	1.9	5.6	2.5	5.0
1952—1978 年 A	9.0	2.0	7.0	2.6	6.4
B	7.9	2.0	5.9	2.6	5.3

Note: simulated circumstance means the economic growth performance without occurrence of the "Great Leap Forward" during 1958 -1961 and the "Cultural Revolution", Scenario A assumes the economic growth rate during 1957 -1978 to be 9.0%, which is high scenari o; Scenario B assumes that the economic growth rate during 1957 -1978 to be 7.5%. The simulated circumstances ranges between A and B.

Source: State Statistics Bureau: *Compilation of Statistic Materials of New China in Fifty Years*, China Statistics Press, 1999 Edition, Page 1-4.

实际情况	Actual circumstance	模拟情况	Simulated circumstance
GDP 增长率	GDP growth rate	人口增长率	Population growth rate
人均 GDP 增长率	Per capita GDP growth rate	就业增长率	Employment growth rate
劳动生产率增长率	Growth rate of labor productivity		

Communism as soon as possible.[14] But any system change is historical change, since there is living, there must be death, their life cycle depends on the suitability and adjustment of history. Only the system that encourages people to create fortune could survive, while the system encourages people to "take free ride" would be difficult to survive.

The decision making of "News Blackout" after the "Tangshan Earthquake" was also an important reflection of political decision-making mistake.[15] After the

---

[14] Hu Angang, *History of China's Politics and Economy (1949–1976)*, Tsinghua University Press, 2008 Edition, Page 314.

[15] On Jul 28, 1976, the major earthquake that shocked China and the world occurred in Tangshan City of Hebei Province, in accumulation, over 242 thousand people died and 164 thousand people

occurrence of Tangshan Earthquake, the Chinese Government back then not only refused to release actual information of the earthquake and disaster within the nation, but also denied any international humanitarian aid and economic aid, which caused great delay and loss to the earthquake relief and reconstruction work. This fully reflected the severe drawbacks of the system back then, including "holding back unpleasant information", "news blackout", lack of transparent information, non-public politics, "black box decision making" by few people, etc. which was very similar to the method adopted after the "big famine" after the failure of China's "Great Leap Forward" during 1959–1961, or was another repeating.[16]

After getting rid of individual decision making and the political decision-making model of "black box decision making" by few people, China currently still hasn't got rid of the huge loss of economy caused by political decision-making mistakes.[17] Comrade Li Jinhua, Chief Auditor of State Auditing Administration (2010) once expressed at national auditing work meeting that "some leaders got sentence for embezzling millions of Yuan, people cheered for that, but some leaders made major decision-making mistakes that could cause over a billion Yuan of loss, which is even worse than corruption",[18] therefore, it can be said without extravagance that decision-making mistake is the biggest mistake in current China, the consequences caused by political decision-making mistakes is very horrible, and we can't waste any time to overcome political decision-making mistakes.

(Footnote 15 continued)
severely injured. Until Aug 18th, the CCP Central Committee issued the *News Report Regarding the Earthquake Disaster Relief at Fengnan of Tangshan*, but the document was only reviewed by few senior cadres.

[16] Hu Angang, *History of China's Politics and Economy (1949–1976)*, Tsinghua University Press, 2008 Edition, Page 494.

[17] As estimated by World Bank: during the "Seventh Five Year Plan" period and "Ninth Five Year Plan" period, the mistake rate of Chinese Government's investment decision making was around 30%, and the waste of funds and economic losses were about RMB 400 billion–500 billion, if calculated by the 70% of success rate of the investment decision making of the entire society, the loss caused from decision-making mistakes in each year is about RMB 120 billion, over the 20 years, the loss should be around RMB 2.4 trillion. For oil and chemical industry alone, during the 20 years from 1979 to 1999, the loss caused from decision-making mistakes was no less than RMB 80 billion. Refer to Chen Lifang and Chen Liuyu: *Causes and Countermeasures of Government's Administrative Decision-making Mistakes,* Journal of Chongqing University of Industry and Commerce (Social Science Edition), 4th Issue in August, 2007.

[18] Chen Lifang and Chen Liuyu: *Causes and Countermeasures of Government's Administrative Decision-making Mistakes,* Journal of Chongqing University of Industry and Commerce (Social Science Edition), 4th Issue in August, 2007.

10.2.3 Improving Cost of Political Decision-Making System

Continuously promoting the improvement of China's political decision-making system is an issue closely attended by the party, the government and the common people. Decision making by officials relates to the public's own interests, in turn, the public's behavior and will also affect the decision making of the officials. Therefore, the broad public plays irreplaceable role in the process of China's improvement of political decision-making system. The improving process of political decision-making system is a long and winding process, many people paid price or even lives during this process. These times, money and even lives contributed for the improvement of a scientific, democratic and legal decision-making system are all the improving cost of political decision-making process.

A nation is a huge and complex system, improvement of China's political decision-making system requires the government to make efforts to realize the scientification, democratization and legalization of decision making, meaning that each decision making shall realize legal and standard procedure, democratic and open process as well as scientific and just result. First of all, a scientific decision-making procedure is the procedural guarantee to realize decision-making scientification and prevent arbitrary decision making. A scientific decision-making procedure shall at least include four processes, open decision making, survey and expert reasoning, collective discussion and decision making be administrative chief, and feedback (refer to Chart 10.1). Although China hasn't got an unified Administrative Procedure Law, but the *Administrative Litigation Act* and *Administrative Permission Law* made regulations on administrative decision-making procedure, the scientific decision-making procedure has preliminarily formed and has been adopted in actual decision making.[19]

Secondly, democratization of government decision making is the active result of political civilization. Government decision making, especially the decision making that involves people's interest, if not fully inquiring the opinions of the people, the result of decision making must be lack of scientific nature and reasonability, and it would be difficult to generate expected effect in practice, not to mention efficiency. Therefore, scientific and reasonable government decision making depends on the government's insistence on the democratization of decision making.[20] However, the democratic degree of government decision making in reality is not high, the "Chief

---

[19]The first national-level price hearing—the railway ticket price hearing of the National Development and Planning Commission was held in Beijing on Jan 12, 2002, which opened the gate of government price decision making to the society, enabled common people to know intelligence and participate in politics, and created the first case of hearing in Chinese Government's decision making.

[20]Currently, China's Constitution and Laws have established a set of democratic system, which created a basic system guarantee for democratic decision making.

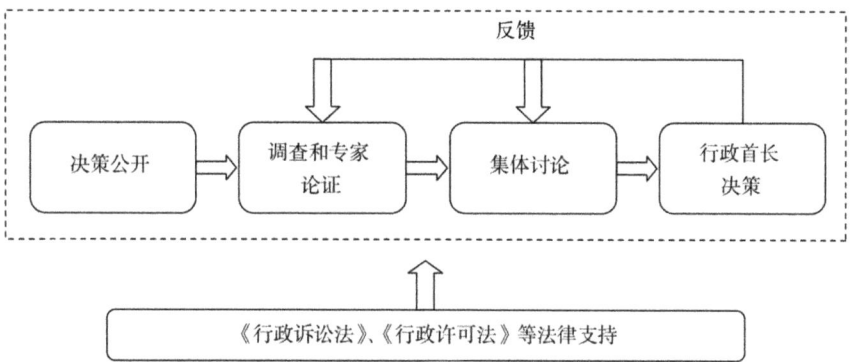

反馈	Feedback	决策公开	Open decision making
调查和专家论证	Investigation and expert reasoning	集体讨论	Collective discussion
行政首长决策	Decision making by administrative leader	《行政诉讼法》、《行政许可法》等法律支持	Support by laws such as *Administrative Procedure Law* and *Administrative Permission Law*

Chart 10.1 Procedure of government decision-making sicentification

Syndrome"[21] is a typical reflection. Therefore, the actual situation urgently requires the government to improve the democratic decision-making system, so as to promote the enhancement of democratic level of decision making.

Thirdly, the realization of the legalization of government decision making mainly include two aspects, the first is that government decision making shall have law to abide by, the second is that the government shall make decisions strictly according to law. Compliance with law is the key and core of the legalization of government decision making.[22] We shall strengthen the legal education of government decision makers and enhance their awareness of making decisions according to law.

---

[21]"Chief Syndrome" means the comprehensive syndrome of officials after becoming "Chief" have authoritative behaviors due to lack of effective restriction and supervision on power, the most basic syndrome is usually authoritative behaviors, and generally could be summarized as "One man's talking during decision making, final say in personnel appointment and simple decision in spending". The "Chiefs" with such syndrome usually tend to personally make major decisions, so as to make those procedures that were designed to guarantee scientific and democratic decision making become in name only. Refer to Li Xuehui: *The Authoritative Behaviors of "Chiefs" must be Corrected*, Party Construction, 4th Issue in 2004.

[22]Jiang Zemin pointed out in the party's report at the 15th National Congress that the government must "maintain the dignity of the Constitution and laws, adhere to everyone being equal before law, no one and no organization shall have privilege of being above the Constitution and Law. All government agencies must govern according to law."

10.3 Forming Causes of Chinese Political Decision-Making System Cost

10.3.1 Immaturity of Political Decision-Making System

Since the establishment of New China, the division of power, duties and scope of party and political decision making hasn't been very clear, the relationship hasn't been completely smoothed yet. The structure of some government's decision-making system is unimproved, incomplete, and hasn't formed a complete decision-making system that is composed of subsidiary systems of information, consultation, decision making and supervision. The division of work within government departments are also not reasonable, the same civil servants group are not only the maker of decisions, but also the executor of decisions, the decision-making powers are unreasonably distributed among the members of decision-making group, and it is difficult to form complementation and balance between the members of decision-making group. Secondly, some decision makers don't make decisions strictly according to procedures[23] and are lack of supervision and restriction on the power of decision maker, the lack of decision-making responsibility system and weak accountability caused abuse of individual decision making.[24] The change and development of decision-making system caused from immature political decision-making system also caused different degrees of decision-making mistakes and made the nation to pay corresponding cost during development.

10.3.2 Lack of Decision-Making Knowledge, Experience, Ability and Quality of Decision Makers (Group)

Many people at leader positions in Chinese governmental departments don't have sufficient scientific and cultural level, their professional knowledge is lacking, and even less about decision-making knowledge, their knowledge structure, ability and quality are unable to meet the requirements of modern political decision making. The decision-making mistakes caused by individual quality would cause loss to the nation, which need high attention.

For the forming causes of Chinese political decision-making system cost, please refer to Chart 10.2.

---

[23] Such as implementation without making survey of people's opinion and social opinion, without holding hearing, without making scientific reasoning and without making pilot project.

[24] A vivid metaphor was that "Pad the head before decision making, pad the chest during decision making, and pad the butt and leave after decision making", these three "pads" vividly described the random decisions makings caused from lack of binding of responsibility system beforehand and lack of accountability of responsible person in decision making mistakes after decision making.

10.3 Forming Causes of Chinese Political Decision-Making System Cost

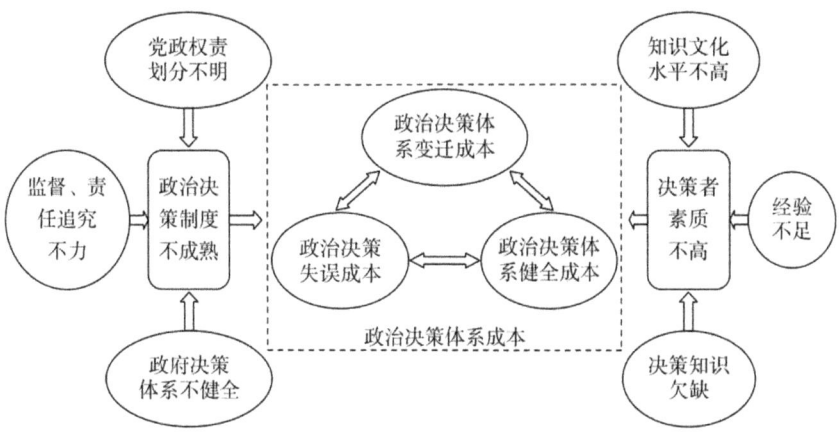

党政权责划分不清	Unclear classification of power and duties between party and politics	知识文化水平不高	Insufficient knowledge and cultural level
监督、责任追究不力	Insufficient accountability of supervision and responsibility	政治决策制度不成熟	Immature political decision-making system
政治决策体系变迁成本	Cost from change of political decision-making system	政治决策失误成本	Cost of political decision-making mistakes
政治决策体系健全成本	Cost from improvement of political decision-making system	决策者素质不高	Insufficient quality of decision maker
经验不足	Insufficient experiences	政府决策体系不健全	Incomplete government decision making system
政治决策体系成本	Cost of political decision-making system	决策知识欠缺	Lack of decision-making knowledge

Chart 10.2 Forming causes of Chinese political decision-making system cost

10.4 Countermeasures for the Cost of Chinese Political Decision-Making System

10.4.1 Establish and Improve Political Decision-Making Regulations and Systems, Promote the Theorization, Scientification, Democratization and Legalization of Decision Making According to Law

The establishment and improvement of political decision-making regulations and systems mainly include the improvement of government decision-making system, providing the scientification of government decision making with organizational guarantee; strengthening the construction of government decision-making system, providing systematic guarantee for the scientification of government decision making; improving democratic centralization system, establishing the standard government leadership collective discussion system; opening the process of government decision making, enabling government decision-making work to be effectively supervised by the broad public; establishing and improving the system to let citizens to widely participate in selection of decision makers, realizing the unification of upwards and downwards accountability; establishing the management system of government decision-making legalization, and enabling the government decision-making according to law to be truly implemented.[25]

10.4.2 Well Selection, Use and Education of People to Establish Highly Efficient Decision-Making Team

Use scientific, democratic and legal decision making to reduce decision-making mistake and waste. Meanwhile, we need to establish decision-making accountability system, which is to hold those people who make irrational decision, individual decision or cause major mistake or waste due to "Rent Seeking" behavior accountable for their legal liabilities and economic liabilities.

---

[25]Liao Xiongjun: *Discussion of Reform and Improvement of Government Decision-making Mechanism*, Journal of Chengdu Administrative College, 4th Issue in August, 2003.

10.4 Countermeasures for the Cost of Chinese Political …

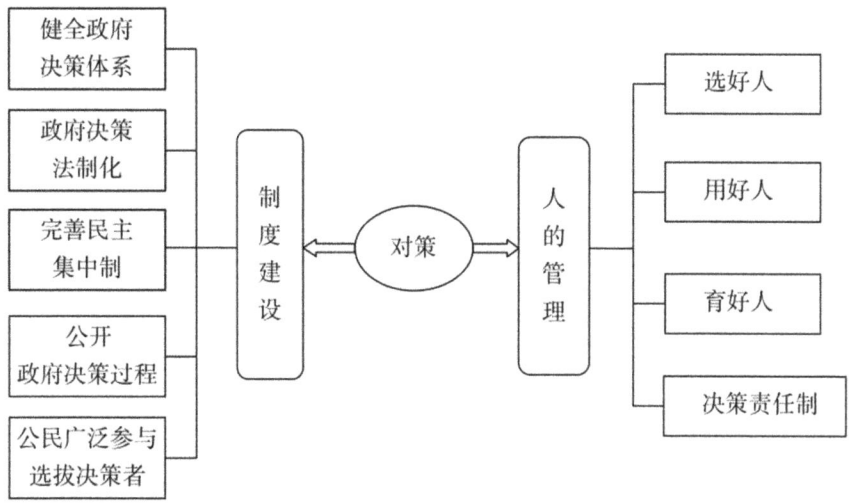

健全政府决策体系	Improve government decision-making system	政府决策法制化	Legalization of government decision making
完善民主集中制	Improve the democratic centralization system	公开政府决策过程	Open government decision making process
公民广泛参与选拔决策者	Citizens broadly participate in selection of decision makers	制度建设	System construction
对策	Countermeasures	人的管理	People's management
选好人	Properly select people	用好人	Properly use people
育好人	Properly nurture people	决策责任制	Decision-making responsibility system

Chart 10.3 Countermeasures for cost of Chinese political decision-making system

For the countermeasures for the cost of Chinese political decision-making system, please refer to Chart 10.3.

Chapter 11
Cost of Chinese Ruling Party Construction

11.1 Connotation of Chinese Ruling Party Construction

Ruling party means the party that controls the political power of a nation. Different with ruling parties in the west, ruling party of socialist country is the only legal ruling party, and the party in the "Leadership Core". In China, Chinese Communist Party is the only legal ruling party.

For any party, in order to consolidate its ruling status and practice its policy advocates, it must continuously strengthen party construction, promote itself to adapt to social advancement and development, so as to enable itself to be continuously consolidated and developed. For Chinese Communist Party that is at the status of only ruling party, the strengthening of party construction is particularly important. Only continuous strengthening of party construction could maintain the party's vitality, effectively lead the socialism construction and realize its own historical mission.

The construction of Chinese ruling party as studied in this book means the theoretical and practical activities of active and continuous self improvement and self development by the Chinese Communist Party in order to consolidate its ruling status and practice ruling strategy, so as to realize the strategic goal of promoting social development. For the Communist Party, as a proletarian party, after it obtains the ruling status, the party's construction should be developed around how to consolidate the socialism national political power, build a socialist society and finally realize communism. Based on the existing consensus of the academic circle, the contents of party construction mainly include ideological construction, organizational construction and working style construction.[1]

---

[1] Wang Jianguo and Wang Hongjiang: *History, Theories and Practices of the Ruling Parties in Socialist Countries*, China Social Science Press, 2008 Edition, Page 5.

Table 11.1 Forms of Chinese ruling party construction cost

Classification of cost	Forms	Specific contents
Cost of ruling party construction	Cost of ideological and cultural construction	Input of ideological and cultural construction (including the cost of the Chinese localization of the Marxism and Leninism)
	Cost of organization construction	Cost from the establishment of basic-level party organization and the development and training of party members
	Cost of working style construction	Input and price of rectification events within the party in history; necessary input of modern party nature and party working style education, and system construction
	Cost of democratic construction within the party	Input from strengthening of democratic construction within the party and price paid for promotion of the realization of democracy
	Forming cost of political civilization	Necessary input and price paid for the construction of the socialist political civilization that is suitable for China's national situation

11.1.1 Forms of Chinese Ruling Party Construction Cost

Chinese ruling party construction cost referred to in this book, as its name indicates, means the cost for the Communist Party to continuously improve in each aspect in order to seek permanent life during the ruling process, specifically it is classified into five forms, ideological and cultural construction cost, organizational construction cost, working style construction cost, construction cost of democracy within the party, and forming cost of political civilization, for specific contents, please refer to Table 11.1.

11.1.2 Ideological and Cultural Construction Cost

Our party has always been highly regarding ideological work, the work in this aspect will directly affect the success or failure of the socialism cause. Since the beginning of the party's first generation of leadership group, they focused on the development of ideological and cultural work. The Party Central Committee believed that "Besides mastering the guidelines, policies and deciding the use of important cadres, the party's leadership departments shall make time and efforts to do ideological and political work, do people work, do citizens work, well educate

11.1 Connotation of Chinese Ruling Party Construction

the communist party members, and well educate the people and youth".[2] To the later "Three Emphases",[3] "Three Represents Theory",[4] "Eight Honors and Eight Disgraces",[5] as well as the proposition and development of education of keeping the party members advanced nature and other slogans and activities, which enabled the ideological and educational work of party members to be constantly maintained at priority of ruling party construction, and continuously innovated. This book will consider all the cost paid by ruling party in ideological and cultural field as ideological and cultural construction cost.

In which the cost of localization of Marxism in China was all the practical cost paid by China in the process of walking the socialism path with Chinese characteristics, making efforts to enable Marxism theory to adapt to China's reality in society, economy, politics, culture and life, and seeking new methods and new measures, creating new experiences. The essence and substance of Marxism as well as its new development after integration with Chinese reality will promote the comprehensive development of the cause of socialism with Chinese characteristics.

11.1.3 Organizational Construction Cost

The party's organizational construction cost includes the establishing cost of basic-level party organization and the developing and cultivating cost of party members. The establishment and improvement of basic-level party organization is the key link of party organization construction, and also the foundation of the development and expansion of party organization. The Organization Department of the Communist Party of China Central Committee declared to Chinese and foreign reporters that "As of the end of 2010, there are 80.269 million members of Chinese

---

[2] Chief Editor: Xue Jianzhong: *Theoretical Guideline of Ruling Party Construction of the Three Generations of Central Leadership Group*, CCP Central Party School Press, 2008 Edition, Page 99.

[3] The "Three Emphases" education started since 1995 means the education developed throughout the party for emphasizing studying, emphasizing politics and emphasizing correct style.

[4] The contents of "Three Represents Theory" includes that Chinese Communist Party always represents the development requirements of the advanced productivity, represents the advance direction of advanced culture and represents the basic interest of the widest people.

[5] The "Eight Honors and Eight Disgraces" socialist view of honor and disgrace was proposed by the General Secretary Hu Jintao of the CCP Central Committee in the afternoon of Mar 4, 2006 at the joint meeting of China Democratic League and China Association for Promoting Democracy of the Fourth Session of the 10th Chinese People's Political consultation Conference. The specific contents include "Honor from loving the motherland, disgrace from harming the motherland, honor from serving the people, disgrace from betraying the people, honor form respecting science, disgrace from ignorance, honor from hard work, disgrace from laziness, honor from unity and mutual assistance, disgrace from being selfish, honor from honesty and creditability, disgrace from forgetting friendship for profit, honor from observing the law, disgrace from breaking the law, honor from hard struggle, and disgrace from extravagance".

Communist Party; 3.892 million basic-level party organizations, including 187 thousand basic-level party committees, 242 thousand general branches, 3.463 million branches; 3222 local party committees across China, including 31 party committees on provincial (district, municipality) level, 396 party committees on city (prefecture or district) level, and 2795 party committees on county (city or district) level.[6] The Party Central Committee always stressed the strengthening of the auxiliary construction of village-level organization with village party organization as the core, take care of party organizations in state-owned enterprises and collective enterprises, strengthen the construction of party organizations in non-public enterprises, highly regard the construction of party organizations in urban communities, comprehensively take care of party construction work in public institutions and increase the working efforts of establishing party organizations in schools, scientific research institutions, social organizations and social intermediary organizations.[7]

Since the establishment of New China, the team of Chinese Communist Party members has been continuously developing and expanding. Chinese Communist Party had 4.488 million party members when New China was established in 1949, before the reform and opening up, there were 36.981 million communist party members, as of the end of 2010, there were 80.269 million communist party members across China, which was nearly 18 times of that at the establishment of New China. During the process of developing party members, party organization shall firstly strengthen the promotion and education of party organization, fully mobilize the activeness of the broad Youth League members and common people, find the excellent objects who are able to become active members for joining the party, these early-stage work of party member development will need the party organization to input certain cost. Later in the cultivation and investigation stage before the active members to join the party, and during the long-term cultivation and education process after becoming party members, party organization needs to continuously input cost to ensure the qualification and advancement of party members.

11.1.4 Working Style Construction Cost

The working style of ruling party relates to the life or death of the party. As for the communist party's strengthening and improvement of working style construction, the core issue is to maintain the flesh and blood bond between the party and the

---

[6]Chief Editor of the Editing Department of the *Party Construction Research* Magazine: China has 80.269 million of CCP members and 3.892 million of basic-level party organizations, 7th Issue of *Party Construction Research* Magazine in 2011.

[7]Chief Editor Xue Jianzhong: *Theoretical Guidelines of Ruling Party Construction of the Three Generation of Central Leadership Groups,* CPC Central Party School Press, 2008 Edition, Page 227.

11.1 Connotation of Chinese Ruling Party Construction

Table 11.2 Main measures of the three generations of party leadership group in working style construction

	Important measures of party's working style construction
First-generation leadership group	Rectification movement, anti-rightist struggle
Second-generation leadership group	Renovation of special treatment to senior cadres, "three emphases" education
Third-generation leadership group	First dependence on education, second dependence on system, and implement to solving of actual problems

people. Targeting the negative tendency after the Chinese Communist Party becoming the ruling party, including deprivation from the reality, deprivation from the people, subjectivism, bureaucracy, dictatorship, the special privilege's mentality, the three generations of leadership group adopted different measures based on actual situation to rectify and strengthen the party's working style, please refer to Table 11.2.

The first-generation leadership group with Comrade Mao Zedong as the core launched two times of large-scale rectification movement within the party during 1950–1953 and 1957–1960, which later evolved into the "Struggle against Rightist". During the "Cultural Revolution", there were continuously criticizing and rectification movements in large and small scales. The purpose of rectification movement was to help party members and cadres to overcome the mistakes made in work, overcome the pride and complacent sentiment, overcome bureaucracy and authoritarianism, and improve the relationship between the party and the people. However, the expansion of the rectification movement and the "Struggle against Rightist" made a large batch of intellectuals, patriots and cadres within the party wrongly classified as "Rightists", which caused unfortunate consequences, and brought immeasurable loss to the nation. The unnecessary cost and loss caused by the rectification movement and the "Struggle against Rightist" was the painful price paid by the Chinese Communist Party in the history of working style construction.

The second-generation leadership group decided, targeting the worsening problem of special treatment of some cadres after recovery at the beginning of the reform and opening-up, to start from senior cadres, rectify the party's working style and take care of the public working style. The "Three Emphases" education since 1995 was the education of party spirit and party working style in the new era, the emphasizing moral rectitude was to inherit and develop the good tradition, good working style formed by our party during the long-term revolution and construction causes, adhere to truth, adhere to principles, and insist on fighting all evil spirits and various corruption phenomenon. The third-generation party leadership group emphasized that in order to develop working style construction, we shall depend firstly on education and secondly on system, and implement to the solving of actual problems. Firstly, we need to strengthen the ideological and political construction, enhance the party members and cadres' activeness and resolve to develop the party's excellent tradition and working style, secondly, we need to establish a set of complete system and mechanism, so as to promote the systemization and standardization of working style construction. The fore-mentioned strengthening of the party's working style construction by the party's central committee through rectification and education is the necessary cost and input for the construction of party working style.

11.1.5 Cost of Democracy Construction Within the Party

The cost of democracy construction within the party is the necessary input to strengthen the democracy construction within the party and the price paid to promote the realization of democracy within the party.

Democracy within the party is the life of party. The essence of democracy within the party is to develop party members' activeness and initiative, enhance party members' sense of responsibility to the party's cause, mobilize party members or the representatives of party members to express opinions within the scope regulated by party constitution, so as to actively participate in the leadership work of the party for the people's cause.[8] The construction of democracy within the party needs guarantee of system, the destruction of democracy system also brought certain loss to democracy construction. The democratic centralism within the party that the party group rather than individuals shall decide major issues[9]" implemented since 1948 was the basic system and core system for the Chinese Communist Party to be able to correctly rule and obtain the support by the people and obtain the political legitimacy. Any practice that can't listen to different opinions, strengthen individual power and deciding everything by one man's say is destruction of democratic decision making and would bring damage to the party and the nation.

11.1.6 Forming Cost of Political Civilization

The forming cost of political civilization is the necessary input for China to construct the socialism political civilization that is suitable to Chinese reality and the price paid for the purpose.

---

[8] Chief Editor Xue Jianzhong: *Theoretical Guidelines of Ruling Party Construction of the Three Generation of Central Leadership Groups,* CPC Central Party School Press, 2008 Edition, Page 163.

[9] Deng Xiaoping: *Report Regarding Modification of Party Constitution, Selected Works of Deng Xiaoping,* Volume 1, People's Publishing House, 1994 Edition, Page 229.

In the "5.31 Talk" (2002), Jiang Zemin made the proposal of "Building the socialism political civilization".[10] With China's reform and opening-up and the continuous advancement and development of the socialism modernization construction cause, people have increasingly urgent need for the construction of socialism democratic politics. Since the Fifteenth National Congress, the party had continuously deepening recognition of this issue, and formed a series of guidelines, policies and rules, adopted many feasible steps and methods, so as to ensure the gradual realization of socialism democratic construction under the precondition of stability. The construction of socialism political civilization includes several rich contents, its core and essence is to build high-level socialism democracy, and guarantee the realization of the ideal and goal of the mass people as the masters of their own country. The proposition of this goal was a leap forward of the construction of Chinese ruling party. For China to develop socialism democratic politics, we must root in Chinese reality, insist on the leadership of Chinese Communist Party, insist on the people as the masters of their own country, insist on the characteristics and advantages of the socialism system. For all the benefiting results of political civilization created by human society, we shall actively refer to them but can't simply copy them, we shall make innovations based on Chinese reality. This requires generations of people to continuously make efforts or even price, so as to promote the true realization of Chinese socialism political civilization.

11.2 Forming Causes of the Cost of Chinese Ruling Party Construction

11.2.1 The Need for Chinese Communist Party to Continuously Develop and Strengthen, and to Consolidate the Ruling Foundation

The establishment and improvement of the basic-level party organization system and the cultivation of party members and cadres require input of certain cost. For Chinese Communist Party as the ruling party of China, its primary task is to consolidate the ruling foundation of the party, the large scale of organization and personnel is the necessary requirement for the ruling party to consolidate its foundation. In one aspect, the organization construction shall enable party organization to realize wide coverage of network system, for example, continuously improving the construction of basic-level party organization in rural areas, private enterprises and urban communities; in the other aspect, continuously developing the

---

[10]Jiang Zemin pointed out in his "5.31" (2002) Speech that "development of socialist democratic politics and construction of socialist political civilization are important goals of socialist modernization construction".

team of party members, focus on development of high-quality party members, especially cultivating and developing the excellent students with all-around development of morality, intellectuality and physical quality in colleges and universities to join the party, so as to inject strong vitality to party organization.

11.2.2 Input of Party Style Rectification and Loss of Expansion

In the process of Chinese Communist Party transforming from a revolutionary party to ruling party, there were some unhealthy practices and evil phenomena within party organization, and they tended to spread. Rectification of party style, punishment of offensive party cadres and establishment of good party style within the party and the solving of a series of working style problems require input of labors, materials and funds. Besides, the intensification of struggle within the party caused expansion of rectification within the party, which caused unnecessary loss.

11.2.3 Necessary Input to Prevent Occurrence of Crisis Within the Party

Ideological education and system construction are effective means to prevent occurrence of crisis within the party. Developing continuous and regular party spirit and honest government education within the party, continuously enhancing the political ideological level of party members, strengthening the crisis awareness of party members, and use system construction as guarantee, only these measures could well prevent occurrence of crisis within the party, so as to reduce the damage to the construction of ruling party. Therefore, necessary input is the coverage of even bigger potential loss.

For the conclusion of the forming causes of the cost of Chinese ruling party construction, please refer to Chart 11.1.

11.3 Countermeasures for the Cost of Chinese Ruling Party Construction

11.3.1 Strengthening of the Ideological and Moral Construction

In order to seize the highlight of construction of ruling party from the origin, we need to make necessary input in ideological education, strengthen the ideological

11.3 Countermeasures for the Cost of Chinese Ruling Party Construction

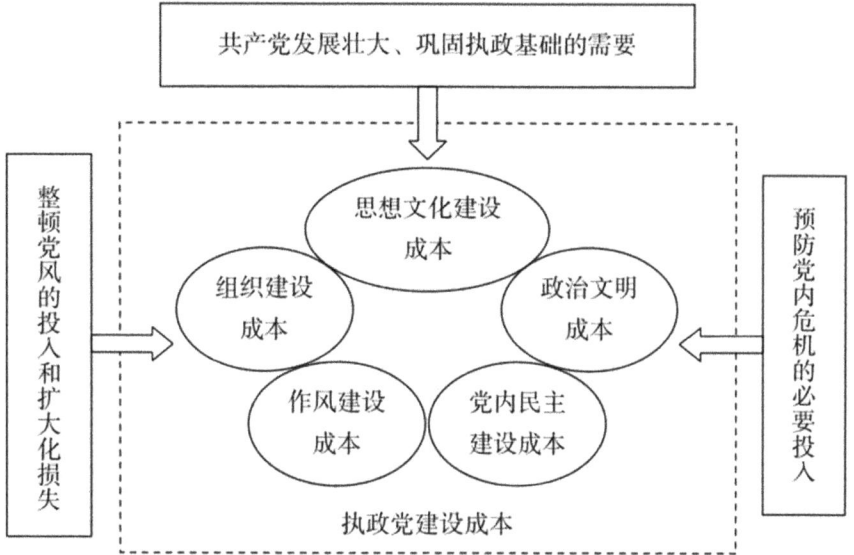

共产党发展壮大、巩固执政基础的需要	The need for the basis of CCP's development, expansion and consolidation of ruling
整顿党风的投入和扩大化损失	Input of party working style renovation and loss of expansion
思想文化建设成本	Cost of ideological and cultural construction
组织建设成本	Cost of organization construction
政治文明成本	Cost of political civilization
作风建设成本	Cost of working style construction
党内民主建设成本	Cost of democratic construction within party
执政党建设成本	Cost of ruling party construction
预防党内危机的必要投入	Necessary inp9ut to prevent crisis within the party

Chart 11.1 Forming causes of the cost of Chinese ruling party construction

and political education, which could not only enhance the idea of wholeheartedly serving the people, but also strengthen the awareness of honesty and self-discipline of party members, as well as good for the ruling party's construction of working style. Therefore, the necessary input of ideological and moral construction is the effective measure to reduce the cost of ruling party construction.

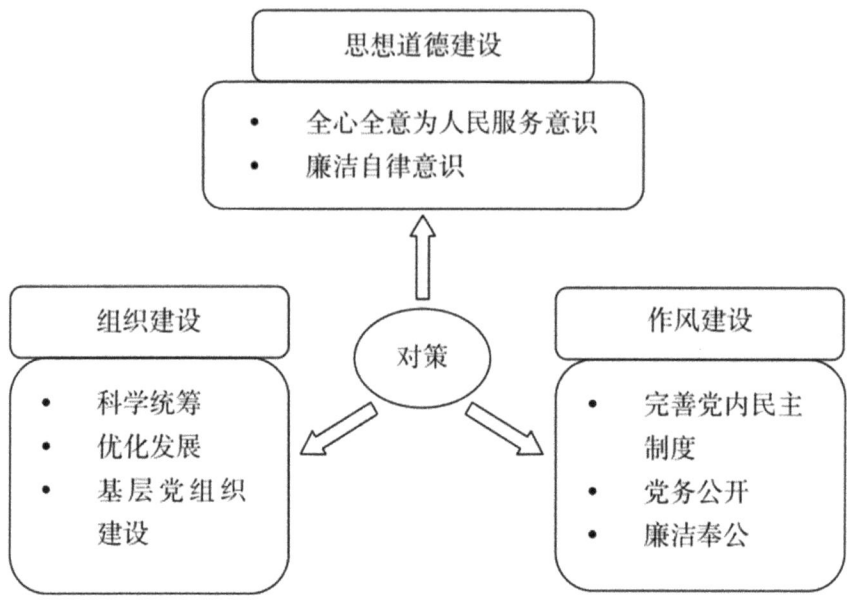

思想道德建设	Ideological and moral construction	●全心全意为人民服务意识 ●廉洁自律意识	●Awareness of all-hearted service to people ●Honest and self-disciplined awareness
组织建设	Organization construction	●科学统筹 ●优化发展 ●基层党组织建设	●Scientific coordination ●Optimized development ●Basic-level party organization construction
作风建设	Working style construction	●完善党内民主制度 ●党务公开 ●廉洁奉公	●Improve democracy system within the party ●Opening of party affairs ●Have integrity and be public-spirited
对策	Countermeasures		

Chart 11.2 Countermeasures for the cost of Chinese ruling party construction

11.3.2 Improve the Democracy Within the Party System and Reduce the Cost of Decision-Making Mistakes

In order to develop democracy within the party, we need to guarantee the democratic rights of party members, promote the opening of party affairs, create a democratic discussion environment within the party; strengthen the system construction, enable party members' opinions, suggestions and criticisms to be able to timely reflected; based on the principle of group leadership, democratic centralization, individual consideration and decision making through meeting, improve the discussion and decision-making mechanism within party committee; reform and improve the party's election system, establish and improve the circulation system within party, the circumstance reflection and major decision-making opinion requisition system, carefully punish the people and affairs that infringe party members' democratic rights, oppress party members' criticisms, retaliation or making false accusation against others.

11.3.3 Realize Scientific Planning and Optimized Development for the Establishment of Basic-Level Party Organization and Development of Party Member Team

Scientific and optimized party organization construction is an important aspect to effectively control the cost of ruling party construction cost. Although large scale of the quantity of basic-level party organization and party members is good for consolidating the ruling foundation of the ruling party and establishing good people foundation, the establishment of basic-level party organization and cultivation and development of party members shall meet the principle of "Scientific Planning and Optimized Development", quality outweighs quantity, we shall focus on quality but quantity, only this could meet the requirement for the ruling party to maintain advancement.

For the conclusion of the countermeasures for the cost of Chinese ruling party construction, please refer to Chart 11.2.

Chapter 12
Cost of Chinese Political Consultation and Crossing of Information Gap

12.1 Connotation of Chinese Political Consultation

Political consultation is an important part of the system of multi-party cooperation and political consultation led by Chinese Communist Party.[1] It contains very abundant democratic resources, and has great significance for the path of socialism political development with Chinese characteristics.[2]

The political consultation system, at the beginning of establishment in 1949, used the specific methods of absorbing democratic parties and patriots without party affiliation to participate in political power, and strengthening the democratic parties' supervision on the ruling party and joint discussion of national affairs, formed the democratic political situation that under the leadership of Chinese Communist Party, different political parties, different ethnic groups and different classes make joint efforts to make contribution to national construction, it developed its system

---

[1] The political negotiation, as an important part of the multi-party cooperation and political consultation system led by Chinese Communist Party and important form of socialist democratic politics, includes two different levels and two different properties of political negotiation: "political consultation between CCP and various democratic parties" and "the negotiation in CPPCC between CCP and various democratic parties, patriots without party affiliation and representatives of different circles". Song Jian: *Discussion on the Connotation and Basic Forms of Political Negotiation*, Journal of Shayang Normal Junior College, 1st Issue in 2008.

[2] Document of CCP Central Committee pointed out that "Under the new historical condition, in order to develop socialist democratic politics and build socialist political civilization, an important part of it is to adhere to and improve the multi-party cooperation and political consultation system led by CCP, expand the orderly political participation by people of various circles, explore the expression channel of social interests, promote the harmonious development of society, and realize the organic integration of CCP's leadership, the people are the masters of their own country and law-based governance". Refer to edited by CCP Central Committee Literature Research Office: *Compilation of the Important Literatures since the 16th National Congress of CCP*, Central Literature Press, 2006 Edition, Page 674.

advantage in the construction history of the Republic. In the historical process of the relationship with various democratic parties as political participation parties and the implementation of multi-party cooperation and political consultation system, Chinese Communist Party not only accumulated successful experiences, but also had lessons of failure, especially committed overall and directional mistakes such as the "Cultural Revolution". The expansion of the "Struggle against the Rightists" in 1957 and the Third Plenary Session of the 11th Central Committee of the Chinese Communist Party at the end of 1978 were two very different historical turning points of the political consultation system. The former symbolized the success-to-failure process of the multi-party cooperation and political consultation system, while the latter symbolized the failure-to-success process of the multi-party cooperation and political consultation system; the former symbolized the strong-to-weak process of the ruling party's ruling capability and the political participating parties' political participation ability, while the latter symbolized the weak-to-strong process of the ruling party's ruling capability and the political participating parties' political participation ability. In 1989 and 2005, the Central Committee of Chinese Communist Party formed the documents twice regarding strengthening the multi-party cooperation and political consultation system under the leadership of Chinese Communist Party, which indicated that this system has important position in the structure of political system in modern China.[3]

Multi-party cooperation, political negotiation, ruling by Chinese Communist Party, political participation by various democratic parties, these are not only the advocates of the Chinese Communist Party in long term, but also the political party system that is jointly recognized by different democratic parties and suitable for the special situation of China, as well as a basic political system that is recorded in China's Constitution. The over 60 years of ruling practice of Chinese Communist Party has proved that "the multi-party cooperation and political consultation system under the leadership of Chinese Communist Party is suitable for the state system of China, receives heartfelt support by different parties, different organizations, different ethnic groups and circles, and reflects strong vitality".[4]

---

[3]For example, the new positioning of considering democratic party as political-participating party, the new explanation of political consultation system reflecting the development requirement of the united frontier of new era, the new thinking of stressing the systemization, standardization and procedure of political negotiation, new explanation of the working theme, form, task, function and other aspects of CPPCC, and new recognition of the correlation between the political consultation system and socialist democratic politics, etc.

[4]Zheng Wantong: Adhere to and improve the multi-party cooperation and political consultation system led by CCP and consolidate and develop the broadest united frontier of patriots, refer to the guiding textbook of *The Decision of CCP Central Committee Regarding Strengthening the Party's Ruling capacity Construction*, People's Publishing House, 2004 Edition, Page 188.

12.2 Forms of the Cost of Chinese Political Consultation and Crossing of Information Gap

The development process of political consultation system reflected the characteristics of advancing with winding, its price paid during the process of encountering destruction and regaining recovery was the evolving cost of political consultation system. Because political consultation represents the "political consultation between Chinese Communist Party and various democratic parties" and the "negotiation between Chinese Communist Party and different democratic parties, patriots without party affiliation and representatives of different circles within the CPPCC",[5] the political consultation and information communication between Chinese Communist Party and different democratic parties, and between Chinese Communist Party and the patriots without party affiliation and representatives of different circles, there must be miscommunication and existence of information gap. Besides, there must also be miscommunication and information gap between central government and local government, between government and common people, there must also be miscommunication and information gap among the five state systems (NPC, Party Committee, CPPCC, Government and Military). The cost of information gap is an important part of political consultation cost. For the forms of the cost of Chinese political negotiation/information gap, please refer to Table 12.1.

12.2.1 Cost of Political Consultation Evolution

The cost of political consultation evolution main means the price paid for political consultation system from establishment to developing with winding (encountering destruction), and to recovery again.

The political consultation system in modern China was established based on the Chinese People's Political Consultative Conference at the end of September, 1949. The role and positioning of the Chinese People's Political Consultative Conference in Chinese political structure was very clear since the beginning, it is not an organ of power and doesn't have the nature of political power.[6] But under the

---

[5] Song Jian: *Discussion of the Connotation and Basic Forms of Political Negotiation*, Journal of Shayang Normal Junior College, 1st Issue in 2008.

[6] The "Common Program" passed in September, 1949 had such regulation regarding the issue of national political system: "The highest national political organ is the National People's Congress", "the Chinese people's political consultation meeting is the organizational form of people's democratic unification", "After the holding of the general election of national people's congress, the Chinese People's Political consultation Meeting shall, regarding the fundamental guidelines of national construction and other important measures, make proposal to the National People's Congress and the Central People's Government". Edited by CCP Central Committee Literature Research Office: *Compilation of Important Literatures since the Third Plenary Meeting*, Volume 1, People's Publishing House, 1992 Edition, Page 3.

Table 12.1 Forms of the cost of Chinese political negotiation/information gap

成本分类	表现形式	具体内容
Cost of political consultation and information gap filling	Cost of the evolution of political consultation system	Cost of the construction–damage–recovery of political consultation system
	Information gap cost	Information communication gap between CCP and other parties; unsmooth information communication between the central government and local governments, and between government and common people; excessive size of national administration system, and information communication gap of five systems

circumstance that the National People's Congress didn't have the time to hold, Chinese People's Political Consultative Conference, as the representative of political power, played the role of authority for several years.

The political consultation system experienced winding development history since 1954. After the establishment of the National People's Congress system, the Chinese People's Political Consultative Conference returned to its own role, which was to focus on the negotiation of major political issue of the country. There were discussions about the necessity of the existence of democratic parties and CPPCC. Comrade Mao Zedong stressed that the national power must absorb certain number of personnel from democratic parties, and criticized some people for neglecting the role of democratic parties and requiring the establishment of a "Pure" national power, and timely explained the necessity and importance of CPPCC.[7] In late 1950s, with the breeding of "Leftist" thinking, the role of political consultation system was gradually weakened and destructed in the national political life, the "Decade-long Cultural Revolution" made the political consultation system to trap in the state of paralysis.

After the Third Plenary Session of the 11th Central Committee of the Chinese Communist Party, the political consultation system was recovered. The reform and opening-up promoted Chinese society to have deep transformation, the promotion of democratic political construction provided broad room for the growth of political consultation system. In 1987, Deng Xiaoping began to connect "Political

---

[7]In 1956, Comrade Mao Zedong proposed "Long-term coexistence and mutual supervision" between CCP and democratic parties", and the accompanying guideline of "Two Long Lives". "Two Long Lives": Long Live the CCP, and Long Live the Democratic Parties. In the view of Comrade Mao Zedong, the measure of "Two Long Lives" was to prevent the problem of "less democracy and more concentration" phenomenon occurred in one-party socialist countries.

Negotiation" and "Multi-party Cooperation" and called it Chinese political party system. In 1989, CPC Central Committee formed the opinion regarding adherence and improvement of Chinese political party system, and formally proposed the standard representation of "Multi-party cooperation and political consultation system under the leadership of Chinese Communist Party". The Constitution Amendment passed in 1993 recorded that "the multi-party cooperation and political consultation system under the leadership of Chinese Communist Party will exist and develop for long term." The Amendment of the Constitution of CPPCC passed in 1994 made further supplementation to the nature of CPPCC based on this and clearly regulated that the "Chinese People's Political consultation Conference is the organization of Chinese People's Patriotic United Front, and an important institution of multi-party cooperation and political consultation under the leadership of Chinese Communist Party." The Amendment of the Constitution of CPPCC added that "It is an important form to develop socialism democracy in China's political life".

12.2.2 Cost of Crossing Information Gap

The cost of crossing information gap needs to cross various gaps, including the information exchange gap between Chinese Communist Party and other parties; the miscommunication between central government and local governments, and between the government and people; due to the huge national administrative departments, the information exchange gap among the five major systems (NPC, Party Committee, CPPCC, Government and Military).

Information communication in political field mainly involves administrative information communication. For a long period of time, Chinese government has been implementing a relatively centralized administrative system, which system reflected obvious tendency of bureaucracy, the boundary of organizational level is clear, the administrative operation procedure is rigid and complex, the administrative style of doing everything by book and each performing its own functions make the government system to be relatively separated from the external environment, and taking care of its own business, so the communication between government and outside world becomes inflexible, or even difficult. Because government itself generates considerable amount of policy information, so it is easy to become monopolist of policy information, and cause the asymmetrical information between government and policy counterparty.

Government monopolizes administrative information in administrative communication process, the disclosure of policy information is highly arbitrary, and the transparency is lacking, the information quantity and contents delivered in communication usually depends on the unilateral will of government and lack of restriction of system, which caused unsmooth information communication between the central government and local governments, between governmental institutions and common people.[8] Plus Chinese administrative system is huge, the five systems, NPC, Party Committee, CPPCC, Government and Military, have their own systems, there would also be certain communication barrier between systems and within systems. The information miscommunication brought by the fore-mentioned information monopoly and the result of information asymmetry must bring certain obstacle to administrative work, and would increase the burden of cost of administrative work.

Besides, China's vast number of administrative divisions would enable administrative information communication to have horizontal and vertical communication gap. So far as concerned, China has 31 provinces (directly governed municipalities and autonomous regions), two special administrative regions of Hong Kong and Macao as well as Taiwan Province; there are 332 district-level administrative divisions in Mainland China, including 284 "District-level Cities"[9]; and 369 county-level cities,[10] China has "Five and Half" levels of government[11] (refer to Chart 12.1). Such large number and levels of administrative divisions not only caused severe information gap phenomenon on horizontal level in China, the vertical administrative communication cost is also very high.

---

[8]The unsmooth information communication between the Central Government and local governments are mainly reflected as: the distortion of the central government's spirits and decisions in the passing process to local governments, and the information of local areas couldn't be timely and accurately reported to the related departments of central government. The unsmooth information between government and common people are mainly reflected as: when making some decisions, the government is unable to widely acquire opinions from the common people, causing real appeals of the mass unable to be reflected to the top government; on the other hand, enforcement of some government policies couldn't be effectively explained among people and would cause the generation of mistakes.

[9]As of November, 2011, Mainland China had 332 provincial administrative units, including 284 prefecture-level cities, 15 districts, 3 counties and 30 autonomous prefectures. If adding the Taibei City, Xinbei City, Taizhong City, Tainan City and Gao Xiong City of China's Taiwan Province, China has 337 provincial prefecture-level administrative units, including 289 prefecture-level cities.

[10]As of November, 2011, China has 369 county-level cities. The Jilong City, Xinzhu City, Taizhong City, Jiayi City and Tainan City of China's Taiwan Province can be considered as county-level cities.

[11]On the hierarchy structure, China has the phenomenon of "One China and Five and Half Levels of Government", meaning besides the five levels of central government–province (autonomous region, directly-governed city)–district–county–township (town), there are 13 deputy-provincial-level cities, such government structure has the most levels in the world.

12.2 Forms of the Cost of Chinese Political Consultation …

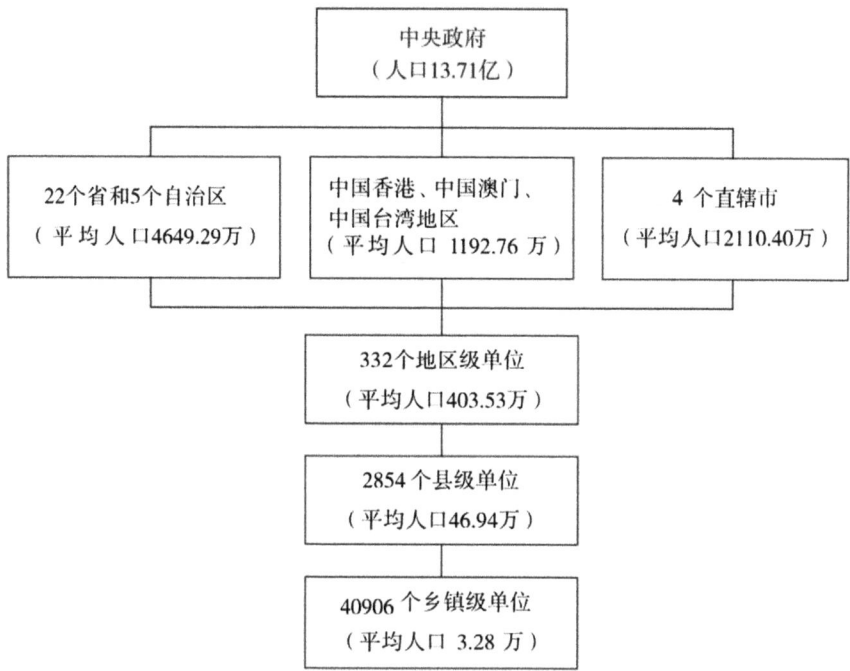

中央政府(人口 13.71 亿)	Central government (Population: 1.371 billion)
22 个省和 5 个自治区(平均人口 4649.29 万)	22 provinces and 5 autonomous regions (average population: 46.4929 million)
中国香港、中国澳门、中国台湾地区(平均人口 1192.76 万)	Chinese Hong Kong, Chinese Macao and Chinese Taiwan regions (average population: 11.9276 million)
4 个直辖市(平均人口 2110.40 万)	4 directly-governed cities (average population: 21.104 million)
332 个地区级单位(平均人口 403.53 万)	332 district-level unit (average population: 4.0353 million)
2854 个县级单位(平均人口 46.94 万)	2,854 county-level units (average population: 469.4 thousand)
40906 个乡镇级单位(平均人口 3.28 万)	40,906 town and township units (average population: 32.8 thousand)

Chart 12.1 Chinese administrative level structure and governed population (*Source* Calculated based on the data from the *Main Communique of the 6th Nationwide Census in 2010* issued by the State Statistics Bureau of the People's Republic of China)

12.3 Forming Causes of the Cost of Chinese Political Consultation and Information Gap Crossing

12.3.1 Historical Cause of the Evolving Cost of Chinese Political Consultation System

The political consultation system was born from historical cause and sustained due to realistic demand, it is a democratic system native born in China. Since China's socialism construction got rid of the restraint of traditional model and walked on the exploration of brand-new path, the in-depth changes occurred in the society promoted the changes of political life, new circumstances and new problems kept occurring. On one hand, political system began to gradually enter the track of reform, people's democratic awareness continuously strengthen, and their will of political appeal becomes increasingly stronger. On the other hand, with the change of economic system and distribution mechanism, social class, group interest, way of life and value orientation reflected diversification, the temptation of conflicts among people increased, and the ways of breakout became complex. The political consultation system, during the process of connecting with the continuously changing Chinese situation since the establishment of New China, continuously met the interest appeal and political participation demand of the broad people represented by different parties. Evolution of such political system must generate costs.

12.3.2 Existence of Information Gap in the Process of Communication in Different Organizations or Systems

Every political party is organizationally independent. The leadership role played by the Chinese Communist Party in multi-party cooperation is leadership of political line rather than leadership of organization. In organization, democratic parties have their own vertical systems, the subordinates shall obey the superior, the local organization shall obey the central committee. The factors including too many organizational levels, overlapped institutions, unsmooth relationships and excessive red tapes and meetings would cause barrier to information communication. Therefore, there is cross-organization and cross-system nature of the information communication between the Chinese Communist Party and other parties, the distortion or loss of information in delivery process would cause information gap cost.

For the conclusion of forming causes of Chinese political negotiation/information gap cost, please refer to Chart 12.2.

12.3 Forming Causes of the Cost of Chinese Political …

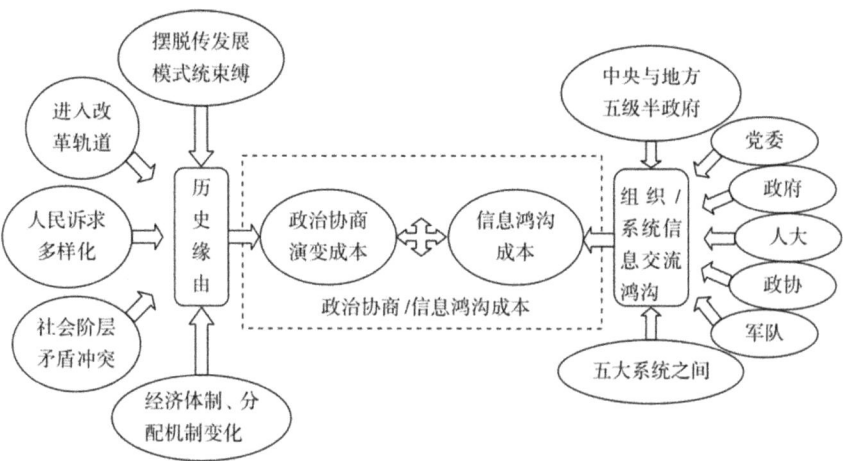

摆脱传统发展模式束缚	Get rid of binding of traditional development model	进入改革轨道	Enter the track of reform
人民诉求多样化	Diversification of people's appeals	社会阶层矛盾冲突	Conflict and contradiction of social classes
经济体制、分配机制变化	Change of economic system and distribution mechanism	历史缘由	Historical causes
政治协商演变成本	Cost of political consultation evolution	信息鸿沟成本	Cost of information gap
政治协商/信息鸿沟成本	Cost of political negotiation/information gap	组织/系统信息交流鸿沟	Communication gap of organization/system information
中央与地方五级半政府	Five-and-half-level central government and local governments	党委	Party committee
政府	Government	人大	NPC
政协	CPPCC	军队	Military
五大系统之间	Between five systems		

Chart 12.2 Forming causes of the cost of Chinese political consultation and information gap crossing

12.4 Countermeasures for the Cost of Chinese Political Consultation and Information Gap

12.4.1 Improve Working Method of CPPCC and Fully Play the Role of Information Democracy

The CPPCC of multi-party cooperation centralizes representatives from different sectors,[12] many works of CPPCC can't be done without the effect of these sectors. CPPCC organization on each level shall encourage members of different democratic parties to strengthen the connection with the people of their own sectors, actively reflect the interest and requirement of such sectors, pay attention to meeting speech by representatives of the sectors, carefully handle the proposals proposed by different parties and organizations, emphasize using sectors to coordinate relationship and solve conflicts, make efforts to establish a democratic information exchange channel among different parties and different sectors. So as to reduce unnecessary information gap cost among governmental systems and between government and the people.

12.4.2 Establish Information System and Implement Scientific Management to Reduce Barrier of Information Delivery

Standard and swift communication of administrative information is the precondition for normal operation of modern administration. Administrative system must establish a well connected, sensitive and accurate administrative information

---

[12]"Sector" means classification and division. As the organization of patriotic united front, the 10th National CPPCC has 34 sectors, including Chinese Communist Party, Revolutionary Committee of the Chinese Kuomintang, China Democratic League, China Democratic National Construction Association, China Association for Promoting Democracy, Chinese Peasants' and Workers' Democratic party, China Zhi Gong Dang, Jiusan Society, Taiwan Democratic Self-Government League, public figure without party affiliation, Chinese Youth League, All China Federation of Trade Union, All-China Women's Federation, China Youth Federation, All-China Federation of Industry and Commerce, China Association for Science and Technology, All-China Taiwan Federation, All-China Federation of Returned Overseas Chinese, cultural and art circle, science and technology circle, social science circle, economics circle, agricultural circle, education circle, sports circle, news and publishing circle, medical and health circle, foreign friendly relations circle, social welfare and social security circle, minorities circle, religion circle, specially invited people from Hong Kong, specially invited people from Macao and specially invited people.

12.4 Countermeasures for the Cost of Chinese Political …

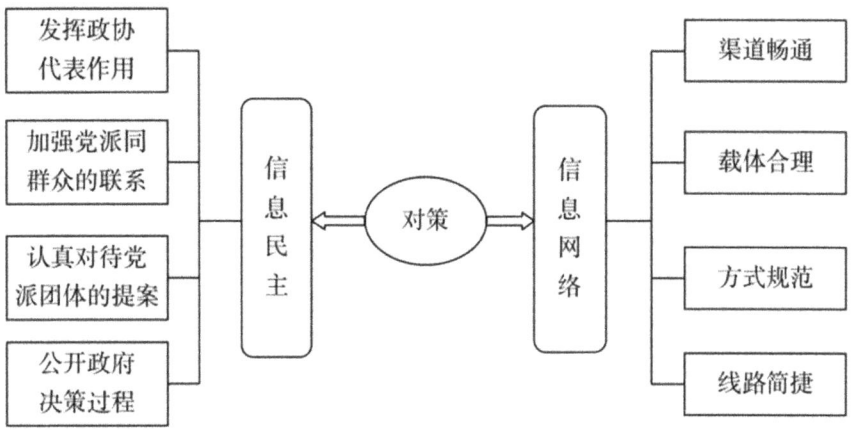

发挥政协代表作用	Play the role of CPPCC representative	加强党派同群众的联系	Strengthen the contact between parties and common people
认真对待党派团体的提案	Carefully treat the proposals of parties and organizations	公开政府决策过程	Open decision-making process of the government
信息民主	Information democracy	对策	Countermeasures
信息网络	Information network	渠道畅通	Smooth channel
载体合理	Reasonable carrier	方式规范	Standard method
线路简捷	Simple and convenient route		

Chart 12.3 Countermeasures for the cost of Chinese political consultation and information gap crossing

network, so as to frequently and swiftly exchange the information of different departments, and realize effective and scientific management. Adhere to the principle of scientific communication of administrative information,[13] such principles include: firstly, smooth communication channel of administrative information; secondly, reasonable carrier of information to facilitate communication; thirdly, the selection of communication method shall be standard and having certain rationality; fourthly, the route of information communication shall be simple; fifthly, adhere to bilateral communication and avoid unilateral communication. Through

---

[13]Yuan Zaijun: *Obstacle in Exchange of Administrative Information and Its Elimination*, Theoretical Discussion, 2nd Issue in 2004.

communication in modern method, we could organize the administrative activities of all government, form an effectively controlled entirety, and enable all administrative activities to maintain dynamic balance with outside world, so as to reduce information gap cost while increasing information delivery efficiency.

For the countermeasures for the cost of Chinese political negotiation/information gap, please refer to Chart 12.3.

Chapter 13
Cost of Chinese Democracy Construction

13.1 Connotation of Chinese Democracy Construction

Democracy is a kind of civilization result formed in the process of long human history, a value of commonly sought after by human being, and also a significant symbol of modern civilized nation. By analyzing the "Democracy" worldwide, it can be concluded that there are multiple forms of "Democracy".[1] The "Democracy" implemented in China is the "Democracy" of the democratic centralization system,[2] which is a organizational principle, working principle and working style that focus on fully expression of opinions and view points on the party's issues based on equality and freedom.[3] From vertical perspective, the during the period since the establishment of New China and the New-democratic Revolution, Chinese Communist Party fought for democracy, and during the period of socialism construction, the party led the people to implement democratic construction and exploration; from horizontal perspective, democratic construction includes various aspects, including the democracy within Chinese Communist Party, the National People's Congress System, the Multi-party Cooperation and Political Negotiation,

---

[1] Various forms of democracy in the world could be basically classified into three types: the first is electoral democracy, its characteristics are to, through fair voting and based on the obedience of minority, form resolutions, laws or elect the candidates or political party based on the interest appeals of the majority participants to govern the country; the second is negotiation democracy, its characteristics are to distribute interest through negotiation, so as to enable all parties to be relatively satisfied with their interest appeals; the third is consultative democracy, its characteristics are to let citizens across the society fairly participate in decision-making of public policies, through broad discussion and talks, form consensus or find the maximum common ground or common interest, make the decisions with collective binding force. Refer to Li Junru: *Which Form of Democracy China could Realize, Beijing Daily,* Sep 26, 2005.

[2] "Democracy" is not anarchism, but a ruling order and system arrangement, there is no "Democracy" without following-up by "centralization", democracy itself includes authoritativeness and centralization.

[3] Zheng Keyang and Li Zhongjie: *Research of the Important Thought of "Three Represents"*, Sichuan People's Publishing House, 2002 Edition.

administrative management system and institutional reform, cadre system reform, judicial system reform, democratic supervision construction, basic-level democracy and villagers autonomy.[4]

At the beginning since the establishment of New China, the democratic rule of law system achieved preliminary effect, to the 1960s and 1970s, democratic construction suffered great destruction, China's democratic construction had a very difficult and winding path until it entered the right track after the Third Plenary Session of the 11th Central Committee of the Chinese Communist Party. Since the reform and opening-up, Chinese Communist Party has been increasingly regarding the construction of democratic politics, and made significant achievements in the aspects of democratic system reform, construction of democratic politics, construction of rule of law, government by law, development of political civilization, etc.

Currently, acceleration of democratic development and enhancement of Chinese political modernization level become an urgent need. On one hand, Chinese political development still lags behind economic development, and has become a barrier of further economic development. Over 30 years, Chinese socialism market economy system has been preliminarily established, the economy and society have realized extensive transformation; the interests of social groups have been segmented, the political requirements to maintain group interests have increased; citizens have stronger awareness of independence and higher enthusiasm in political participation. However, the society also has many negative factors, such as the lasting and incurable corruption among officials, which caused extreme dissatisfaction among common people on the reality of corruption. Under such circumstance, each aspect of Chinese society has extremely high expectation on political democracy, energetic promotion of political system reform and promotion of political democratization have become the urgent need to comprehensively promote socialism reform. On the other hand, under the background of rising globalization, political democratization also faces huge external pressure. Currently, economic mutual benefiting and competition, cultural integration and conflict, political democracy and stabilization have become a worldwide development trend. Any political community that wants survival and development can't stand outside this developing world trend. The international community commonly considers the situation of democracy, freedom and human rights protection as an important indicator to evaluate a country's international image, how much soft power a country has also depends on this. Therefore, acceleration of democratization process has become an issue that China must face in order to integrate into the world.

---

[4]Gao Fang: *Respond to New Wave of Globalization, Develop Socialist Democracy,* Journal of Ningxia Party School, 3rd Issue, Volume 7, May, 2005.

13.1 Connotation of Chinese Democracy Construction

Table 13.1 Forms of Chinese democratic construction cost

Classification of cost	Forms	Specific contents
Cost of democracy construction	Cost of democracy explanation	Making democracy explanation to international community, including active ones and passive ones
	Cost of democracy development	Constant input and price of democratic construction, history–present–future
	Cost of political fluctuation	Loss caused from political disturbance relating to democracy

Chinese democratic politics model is based on Chinese reality, and is the political operation form that is suitable to Chinese social goals at current stage.[5] Rooting in national reality and establishment of scientific Chinese political development view could avoid risks, promote democratic development, realize the socialism modernization and the great revival of Chinese nation by resolutely adhering to the socialism path with Chinese characteristics.

13.2 Forms of Chinese Democracy Construction Cost

Chinese democracy construction cost studied in this book, as its name indicates, means the cost of a series of measures implemented in order to realize socialism democratic politics in China, including democratic explanation cost, democratic development cost and political disturbance. Refer to Table 13.1. Democratic explanation cost involves the input for media explanation and publicity towards the doubts, misunderstanding and distortion of the international community on Chinese democracy issues, including democratic explanation in active and passive manners; democratic construction cost means the input and loss of China in a series of democratic construction process since the establishment of New China, including the preliminary establishment of democratic system at the beginning of New China, the destruction of democratic system during the "Cultural Revolution", the accelerated steps of democratic construction and achievements made since the reform and opening-up; political disturbance cost means the loss caused from political disturbance events.

---

[5]In September, 2006, Wen Jiabao expressed, as accepting an interview from European media, that "Democracy is a value jointly pursued and civilization result commonly sought after by human being, but its realization forms and paths vary at different stages and in different countries, there is no unified model".

13.2.1 Democracy Explaining Cost

In news coverage on Chinese politics, the ideological prejudice among western media is still not changed, when facing the misunderstanding among international media towards Chinese democratic politics, it requires China to adopt appropriate measures to make democracy explanation, such passive democracy explanation lies in the saving of image in crisis PR, and requires certain cost. But it can't be denied that the Chinese government should actively image of a civilized, modern, just and responsible democratic political power as a relatively clear positioning in the molding of Chinese national image. Chinese idea of "Harmonious World" is a major contribution to the harmonious development of international politics, and is good for vivid reflection of the image of modern China as a civilized, modern, just and responsible democratic power. China should abandon simple argument of ideology, hold hands with the international community, adhere to the ideas of peace, development, cooperation, win–win situation and tolerance, in the exchange and cooperation with countries in the world, China should actively project China's image as a democratic political power, which would need active democracy explaining cost.

13.2.2 Democracy Development Cost

Democracy development cost means the price that China has paid in the historical process of democracy construction under the leadership of Chinese Communist Party.

To review China's history of democratic politics construction, it must be seen that Chinese democracy today was started from the holding of Chinese People's Political consultation Conference in 1949. In other words, after Chinese Communist Party obtained decisive victory on the battlefield of Liberalization War, it didn't monopolize the political power, but discussed national affairs with different democratic parties and the patriots without party affiliation, adopted the form of political negotiation, formed the Common Program. On this basis, the Central People's Government of the People's Republic of China was generated through election. Therefore, the ruling status of Chinese Communist Party was established based on the combination of negotiating democracy and elective democracy. This is the source of legitimacy of Chinese Communist Party's ruling. Later the Constitution was formed and the National People's Congress was held according to the Constitution. The ruling status of Chinese Communist Party had further legal foundation. This is a kind of democratic form created by history and with connotation of Chinese culture.

Chinese people also paid dear price in the exploration of how to realize people's democracy. At the preliminary period of democratic construction, Comrade Mao Zedong referred to the lessons of democratic construction of Soviet Union, and considered the expansion of democracy and development of democracy as the fundamental measure to oppose and prevent revisionism, so as to enable the broad basic level to widely, directly and regularly participate in the nation's political life. The democratic practice promoted by Comrade Mao Zedong reached an extreme during the Cultural Revolution, formed "Grand Democracy", which caused "Major Turmoil" of the society, people could hardly guarantee their lives during turmoil, let alone freedom and civil rights. "Grand Democracy" reflected the beautiful wish of Comrade Mao Zedong to realize People as masters of their own country, but destructed the original and important political foundation of socialism democracy—the Leadership of Communist Party and Socialism Rule of Law. After losing the Party's leadership and regulation of law, people's direct political participation and autonomy quickly evolved into intensive explosions of various conflicts.

Since the reform and opening-up, while maintaining social harmony and stability, China realized long-term high-speed economic development and huge advancement of democratic construction through gradual reform of economic system and political system. Currently, China is already the second biggest economy in the world,[6] the quality of Chinese democracy keeps improving, the people's living standard has extensive advancement, and feasibly enjoys increasingly abundant human rights and civil rights of different kinds. China has created a new path that is different from the "Washington Consensus" promoted by the west, China's development experience is also referred to as "Beijing Consensus".[7] In order to make Chinese democratic system to be able to continuously improve and mature, China needs to continuously input certain cost to guarantee the smooth operation of democracy construction.

Special Column 13-1 How the Democracy System within the Party failed (1957–1965)

Analyzing from the party's national congresses and plenary sessions of the central committee during 1957–1965:

Firstly, the actual term of the party's 8th National Congress and 8th Central Committee was as long as 13 years (from 1956 to 1969), which was equivalent to two and half normal terms of Central Committee, it violated the regulations in Article 31 and Article 33 of the Party Constitution.

Secondly, during 1957–1965, the party's national congress was held only twice (the 8th in 1965 and the 2nd Session of the 8th Party's Congress in 1958), which violated the regulation in Article 31 of the Party Constitution.

---

[6]According to the data released by State Statistics Bureau of China, Japan's GDP in 2010 was lower than Chinese GDP of USD 5.8786 trillion released in January, China has become the second biggest economy in the world.

[7]After well-known US scholar Shuja Raymer proposed "Beijing Consensus" in 2004, it has drawn wide attentions from scholars across the world. Refer to Edited by Yu Ping: *Chinese Model and "Beijing Consensus"; Beyond "Washington Consensus"*, Social Science Literature Press, 2006 Edition, Page 413–415.

Thirdly, besides the two plenary sessions of the central committee (the 5th and 6th Plenary Session of the Central Committee), there was only one plenary session held in history, in 5 years of which (1960, 1963, 1964 and 65), there was no plenary session of the central committee. It violated the regulation in Article 36 of the Party Constitution.

Even before Comrade Mao Zedong launched the "Cultural Revolution", the party's regulations and rules were severely damaged, the system was already "Existing in Name only". So it can be said that the party's regulations and rules "failed" first, then Comrade Mao Zedong launched the "Cultural Revolution". If the party's regulations and rules could play its effect, and the entire party do things according to the (party) Constitution", then the party's National Congress or Plenary Session of the Central Committee would not pass the decision of "Cultural Revolution" proposed by Comrade Mao Zedong.

Source: Hu Angang: *Chinese History of Economy and Politics* (1949–1976), Tsinghua University Press, 2008 Edition, Page 372.

13.3 Forming Causes of the Cost of Chinese Democracy Construction

13.3.1 Necessary Requirement for China to Actively or Passively Make Democracy Explanation and Promote National Image in International Community

Due to the restriction of international broadcasting channel and cultural exchange, cultural and language difference as well as the mistake or possible distortion of fact in information communicating process, the national image is often subject to various forms of misinterpretation, as a result, it would inevitably affect the effect of this country's external promotion, so as to increase the cost for the country to maintain and explore national interest. The cost of democracy explanation in international community is one of such forms.

13.3.2 Chinese Democracy Construction Development Faces Restriction of Historical Factors and Basic National Situation

First of all, there is deep historical cause for the difficult steps of Chinese democratic politics process. China is a nation with over 2100 years of feudal autocracy history, vast territory, many ethnic peoples and extreme imbalance in development.

The cultural tradition of Feudal Autocracy passes from one generation to another, penetrates into every aspect of social life, to this day, it is still a heavy burden. China's territory is equivalent to that of the entire Europe, the population equals that of Europe and America combined, the differences among different regions in language, culture, productivity level, or even life style are huge. In order to ensure such a multinational country to maintain unification and realize long-term peace, its political operation form must also have particularity. The democratic forms that are effective in many countries are not necessarily suitable for China, China needs to gradually create its own democratic form based on its momentum of development. Secondly, for over 30 years since the reform and opening-up, China's basic national situation of being at preliminary stage of socialism hasn't been fundamentally changed, it still restrains the degree and speed of democratic politics development. Although in economically developed coastal regions, the modernization degree of major cities is already very impressive, the middle-income class keeps expanding and makes appeals for democracy from time to time, as a whole, China's society and politics are still not separated yet, an independent social system is not composed yet, not to mention the wide rural areas, especially the poor and backward undeveloped regions. Many places just eliminated poverty or realized moderately well-off, they have even less economic foundation to realize political democratization. Therefore, for the democracy construction with Chinese characteristics, there is no existing pattern to copy, on one hand, we need to refer to the international advanced experiences and pay the tuition; on the other hand, we need to, based on the reality, continuously explore and advance, constantly innovate, so the necessary input and price of mistakes are inevitable.

13.4 Countermeasures for the Cost of Chinese Democracy Construction

13.4.1 Actively Expand Foreign Exchange and Cooperation and Demonstrate Chinese Image of Democracy to Reduce Passive Democracy Explaining Cost

We should adopt broad, multi-directional and network-shaped communication method and build a set of system to systematically demonstrate China's cultural, political and economic lives to foreign countries, so as to enable foreigners to be able to see both the good side and bad side of China, eventually it will form the understanding method of the international community towards China. The public good of national image optimization would not only need planning and organization by government, but also need wide participation by all levels of society. We

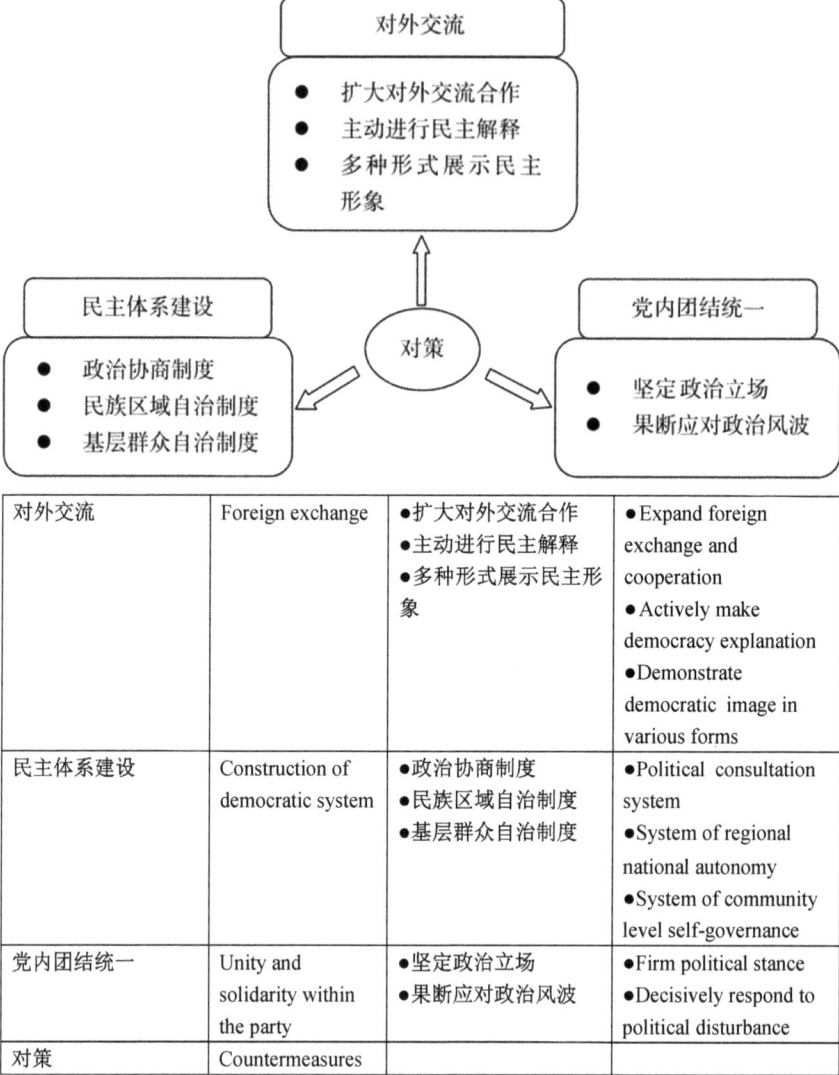

Chart 13.1 Countermeasures for the cost of Chinese democracy construction

should encourage private groups and private organizations to strengthen exchange with foreign groups or even world organizations, through such private exchanges, the people in the world could see the open and democratic side of China, so as to be willing to or even looking forward to exchange and cooperation with China.

13.4.2 Continuously Improve China's Democracy Construction System and Enhance China's Democratic Level

As indicated in the Report of the Party's 17th National Congress: "We shall adhere to the organic unification of party's leadership, people as master of their own country and rule by law, adhere to and improve the People's Congress System, the Multi-party Cooperation and Political consultation System, the System of Regional National Autonomy and the System of Community Level Self-governance under the leadership of the Chinese Communist Party, continuously promote the self-improvement and development of the socialism political system.[8] China shall continue to advance and innovate on the path of developing the socialism democratic politics.

For the countermeasures for the cost of Chinese democracy construction, please refer to Chart 13.1.

---

[8]Hu Jintao: *Report on the 17th National Congress of Chinese Communist Party,* People's Daily, Oct 25, 2007.

Summary of Chinese Political Development Cost

The cost of political development includes not only normal input but also loss and price, the latter accounts for majority of political development cost. The political reform cost before 1978 was relatively high, the corruption cost after the reform and opening-up reflected situation of radical increase.

For the construction of the ruling party, Chinese Communist Party itself, including the nearly 17 times of scale increase of party members from 1949 to 2010, plus the ideological and style construction, large cost of input is needed. Since 1978, the political system reform that China has been gradually developing to extensively streamline government organs, and consolidate ministries and commissions enabled the departments of the State Council to decrease from 100 to 25 (2013), which is also good for reducing China's information gap cost. The fore-mentioned normal input in political development process is the necessary cost for China to realize democratic politics.

Political reform cost (mainly cost of political struggle) brought high extent of influence and damage to China's economic construction and social development, and impeded China's development of productivity. The occurring times of political struggle events reached two peaks around 1950 and at the early period of the "Cultural Revolution". Especially the "Cultural Revolution" made China loses the golden development period, and reduced the potential human capital stock by 14.3%. Besides, from the perspective of political decision-making cost, the economic loss caused by two most important political decision-making mistakes, the "Great Leap Forward" and the "Cultural Revolution" accounted for 1/4–1/3 of the economic growth rate, accounting for 1/2–1/3 of the simulated gross GDP in 1978. The cost of social turmoil risk reflected through GINI Coefficient generally reflected growing trend, which indicated that China's rich-poor gap is widening, the risk of having social turmoil is increasing. The corruption cost reached peak during 1993–1997, it basically maintained at high level after that and peaked in 2002, and tended to decline since 2006. The high price China paid in political development process is the valuable experience and lesson in future development. The core and essence of the construction of socialism political construction is to build high-level socialism democracy, and guarantee the realization of the ideal and goal of people being masters of their own country.

Part IV
Cost of Chinese Social Development

Chapter 14
Chinese Social Livelihood Issues and Their Cost

Since ancient times, China has words representing care for "Livelihood" issues, such as "National Economy and the People's Livelihood", "the people are the foundation of the state, the state serves its people",[1] etc., which not only reflected the inseparable relationship between "Livelihood" and the existence and development of state, but also reflected the high regard of livelihood issues by worthy predecessors.[2]

Over the 30 years since the reform and opening up, China has made achievements in economic construction that amazed the world, the market economy system has been preliminary established, the income level of urban and rural citizens has been steadily growing, people's living standard realized extensive enhancement, major breakthroughs have been made in the construction of social security system, etc. However, while the economy and society have been rapidly developing, unequal resource distribution and other realistic conditions have also caused a series of social livelihood problems in the fields such as education, employment, income distribution, social security, etc., for example, the growth of resident income is relatively slow, the pricing level is unstable, the income distribution is not reasonable enough, the general level of social security is not high, the distribution of education resource is not fair, these negative factors have caused negative influence on the establishment of China's socialism harmonious society and China has paid price for it, therefore, studying the livelihood construction and cost of current Chinese society, analyzing the existing problems and causes, and exploring effective countermeasures have great theoretical and practical significance in the promotion of Chinese social advancement.

---

[1] Quoted from *Book of History—Wu Zi Zhi Ge*, meaning "the people are the foundation of the state, the state serves its people".
[2] *Selected Works of Sun Zhongshan*, People's Publishing House, 1981 Edition, p. 802.

14.1 Definition and Connotation of Livelihood Issue

The word "Livelihood" was originated from "Livelihood depends on hard work, there will be no shortage after hard work" cited from *Zuo Zhuan—The 12th Year of Duke Xuan*. The first person who made detailed interpretation on "Livelihood" for the first time in modern times was Sun Yat-sen, his classic interpretation of livelihood issue was that "Livelihood is people's lives—meaning the existence of society, livelihood of people and lives of the mass."

Chinese Communist Party has always been highly regarding livelihood, and always considered the solving of livelihood issues as an important means to seek interests for the people. It was emphasized in the report of the party's 17th Congress that "We must, based on economic development, pay higher regards on social construction, focus on the guarantee and improvement of livelihood, promote the social system reform, expand public service, improve social management, promote social fairness and justice, make efforts to enable the entire people to have education, to have income for their work, to have medical service for diseases, to have pension after getting old and to have houses for sheltering."

In modern sense, "the so-called livelihood mainly means the basic existing and living state of the people, as well as the state of the people's basic development opportunity, basic development ability, and protection of basic rights and interests, etc.[3]" Specifically, it mainly includes six aspects, employment, education, income distribution, social security, medical condition and social stability (see Chart 14.1).

Since the reform and opening-up, Chinese government has been actively developing livelihood causes, continuously expanding the degree of fiscal investments, policies and subsidies, the livelihood issues have been prominently improved. "During the last five years, the development of various social undertakings has been accelerated, and people's lives have been prominently improved. The undertakings in education, science and technology, culture, health and sports have comprehensively advanced. 57.71 million of additional employments in urban areas were realized, 45 million of rural labors were transferred; the average distributable income of urban residents and average net income of rural residents actually grew by 9.7 and 8.9% respectively; the social security system that covers urban and rural areas have been gradually completed.[4] But on the other hand, the "Marketization" also caused continuous prominence of China's livelihood problems, the social phenomenon occurred, including people can't afford education, seeing doctors, widening gap between the rich and poor, etc., globalization also continuous amplify Chinese

---

[3]Wu Zhongmin: *Basic Definition and Characteristics of Livelihood*, Forum of Chinese Party and Political Officials, 5th Issue in 2008.

[4]Wen Jiabao: *2011 Government Work Report*, Wen Hui Bao, Mar 6, 2011.

14.1 Definition and Connotation of Livelihood Issue

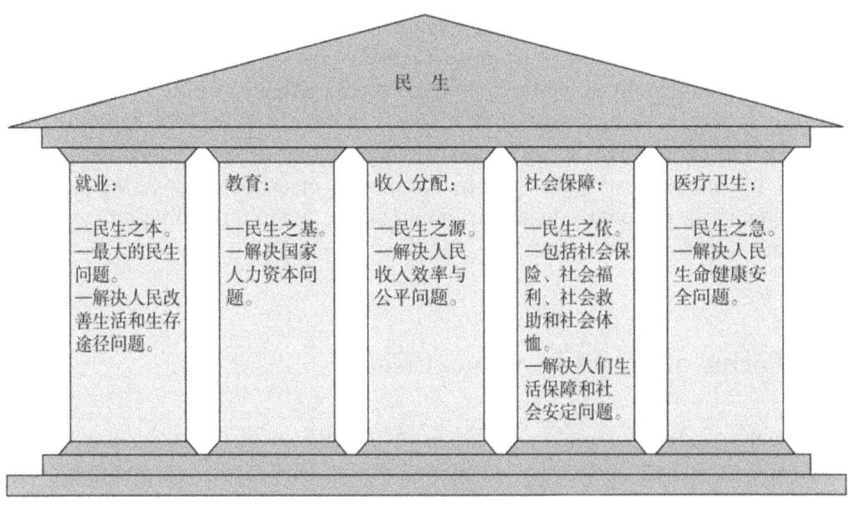

民生	Livelihood
就业： --民生之本。 --最大的民生问题。 --解决人民改善生活和生存途径问题。	Employment: -- Basis of livelihood. -- The biggest livelihood issue. -- Solve the issues of people's livelihood improvement and living means.
教育： -- 民生之基。 --解决国家人力资本问题	Education: --Basis of livelihood. --Solve the issue of a country's human capital
收入分配： --民生之源 --解决人民收入效率与公平问题。	Income distribution: --Source of livelihood --Solve the issue of people's income efficiency and fairness.
社会保障： --民生之依。 -- 包括社会保险、社会福利、社会救助和社会体恤。 --解决人们生活保障和社会安定问题。	Social security: --Dependence of livelihood. --Including social insurance, social welfare, social aids and social compensation. --Solve the issues of people's living security and social stability.
医疗卫生： --民生之急。 --解决人民生命健康安全问题。	Medical and health: --Urgent appeal of livelihood. --Solve the issues of people's life and health security.

Chart 14.1 Definition and connotation of social livelihood issue

livelihood problems.[5] Livelihood problems have become the bottleneck that restrains further development of Chinese economy and society.

How to fundamentally solve the livelihood problem that is caused by systematic flaws, this is the important issues that Chinese government actively explores and solves. This to certain extent constitutes another content of social livelihood cost.

Social livelihood cost means the overall social investments covering various aspects, including national education, employment, income distribution, medical service and social security.

14.2 Forms of Social Livelihood Cost

Based on the analysis in previous section, it is believed in this book that social livelihood cost includes education development cost, employment promotion cost, social security lacking cost, income distribution imbalance cost, medical service development cost, etc. For more detailed contents, please refer to Table 14.1.

14.2.1 Education Development Cost

Education investment is the fundamental and strategic investment that supports the long-term development of a nation, and an important part of government investment in livelihood. For many years, Chinese fiscal departments on each level have always been putting education at prominent position of public finance, and continuously increasing investment. During 2001–2010, China's education investment from public finance increased from about RMB 300 billion to about RMB 1.42 trillion with average annual growth rate of 19.2%; the ratio of education expenditure of public fiscal budget in fiscal expenditure increased from 14.3 to 15.8%, becoming the biggest expenditure of public finance; its ratio in GDP also increased from 3.14% to over 3.57% (refer to Table 14.2). But this number is still far from the average level in the world, the world average is 4.9%, and the level in developed nations is 5.5%.

In the era of planning economy, in order to create talents in fast and early manner, based on the principle of "Efficiency first and fairness into account", the

---

[5]While making the international market prosperous, economic globalization also makes Chinese economy more vulnerable, causes relatively severe social problems, and make China's employment, income and accumulation of wealth subject to certain degree of influence. Globalization expands the inequality between those receiving more education and those receiving less education, which accelerated the uneven income between urban and rural areas, and increasingly expanded the gap between rich and poor. Around the change of century, from Southeast Asian Financial Crisis, SARS crisis to US Financial Crisis, these globalization risks that broke the restriction of time and space have seriously affected the security and solving of China's livelihood.

Table 14.1 Cost of social livelihood issues

Classification of cost	Forms	Measurement indicator
Cost of solving social livelihood issues	Cost of education development	Fiscal expenditure on education, cost of lack of education equality
	Cost of employment promotion	Fiscal expenditure on employment
	Cost of lack of social security	Moderation level of social security, actual level of social security
	Cost of income distribution balance	Urban-rural income ratio, income change coefficient
	Cost of medical and health development	Health input, infant mortality rate, maternal mortality rate

government adopted the measure of centralizing fiscal power to focus on a batch of key schools, and created a "Sliding" system for key schools in the aspects of teacher team construction, expense guarantee, education conditions, etc., which gradually widened the gap of resource distribution in Chinese middle schools and primary schools. Currently, the problem of unfair education becomes increasingly severe, the government is already deeply aware of it and has formed a series of solutions[6] to enlarge investments in the education among vulnerable groups.[7]

In 2010, the total amount of subsidies for regular colleges and universities, secondary vocational schools, and regular high schools as well as the living subsidies for resident students at the stage of compulsory education was RMB 87.65 billion, increased by RMB 20.595 billion compared with that in 2009, or 30.71% of growth. In which the fiscal expenditure was RMB 65.804, increased by RMB 18.181 billion compared with that in 2009, or 38.18%. In 2010, over 15.9 million resident students with difficult economic conditions in their families at the stage of compulsory education received living subsidies, increased by 1.35 million students compared with that in 2009, the total subsidy fund was RMB 13.28 billion, increased by RMB 4.22 billion compared with that in 2009, or 46.6%. In 2010, based on exemption of the tuition and fees of 150 million students at the stage of compulsory education across China, the national government provided free textbooks to 136 million students, including all rural students and urban students with difficult family economic conditions. According to the data released by the Ministry of Education in 2010, China's regular colleges and universities

---

[6]Such as implementation of the "Two Exemption and One Subsidy" measure to the rural poor students at compulsory education stage, enlarge the support to poor students in colleges and universities, the standard of state stipend in regular institution of higher learning for poor students has currently reached RMB 3000 per student; the standard of state stipend in regular high schools for poor students has currently reached RMB 1500 per student, the coverage accounts for 20% of total students at regular high schools.

[7]Vulnerable group mainly includes poor farmers, laid-off workers, migrant workers, disable people, minorities, etc. Education of vulnerable group also include the two types, vulnerable group itself and education of their children.

Table 14.2 Total education expenditure of chinese public finance and ratio in GDP during 1992–2010

年份	财政性教育经费（亿元）	财政性教育经费占 GDP 的比重（%）
1992	729	2.70
1993	868	2.51
1994	1175	2.51
1995	1412	2.41
1996	1672	2.46
1997	1863	2.50
1998	2032	2.59
1999	2287	2.79
2000	2563	2.86
2001	3057	3.14
2002	3491	3.41
2003	3851	3.28
2004	4466	2.79
2005	5161	2.82
2006	6348	3.00
2007	8082	3.32
2008	10450	3.48
2009	12224	3.59
2010	14200	3.57

Note: National fiscal education expenditure includes education expenditure of national fiscal budget, taxes and fees levied by governments on each level for education, education expenditure of the schools established by enterprises, and education expenditure from school enterprises, study on a work-study basis and social service income.

Source: State Bureau of Statistics: *Chinese Development Report 2011*, China Statistics Press, 2011 Edition, Page 615.

年份	Year
财政性教育经费（亿元）	Fiscal expenditure on education (RMB 100 million)
财政性教育经费占GDP的比重（%）	Ratio of fiscal expenditure on education in GDP (%)

released RMB 11.357 billion of state-subsidized student loans in 2010, increased by 21.37% compared with the same period in previous year. In order to guarantee the smooth enrolling of poor students, colleges and universities across China have generally established the "Green Pass" system.

With the acceleration of China's urbanization process, the education issue of the children of migrant workers in cities becomes gradually prominent. "According to statistics, the number of children of migrant workers at the stage of compulsory

14.2 Forms of Social Livelihood Cost

education reached 9.971 million in 2009.[8]" In order to solve this problem, since 2001, the national government formed the system of "Putting the migration-in government and public schools in priority", and appropriating special funds to supplement and enhance the education of the children of migrant workers. However, due to the existence of household registration system, the children of migrant workers are still unable to receive the "equal treatment" and suffer education discrimination.

14.2.2 Employment Promotion Cost

Employment issue is a world problem that is related to social stability and development, nations across the world are actively exploring the ways to solve employment problems. In 1990s, China had large scale of "Lay-off Flood" and "Unemployment Flood", during this period, nearly 60 million regular employment positions in state-owned units and urban collective units were destroyed in accumulation, if we extend the laid-off and unemployed people to their families, calculating based on the average family members in urban area, about 190 million urban population suffered influence to different extent. That "Lay-off Flood" also caused various social instability events, which is sufficient to reflect the complexity and importance of employment issue.

The national employment cost means the cost and efforts made by the central government in the aspects of taxation, fiscal and industrial policies in order to promote employment and reduce unemployment.

Before the liberation, due to years of wars, social turmoil and economic depression, most people in China were unemployed, in 1949 when New China was established, the total employed and working population was only 180 million people, while the population at labor age back then was 340 million, the urban unemployment rate reached 7%, while the employment population of China back then accounted for 20% of that in the world,[9] which was obviously lower than the ratio of working population in that of the world. At the preliminary stage of New China, in order to maintain social stability, the government implemented "Full Coverage" policy on all urban people at laboring ages, within 3 years, the working people in the society had net increase of 26.47 million. After 1952, China was transferred into the large economic construction, which was the stage of fast employment growth. In 1958, China launched the "Great Leap Forward" Movement, the government streamlined employment approving authority, and allowed the recruitment of farmer workers in cities, the urban employment had net

---

[8]CCP Central Committee Publicity Department Theoretical Bureau: *Seven How to Look at It*, Studying Press and People's Publishing House, 2010 Edition, p. 62.

[9]As estimated by IMF, the global employed people in 1950 was 900 million, World Economic and Financial Surveys, World Economic Outlook Database, September 2006 Edition.

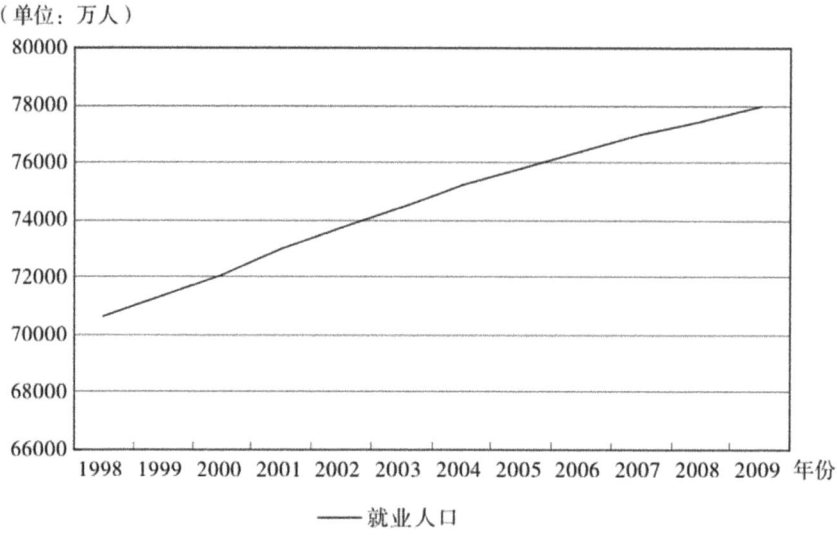

(单位：万人)	(Unit: 10 thousand people)	年份	Year
就业人口	Employed population		

Chart 14.2 Change of Chinese employment population (1998–2008) (*Source* State Bureau of Statistics: *Chinese Development Report 2011*, China Statistical Press, 2011 Edition, p. 595)

increase of 20.95 million people in that year, the excessively fast growth of employment caused more personnel than work available and pressure of food supply shortage, so streamlining had to be made later.[10]

Entering 1990s, the reform of state-owned enterprises caused large-scale destruction of employment, in order to place the employment of laid-off personnel, the national government implemented the "Re-employment" project, which encouraged urban employment to transform from regular employment to irregular employment. Through more flexible employment mechanism, China realized transformation of employment structure (see Charts 14.2 and 14.3).

In order to promote employment, the national government made energetic preferential policies in fiscal and taxation. During 2000–2009, China's fiscal expenditures in social security and employment increased from RMB 100 billion to RMB 760.668 billion, the total employment population in the society increased from 706.37 to 77.995 million. Through active fiscal policies, the economy realized rapid growth, which in turn promoted the growth of general employment.

---

[10]Hu Angang: China: *Livelihood and Development,* China Economy Press, 2008 Edition, p. 171.

14.2 Forms of Social Livelihood Cost

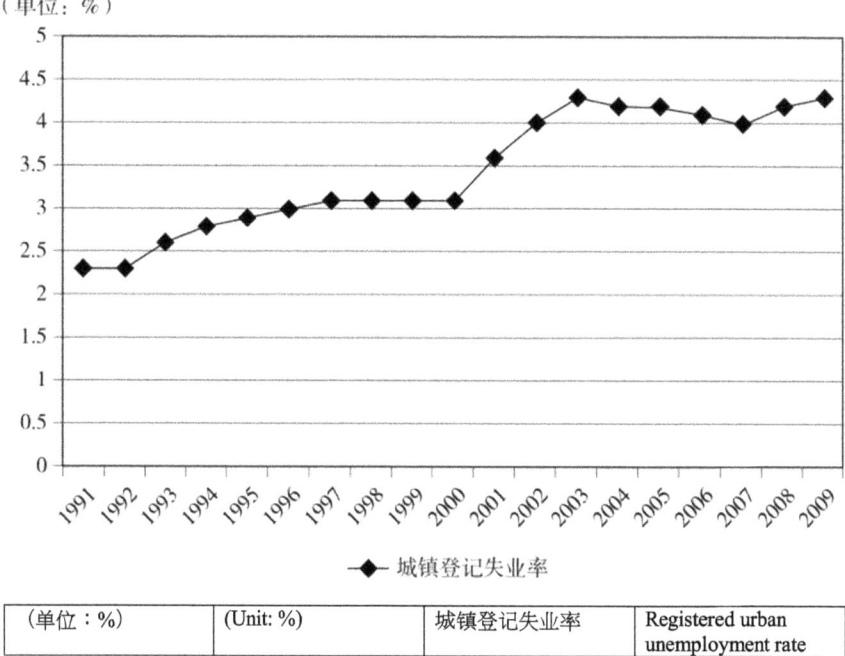

Chart 14.3 Registered unemployment rate in urban areas (1991–2009) (*Source* Concluded from the Statistic Year Book of State Bureau of Statistics and Related Data)

In 2009, the fiscal expenditure of central government in employment increased from RMB 26 billion to RMB 42.6 billion, or 59% of growth, in 2010, the national government appropriated RMB 39.4 billion of special fund of employment subsidy, which was mainly used for occupational introduction subsidy, occupational training subsidy, social insurance subsidy, public benefiting position subsidy, occupational skills evaluation subsidy, special employment policy subsidy, basic living subsidy for internship of graduates of colleges and universities, etc. In 2010, in the *Notice about Supporting and Promoting Employment-related Taxation Policies* issued by the State Administration of Taxation, it clearly proposed to grant preferential taxation policy to the start-ups by unemployed personnel and the enterprises recruiting unemployed and laid-off personnel.

14.2.3 Social Security Lacking Cost

From the establishment of New China to present time, China's construction of social security system has a history of nearly 60 years, especially after the reform and opening-up, China's social security work realized rapid development. The

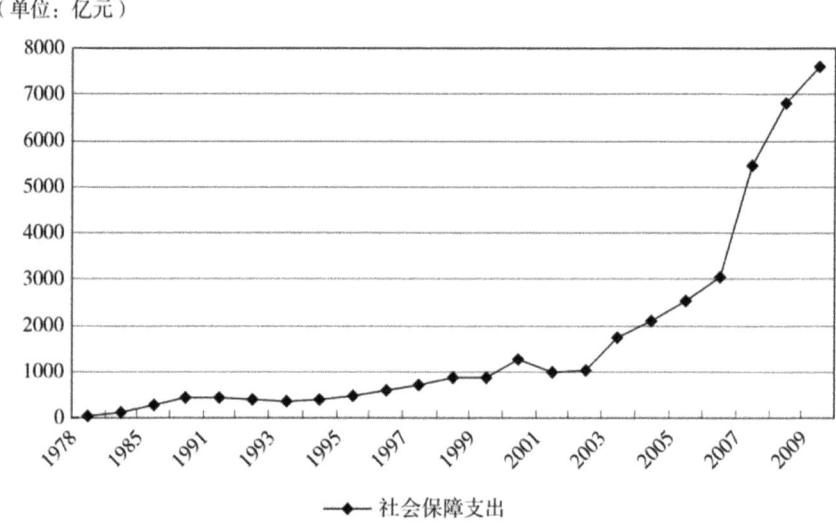

| (单位：亿元) | (Unit: RMB 100 million) | 社会保障支出 | Social security expenditure |

Chart 14.4 Chinese expenditure in social security (1978–2009) (*Note* The data of 2007–2009 was the total amount of national-level fiscal expenditure in social security and employment. *Source* State Administration of Statistics: *Chinese Development Report 2011,* China Statistical Press, 2011 Edition, p. 615)

social security expenditure increased in each year. In 1978, China's national-level fiscal expenditure in social security was only RMB 3.005 billion, it increased to RMB 13.802 billion in 1980, in 2000, the national-level fiscal expenditure in social security reached RMB 125.531 billion. After 2005, the expenditure in social security has been rapidly increasing with annual average growth higher than 20%.[11] (see Chart 14.4).

But on the other hand, with the large-scale increase of the people in Chinese social vulnerable groups and marginal groups, the social security fund can't make ends meet, the coverage of social security and public health are relatively low, the social security capacity is severely lacking. As of the end of 2010, the number of people who had participated in the basic endowment insurance across China was 257.07 million, the number of people who had participated in unemployment insurance was 133.76 million, and the number of people who had participated in medical insurance was 432.63 million, their ratios in China's 779.95 million of total employed population (statistic data in 2009) were 32.9, 17.1 and 55.5%

---

[11] Hu Angang: *China: Livelihood and Development,* China Economics Press, 2008 Edition, p. 160.

14.2 Forms of Social Livelihood Cost

respectively.[12] In individual regions of China, there are still abandoning of deformed or severely ill babies, and there are still severely ill old people who are left in their homes without people looking after them. All these are because of the lack of social security system in China.[13]

Social security level means the degree of social security enjoyed by social members of a nation (region) at certain stage, its main measurement indicator is the ratio of social security expenditure in GDP. There is an "Appropriate" range of social security level in objectivity, social security level being too high or too low would generate negative influence on the operating mechanism of social security system itself and social and economic development.[14] Adopting the research results of Gao (2002),[15] the formula in this book to calculate the loss or cost of social security is as follows:

$$\text{Social security lacking cost} = (\text{Social security fitness} - \text{Actual level of social security}) \times \text{GDP} \tag{14.1}$$

Calculating based on the 10–12% of social security fitness level in China and the 5–6% of the actual level of social security, China's social security lacking cost has reached about RMB 2 trillion.

14.2.4 Income Distribution Imbalance Cost

Since the reform and opening-up, in order to get rid of the restriction of planning economy and absolute "Equalitarianism", China adopted the income distribution system of "Efficiency first and took fairness into account". This system has to certain extent liberated and developed productivity, and promoted the establishment of socialism market economy system and rapid economic development. But it emphasized more on "Efficiency" while neglected "Fairness", and caused continuous widening of unequal degree of fortune distribution among Chinese groups. Despite the year-on-year enhancing trend of the income level among urban and rural residents, but the urban-rural gap is widening, China has quickly changed from a country with relatively equal fortune distribution to one of the countries with

---

[12] State Statistics Bureau: *China Development Report 2011*, China Statistics Press, 2011 Edition, p. 136.

[13] According to China's population data in 2008, China has 109.56 million people at 65 years old or older, accounting for 8.3% of the national population, and has exceeded the international standard of 7%. Aging society also proposes higher requirements on the improvement of social security system.

[14] Gao Liping: *Social Security Level of Shandong Province and Its Proper Selection*, Population and Economy, 5th Issue in 2002.

[15] Gao Liping: *Social Security Level of Shandong Province and Its Proper Selection*, Population and Economy, 5th Issue in 2002.

the most unequal fortune distribution in the world. In 1978, China's average income ratio between urban and rural areas was 2.4:1, in 1983, it was reduced to 1.7:1, which reflected obvious trend of reduction, but after 1997, the income ratio among urban and rural residents quickly widened, during 2000–2009, the income ratio among urban and rural residents expanded from 2.85:1 to 3.57:1.[16] The income gaps among regions, urban and rural areas, industries and social members have been continuously expanding, which is in conflict with the essence and objective of socialism common prosperity, reducing people's sense of happiness and causing extreme imbalance of human development in different regions, or even cause lower recognition of underdeveloped regions towards the policies and systems of central government, so as to increase potential hazards for social instability. This already has related data support in economic transformation cost, so it will not be repeated in this chapter.

In order to relieve the conflict of income distribution imbalance among Chinese groups, the party and government have made continuous improvements on the national income distribution system. The party's 17th Congress proposed that "We need to well handle the relations between efficiency and fairness in both primary distribution and secondary distribution, and focus more on fairness in secondary distribution", and begin to try tertiary distribution mechanism.[17] However, due to the lack of executable mechanism of the system itself and the incomplete national fiscal system, the regulatory function of distribution is still very limited.

14.2.5 Medical Service Development Cost

With the continuous enhancement of economic development and the people's demand for medical service, Chinese government has also extensively increased fiscal investments in the aspect of medical service. In 1978, the expenditure by Chinese government in medical service was only RMB 3.544 billion, in 1985, the number increased to RMB 10.765 billion, since 2000, the growth speed obviously accelerated, during 2000–2008, the government expenditure in health increased from RMB 70.952 billion to RMB 359.394 billion with annual average growth of 20.3%[18] (see Chart 14.5).

Prior to 1990, China had made outstanding performance in health, the key indicators such as child mortality and maternal mortality had extensive reduction. But since 1990, China had both good news and bad news in the progress of health

---

[16]State Statistics Bureau: *China Development Report 2011*, China Statistics Press, 2011 Edition, p. 117.

[17]Tertiary distribution mechanism means that primary distribution is mainly formed by market mechanism, secondary distribution is mainly dominated by government regulatory mechanism, while tertiary distribution means supplementation of social force on government regulation.

[18]Hu Angang: *China: Livelihood and Development,* China Economics Press, 2008 Edition, p. 150.

14.2 Forms of Social Livelihood Cost

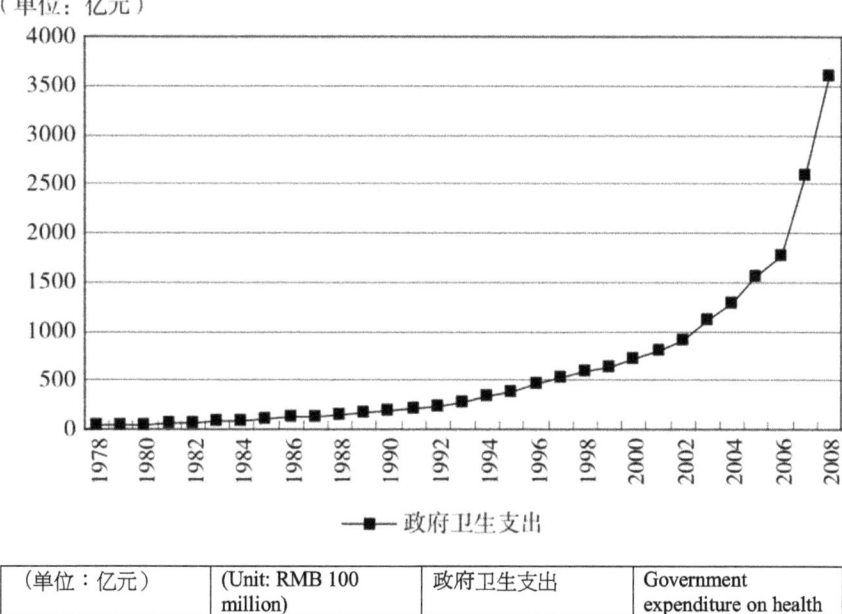

| (单位：亿元) | (Unit: RMB 100 million) | 政府卫生支出 | Government expenditure on health |

Chart 14.5 Government expenditure in health over the years (1978–2008) (*Source* Calculated and sorted out based on the data from *China Statistical Yearbook in 2011*)

indicators, while making certain advancements, China also faced huge challenges, which were mainly reflected in: the progress of child mortality reduction was relatively slow; the reduction of the prevalence rate and mortality rate of tuberculosis lagged behind neighboring countries; the spreading rate of AIDS was astonishing. Even in the relatively good aspects, such as reduction of malnutrition population and reduction of maternal mortality and prevention of malaria, etc. there is still crisis.

Based on the statistical and monitoring data of the "Millennial Development Goals (MIDGs)" of the World Bank for China in 2004: in 1990s, China's speed in reducing child mortality rate was only 2%, which was lower than the average level (2.5%) among developing nations. Compared with the reduction of child mortality rate, China's achievement in reducing maternal mortality rate was relatively big. In 1990s, China's maternal mortality rate reduced by 4.8% on average in each year, which extensively exceeded the average annual reduction by 3.2% among developing nations (Table 14.3).

Table 14.3 New-born baby mortality rate of maternal mortality rate of China over the years

年份	新生儿死亡率 (‰)			孕产妇死亡率 (1/10 万)		
	合计	城市	农村	合计	城市	农村
1991	33.1	12.5	37.9	80.0	46.3	100.0
1992	32.5	13.9	36.8	76.5	42.7	97.9
1993	31.2	12.9	35.4	67.3	38.5	85.1
1994	28.5	12.2	32.3	64.8	44.1	77.5
1995	27.3	10.6	31.1	61.9	39.2	76.0
1996	24.0	12.2	26.7	63.9	29.2	86.4
1997	24.2	10.3	27.5	63.6	38.3	80.4
1998	22.3	10.0	25.1	56.2	28.6	74.1
1999	22.2	9.5	25.1	58.7	26.2	79.7
2000	22.8	9.5	25.8	53.0	29.3	69.6
2001	21.4	10.6	23.9	50.2	33.1	61.9
2002	20.7	9.7	23.2	43.2	22.3	58.2
2003	18.0	8.9	20.1	51.3	27.6	65.4
2004	15.4	8.4	17.3	48.3	26.1	63.0
2005	13.2	7.5	14.7	47.7	25.0	53.8
2006	12.0	6.8	13.4	41.1	24.8	45.5
2007	10.7	5.5	12.8	36.6	25.2	41.3
2008	10.2	5.0	12.3	34.2	29.2	36.1
2009	9.0	4.5	10.8	31.9	26.6	34.0

Source: Calculated and sorted out based on the data from *China Statistical Yearbook in 2011*.

年份	Year	新生儿死亡率 (‰)	Neonatal mortality rate (‰)
合计	Total	城市	Urban
农村	Rural	孕产妇死亡率 (1/10 万)	Maternal mortality rate (1/100 thousand)

14.3 Forming Causes of Social Livelihood Cost

The fundamental cause for generation of livelihood issues is system issue. On macro system level, China has problems in the aspects of deviation of value orientation, insufficient coordination of system, insufficient fairness and justice in systems, insufficient enforcement of systems, etc.; on micro system level, there are flaws in the aspects of employment, education, income distribution, social security

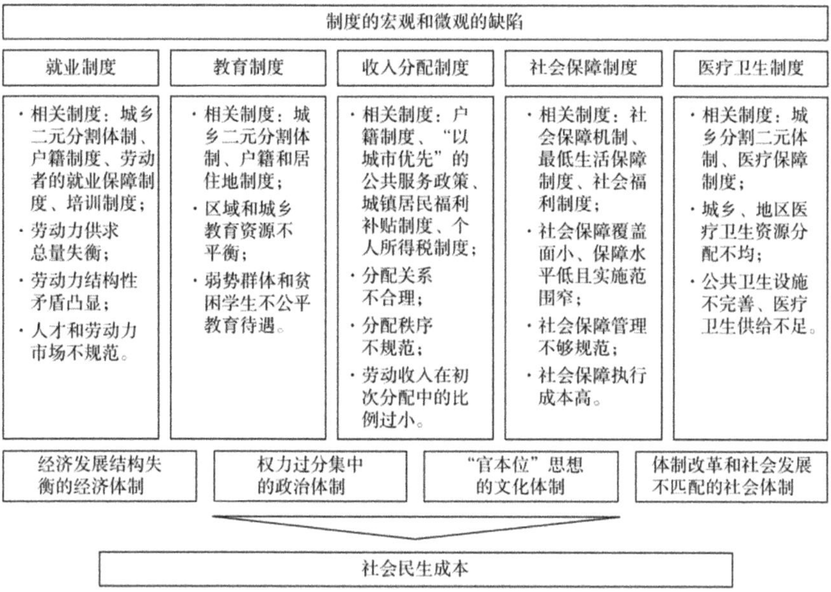

14.3 Forming Causes of Social Livelihood Cost

```
┌─────────────────────────────────────────────────────────────────┐
│                      制度的宏观和微观的缺陷                            │
└─────────────────────────────────────────────────────────────────┘
┌──────────┬──────────┬──────────┬──────────┬──────────┐
│ 就业制度  │ 教育制度  │ 收入分配制度│ 社会保障制度│ 医疗卫生制度│
├──────────┼──────────┼──────────┼──────────┼──────────┤
│·相关制度：│·相关制度：│·相关制度： │·相关制度： │·相关制度： │
│城乡二元分 │城乡二元分 │户籍制度、 │社会保障机 │城乡分割二 │
│割体制、户 │割体制、户 │"以城市优先"│制、最低生 │元体制、医 │
│籍制度、劳 │籍和居住地 │的公共服务 │活保障制度、│疗保障制度；│
│动者的就业 │制度；     │政策、城镇 │社会福利制 │·城乡、地区│
│保障制度、 │·区域和城乡│居民福利补 │度；       │医疗卫生资 │
│培训制度； │教育资源不 │贴制度、个 │·社会保障覆│源分配不均；│
│·劳动力供求│平衡；     │人所得税制 │盖面小、保 │·公共卫生设│
│总量失衡； │·弱势群体和│度；       │障水平低且 │施不完善、 │
│·劳动力结构│贫困学生不 │·分配关系  │实施范围窄；│医疗卫生供 │
│性矛盾凸显；│公平教育待 │不合理；   │·社会保障管│给不足。   │
│·人才和劳动│遇。       │·分配秩序  │理不够规范；│          │
│力市场不规 │          │不规范；   │·社会保障执│          │
│范。      │          │·劳动收入在│行成本高。 │          │
│          │          │初次分配中 │          │          │
│          │          │的比例过小。│          │          │
└──────────┴──────────┴──────────┴──────────┴──────────┘
┌──────────┬──────────┬──────────┬────────────────┐
│经济发展结构│权力过分集中│"官本位"思想│体制改革和社会发展│
│失衡的经济体│的政治体制 │的文化体制 │不匹配的社会体制 │
│制        │          │          │                │
└──────────┴──────────┴──────────┴────────────────┘
                            ▼
                  ┌─────────────────┐
                  │   社会民生成本    │
                  └─────────────────┘
```

制度的宏观和微观的缺陷	Overall and micro flaws of system
就业制度	Employment system
教育制度	Education system
收入分配制度	Income distribution system
社会保障制度	Social security system
医疗卫生制度	Medical health system
●相关制度:城乡二元分割体制、户籍制度、劳动者的就业保障制度、培训制度； ●劳动力供求总量失衡； ●劳动力结构性矛盾凸显； ●人才和劳动力市场不规范。	●Related systems: urban-rural dual system, household registration system, labor employment security system, training system; ●Imbalance of total labor supply and demand; ●Prominent labor structural conflict; ●Irregularity between talent and labor market
●相关制度:城乡二元分割体制、户籍和居住地制度； ●区域和城乡教育资源不平衡； ●弱势群体和贫困学生不公平教育待遇。	●Related systems: Urban-rural dual system, household registration and residence place system; ●Imbalanced education resources between regions and between urban and rural areas; ●Unfair education treatment of vulnerable groups and poor students
●相关制度:户籍制度、"以城市优先"的公共服务政策、城镇居民福利补贴制度、个人所得税制度； ●分配关系不合理；	●Related system: household registration system, "city in priority" public service policy, urban residents welfare subsidy system, individual income tax system;

Chart 14.6 Forming causes of social livelihood cost

●分配秩序不规范; ●劳动收入在初次分配中的比例过小。	●Unreasonable distribution relationship; ●Irregular distribution order; ●Ratio of labor income in primary distribution is too small.
●相关制度:社会保障机制、最低生活保障制度、社会福利制度; ●社会保障覆盖面小、保障水平低且实施范围窄; ●社会保障管理不够规范; ●社会保障执行成本高。	●Related systems: social security mechanism, minimum living security system, social welfare system; ●Small social security coverage, low security level and narrow implementation scope; ●Social security management is not standard enough ●High social security implementation cost.
●相关制度:城乡分割二元体制、医疗保障制度; ●城乡、地区医疗卫生资源分配不均; ●公共卫生设施不完善、医疗卫生供给不足	●Related systems: urban-rural dual system, medical security system; ●Uneven distribution of medical and health resources between urban and rural areas and between different regions; ●Incomplete public health facilities and insufficient medical and health supply
经济发展结构失衡的经济体制	Economic system with unbalanced economic development structure
权力过分集中的政治体制	Political system with excessively concentrated power
"官本位"思想的文化体制	Cultural system with "Official Standard" thinking
体制改革和社会发展不匹配的社会体制	Social system with unmatched system reform and social development
社会民生成本	Social livelihood cost

Chart 14.6 (continued)

system, etc. The backward urban-rural segmentation system, household registration system, education, unemployment insurance system, etc. all affect the solving of Chinese livelihood issues (see Chart 14.6).

Statistics indicate that China's development level in social livelihood undertakings such as science, education, culture, health, social security, etc. is 5–8 years behind the economic development level. The cause of lagged development is no longer the problem of weak foundation, but uneven resource distribution caused from system flaws. On one hand, the government's fiscal investments in social livelihood undertakings are insufficient, on the other hand, the internal resource distribution among social undertakings is also irrational, the demand-supply conflict in the fields of education, health service, social security and other public services is particularly prominent, people have strong appeals over these matters and even called them the "New Three Mountains".

14.3 Forming Causes of Social Livelihood Cost

For Example, the fundamental cause for the generation of education issue lies in the low scale of fiscal expenditure in education and imbalance of distribution structure.[19] Statistics indicated that tuition of colleges and universities in recent years has been increasing by 20% on average in each, the burden of higher-education cost on students and their families grew from 1.65% of per capita GDP to 51.87% of per capita GDP; 77% of education investments were used in cities, rural population, which accounts for 60% of total population, only obtained 23% of education investments[20]; fiscal expenditure in education has preferential policy to key schools, which made the originally limited education resources to be even more centralized, the segmentation between key schools and normal schools becomes too excessive; the unreasonable college entrance exam system among different regions and the lack of long-term education management system made the problem of "Post-junior high school education" of the children of migrant workers becomes increasingly prominent. These rigid household registration system and education system caused the "Segmentation" and discrimination phenomenon of the children of migrant workers, which severely impeded the solving of the education fairness problem of the children of migrant workers.

The drawbacks of Chinese medical management system become increasingly prominent in social development. The uneven distribution of health service resources between urban and rural areas and among different regions, the incompletion of public health service facilities, and severe lack of medical service supply have caused severe difficulties for the people's medical demands. The "SARS" erupted in China in 2003 fully exposed this problem. According to statistics, the beds, equipments, medical personnel and other medical resources owned by Chinese public hospitals still account for over 90% of medical resources across China, and 30% of which are concentrated in big hospitals. The official data released in 2004 indicated that 44.8% of urban population and 79.1% of rural population don't have any medical security.

With the continuous deepening of reform and the establishment of market economy system, the original social security system is no longer able to meet the demand of economic development. China's social security capacity is severely lacking, and insufficient to resist the risks of economic marketization and internationalization. The social security system hasn't fully covered the entire urban employees, some foreign invested enterprises, individual private enterprises and public institutions are not included in the scope of coverage. The ratio of rural population with insurance participation is even lower. According to the statistics of the Rural Social Security Department of the Ministry of Labor and Social Security,

---

[19]Distribution structure of education expenditure is mainly reflected in the investment in compulsory education and non-compulsory education, distribution structure of education expenditure among different governments, distribution structure in higher education, medium education and junior education, and distribution structure in geographic location.

[20]Jia Pinrong: *Lagging Reform and China's Education Reform, the Eight Major Unjust Confuse the Vulnerable Groups,* China Economic Times, Sep 11, 2007.

as of the end of 2005, there were only 54.42 million Chinese farmers participating in the rural social endowment insurance, accounting for 5.8% of total farmers, the accumulated fund was RMB 31 billion.

14.4 Countermeasures for Social Livelihood Cost

14.4.1 Establish Public Fiscal System, Expand Social Livelihood Investments and Build Service-Type Government

The national and local governments shall expand the investments in social livelihood undertakings, establish the system to fairly distribute public education resources, form the training policies that could enhance the competitiveness and skills of common social members, particularly the difficult groups; guarantee social safety net with public fiscal means; establish public income system; establish nationally unified basic social security system. The government shall further strengthen the functions of social management and public service, strengthen the transparency and public participation degree in forming public policies, improve the national macroeconomic regulating system, as well as enhance scientific nature and predictability of macroeconomic regulation, strengthen the pertinence and flexibility, put more fiscal force on the guarantee and improvement of livelihood, and build a service-type government.[21]

14.4.2 Improve Social Redistribution System Policies and Enhance Social Equity and Efficiency

China shall improve the redistribution policy with the core of tax regulation and social security system in order to prevent excessive concentration of fortune to few social members and excessive loss of interests of partial social members. The government shall strengthen the degree of income redistribution, accelerate medical insurance, endowment insurance, unemployment insurance and other social security systems, and strengthen the supervision of various enterprises in China in their fulfillment of social security obligations. China shall establish fair and cooperative labor-capital relationship.[22] Chinese government shall accelerate the legislation of

---

[21]Service-type government means the government that is under the guidance of the citizen standard and social standard idea, within the framework of overall social democratic order, through legal procedures, based on the will of citizens, for the aim of citizen service, and bear the responsibilities of service.

[22]Distribution rate means the ratio of total labor compensation in GDP.

minimum wage, include education and medical expense or other labor capital investment expenses into the scope of minimum wage security, determine employees' minimum wages based on the factors such as family members, living expense, price change, education and medical expense, guarantee the basic living demand of workers' families and the demand for human capital accumulation such as basic education, basic medical service, etc.; China shall also improve the social insurance system for the endowment, unemployment and medical service of all labors, enlarge the labor supervision degree of labor-capital relationship management, so as to feasibly protect the rights and interests of labors.

14.4.3 Enlarge Investments in Fiscal Education Expenditure and Guarantee the Balanced Distribution of Education Resources in Urban and Rural Areas

Based on the various problems existing so far, the national government shall further enlarge investments in education expenditure, more importantly, the current distribution system of education expense shall be improved. In order to realize the 4% target on time,[23] based on strict implementation of the mandatory growth requirements of education expenditure,[24] we shall also further enhance the ratio of fiscal education expenditure in public fiscal expenditure, meanwhile, we shall actively explore the sources and channels of fiscal education expenses, including the unification of domestic and foreign enterprises and individual education surcharge system, comprehensively levy local education surcharge, and proportion education fund from land transfer incomes. Besides, the education system should get rid of the restriction of household registration system to guarantee the balance of education resources in urban and rural areas. China shall gradually eliminate urban-rural difference through the reform of teacher personnel system, use the talent selection and diversified talents evaluation system reform to provide the students with knowledge and abilities with more fair opportunities; meanwhile, China shall realize the balanced development of compulsory education, energetically develop occupational education, accelerate the implementation of national student financial aid system, guarantee the children of migrant workers in cities to receive compulsory education as well as other means to promote education equity, so as to solve the livelihood issue of education.

---

[23]The *Planning Outline of National Long-term Education Reform and Development (2010–2020)* clearly indicated that China shall realize the goal in 2012 that the ratio of national fiscal expenditure on education accounts for 4% of the GDP.

[24]The *Education Law* stipulated that the growth of government fiscal appropriation on education on each level of People's Government shall be higher than the growth of current fiscal income.

14.4.4 Improve Market Employment System and Create Fair Employment Environment

Based on the characteristics and national situation of labors in China, China shall energetically develop labor-intensive industry and service industry, develop flexible employment forms, use the improvement of labor employment training system to encourage various forms of employment and start-ups. The government shall consider the promotion of employment growth and reduction of high unemployment rate as the most important economic growth goal and social stability goal. The government shall feasibly strengthen the employment aid to unemployed personnel in urban area, provide them with position subsidy, training subsidy, etc.; create a fair employment environment, eliminate employment discrimination, so as to provide fair employment environment for the labors transferred from rural areas and graduates from colleges and universities.

14.4.5 Enlarge Government Investments in Public Health Undertakings and Establish a Civil and Integrated Medical Service System

First of all, we shall ensure government investment in public health undertakings, based on the difference of the uncertainty of individual disease risks and individual economic abilities, the government shall realize social mutual aid through fiscal investment; secondly, the government shall break various boundaries of urban-rural area and ownership in order to establish a civil and integrated medical service system, avoid the excessive concentration of medical resources to cities and developed regions, and increase the protection of farmers' rights and interests through urban-rural integrated medical service system construction. The government shall establish the appropriate development ratio of public and profiting medical institutions, the public medical institutions shall directly held by the government, which shall also bear the functions of public health service and basic medical service, which shall not have goal and action of profiting, and their income and expenditure shall be strictly separated.

Chapter 15
Cost of Population Change in Chinese Society

15.1 Definition and Connotation of Population Change

Population change means the change of population in quantity and structure along with change of space and time. The process of population development is a process that is closely connected with the change of social and economic changes.[1] In the 20th Century, China's population development could be basically divided the two main stages of first 50 years and second 50 years, these two stages reflected two changes of Chinese population: firstly, the transformation from old population[2] to new population[3]; secondly, the transformation of inherent population[4] to active

---

[1] Development of productivity will inevitably cause a series of changes in social, economic, political and cultural conditions, and eventually cause the changes in population birth rate, mortality rate, and population natural growth rate, so as to form the population reproduction type of different societies, histories and different development stages. The effect of population transition on economic development is generally including three stages: Stage I is the stage of population burden of young people; stage II is the stage of population bonus; and Stage III is the stage of the burden of aging population.

[2] The so-called old population means the population in the first 50 years, which basically continued the mode of high birth rate, high mortality rate and low natural growth rate. According to studies, old China had birth rate of 38‰, mortality rate of 33‰ and natural growth rate of about 5‰.

[3] The so-called new population means the population in the later 50 years, which gradually completed the transition of high birth rate, low mortality rate, high natural growth rate and low birth rate, low mortality rate and low natural growth rate.

[4] The so-called inherent population means that during the first 50 years, the birth behaviors of population were in inherent state and subject to higher influence by the social and economic environment.

population.[5] In worlds' history, China has always been the country with the most population, large population is not only the basic national situation of Chinese society, but also the basic factor of Chinese modernization. China spent 60 years to quickly realize the transformation of modern population at extremely low income and low income conditions, and transformation from "high birth rate, high mortality rate, low natural growth rate" to "low birth rate, low mortality rate and low natural growth rate", such population change is the result of integration between compulsory change and economic development and social advancement, which created "Population Bonus" for economic take-off.

In the process of China's fast population change, China not only paid the social turmoil cost that was caused from fast population increase since the establishment of New China, but also faced the development risks at current stage in population structure imbalance, and approaching of aging times. No matter in female-sum fertility rate and other main population indicators, China has quickly transformed from a typical underdeveloped nation type to a relatively developed or developed nation type. This determines the trend of Chinese population structure change: aging population, low birth rate, gender imbalance, etc. Therefore, the cost of population change in Chinese society could be concluded as population increase cost, aging cost and population management cost.

An issue that China should highly regard today is to prevent population change from forming strong negative impact on the economic and social development, this needs us to rethink the section of previous population policy and national development strategy, so as to enable policy adjustment to effectively adjust changes of population, economy and society.

15.2 Forms of the Cost of Population Change in Society

It is believed in this book that the cost of population change in society includes population increase cost, aging cost and population management cost. For the detailed contents, please refer to Table 15.1.

---

[5]The so-called active population means that the population's birth behaviors are always under the influence of certain theory, such as the "human hand theory" in 1950s that stressed the infinite creativity of human, which accelerated the operation of the train of Chinese population; the "population theory" proposed in 1970s that saw the side of human as consumers, which set theoretical foundation for the practice of population control; the "human brain theory" proposed in 1990s that proposed to "reduce population quantity and enhance population quality", while controlling the quantity of population, it emphasized the enhancement of internal population quality.

15.2 Forms of the Cost of Population Change in Society

Table 15.1 Cost of Chinese population change

Classification of cost	Forms	Specific contents
Cost of social population change	Cost of increasing population	Pressure of population growth and family planning
	Cost of aging	Cost of increasing total endowment of society, cost and decreasing ratio of national fiscal investment
	Cost of population management	Cost of gender imbalance and migrant workers flow

15.2.1 Population Increase Cost

China is the most populous nation in the world. As estimated, China's total population in 2005 (1.30756 billion) was equivalent to 20.4% of the world's total population in 2004 (6.345 billion).[6] Such a large population, on one hand, provided abundant labor resources and huge domestic consumer market for the economic and social development of China, and on the other hand, indeed brought great and constant pressure on China's resources and environment, as well as caused many social problems.

During certain period after the establishment of New China in 1949, China once encouraged more birth. In 1950s and 1960s, China's population surged due to this reason, it caused economic and social influence was reflected as employment difficulty, "Send down youth to the countryside" movement and political turmoil in the 1960s and 1970s, the vast and surging population formed sharp conflict with resource shortage and backward economic development, the great population pressure and low population quality all became the biggest restraining condition for the launch of industrialization.[7]

During 1959–1961, China encountered severe attack of natural disasters, plus some mistakes of work policies, a disastrous nation-wide great famine was erupted, countless people died due to lack of food. According to statistics, millions of people died for this reason.

At the beginning of 1970s when the family planning policy was comprehensively promoted, the total population in China had reached 830 million, which increased by 53.2% within 20 years with annual average growth rate of about 2.1%. Since 1970s, due to the implementation of effective family planning policy, the population growth rate continuously declined, the total population in Mainland

---

[6] World bank, 2006, *World Development Report 2006*, Oxford University Press, pp. 292–293.
[7] On Jun 30, 1953, China implemented the first national census since the establishment of New China. The census result indicated that the Chinese population was over 580 million, which was much higher than the population scale of developed nations at preliminary stage of industrialization, and was 4.8 times of the total population of the 12 Western European countries in 1820, 52 times of the total population of the US in 1820 (10 million people), and 15.7 times of the total population of Japan in 1870 (34.44 million people).

China reached 1.328 million in 2008, accounting for about one fifth of the world's total population. The excessively fast population growth not only affected the enhancement of per capita income, the supply and food and residence, the meeting of demands in education, labor and employment, and environmental problems all became the social problems that urgently needed solving. Because the increase of population, the excessive land usage destructed the originally vulnerable ecological balance, despite Chinese government has made great efforts to restore ecological environment, such as the creation of "Three North" protective forest, the trend of deteriorating ecological environment still hasn't been effectively contained.[8]

Since 2000, China had entered the peak period of population bonus, the occurrence of population bonus is closely connected with China's population policy since 1970s,[9] low population growth rate made China enter the population bonus period in advance, but this also comes with great economic cost, it made Chinese population bonus period to end in advance and enter the population burden period in advance, made China "Aging before getting rich", enter the aging society at the stage of medium and low income, and enter the relatively severe aging society at the stage of medium income.

During the "Tenth Five Year Plan" period, the national per capita population and family planning expenditure reached RMB 10, in 2010, the national per capital population and family planning expenditure increased to RMB 22. In 2010, Chinese government invested RMB 3.46 billion in total to reward and support rural family planning families, and coordinate and promote the preferential treatment to the family planning families in the policies such as new agricultural insurance, collective forest right reform, land acquisition compensation, poverty elimination and development, so as to enhance their welfare and treatment. This symbolized the preliminary formation of China's interest-oriented policy system in population and family planning.

---

[8]Chen Jianping: *China's Issue of Population Scale and Structure and Related Policy Adjustment*, Population and Development, 2nd Issue in 2009.

[9]A major characteristic of the evolution of Chinese population system is that the trend of excessively fast total population growth was effectively controlled, and entered the "Three Low" stage of "low birth rate, low mortality rate and low growth rate" on the condition of low income; the ratio of working-age population (meaning the population at 15–64 years old) in total population reached the highest level in history (72.0% in 2005), so it formed the historic stage of rapid growth of human resource, which is referred to as "Population (Birth Control)" Bonus.

15.2.2 Aging Cost

Since 1980s, China's population birth rate has been extensively reduced. It made China become an aging society. In 2000, the ratio of Chinese population above 60 years old was 6.96%,[10] which approached the standard of aging society at 7%; in 2005, the ratio of old people was 7.5%, equivalent to the ratio of Japan in 1973[11]; in 2010, this ratio increased to 8.87%.[12] Besides, according to the estimation of Chinese old population in coming decades by the *National Population Development Strategy Research Report*, in 2020, the number of old people above 60 years old will reach 234 million, and the ratio will increase from 9.9% in 2000 to 16.0%; the population of old people above 65 years old will reach 164 million, and the ratio will grew from 7.0% in 2000 to 11.2%. It is estimated that the aging population's peak platform will form in late 2040s, the population of old people above 60 years old will reach 430 million, accounting for 30%; the population of old people above 65 years old will reach 320 million, accounting for 22%, by then, there is 1 old person in every 3–4 people (Refer to Chart 15.2). In 2020 and 2050, the population of old people above 80 years old will reach 22 and 83 million respectively (Chart 15.1).

The aging of working-age population caused by aging population is not good for adjustment of industrial structure. In large-scale industrial transfer, it requires many young labor forces with new-type technical structure, if the young labor force is unable to meet the demand, we could only re-train and re-educate the aged labor force. Due to the slow speed of old labor force in accepting new technology, the expense of re-training will be high and difficult to adapt to the requirement of industrial structure adjustment, therefore, it will definitely result in structural unemployment and restrain the adjustment of industrial structure.[13]

Aging population will increase the economic burden of China's total population, the elderly dependency coefficient will quickly enhance. Since 1980s, China's growth of endowment fund has been relatively fast, in 1979, there were 5.96 million of veteran, retired and resigned employees, in 1989, the number increased to 22.015 million, it increased by 2.7 times within 10 years, the insurance welfare expenditure for the veteran, retired and resigned employees accordingly increased from RMB 3.25 billion to RMB 38.26 billion, increasing by 10.8 times.[14]

---

[10] Zheng Jingping: *Impact of "Silver Wave" on China's Endowment Insurance System and Countermeasures*, Statistical Research, 1st Issue in 2002.

[11] Hu Angang: *China's Mid-and-long-term Population and Comprehensive Development Strategy*, Tsinghua University Press, 5th Issue in 2007.

[12] State Statistics Bureau: *China Development Report 2011*, China Statistics Press, 2011 Edition, Page 86.

[13] Lu Zhiguo: *Brief Discussion of the Influence of Aging Population on the Adjustment of China's Industrial Structure*, Journal of Shenzhen University, 2nd Issue in 2001.

[14] Ding Junqiang: *Analysis and Countermeasure of China's Aging Population in the 21st Century on Sustainable Development*, Theoretical Monthly, 10th Issue in 2002.

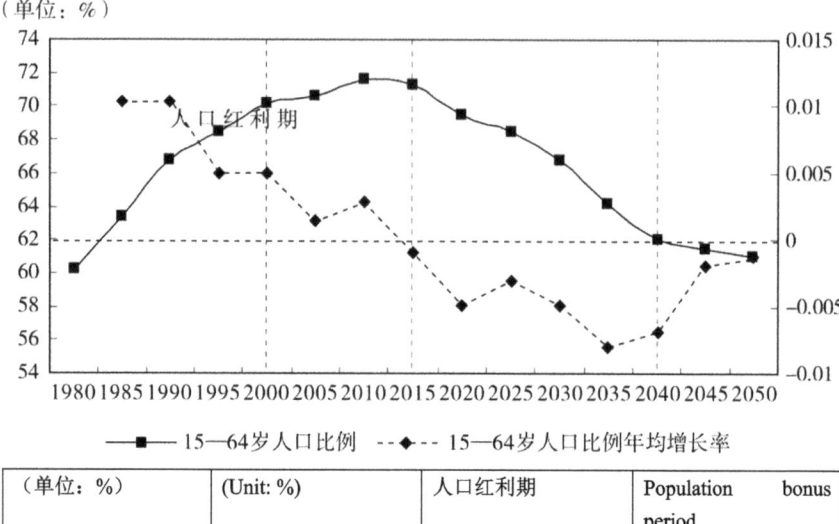

（单位：%）	(Unit: %)	人口红利期	Population bonus period
15-64岁人口比例	Ratio of population at 15-64 years old	15-64人口比例年均增长率	Annual growth rate of the ratio of population at 16-64 years old

Chart 15.1 Change of working-age population in China (1980–2050)

In 2000, 9.1 labors supported one old person on average, in 2020, 5.9 working-age people support one old person on average, in 2050, 2.7 working-age people support one old person on average.[15] It can be said that China's population liability period is quickly approaching (Chart 15.2).

Due to the deepening of aging degree, the general dependency ratio of the society is increasing, the endowment and medical expenses used on old population increases, the aging society would increase the part of national income used in consumption, which would definitely cause the expansion of consuming fund in national income and decrease of accumulation fund, so as to make the fiscal capacity that supports the long-term national development to be restricted, which would directly affect the sustainability of economic growth. From 1978 to 1993, the social security and welfare expense of Chinese retired employees increased from RMB 1.73 billion to RMB 91.37 billion. While estimated by the Ministry of Labors, in 2050, the total number of retired employees in China will surpass 100 million, the retirement expenses paid in each year will reach RMB 18 trillion, which is over 20 times of that in 1993.[16] Surveys in different countries indicated

---

[15] Hu Angang: *China's Mid-and-long-term Population and Comprehensive Development Strategy*, Tsinghua University Press, 5th Issue in 2007.

[16] Xiao Zili and Zhou Shuangchao: *Chinese Population and Sustainable Development*, Chinese Population Press, 1998 Edition, pp. 164–166.

15.2 Forms of the Cost of Population Change in Society

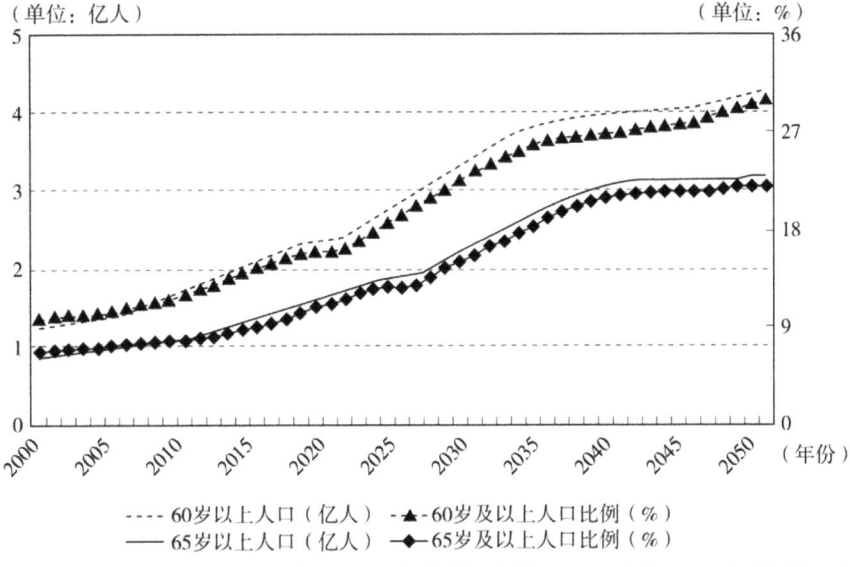

(单位: 亿人)	(Unit: 100 million people)	(单位: %)	(Unit: %)
（年份）	(Year)	60岁以上人口（亿人）	Population older than 60 (100 million people)
60岁及以上人口比例（%）	Ratio of population at no less than 60 years old (%)	65岁以上人口（亿人）	Population older than 65 (100 million people)
65岁及以上人口比例（%）	Ratio of population at no less than 65 years old (%)		

Chart 15.2 Estimation of Chinese aging population in the future. *Source* World Bank, World Development Indicator 2006; sorted out from the Statistical Yearbook data of the State Administration of Statistics

that the medical expense of old population is usually about 3–5 times of that of adults, besides, with the increase of people's living standard, the average consumption fund of old people also increases accordingly. From 1986 to 1997, the number of veteran and retired people in China increases by 5.97% in each year, while the average growth rate of their insurance and welfare expenses increases by 23.96%.[17] (Table 15.2).

---

[17]Yang Zhongxin: *Aging and Industrial Structure Adjustment*, Guangxi People's Publishing House, 2000 Edition, pp. 30–33, 37–47.

Table 15.2 Age structure of chinese population (1990–2040, ratio in total population)

(Unit: %)

年份	0—14 岁	15—64 岁	劳动力人口	老年抚养人口 a	老年抚养人口 b	抚养人口 a	抚养人口 b
1990	27.69		60.06		10.34		38.02
1995	26.70	67.01	60.52	6.30	13.01	32.91	39.64
2000	25.31	67.67	60.95	7.03	13.81	32.25	39.05
2005	22.40	69.97	62.55	7.63	15.09	29.96	37.45
2010	20.74	71.08	62.40	8.18	16.90	28.85	37.60
2015	19.75	70.90	61.49	9.35	18.81	29.03	38.51
2020	19.30	69.26	59.60	11.44	21.15	30.68	40.40
2030	17.57	67.09	54.72	15.34	27.76	32.84	45.28
2040	16.51	63.43	53.43	20.05	30.10	36.49	46.57

Note old population with financial support a means the population above 65 years old; odl population with financial support b means the population of male above 60 years old and female above 55 years old; the population with financial support a means the sum of population at 0-14 years old and the population above 65 years old; the population with financial support b means the sum of the population at 0-14 years old and the population of male above 60 years old and female above 55 years old. The national population data in 1990 was the data from the fourth national census

Source Sorted out and calculated from the related data in the *2011 Chinese Statistical Yearbook*

年份	Year	0-14 岁	0-14 years old
15-64 岁	15-64 years old	劳动力人口	Population of labors
老年抚养人口 a	Supported older population a	老年抚养人口 b	Supported older population b
抚养人口 a	Raised populated a	抚养人口 b	Raised populated b

15.2.3 Population Management Cost

Chinese population is transforming from traditional agricultural population to modern industrial and commercial population, large quantity of agricultural population is entering cities,[18] which made great contribution to the relief of the employment pressure of large excessive labor force in rural China, supplementation of the working-age population in cities, coverage of shortcomings in urban industrial structure and acceleration of urban construction. But on the other hand, the vast scale of floating population also brings vast pressure on the infrastructure,

---

[18] According to statistics, currently China has about 210 million floating population, including 150 million of migrant workers and 60 million of other floating population.

transportation, environment, employment and security of cities, and brings great challenge to urban population management, especially the "Marginal" state[19] of "Migrant Workers" in cities makes them become important unstable factor of the society. The chaotic increase of floating population enhances the probability of illegal and criminal activities that jeopardizes social security. The places with concentration of floating population are usually the places with frequent occurrence of criminal cases, according to statistics, in China's economically developed regions, the ratio of criminal cases committed by floating population accounts for over 60% of all cases, this number is as high as 80–90% in certain regions.[20]

The social price brought by Chinese's gender imbalance also constitutes one of the factors of population management cost. Since 1980s, the gender ratio of new-born Chinese has been continuously enhancing, according to the Fifth National Consensus in 2000, during 1990–1999, the gender ratios of China were: 111.4, 113.5, 114.6, 115.2, 116.6, 117.8, 118.5, 120.4, 122.1, 117.8 respectively.[21] In 2005, China's male-female gender ratio was 106.30, the birth gender ratio was 118.88, which had severe deviation from the normal value of 104–107. The biggest influence of gender imbalance is that it could cause crimes and social turmoil. As indicated by statistical materials of the Ministry of Public Security of China, there were 65,236 cases of apprehended criminals recorded for women trafficking during 1990–1991.[22] Gender ratio imbalance brought the problem of "Marriage Squeeze", which increases the employment pressure of female, enlarges the pressure of social endowment and causes impact to family and society.

---

[19]"Marginal" state means that farmers enter city and engage in non-agricultural works, but never changed the identity as farmer, they work in city, but don't have urban household registration, so they don't receive social security; they are looking forward to cities, but they haven't been recognized and accepted by cities, they are at margin of industry, margin of urban and rural areas and margin of system.

[20]Chen Shangkun: *Law Breaking and Crime Problems of Floating Population and Thinking of Countermeasures*, Academic Journal of Population, 5th Issue in 2004.

[21]Yu Jinghui: *Causes and Countermeasures of the Gender Unbalance of Newborn Population in China*, Academic Exchange, 1st Issue in 2008.

[22]Chen Shangkun: *Law Breaking and Crime Problems of Floating Population and Thinking of Countermeasures*, Academic Journal of Population, 5th Issue in 2004.

15.3 Forming Causes of the Cost of Population Change in Society

The most important factor of the severe imbalance of Chinese population structure is the rapid decrease of birth rate caused by birth policy. China has implemented family planning policy for over 30 years, which controlled the excessively fast growth of population, but the decrease of birth rate directly caused the severe imbalance of reproduction of population itself, caused imbalance of age and gender structure, so as to result in a series of problems in social and economic development.

In 1950s, China obviously underestimated the complex influence of population growth, the policy of encouraging birth back then was because the government didn't realize that China at that time was still in agricultural society, the excessively fast population production would cause shortage of material production. While in 1960s and 1970 s, after the government felt the pressure of excessively fast population growth, China began to implement population control policy in early 1970s, ten years later, it evolved into the family planning policy that we have been adopting in recent twenty years. China's population policy adjustment in the second half of 20th Century also lagged behind the change of population, economy and society.

Urbanization is an important content of national development and an important way to reduce the urban-rural gap. However, in China's urbanization process, the separated urban-rural dual structure severely affected the citizen interest of farmers. Because of China's current dual structure social system, the floating population are living outside of system, they don't have equal position in the labor-capital relationship and the relationship with locals, the urban-rural separated employment management system, household registration system, social security, etc. made them unable to effectively guarantee their economic, political and cultural rights and interests, so as to result in more social conflicts and social confrontations. As for the system, China's laws and regulation on population management are still incomplete, although different places have formed some local regulations based on their reality, but the updating of system still lags behind the demand of labor flowing, and there is even omission of management.

15.4 Countermeasures for the Cost of Population Change in Society

The detailed countermeasures are concluded as follows Chart 15.3.

15.4 Countermeasures for the Cost of Population Change in Society

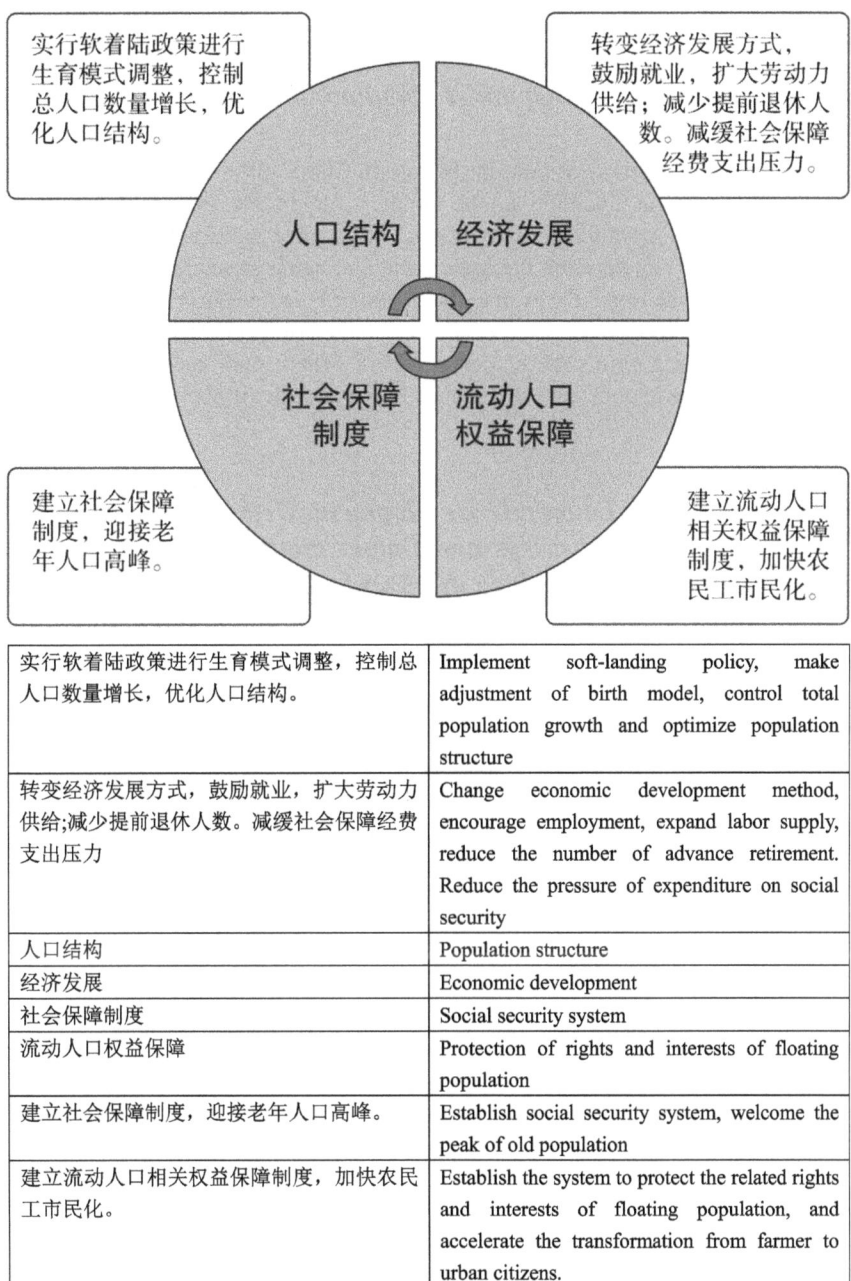

实行软着陆政策进行生育模式调整，控制总人口数量增长，优化人口结构。	Implement soft-landing policy, make adjustment of birth model, control total population growth and optimize population structure
转变经济发展方式，鼓励就业，扩大劳动力供给；减少提前退休人数。减缓社会保障经费支出压力	Change economic development method, encourage employment, expand labor supply, reduce the number of advance retirement. Reduce the pressure of expenditure on social security
人口结构	Population structure
经济发展	Economic development
社会保障制度	Social security system
流动人口权益保障	Protection of rights and interests of floating population
建立社会保障制度，迎接老年人口高峰。	Establish social security system, welcome the peak of old population
建立流动人口相关权益保障制度，加快农民工市民化。	Establish the system to protect the related rights and interests of floating population, and accelerate the transformation from farmer to urban citizens.

Chart 15.3 Countermeasures for the cost of population change in society

15.4.1 Implement Soft-Landing Policies to Undertake Birth Model Adjustment, Control the Growth of Total Population and Optimize Population Structure[23]

Over a considerable period of time in the future, China still needs to consider the control of population quantity as the primary target. As expressed by Zhai Zhengwu: "Just in the comprehensive analysis of quantity and structure, we believe that in the comparison between the scale issue and aging structure issue of population, the scale issue is still the primary issue currently and in the coming ten years, while structure is the secondary issue."[24] We should strictly control the increase of population from long-term strategic perspective.[25] When conditions allow, we will enter negative growth and gradually reduce the total scale of population.[26]

15.4.2 Transform Economic Development Method, Actively Encourage Start-Ups and Employment, Expand Labor Supply, Reduce the Number of Advanced Retirees, and Relieve the Pressure of Social Security Expenditure

Macroeconomic policies shall put the increase of employment as the key, optimize industrial structure, open new employment channels, continuously implement preferential distribution policies. China shall actively develop the tertiary industry, accelerate the construction of small towns, expand the employment space of surplus labor forces in urban and rural area, and promote the effective transfer of surplus labor force in the primary industry.[27] China shall reform the retirement policy, implement more flexible employment system, including independent employment,

---

[23]From the goal of population development, we need to use China's system advantage to actively and forwardly design the national goal of population development, including the following three goals: firstly, moderate population scale, secondly, rational population structure to maintain the population of labors to be no less than 50% for a long term, thirdly, population of high quality and higher productivity.

[24]Zhai Zhenwu: *Analysis of Chinese Population Scale and Age Structure Conflict*, Population Research, 3rd Issue in 2001.

[25]Song Haiyuan: *Strictly Controlling Population Growth is an Important Guarantee to Shorten the Gap with the Eastern Region*, Economic Reform, 6th in 1996.

[26]Zhai Zhenwu: *Chinese Population Development: New Challenges and Choices*, Theoretical Vision, 9th Issue in 2007.

[27]In the adjustment of industrial structure, we should support and develop the industries with great potential; in the adjustment of investment structure, we should enlarge the investment ratio of these industries, energetically develop the small and medium-sized enterprises, open private investment field; enable these industries to become new economic growth point and new employment base.

irregular employment, contractual employment, family employment, etc., as for retirement age, we shall consider the suitability to the trend of aging society.[28]

15.4.3 Establish Social Security System and Face the Peak of Elderly Population

China shall establish a comprehensive, multi-layered and unified security system that is based on family security, integrating social security and social aid, and using social and community service as auxiliary means. China shall further expand the coverage of endowment security, and establish the "Safety Net" of national and people's development. The government shall strengthen the management of endowment fund, and enhance the income efficiency of endowment fund. We shall fully develop labor resources, enhance the utilization rate of labor resource, enlarge the accumulation degree of labor force, and strengthen the flowing of labor force.[29] China shall actively develop aging industry, with the acceleration of aging degree, increase of elderly people as well as the expansion of demanding scope of old people and their demanding layers, the market share of aging population will continuously enhance, therefore, aging industry is likely to become new growth point of the national economy, so as to promote national economic development.[30]

15.4.4 Establish the System to Protect the Related Rights and Interests of Floating Population and Accelerate the Citizenization of Migrant Workers

To solve the problems of farmers and floating population, its fundamental reform thinking direction is to break the management and public service system with household registration as the boundary, the national government shall plan as a whole, comprehensively coordinate, and form the social policy system as soon as possible that is cross-departments, cross-regions and comprehensively covering the group of migrant workers. China shall break the urban-rural dual system and create conditions to promote the conversion of migrant workers from depending on rural

---

[28]Since 1950s, China's retirement policy regulated that the legal retirement age is 60 years old for male and 55 years old for female. With the continuous increase of average expected life expectancy of population, we should consider modifying this outdated retirement policy, so as to respond to the quickly aging population.

[29]A simulation by World Bank indicated that under the presumption of transferring 1, 5 and 10% of the agricultural labor force, the total GDP will be increased by 0.7, 3.3 and 6.4% respectively.

[30]The so-called aging industry means the industry formed from the growth of old people's consumer market demand, including all the production, operation, service, facilities and economic activities regarding the special demands of old people.

land security to depending on urban public service security. Including the establishment and improvement of floating population social security system, establishing work-related injury insurance system, critical illness insurance system and endowment insurance system[31]; creation of excellent policy environment for population flowing, migration and residence, grant them the rights to freely select works and choose resident places, and to equally and friendly treat each other.

---

[31]Currently, many big cities have established social security system of floating population, which has to great extent relieved the development imbalance caused from urban-rural dual system and social and economic system, and it could also solve the concerns of migrant workers and reduce their tendency to commit crimes.

Chapter 16
Chinese Social Management Cost

16.1 Definition and Connotation of Social Management

Social management, as one of the functions of government, means that a government, through forming specific, systematic and standard social policies and regulations, manages and normalizes social organizations, cultivates rational modern social structure, adjusts social interest relationship, responds to social appeals, solve social conflicts, maintains social justice, social order and social stability, creates rational, tolerant, harmonious and civilized social atmosphere, and establishes a social environment for coordinated development of economy, society and nature.[1] The objective of social management is to promote social development, focusing on the design and creation of social order, it involves more on social relationships and involves mutual relationships between individuals, and between individual and society (organization).

Since the reform and opening-up, Chinese government has been highly regarded the strengthening of the government's social management function, continuously reforming and improving the social management system. The *Explanation about the Institutional Reform Solution of the State Council*[2] issued at the First Session of the 9th NPC in 1998 and the 16th National Congress of the Chinese Communist Party[3] both clearly considered the strengthening of social management as a target of governmental function transformation. In 2004, the State Council Premier Wen Jiabao stressed "Focusing more on the fulfillment of social management and public

---

[1] Wu Shouyong: *How Enterprises Enhance the Financial Fine Management Capability*, Economic and Trade Time, 11th Issue in 2007.
[2] On Mar 6, 1998, Luo Gan, the Secretary General of the State Council, made a report on the First Session of the 9th National People's Congress, the *Explanation about the Institutional Reform Solution of the State Council*, which proposed to "feasibly transform government functions into macroeconomic adjustment, social management and public service".
[3] The 16th National Congress of the CCP proposed to "improve the government's functions in economic regulation, market supervision, social management and public service".

service functions, and putting more energy on the development of social undertakes and the solving of people's livelihood issues[4]". The 3rd Plenary Session of the 16th National Congress of the Party positioned the government's function as four aspects, including "economic regulation, market supervision, social management and public service", and clearly defined social management as one of the main functions of the government, expressed the prevention and treatment of social issues, ensuring social security and coordinated development, it has gradually upgraded into a strategic choice with very great economic significance and political significance.

On one hand, the government has paid huge administrative management cost in order to strengthen social management; on the other hand, the flaws of social management system have also caused some social problems, such as widening gap between rich and poor, excessively low coverage of social security, deterioration of employment situation, slow development of social undertakings, etc., which all constitute the social management cost of China. In summarization, social management cost could be classified into social harmony cost and administrative management expense (Chart 16.1).

16.2 Forms of Social Management Cost

Based on the analysis in previous section, social management cost mainly includes the following forms, social harmony cost and administrative management expenditure. For more details, please refer to Table 16.1.

16.2.1 Social Harmony Cost

At current stage of China, traditional types of social problems, such as epidemic diseases, natural disasters, etc., still constitute threat to people's lives and social security, while in the process of modernization with symbol of industrialization and urbanization, some social problems are also constantly emerging and intensifying, such as unemployment, widening gap between rich and poor, production accident, labor-capital conflict, crime, etc. Meanwhile, when China's social transformation is not completed yet, the old social resource distribution system, control mechanism and integration mechanism are tending to dissolve, while new system and mechanism haven't been completed nor fully effective, so it induced and intensified some

---

[4]On Feb 21, 2004, the Premier Wen Jiabao indicated, at the education ceremony of the special research class of the subject of "Establishing and Implementing the Outlook of Scientific Development" of the Main Leaders and Cadres of Provincial and Ministry Level", that we need to "pay more attention to the fulfillment of social management and public service functions, and put more force on the development of social courses and solving of people's livelihood issues".

16.2 Forms of Social Management Cost

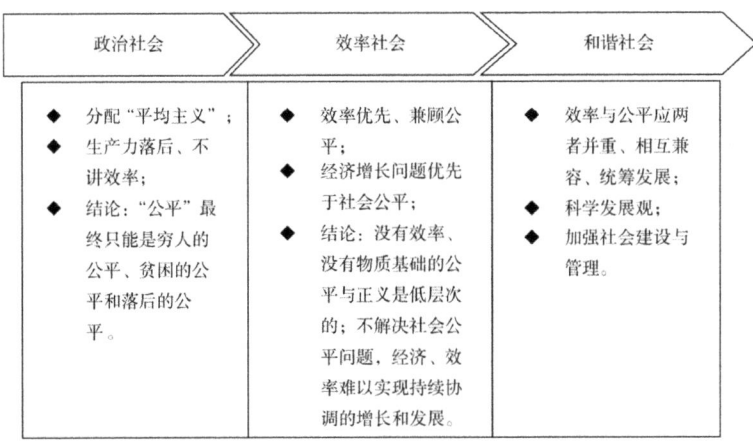

政治社会	Political society	效率社会	Efficient society
和谐社会	Harmonious society	◆分配"平均主义"; ◆生产力落后、不讲效率; ◆结论:"公平"最终只能是穷人的公平、贫困的公平和落后的公平。	◆"equalitarianism" of society ◆Backward productivity, pay no attention to efficiency; ◆Conclusion: such "equality" will eventually be the equality of poor people, poor equality and backward equality
◆效率优先、兼顾公平; ◆经济增长问题优先于社会公平; ◆结论:没有效率、没有物质基础的公平与正义是低层次的;不解决社会公平问题,经济、效率难以实现持续协调的增长和发展。	◆Efficiency first and take fairness into account; ◆Economic growth before social equity ◆Conclusion: the equity and justice without efficiency and without material foundation is low-level; it will not solve the issue of social equity, and the economy and efficiency would be difficult to realize continuous and coordinated growth and development.	◆效率与公平应两者并重、相互兼容、统筹发展; ◆科学发展观; ◆加强社会建设与管理。	◆Efficiency and equity shall be emphasized equally, complementing each other and planned as a whole; ◆Outlook of scientific development; ◆Strengthen social construction and management

Chart 16.1 Three stages of Chinese social management model

Table 16.1 Chinese social management cost

Classification of cost	Forms	Specific contents
Cost of social management	Cost of social harmony	Cost of social security and cost of public security
	Expenditure of administration	Administrative cost

special types of risks, such as widening gap between rich and poor, deviation from the society, intensifying crimes, higher difficulty in control of epidemic diseases, intensifying racial conflicts, demoralization, trust crisis and control failure, etc.[5] These all cause social harmony cost when the government plays social management function.

The economic loss of China caused from social and public security issues in each year is about RMB 650 billion, accounting for about 6% of the GDP, including RMB 250 billion of loss from production safety accidents, RMB 150 billion of loss from social security accidents, RMB 200 billion of loss from natural disasters, RMB 50 billion of loss from biological violation. In each year, public security issues take away the lives of 200 thousand people.[6] In recent years, China's social security situation has been relatively severe. The crime rate of many cities is continuously increasing. According to statistics of the Ministry of Construction, the appealing volume about land acquisition and demolition in the first half of 2004 alone exceeded the total volume in 2003.[7] In 2009, as of October, the number of criminal cases in China grew by over 10%, the number of public security cases grew by about 20%, the number of criminal cases filed in China reached 5.3 million, the number of public security cases reached 9.9 million, such growth trend broke the constant stable trend of law violation and crimes since 2000. On the other hand, the number of crimes in main crime types also had prominent growth. It was mainly reflected as prominent growth of crimes in violence and infringement of property, mob crimes are in active period, mass disturbance events continue to increase.[8] All these symbolize the intensification of social harmony risks at the same time of economic development in China.

---

[5] Zheng Hangsheng and Hong Dayong: *Potential Social Security Hazards and Countermeasures in China's Transformation Period*, Journal of China Renmin University, 2nd Issue in 2004.

[6] Pu Shurou: *Treating Public Security as Basic National Policy*, Guidance of Chinese Society, 4th Issue in 2005.

[7] Sun Xiuyan: *Preliminary Analysis of the Origin and Challenges of Strengthening Social Construction and Management*, Journal of CPC Fujian Provincial Committee Party School, 12th Issue in 2005.

[8] Law Institute of the Chinese Academy of Social Science: *Chinese Rule of Law Development Report No. 8 (2010)*, Social Science Literature Press, 2010 Edition, pp. 177–189.

16.2.2 Administrative Management Expenditure

Administrative management expenditure means the consumption of economic resources from governmental administrative activities. Normal operation of governmental institutions needs consumption of resources, such as buildings, office facilities, personnel remuneration, etc. Since the reform and opening-up, the scale of the Chinese administrative management expenditure, no matter in absolute value or relative value, has been growing at astonishing speed, in recent years, the growing trend didn't decrease and reflected multiple growth, its ratio in fiscal expenditure and GDP have been continuously increasing, and even exceeded the level of developed nations.

In 1978, China's administrative management expenditure was only RMB 5.29 billion, which increased to RMB 651.234 billion in 2005, the total amount increased by 123 times, the per capita annual administrative management expense increased by negative 9.22 times.[9] From relative scale, the ratio of administrative management expense in fiscal expenditure also increased bigger and bigger, in 1978, the ratio of administrative management expenditure in fiscal expenditure was only 4.71%, in 2003, the ratio was as high as 19.03%, which far exceeded the 2.38% in Japan, 4.19% in the UK, 5.05% in Korea, 6.5% in France, 7.1% in Canada and 9.9% in the US during the same period (Table 16.2).[10]

Reasonable administrative management expenditure is the necessary expense to guarantee the normal operation of government on each level, effectively provide public goods and cover market flaws, besides, with the enhancement of economic development level, the continuous growth of its absolute scale is inevitable.[11] China is currently at the stage of industrialization, urbanization and economic transformation, relatively high administrative cost has certain objective factor, but it is already too high compared with that in foreign countries, and from the examples in different countries in the world, the ratio of administrative management expenditure

---

[9]During the period of transition, China's government revenue was not regular, besides income within budget, there are vast amount of incomes outside the budget and the system. The preliminary cause of forming incomes outside budget is the independent income and expenditure and independent management of each local government and unit allowed by the state to cover the inefficient expenditure of government agencies, although its scale has been effectively controlled in recent years, but its ratio in income within budget is still maintained above 20%. Incomes outside system means the income independently received and spent and in different names of different regions and departments, most of which are irregular charges, sharing and fines, due to its complex sources, hidden nature, high un-transparency, plus great difference between different fiscal levels and different regions, there is basically no open statistic data. Therefore, if counting the income outside budget and outside system, China's administrative cost would be even bigger.

[10]Dong Zaiping: *Current Status of China's Administrative Management Expenses and Its Control*, Administrative Forum, 1st Issue in 2008.

[11]Famous German fiscal expert Wagner proposed the "Wagner's Law", which attributed expansion of fiscal expenditure to the enhancement of industrialization level and acceleration of urbanization process, which results in increasingly complex economic activities and more public affairs.

Table 16.2 Administrative management expenditure and ratio in fiscal expenditure during 1978–2009

年份	行政管理费（亿元）	财政支出总额（亿元）	行政管理费占财政支出比重（%）
1978	52.9	1122.09	4.71
1980	75.53	1228.83	6.15
1985	171.06	2004.25	8.53
1990	414.56	3083.59	13.44
1995	996.54	6823.72	14.60
1996	1185.28	7937.55	14.93
1997	1358.85	9233.56	14.72
1998	1600.27	10798.18	14.82
1999	2020.6	13187.67	15.32
2000	2768.22	15886.5	17.42
2001	3512.49	18902.58	18.58
2002	4101.32	22053.15	18.60
2003	4691.26	24649.95	19.03
2004	5521.98	28486.89	19.38
2005	6512.34	33930.28	19.19
2006	7564.70	40422.73	18.7
2007	8514.24	49781.35	17.1
2008	9795.52	62592.66	15.6
2009	9164.2	76299.9	12.0

Source State Statistical Bureau: *Chinese Development Report 2011*, China Statistics Press, 2011 Edition, Page 615

年份	Year
行政管理费（亿元）	Administrative cost (RMB 100 million)
财政支出总额（亿元）	Total fiscal expenditure (RMB 100 million)
行政管理费占财政支出比重（%）	Ratio of administrative cost in fiscal expenditure(%)

in fiscal expenditure tends to continuously decrease along with economic development. The unreasonable growth of China's administrative management expenditure has squeezed the funds for education, science, culture, health, social security and other public services, severely affected the government's comprehensive fulfillment of functions and restrained the healthy development of the economy and society.[12]

---

[12]Dong Zaiping: *Current Status of China's Administrative Management Expenses and Its Control*, Administrative Forum, 1st Issue in 2008.

16.3 Causes of Social Management Cost

China faces severe challenge in social construction and management, the uncoordinated economic and social development, the social construction and social management lagging behind economic growth have become prominent conflict, and directly affect the construction of harmonious society. The main causes of social management cost include:

16.3.1 Mistakes in the Positioning of Government Functions and Perspectives

In social management function, Chinese government still inherits the management model of comprehensive government under the planning economy system and undivided government and enterprise, which cause heavy burden of government and low efficiency in social management[13]; on the other hand, due to the long-term neglect towards social development by the government, plus the traditional government's social management system and management method become increasingly unable to adapt to new situations and requirements, many fields and links that should be originally managed by the government become very problematic due to misplaced measures and insufficient policy enforcement.[14]

16.3.2 Backward Social Management System Causes Confusion and Failure in Social Management

China's current social management system is still the static social management system of "Strong Government and Weak Society" formed under the planning economic system, which is lack of diversified participants, there is no buffer zone for social conflicts, the market force and social force depend too much on government, the room for participating in social management is very few, and self-governing capacity is poor. Meanwhile, the differentiated management

---

[13] The currently market management still mainly use administrative means to manage the increasingly diversified and complex social issues, which makes social management mostly reflect as dynamic, momentum, simplified, and lack of effectiveness, regularity, system and uniformity.

[14] Division of departments and multiple management are also a factor to restrain a government from effectively managing various fields of society. Different management departments in the same field would have crossed functions, unclear duties, and all could enforce law, while they don't have clear boundaries and responsibilities, it is very easy to cause law enforcement vacuum, vague duties or shuffling of responsibilities, so as to give criminals opportunities to take advantage of.

treatment in certain issues by social management system, such as the household registration system and the caused urban-rural dual social structure, enlarged the imbalance of development between urban and rural areas, and affected social stability.

16.3.3 The Social Management and Operation System Lacks Complete Management Means and Organizational System

China's current problem of excessively high administrative management expense is on one hand affected by economic development factors[15] and on the other hand also the result of China's administrative system reform severely lagging behind economic reform. The governmental machine is huge, the using efficiency of administrative expense is low, the supervision mechanism for administrative management cost is incomplete. From the ratio of Chinese total population and the number of personnel supported by governmental finance, it was 600:1 in 1950s, 155:1 in 1970s and 40:1 in 1990s.[16] The expansion of governmental institution and swollen personnel will inevitably cause the increase of administrative management expenditure.

China had implemented planning economic system for many years, the development of administrative undertakings had always depended on collective income and collective expenditure. As a direct result, the awareness of administrative cost is weak. Besides, the process of the government providing public service is naturally "Monopoly" in nature, which is exclusive and mandatory. The two characteristics of governmental organization determine the detachment between governmental work input and output and caused low fund using efficiency. Besides, the political performance assessment system of Chinese officials makes the officials lose the internal motive and external pressure of reducing administrative cost. The measurement of political performance of Chinese officials mainly depends on GDP. The indicators such as fiscal income, introduction of foreign investments and export and foreign exchange creation indulged the wasteful administrative idea of not caring about cost (Chart 16.2).[17]

---

[15]With the development of economy, the society also has continuously increasing requirements on administrative management, the acceleration of population migration and urbanization will also inevitably need public sector to input more resources for management. Government, as the provider of public goods, must make all efforts to meet the increasing public demands of the people, therefore, the administrative management that plays important role in public expenditure will inevitably increase.

[16]Yu Linfang: *Causes and Controlling Measures for the Growth of China's Administrative Management Expenses*, Modern Economic Information, 9th Issue in 2010.

[17]Yu Linfang: *Causes and Controlling Measures for the Growth of China's Administrative Management Expenses*, Modern Economic Information, 9th Issue in 2010.

16.4 Analysis of Countermeasures for Social Management Cost

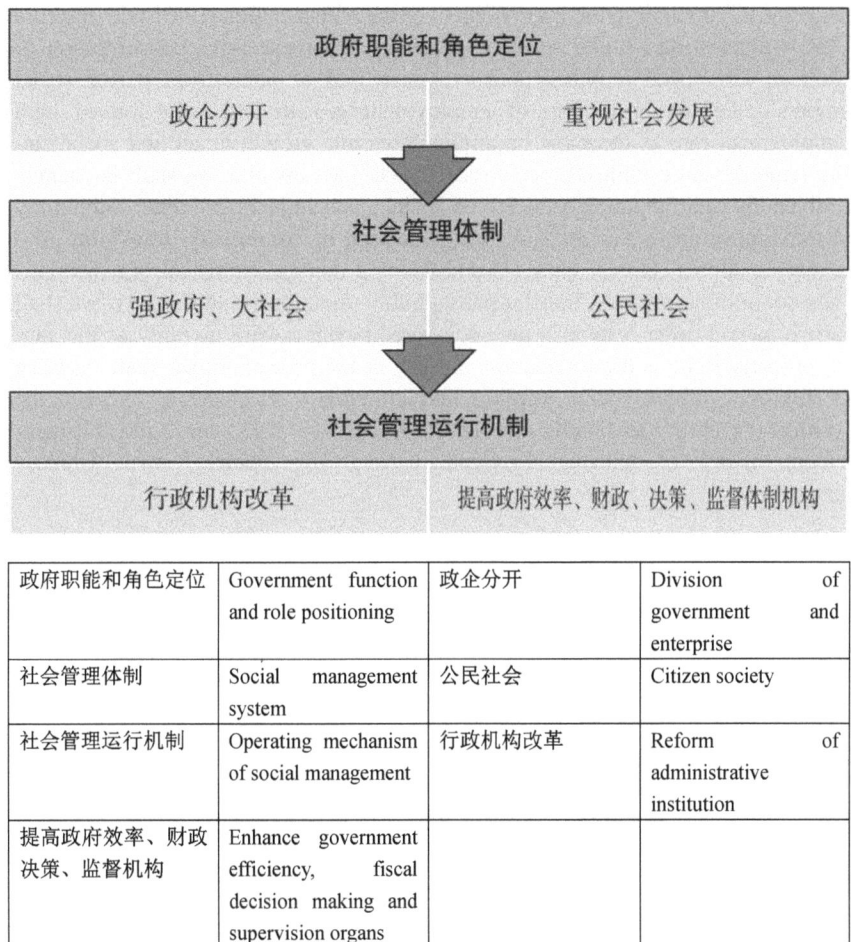

政府职能和角色定位	Government function and role positioning	政企分开	Division of government and enterprise
社会管理体制	Social management system	公民社会	Citizen society
社会管理运行机制	Operating mechanism of social management	行政机构改革	Reform of administrative institution
提高政府效率、财政决策、监督机构	Enhance government efficiency, fiscal decision making and supervision organs		

Chart 16.2 Countermeasures for reduction of social management cost

16.4 Analysis of Countermeasures for Social Management Cost

In order to maintain social harmony and reduce social harmony cost, we must quickly complete the government's social management system and social regulation means, establish the public fiscal system, social policy system, maintain the stable and harmonious social order, strengthen the government's function in social management,[18] including the improvement of the redistribution policy with the core of tax regulation and social security system, prevent excessive concentration of fortune to few social members and excessive loss of interests of partial social members, take the promotion of employment growth and reduction of high unemployment rate as the most important economic growth target and social stability target,[19] and establish a society of rights.[20] Meanwhile, we shall encourage the diversified participating subjects of social management, promote community construction, improve the self-governing function of community, transform government functions, actively cultivate and develop non-governmental organizations or non-profit organizations, transfer partial public functions to the society; we shall improve market order, establish the social credit system with morality as the support, property right as the foundation and law as the protection; we shall establish and improve various social warning and emergency mechanism, enhance the government's ability to handle emergency events and risks, and build the management model of "Strong Government and Big Society" with Chinese characteristics.[21]

---

[18]The 16th National Congress of CCP clearly indicated the basic requirements of deepening China's administrative management system at new stage, to "further transform the government functions, improve management methods, promote electronic government, enhance administrative efficiency, reduce administrative cost, form the administrative system that is standard in behavior, coordinated in operation, fair, transparent, clean and highly efficient", "We must, based on the requirement of the outlook of scientific development, while energetically promoting economic development, pay more attention to acceleration of social development".

[19]Li Junpeng: *Discussion on the Achievements, Problems and Countermeasures of the Social Management by Chinese Government*, Journal of Hubei Administrative College, 1st Issue in 2005.

[20]Society of rights means a kind of social and political state that citizens are entitled to various political, economic, social and cultural rights, the rights of government are deprived from citizen authorization, the fundamental responsibility of government is to safeguard citizens' rights, and implement community cooperation and social mutual assistance. For establishment of a society of rights, the core is to form and implement the social policies oriented to the development of common civil rights. The substance of the society of rights is the people-first social development process.

[21]The basic meaning of "Big Society" is that the power of government is sourced from people, the people have supervision over government, and the people participate in the management of national and social affairs; meanwhile, social organization is not only the object of social public management, but also the entity of social management, social autonomy means the society's autonomous organization and management, autonomous service, autonomous development and autonomous satisfaction. The social development of China is the interactive process that is guided by government, using social organization as intermediary, community as foundation and public widely participating in.

16.4 Analysis of Countermeasures for Social Management Cost

Since the 1960s and 1970s, with the deepening of people's recognition of government failure, the new public management movements[22] with governmental mercerization reform as the main contents have been emerging. As indicated by Peters, "No matter if the richest Western Europe nations or the poorest African nations consider administrative reform, people commonly assume that the best or even the only method to enhance governmental organizational efficiency is to replace traditional bureaucracy system with certain mechanism on the basis of market".[23] Based on the principle of efficient market, only by complying with the reform thinking of the governmental finance doing certain things of refraining from doing other things, and breaking the monopoly of government and other public institutions, we could reach the objective of reducing administrative management cost and enhancing administrative management effect.

To solve the problem of rapid growth of administrative management cost, we must deepen the reform of administrative institution, enhance the using efficiency of administrative expenditure, establish and improve the government behavior supervision mechanism, optimize fiscal expenditure structure; continuously insist on streamlining institutions and personnel, reduce the "Five Levels" governmental hierarchy,[24] change "Vertical Management" into "Leveled Management", scientifically divide and determine the functions of administrative departments, energetically control the scale and positions of administrative departments, focus on solving the problems of institutional overlapping and function crossing[25]; enhance government efficiency, establish effective competitive mechanism of government civil servants ranks, change the previous situation of swollen institution and low efficiency, make efforts to reduce the number of personnel supported by government finance, save fiscal expenditure, improve the budget management system; establish scientific decision-making mechanism and supervision mechanism, enhance the decision-making level of leaders, and enable more government behaviors to be supervised by the public.

---

[22]Due to the difference of specific practices in different countries, they have various definitions over new public management, for example, the definition by the OECD is that (1) adoption of enterprise management technology; (2) service and customer-oriented strengthening; (3) introduction of market mechanism and competition function within public administrative system. Despite the different representations, the basic direction of various definitions are the same, meaning that new public management is a management approach that adopts the theories, methods and techniques of commercial management, introduces market competition mechanism, enhances public management level and public service quality as the characteristics.

[23][US] B. Guy Peters: *Government's Governance Model in the Future*, translated by WuHaiming and Xia Hongtu, China Renmin University Press, 2001 Edition, p. 25.

[24]China's current government organ has five levels of government: central government, provincial government, local government, county government and township government. In other market economy countries, they usually implement three-level government.

[25]Chang Ruixia and Xie Wei: *Empirical Analysis of the Changes of China's Administrative Fees and Expenditures,* Oriental Enterprise Culture, 2nd Issue in 2011.

Chapter 17
Chinese Social Stability Cost

17.1 Definition and Connotation of Social Stability Cost

During 1957–1977, China experienced a period of political "Instability", the national economy was on the verge of collapse. Just in that period, the gap between China and the world was widened, China paid heavy price in social development for this reason.[1]

It was exactly because of the keenly felt pain of "Cultural Revolution", the national leaders highly regarded social stability, in 1989, Deng Xiaoping stressed in his meeting with the US President that: "the overwhelming need to solve Chinese problems is stability".[2]

Maintaining social stability is an important responsibility of Chinese government. Since the reform and opening-up, China's society and politics have been generally stable. Comrade Hu Jintao had in-depth presentation of the great significance of the maintenance of stability for the current Chinese society at the 90th

---

[1]From May, 1966 to October 1976, China erupted the decade-long "Cultural Revolution". The "Cultural Revolution" made a large batch of leaders, cadres and mass, including party and national leaders, subject to slander and persecution, the national economic development was slowed down, the main ratios were out of balance for a long time, and the economic management system became even more rigid. During those ten years, calculating based on the additional benefits for RMB 100 of investment in normal years, the loss of national income reached RMB 500 billion. The 1970s was a period when the international situation tended to ease, and many countries had economic soaring or started continuous development. However, due to the influence of the "Cultural Revolution", China was not only unable to shorten the existing gap with developed countries, but also lost one development opportunity. This "Cultural Revolution" that was originated from cultural field, had particular damage on education, science and culture, and had far-reaching influence, for a period, it caused cultural interruption, technological interruption and talents interruption.

[2]On February 26, 1989 when Deng Xiaoping met the US President Bush, he indicated that "For China's issue, the most important of all is stability. Without a stable environment, everything is pointless, and the achievements already made will be lost. Our country needs reform, which must have a stable political environment, without it, we can accomplish nothing".

Anniversary Conference of the Chinese Communist Party, "Development is the top priority, stability is the top mission; without stability, nothing can be done, and the already obtained achievements would be lost".[3]

The reason why social stability issue was stressed in such way is because China has many unstable factors at current stage. Especially the trend of world integration and the complex international and domestic environment make Chinese government face increasingly great pressure in maintaining stability. It should be said that China from top to bottom has been paying high regard to social stability in recent years, the central government and local government spent great thinking and paid considerable labors, materials and funds in order to maintain social stability, and ensured the social peace and stability in many sensible periods.

China's social stability cost includes not only the national defense expenditure that is used for national defense construction and safeguarding national security, as well as the stabilization cost paid for solving and handling the events that may affect national stability in order to maintain social stability, but also the huge economic and social losses caused to the nation from the events that destructed social stability.

17.2 Forms of Social Stability Cost

Based on the analysis in previous section, the forms of social stability cost mainly include: national defense expenditure, stabilization cost and the price of events that destruct social stability, etc. (Table 17.1).

17.2.1 National Defense Expenditure

National defense expenditure means the military expenditure needed to satisfy the security demand of all social members.[4] The purpose of national defense expenditure is to strengthen national defense construction, establish a modernized national defense power, maintain national independence, protect territorial integration, security and sovereignty from violation. Since the establishment of New China, China's national defense expenditure has been taking important position in

---

[3]In the speech in the celebration of the 90th Anniversary of CCP, Comrade Hu Jintao clearly indicated that "development is the absolute principle and stability is the absolute task; without stability, nothing can be accomplished, and the already made achievements will be lost. This principle shall not only be firmly reminded by all comrades within the party, but also firmly reminded by all the people through guidance". Here, Hu Jintao started from the overall situation of China's socialist modernization construction, proposed "stability is the absolute task" for the first time, and put it at the same "absolute" status with "development is the absolute principle".

[4]Chen Bo: *Long-term Balance and Causal Relationship between National Defense Expenditure and Economic Growth*, China Annual Meeting of Economics, 2005.

17.2 Forms of Social Stability Cost

Table 17.1 Chinese social stability cost

Classification of cost	Forms	Specific contents
Cost of social stability	Expenditure of national defense	Cost of military and national defense expenditure
	Cost of stability maintenance	Cost of public security expenditure
	Price of events damaging social stability	Cost of damage from "Xinjiang Independence" and "Tibet Independence" events

the national fiscal expenditure. Military expenditure has decisive role in the stability and support of national security, and the obtaining and maintenance of the nation's political and economic interest. The contents of national defense expenditure include military expenditure, reserve expenditure, national defense scientific research undertaking expenses and air defense expenditure, etc.[5]

Since the establishment of New China, China's national defense expenditure has generally experienced six stages.[6] At the beginning after the establishment of New China, the ratio of national defense expenditure was relatively high, the military expenditure input supported the elimination of the residual KMT force, bandits suppression and anti-hegemony, the war to resist US aggression and aid Korea, suppression of counterrevolutionary movements and other main military activities of the country, consolidated the new-born people's power, cracked down the foreign and domestic reactionary forces, and played important role in creating a

---

[5]Chen Bo: *Long-term Balance and Causal Relationship between National Defense Expenditure and Economic Growth*, China Annual Meeting of Economics, 2005.

[6]Stage I: 1949–1956, the New China was just established, most of fiscal expenditure of the Central Government back then was for military spending, which was huge, at that stage, the military budget during eight years accounted for 35% of the government fiscal expenditure on average. Stage II: 1957–1966, the stage of socialist economic construction, during such stage, China implemented the guideline of streamlining military personnel and reducing national defense expense, the ratio of national defense budget in central government fiscal expenditure decreased from 42% in 1951 to 10% in 1960. But after 1961, with the change of international situation and surrounding environment, the military personnel and national defense budget expanded again, to 1966, the ratio of national defense budget in national fiscal expenditure returned to the level of 18.1%. Stage III: 1967–1977, the "Cultural Revolution" expanded the military expenditure, the ratio of national defense expenditure in GDP was very stable and always maintained at 4.65–6.5%. Stage IV: 1978–1985, the stage of economic construction as priority at the beginning of the reform and opening-up, at this stage, the ratio of national defense budget clearly declined, and its ratio in national fiscal expenditure declined from 17.37% in 1978 to 10.63% in 1984. Stage V: 1986–1998, the reform and opening-up proceeded and the stage of military endurance, at this stage, the decline of military expenditure continued to accelerate and had the phenomenon of extreme shrinking, the annual declining rate was 4%. China's annual military budget could barely maintain. Stage VI: 1999 to present, the stage of repayment and catching-up, the problem of declining national defense budget appeared, and the military expenditure began to grow year by year.

relatively peaceful domestic and international environment for the economic construction and social reform of New China. At the end of 1970s, the international situation changed, China's strategic focus transferred from the preparation to fight a nuclear war to economic construction. In order to support economic construction, the ratio of national defense expenditure had prominent decease. Its ratio in national fiscal expenditure was reduced from 17.37% in 1978 to 9.47% in 1986. Until 1999, due to the major challenge of national security and territorial integration caused from many years of weakened national defense construction, meanwhile due to the rapid increase of total economy, the people across China had accumulated huge fortune, which also urgently needed a strong military to safeguard, the government began to consider enlarge the input of national defense construction. Since 1999, the national defense expenditure increased year by year, and kept a growth range of 12–20% in each year. During the 11 years from 1999 to 2009, the national defense expenditure increased from RMB 107.6 billion to RMB 495.11 billion, the total amount increased too times or more (Refer to Table 17.2).

From 1999 to 2008, the national defense expenditure had been growing steadily with average annual growth of 15.9%, during the same period, the annual average GDP growth rate calculated based on the prices of current year was 12.5%, the annual average growth rate of national fiscal expenditure was 18.5%. The ratio of military expenditure in national fiscal expenditure generally reflected decreasing trend. On the other hand, China's total military expenditure and average amount per military personnel are both lower than the level of some main political and military powers in the world, China's ratio of military expenditure in GDP still lags far behind the world average level at 3.5%[7] (In 2006, the ratio of US military expenditure in GDP was nearly 5%). In 2007, China's annual national defense expenditure was equivalent to 7.51% of that in the US, and 62.43% of that in the UK. In 2002, China's military expenditure per military personnel was USD 8192, the amount was USD 235.2 thousand in the US, USD 113.9 thousand in the UK, USD 106.1 thousand in France and USD 105 thousand in Germany. The amount of military expenditure per military personnel was equivalent to 4.49% of that in the US, 11.3% of that in Japan, 5.31% of that in the UK, 15.76% of that in France and 14.33% of that in Germany.[8]

---

[7]Hu Angang and others: *Comparison of National Defense Strength of China, the US, Japan and India*, Strategy and Management, 6th Issue in 2003.

[8]Xu Guangyi: *Preliminary Discussion of the Input of China's National Defense Budget*, Economic Horizon, 6th Issue in 2007.

17.2 Forms of Social Stability Cost

Table 17.2 Scale of national defense expenditure and its ratio in GDP

年份	国防开支（亿元）	国防开支占财政支出的比例（%）	国防开支占GDP的比重（%）
1950	28.00	45.00	
1955	65.00	26.10	7.14
1965	87.00	18.30	5.06
1975	142.00	17.50	4.75
1980	194.00	16.70	4.26
1990	290.00	9.87	1.55
1995	636.72	10.20	1.05
1996	720.06	9.72	1.01
1997	812.57	9.40	1.03
1998	934.70	9.47	1.11
年份	国防开支（亿元）	国防开支占财政支出的比例（%）	国防开支占GDP的比重（%）
1999	1076.40	9.40	1.20
2000	1207.54	9.01	1.22
2001	1442.04	8.80	1.32
2002	1694.40	9.04	1.41
2003	1907.90	8.79	1.40
2004	2200.00	8.33	1.38
2005	2475.00	7.82	1.35
2006	2979.00	7.57	1.41
2007	3554.91	6.93	1.44
2008	4178.76	6.81	1.39
2009	4951.10		1.45

Source State Administration of Statistics; *Chinese Development Report 2011,* China Statistic Press, 2011 Edition, Page 615.

年份	Year
国防开支（亿元）	National defense expenditure (RMB 100 million)
国防开支占财政支出的比例（%）	Ratio of national defense budget in fiscal expenditure (%)
国防开支占GDP的比重（%）	Ratio of national defense expenditure in GDP (%)

17.2.2 Stabilization Cost

Stabilization cost is reflected as public security expenditure, which is mainly used for stabilization institutions and personnel, "Stabilization Fund"[9] and various expenditures to prevent and handle mass disturbance events,[10] etc. In recent years, the stabilization cost of Chinese society reflects the trend of rapid increase, the stabilization cost is huge and the society almost can't bear the heavy burden. The per capita expenditure of stabilization has far exceeded the per capita expenditure in education and medical service by the government.[11] During 2000–2006, the average annual growth rate of local fiscal expenditure in armed police was 26%, which was higher than the annual growth rate of fiscal expenditure of the central government and local governments (17 and 19% respectively).[12] Since 2007, the national fiscal expenditure's final account began to separately list "Public Security" in one item. In 2009, the fiscal expenditure of the whole nation in public security was RMB 474.49 billion, increasing by 16.5% compared with the same period in previous year. The actual amount was almost equivalent to the national defense expenditure. In 2010, the budget used for internal security was as high as RMB 514 billion, the fiscal budget of public security will increase by another 8.9%, the growth rate exceeded that of military expenditure (Chart 17.1).[13]

17.2.3 Economic and Social Price of the Events that Destructs Social Stability

Since mid 1990s, the development trend of mass disturbance in China has been increasingly severe, the annual occurrence rate reflected rapid increasing trend, especially after 1997, the occurring frequency accelerated. Prior to 1996, China's annual occurring number of mass disturbance events was under 12 thousand, in 1997,

---

[9]Under the huge pressure of "Zero Indicator" and "One Vote Veto" by superior level, each level of local government, in case of "Sensitive Period" or "Sensitive Event", would make large-scale mobilization, all departments would engage to ensure pace of their areas. "Everything shall make way for stabilization", which is actual cases of administration happened in many places. Such stabilization model requires large quantity of manpower and financial input. So stabilization fund was generated under such circumstance.

[10]Zhang Jinghong: *High Cost Dilemma of "Rights Protection" and "Stabilization"*, Theory and Reform, 3rd Issue in 2011.

[11]Liang Wei: *Dilemma of Stabilization and Lack of Citizen Society,* Lingnan Journal, 3rd Issue in 2011.

[12]Chen Zhirou: *Diversified Group Protest at Post-Reform Stage,* Thesis of the "7th Empirical Research Workshop of Organizational Sociology Meeting.

[13]Zhang Jinghong: *High Cost Dilemma of "Rights Protection" and "Stabilization"*, Theory and Reform, 3rd Issue in 2011.

17.2 Forms of Social Stability Cost

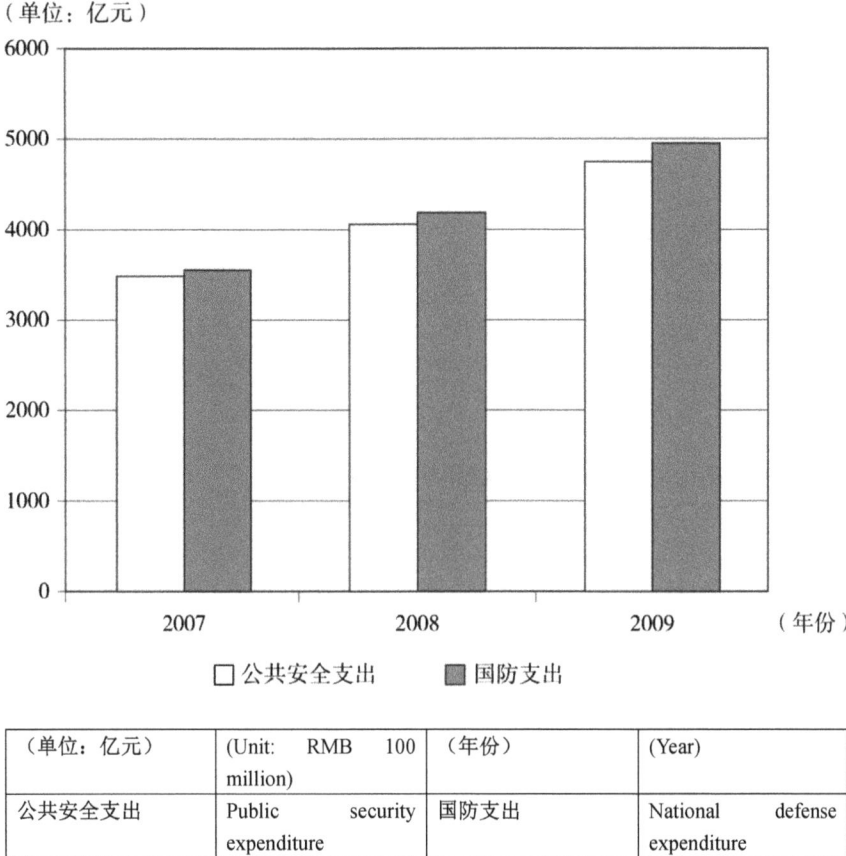

(单位：亿元)	(Unit: RMB 100 million)	(年份)	(Year)
公共安全支出	Public security expenditure	国防支出	National defense expenditure

Chart. 17.1 National public security expenditure during 2007–2009

it rapidly increased to 37 thousand,[14] then increased to 87 thousand in 2005 and 90 thousand in 2006. Although the occurring number of mass disturbance events in China in 2007 and 2008 had no authoritative statistic material to prove, but it can be sure that the specific number can only be more instead of less than that in 2006. Some media called 2008 as the "Breaking Year of Mass Disturbance Events", and made high alert warning on the likely mass disturbance events in 2009.[15]

From the perspective of scale, currently, the number of participants of mass disturbance events in China reflects the trend of increase. The industries involved in these events become more and more, and the subjects are diversified. Mass

---
[14]Wang Meizhi: *Preliminary Discussion of the Transformation from Rigid Stabilization to Flexible Stabilization*, Journal of Party and Political Officials, 4th Issue in 2010.
[15]Zhang Chuanhe: *Research on the Latest Development Trend, Causes and Countermeasures of China's Mass Disturbance Events*, Shandong Social Science, 5th Issue in 2010.

disturbance events have many participants and large involved ranged, once occurred, they would affect the local political, economic and social order. Firstly, they would damage the image of the party and government, affect the party-mass relationship, cadre-mass relationship, cross-region relationship, ethnical relationship, and even affect people's socialism ideas and belief, so they are not good for the formation of normal interest appealing mechanism and social regulatory mechanism; secondly, they would disturb the normal social order, destruct social stability and development; thirdly, they would bring huge economic loss. The frequently occurring mass disturbance events in China have become an important fact that affects social harmony and stability, such events would not only cause direct economic loss, and need large amount of expenses for handling and prevention, which would greatly increase China's social stability cost.[16]

17.3 Forming Causes of Social Stability Cost

I. The lack of public interest expression mechanism causes the increase of legal rights safeguarding events and increases the stabilization pressure of the government.

International experiences indicated, when per capita GDP reaches the level between USD 1000 and USD 40,000, it is often an important "Pass": which is not only an important development window period, but also a period with the most intensive structural changes and most prominent conflicts of various kinds.[17] In such period, economic system has in-depth reform, social structure has in-depth changes, interest pattern has in-depth adjustments, ideas and beliefs have in-depth transformations, which would bring the strengthening of various economic components, classification and restructuring of social classes, strengthening or even intensification of various interest conflicts, and reflection of diversified characteristics of social ideas. China is currently in such a social transformation period, especially since the eruption of the international financial crisis, the intensification of economic difficulty, severe social security issues plus the already existing widening gap between rich and poor, increasingly affect the attitude of various social groups, especially the dissatisfaction of socially difficult groups accumulated under living pressure is easy to cause mass disturbance in the excuse of relevant events with indirect interests.[18]

The root of mass disturbance events is the lack of public interest expression mechanism. When some residents are treated unequally in some events, because

---

[16]Jin Ming: *Characteristics and Handling Strategy of China's Mass Disturbance at Current Stage*, Theoretical Frontier, 24th Issue in 2009.

[17]Zou Hongyi: *Development is the Absolute Principle and Stability is the Absolute Task*, Mass Magazine, 9th Issue in 2011.

[18]Liang Wei: *Dilemma of Stabilization and Lack of Citizen Society*, Lingnan Journal, 3rd Issue in 2011.

they don't have the organizations or channels that truly represent their own interests to express their own interests or dissatisfaction, such inequality couldn't receive reasonable interest expression and appeal, they could only use force of mass to vent the dissatisfaction to the society, and become the blasting fuse of mass disturbance events.

The government's current civil rights idea is usually reflected as extreme emphasis on people's interest in written and oral manner, but neglecting the interest of victim individuals on realistic level, so they couldn't effectively solve the problems faced by people. According to a survey, in the Beijing community of people with the complaint and opinions to the government, the number of cases that were finally solved only accounted for 0.2% of the total cases.[19]

II. The complete and vast stabilization mechanism promoted the excessively swollen stabilization cost.

The increase of social security pressure caused the government to highly regard stabilization work. In order to maintain social stability, the party and government have adopted a series of preventive and control measures: such as the issuing the regulations including the *National General Plan for Emergency Public Events*, the *Emergency Response Law of the People's Republic of China*, etc.; training of officials on each level about the strategies and skills to prevent and handle mass disturbance events; design stabilization departments on each level nationwide, such as complaint divisions on each level, "Stabilization Office", "Comprehensive Handling Office", etc.; setting of "Stabilization Fund", etc.[20]

The governments on different levels have set zero occurrence rate of mass disturbance as the stabilization target and implemented "One-vote Veto System" on the assessment of main leaders. Under the great pressure of stabilization, once there is any occurrence of unstable factor, governments on each level would resort to all resources and means to oppress the situation. Such stabilization model realizes short-term social stabilization through oppression and sacrifice of the interest expression of vulnerable groups, it can only be expedient stabilization. This would eventually increase stabilization cost.

III. With the in-depth change of international relationship, China faces increasing and new external challenges and threats, and the international responsibilities shouldered by China are also increasing, there are more and more international affairs to be participated and solved by China, various large international activities need China to hold, various security tasks keep coming, so the security expenditure is increased.

The all-out stabilization work plays active role in maintaining social stability, but on the other hand, it also caused excessive swollen stabilization cost of China. The high stabilization cost increasingly becomes the heavy burden of local governments

---

[19]Li Changping: *Why Petitioners Increasingly Concentrate in Beijing*, Chinese Agriculture, Rural and Farmer Issue, 4th Issue in 2006.

[20]Zhang Jinghong: *High Cost Dilemma of "Rights Protection" and "Stabilization"*, Theory and Reform, 3rd Issue in 2011.

on each level and the society. For example, the total political and law input of Jinshan District of Shanghai City during 1996–2006 was RMB 1.24609 billion, accounting for 5.6% of the total fiscal expenditure; the social stabilization cost of Guangzhou City in 2007 was RMB 4.4 billion, which far exceeded the social security and employment fund of RMB 3.52 billion (Chart 17.2).[21]

17.4 Analysis of Countermeasures for Social Stabilization Cost

17.4.1 Increase Necessary National Defense Expenditure, Reduce Unnecessary Stabilization Expenditure and Optimize Social Stabilization Cost Structure

Currently, the gap among countries in the world in military power is gradually expanding. The western world led by the US and NATO, due to their powerful economic strength and advanced military technology, continuously widen their gap with third-world countries in military power, which is great threat to the peaceful development of the world. China, as a big power of the third world, should enlarge military investment, the key is to enlarge the investment in high-tech military technology, so as to balance the world's comparison of military power.[22] China's investment in national defense expenditure should approach the world average level, therefore, China should enlarge the ratio of national defense expenditure in GDP and enhance the per capita equipment level of Chinese military personnel.

17.4.2 Strengthen Long-Term Macroeconomic Planning, Determine Correct Stabilization Idea and Rights Idea

The government should change functions and strengthen the government's role as the maker of rules and procedures as well as the regulator and arbitrator of conflicts; gradually transform from pursuing the zero occurrence rate of mass disturbance to the stabilization idea that focus on long-term peach of society, transform from ignoring people's individual interests in practice to the rights idea that emphasizes

---

[21] Wang Meizhi: *Preliminary Discussion of the Transformation from Rigid Stabilization to Flexible Stabilization*, Journal of Party and Political Officials, 4th Issue in 2010.

[22] Hu Angang and others: *Comparison of National Defense Strength of China, the US, Japan and India*, Strategy and Management, 6th Issue in 2003.

17.4 Analysis of Countermeasures for Social Stabilization Cost

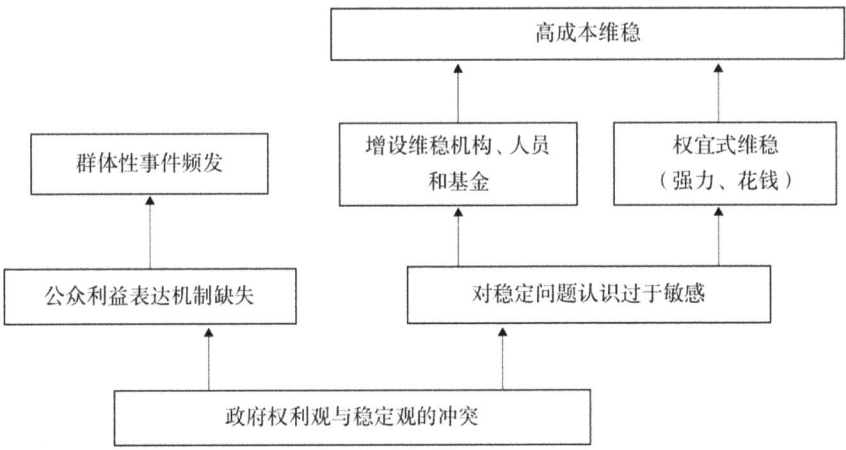

高成本维稳	Stabilization at high cost
群体性事件频发	Frequent occurrence of mass incidents
增设维稳机构、人员和基金	Setting up stabilizing institutions, personnel and funds
权宜式维稳（强力、花钱）	Expedient stabilization (force and money)
公众利益表达机制缺失	Lack of expression mechanism of public interest
对稳定问题认识过于敏感	Too sensitive in recognition to stabilization issue
政府权力观与稳定观的冲突	Conflict between the government's view of power and view of stabilization

Chart. 17.2 Causes of high social stabilization cost

the protection of the realistic rights and interests of individuals, so as to turn the expedient stabilization situation around.[23] The government shall always put the people's interest in priority, feasibly solve the issues relating to the interests of the people, help those in distress and aid those in peril, work for the convenience of the people. The government shall also properly delegate powers to lower levels, delegate partial administrative functions in the fields of production, operation, civil and cultural to the civil organizations in the market, so as to leave room for citizens to participate in more social affairs.

---

[23]Zhang Jinghong: *High Cost Dilemma of "Rights Protection" and "Stabilization"*, Theory and Reform, 3rd Issue in 2011.

17.4.3 Emphasize Social Management Innovation, Establish Effective Expression Mechanism of People's Interest

Establish the diversified interest expression mechanism primarily through judicial means and secondarily through complaint settlement and other administrative means; develop socialism democratic politics, ensure the realization of citizens' rights to know, to participate and to supervise through democratic political construction; establish properly relaxed system environment, cultivate social organizations, develop their active function for social justice and social harmony[24]; strengthen and improve the rule of law mechanism to solve social conflicts, enable rule of law to become the effective systematic means to solve social conflicts; establish the interest balance mechanism under the condition of market economy, change the current situation of severe imbalance in the interest relationship in society, and provide systematic channel for the venting of dissatisfactions in the society,[25] so as to fundamentally solve the mechanism problem of interest imbalance and social justice.

---

[24]Shi Xuehua and Yang Danhua: *Building of Harmonious Society Needs Social Capital Construction*, Journal of Beijing Administrative College, 5th Issue in 2009.

[25]Social Development Research Team of Tsinghua University: *New Thinking of "Stabilization": Systematic Expression of Interest to Realize Long-term Social Peace*, Tsinghua Social Development Forum, 2007.

Chapter 18
Chinese Social Advancement Cost

18.1 Definition and Connotation of Social Advancement Cost

China has always been a big nation of knowledge in history. After experiencing the short-term decline in modern times, it began to quickly emerge. China's scientific and technological force has been continuously and rapidly enhancing and having changed the scientific and technological situation of the world.[1] Scientific and technological innovation is the motive power of social advancement. The development and breakthrough of science and technology are leading to tremendous changes in the world's politics, military industrial structure, production model, economic level, management system and way of life. Just like the first technological revolution resulted from the steam engine and textile machine in the 17th Century, the development of science and technology will inevitably bring new industrial revolution. On one hand, it transforms the traditional industries, on the other hand, it will form new industrial groups, such as electronic information industry, new-type material industry, biological engineering industry, new energy industry, etc., so as to result in the rapid development of social productivity, and extensively enhance labor productivity. The development of science and technology will definitely drive the development of various industries in the society, enrich the variety of products in market and shorten the life of product market.

Since 1950s, the world had entered the times of knowledge economy with characteristics of service industry orientation and computerization, and became the

---

[1]Hu Angang: *National Situation and Development,* Tsinghua University Press, 2005 Edition, p. 120.

third wave in human history after the agricultural revolution and industrial revolution.[2] As estimated, the number of research results achieved during the first 50 years of the 20th Century far exceeded that in the 19th Century, while since 1960s, the number of new discoveries and new inventions in science and technology even surpassed that of the past two thousand years combined.

Social advancement cost referred to in this book mainly means innovation cost, fully information symmetry cost and people's comprehensive development cost.

18.2 Forms of Social Advancement Cost

It is believed in this book that the forms and specific representative indicators of social advancement cost are as follows in Table 18.1.

18.2.1 Innovation Cost

Since the reform, China has rapidly transformed from a big nation of knowledge that mainly introduces innovation and simulates innovations to a big nation of independent innovations. China's comprehensive strength of science and technology has been continuously enhancing, China's ratio in the world in 1980 was 3.8%, the number increased to 4.9% in 1990, and to 7.9% in 2000, in 2009, China's ratio in the world was 17.4%, its scientific and technological strength had surpassed that of Japan and only next to the US.[3] The ratio of China's production and export of high-tech products has also been continuously enhancing. In 1980, China's export amount of high-tech products accounted for 0.03% of the world total, the number reached 1.8% in 1995, in 2002, China began to surpass Japan; in 2005, China surpassed the US; in 2006, China surpassed the EU (27 countries) and became number one in the world. In 2008, China's ratio in world total was nearly 20%, which was equivalent to 1.14 times of that of EU, 1.46 times of that of the US and 2.45 times of that of Japan.[4]

The great achievements made by China in scientific and technological advancement can't separate from the fund investments made by the government in promotion of research and development. As indicated by the data released by the

---

[2]Alf Toffler classified, in his book *The Third Wave* (Alf Toffler: *The Third Wave*, Chinese Edition, CITIC Publishing House, 2006 Edition), the human society into three stages: stage I is agricultural stage, which began from about 10 thousand years ago; stage II was the industrialization stage, which began since the end of 17th Century; stage III is the stage of information (or service industry), which began since the end of 1950s.

[3]Hu Angang: *National Situation and Development,* Tsinghua University Press, 2005 Edition, pp. 125–130.

[4]HIS Global Insight, World Industry Service Database; National Science Foundation (NSF), Science and Engineering Indicators 2010.

18.2 Forms of Social Advancement Cost

Table 18.1 Chinese social advancement cost

Classification of cost	Forms	Representative contents
Cost of social advancement	Cost of innovation	China's input of R&D expenditure
	Cost of full information symmetry	Development level of internet, etc.
	Cost of comprehensive human development	Human development index

State Administration of Statistics, the total R&D expense of China in 2009 was RMB 580.21 billion, which was 6.5 times of that in 2000 with average annual growth of 23.0%. In 2009, China's expense in researching and development (R&D) accounted for 1.70% of GDP, which increased by 0.8% compared with that in 2000. In 2010, China's total R&D expense was RMB 698 billion, and the ratio reached 1.75% (Table 18.2).[5]

From international comparison, China is also the country with the fastest growth of total R&D expenses. The data provided by the Organization for Economic Co-operation and Development (OECD) indicated that China's annual average growth rate of R&D expenses during the ten years prior to 2005 was astonishing and almost reached 18%.[6] Based on the data provided by the National Science Foundation of the US, during 1996–2007, the average growth rate of Chinese total R&D expenses was as high as 21.9%, which far exceeded that of the three major economies, EU, the US and Japan; after 1990s, China's relative gap with the US was continuously reduced, which was 21.2 times in 1990, then reduced to 10.1 times in 2000, in 2009, China had surpassed Japan to become the second biggest country in the world in terms of R&D investment, and China's gap with the US was further reduced to 2.5 times. Table 18.3

18.2.2 Full Information Symmetry Cost

When the world had the first scientific and technological revolution with the application of steam engine as the core and the second scientific and technological revolution with the application of electricity as the core, China was basically a laggard and marginalized. Since 1980s, the world had the wave of the third scientific

---

[5]Compiled by State Statistics Bureau: *China Statistics Abstract 2011*, China Statistics Press, 2011 Edition, p. 173.

[6]The data provided by OECD indicated that the average growth rate of China's R&D during 1995–2005 was far higher than the record of OECD countries, and was the highest among non-OECD countries in terms of R&D expenditure.

Table 18.2 China's investment in scientific and technological research and development and results in science and technology (2005–2009)

指标	2005	2006	2007	2008	2009
研究与试验发展（R&D）投入情况					
R&D 人员全时当量（万人年）	136.5	150.2	173.6	196.5	229.0
#基础研究	11.5	13.1	13.8	15.4	16.3
应用研究	29.7	30.0	28.6	28.9	31.5
试验发展	95.2	107.1	131.2	152.2	181.2
R&D 经费内部支出（亿元）	2450.0	3003.1	3710.2	4616.0	5791.9
#基础研究	131.2	155.8	174.5	220.8	264.8
应用研究	433.5	489.0	492.9	575.2	724.9
试验发展	1885.2	2358.4	3042.8	3820.0	4802.2
#政府资金	645.4	742.1	913.5	1088.9	1329.8
企业资金	1642.5	2073.7	2611.0	3311.5	4120.6
R&D 经费内部支出相当于国内生产总值比例（%）	1.32	1.39	1.40	1.47	1.70
科技产出及效果情况					
发表科技论文（万篇）	94	106	114	119	141
出版科技著作（种）	40120	42918	43063	45296	47826
国家技术发明奖（项）	40	56	51	55	55
国家科学技术进步奖（项）	236	241	255	254	282
专利申请受理数（件）	476264	573178	693917	828328	976686
#发明专利	173327	210490	245161	289838	314573
专利申请授权数（件）	214003	268002	351782	411982	581992
#发明专利	53305	57786	67948	93706	128489
高技术产品进出口及技术市场情况					
高技术产品进出口额（亿元）	4160	5288	6348	7574	6868
高技术产品出口额	2182	2815	3478	4156	3769
高技术产品进口额	1977	2473	2870	3418	3099
技术市场成交额（亿元）	1551	1818	2227	2665	3039

Source Sorted out and calculated from the related data of *Chinese Statistical Yearbook* in previous years.

指标	Indicator	研究与试验发展(R&D)投入情况	Research and test development (R&D) input
R&D 人员全时当量(万人年)	Full-time equivalent of R&D personnel (10 thousand people per year)	#基础研究	#Basic research
应用研究	Application research	试验发展	Test development

(continued)

18.2 Forms of Social Advancement Cost

Table 8.2 (continued)

R&D 经费内部支出(亿元)	Internal expenditure of R&D expenses (RMB 100 million)	#基础研究	#Basic research
应用研究	Application research	试验发展	Test development
#政府资金	#Government fund	试验发展	Test development
#政府资金	#Government fund	企业资金	Enterprise fund
R&D 经费内部支出相当于国内生产总值比例(%)	Ratio of internal expenditure of R&D expense in GDP (%)	科技产出及成果情况	Technological output and results
发表科技论文(万篇)	Published technological papers (RMB 10 thousand)	出版科技著作(种)	Published scientific and technological books (types)
国家技术发明奖(项)	National technological invention awards (units)	国家科学技术进步奖(项)	National award for science and technology progress (units)
专利申请受理数(件)	Number of accepted patent application (units)	#发明专利	#Patents for invention
专利申请授权数(件)	Number of authorized patent applications (units)	#发明专利	#Patents for invention
高技术产品进出口及技术市场情况	High-tech product import/export and technological market	高技术产品进出口额(亿元)	Import/export amount of high-tech products (RMB 100 million)
高技术产品出口额	Export amount of high-tech products	高技术产品进口额	Import amount of high-tech products
技术市场成交额(亿元)	Completed amount of technological market (RMB 100 million)		

and technological revolution, meaning the wave of ICT revolution.[7] But in this revolution, China is still just a follower, the developed nations have already hopped on the first train of this network information revolution, use their "information advantage" and "knowledge advantage" to further create "competitive advantage". At the end of 20th Century, China paid huge development cost of "Digital Gap".

---

[7]ICT means information communication technology, it is a new concept and new technological field formed from the integration of information technology and telecommunication technology.

Table 18.3 Ratio and rank of china's main scientific and technological indicators (2003, 2007–2009)

	2003	2007	2008	2009
R&D 经费支出占世界比重（%）	2.5	4.9	5.8	
R&D 经费世界排名	6	4	4	4
发明专利授权量占世界的比重①（%）	6.3	9.6		
发明专利授权量世界排名	4	4	3	3
SCI/EI/ISTP 收录我国论文占世界比重②（%）	5.1	9.8	11.5	11.6
SCI 收录我国论文占世界比重（%）	4.48	7.5	8.1	8.3
SCI/EI/ISTP 收录我国论文数世界排名	5	2	2	2
EI 收录我国论文数世界排名	3	1	1	1
SCI 收录我国论文数世界排名	6	3	2	2

Source Development Planning Department of the Ministry of Science and Technology: *Compilation of Statistical Materials in Science and Technology*, 2009 Edition, Page 4

R&D 经费支出占世界比重(%)	Ratio of R&D expenditure in the world (%)
R&D 经费世界排名	World ranking in R&D expenditure
发明专利授权量占世界的比重①(%)	Ratio of the number of authorized patents for invention in the world①(%)
发明专利授权量世界排名	Ranking of the number of authorized patents for invention in the world
SCI/EI/ISTP 收录我国论文占世界比重②(%)	Ratio of China's thesis included by SCI/EI/ISTP in the world ②(%)
SCI 收录我国论文占世界比重(%)	Ratio of China's thesis included by SCI (%)
SCI/EI/ISTP 收录我国论文数世界排名	Ranking of China's number of thesis included by SCI/EI/ISTP in the world
EI 收录我国论文数世界排名	Ranking of the number of China's thesis included by EI in the world
SCI 收录我国论文数世界排名	Ranking of the number of China's thesis included by SCI in the world

Full communication of effective information is an important means to break regional development gap, and also a field that requires continuous payment of reasonable cost.

China, as a member of developing countries, become a "Digital Poor Nation" of this information revolution, and situates at the other end of the "Digital Gap",[8]

---

[8]The "Digital Gap" mentioned in this book means the gap between individuals, households, business institutions and different regions of different economic and social development levels in the opportunity of accessing ICTs and using internet to engage in various business activities.

18.2 Forms of Social Advancement Cost

China not only faces the huge gap with the popularization level of internet in the world, but also faces huge internal gap; the "Digital Gap" between different regions and "Digital Gap" between urban and rural areas are collectively referred to as the three gaps faced by China in the times of information.

At the beginning of 21st Century, although Chinese population accounted for 21.15% of the world's total population, but China only has 0.13% of the total number of internet host servers and 6.11% of the total number of internet users in the world, China's development level of internet was not only greatly lower than that of developed countries like the US and Japan, but also greatly lower than the average level of world development. In 2000, the number of internet host servers per 10 thousand people in the US already reached 2419 units, the world average level reached 152 units, but China was only 0.7 unit (Table 18.4).[9]

The internet popularization level among different regions in China was also extremely imbalanced. It can be seen that the Eastern Regions had certain development, while the central and western regions became the "Digital Poor" regions; it can also be seen from the comparison of urban and rural areas that the rural regions had completely become the "Digitally Marginalized" regions.

As calculated from the statistical result of the China Internet Network Information Center (CNNIC) that the number of domain names per ten thousand people in Beijing in January, 2000 was 23 times of that of the national average level, while its average income level in 1999 was 2.45 times of the national average level; while the indicator for Shanghai was 8.9 times and 3.9 times respectively; the number of domain names in 5 regions of Beijing, Shanghai, Guangdong, Zhejiang and Fujian accounted for 64% of the national total, their population accounted for 14% of the national total, their average number of domain names per 10 thousand people was 4.6 times of that of national average level, and their per capita GDP was 1.9 times of the national average level.[10]

During the first ten years of the 21st Century, China's information technology and degree of computerization realized rapid development, but was still a big information nation, it still had a long way to the target of information power. In 1997, China's number of internet users was only 620 thousand, and three quarters of which used internet through dial-up method,[11] but in 2007, the number of Chinese internet users already surpassed that of the US and became the country with the most internet users in the world, in December, 2010, the number already reached 457 million, including 303 million of internet users using smart phones, which both ranked top in the world, the internet popularization rate climbed to

---

[9]Hu Angang: *How China Respond the Increasing Digital Gap*, National Situation Report, 3rd Issue in 2002.

[10]Hu Angang: *How China Respond the Increasing Digital Gap*, National Situation Report, 3rd Issue in 2002.

[11]CNNIC: *The 1st Statistic Report of Chinese Internet Development Situation*, October, 1997.

Table 18.4 Comparison of internet development level between China and developed countries and other regions

Country or region	Ratio of host computers in the world (%)	Ratio of the number of users in the world (%)	Ratio of population in the world (%)	Number of internet host computers per 10 thousand people (unit)	Number of internet users pre 10 thousand people (unit)	Number of internet users borne by each internet host computer (unit)
The US	62.85	31.48	4.58	2836.32	6446.00	2.27
Japan	5.49	10.27	2.14	530.42	4502.63	8.49
China	0.13	6.11	21.15	1.30	271.35	208.37
Africa	0.24	0.71	12.69	3.96	52.20	13.19
Asia	9.11	27.53	60.76	31.04	425.36	13.70
Europe	19.87	29.48	12.35	332.86	2241.28	6.73
Australia	1.80	2.75	0.50	741.89	5145.07	6.94
South America	1.23	3.35	5.69	44.72	553.37	12.37
North America	67.74	36.18	8.01	1750.24	4242.31	2.42
World	100.00	100.00	100.00	206.90	938.93	4.54

Source Sorted out and calculated based on the related data of the *International Statistical Yearbook in 2011*.

34.3%,[12] which was already higher than the world average level. Enhancing the average popularization rate and building a information power are the targets of China's computerization construction in the future.

---

[12] CNNIC: *The 27th Statistic Report of Chinese Internet Development Situation*, Jan 29, 2011.

18.2.3 Comprehensive Human Development Cost

The comprehensive human development theory in Marxism is the objective of human development for socialism nations. The international community usually uses human development index to measure the standard of living and development. Actually the aspects included in comprehensive human development are wider, including the development of all intellectual and physical strength of human.

Human development index (HDI) reflects the human development level of a country or region, and can be used as an index to reflect population quality.[13] Human development index is mainly composed of three basic factors of human lives: longevity, knowledge and decent living standard.[14]

In recent thirty years, China has achieved great advancements in the aspect of human development. The human development index is at the highest level in history and approaches the standard of "Nations with High Human Development". However, major challenges still exist. According to the related data of the United Nations Development Program, the gaps of human development between urban and rural areas, between coastal developed regions and inland poor regions, between different genders, and between the people with urban household registration and people with non-urban household registration are still widening.

Besides, the latest era significance of Chinese Dream is to enable the people to share the opportunity of achieving great things in life, which in fact is to stress the possibility of providing comprehensive human development, this project will be a process of long-term and sustainable enhancement (Chart 18.1).

18.3 Forming Causes of Social Advancement Cost

Each of the three scientific and technological revolutions in the world constituted an opportunity of great national development. But China missed them over and over again due to various reasons, which resulted in the widening gap between China and western nations in modern times in the aspects of scientific and technological and economy, etc., and made China pay great social advancement cost for it.

---

[13]The multiplying between HDI and total population is the total welfare of human development. It is not only correlated with the human development level of a country or region, but also correlated with the total population size and its growth trend of a country or region, so it is an indicator that comprehensively reflects population quality and quantity.

[14]Longevity indicate has relatively high correlation with health, nutrition and other conditions, and could to certain extent reflect the variables of human health, nutrition, etc.; knowledge indicator means education indicator; decent living standard indicate is expressed with per capita GDP.

During the period of the first scientific and technological revolution, China was under the corrupt rule of the late Qing Dynasty. The rotten feudal rule, economic and cultural system impeded the advancement and development of science. Affected by the despotism and the seclusion policy, China's economic development lagged behind and directly caused the capitalism countries to wage the Opium War against China, and made China change from a feudal society to a semi-colonial and semi-feudal society.

During the period of the second scientific and technological revolution, China experienced the political turmoil of the collapse of Qing Dynasty, the rule of the Republic of China and the northern warlords, the long-term domestic strife and foreign aggression made the government unable to seize the second opportunity of major development. The Westernization Movement was just learning advanced western science and technology in order to maintain the feudal rule, and failed very quickly, foreign capitalists were not willing to see a strong China, therefore, they didn't introduce advanced technologies and equipments, but continuously exploited China through wars and capital outputting; China's national capitalism had slow development.

During the period of the third scientific and technological revolution, China was during the beginning period of the liberation war and New China construction. Due to the hostility, block and besiege of imperialist countries as well as the influence by the highly centralized political and economic system of the Soviet Union, the Leftist thoughts overflew, China was busy in making political movements, while the western countries had already used the third scientific and technological revolution to enter the golden age of development.

Economic development level and knowledge development are the main factors that affect the various "Digital Gaps" in China. Chinese average income level, urbanization rate, information infrastructure level, research and development investment still have very big gaps with both the world average level and developed countries such as the US, which greatly restrains the development of China's information technology. In 1998, China's average income was only 1/10 of that in the US and 50% of the world average level; the urbanization rate was 31.14%, which was not only far lower than 76.76% in the US, but also lower than 46.08% of the world average level; there was also relatively big gap in information infrastructure, in 1999, the number of main telephone lines per thousand people in China was 85.8 lines, while the world average level was 157.34 lines, and the US was 644 lines.[15]

---

[15]Hu Angang: *How China Respond the Increasing Digital Gap*, National Situation Report, 3rd Issue in 2002.

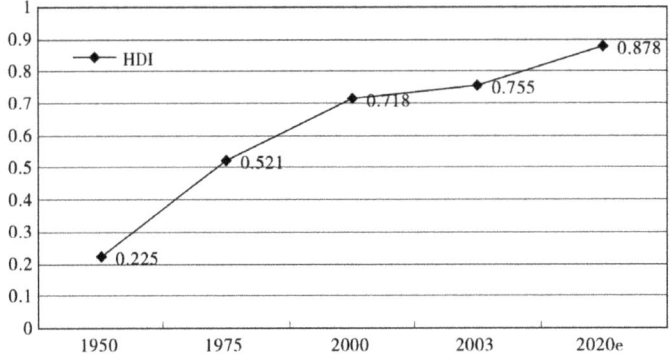

Chart 18.1 Development and Forecast of Chinese HDI (1950–2020). *Source* Concluded from the historical data released by the United Nations Development Program (1950, 1975, 2000 and 2003), the data in 2020 was estimated by the author

18.4 Countermeasures for Social Development Cost

The period of strategic opportunity decides the development destiny of a country and people. Currently the trend of economic globalization in the world is developing rapidly, it is both opportunity and challenge for countries in the world. Peach and development are two major themes of the era, which both closely connected with scientific and technological development. In order to reduce the gap with developed nations in the world, China must well use the development opportunities brought to China by the new round of scientific and technological revolution, and rely on science and technology to promote social advancement.

The essence of science is innovation, we shall form innovation system and mechanism in the entire society, promote scientific and technological innovation and knowledge innovation, encourage independent innovation, establish innovation-type country. China shall energetically develop science and technology, the government must continuously enlarge the degree of investment in scientific researches, and optimize the expense structure among basic research, application research and experiment & development, as well as heavily reward those top-edge scientific and technological talents and strategic scientists who made major contribution in independent innovations of science and technology.

The world is experiencing the new round of information technology revolution that is represented by cloud computation,[16] internet of things,[17] next-generation internet,[18] next-generation mobile telecommunication technology,[19] which has already become the key commonly regarded in society and economic development in the post-financial-crisis period of the world.[20] In this new round of information technology revolution, whoever seizes the commanding height could obtain the dominance of the world in the future.

In this time, we keep synchronized research and development level with the world, this is an important opportunity for us. Currently, both the global internet technological reform and China's new round of information technology development have already entered key period. While the accelerated global economic turmoil and prominent slowdown of economic growth bring major challenges to China's economic development, in order to realize excellent and fast development, China must focus more on the development of new generation of information technology, pay more attention to innovation and using new generation of information technology to reform the traditional industries, promote the transformation of economic development method and the adjustment and optimization of industrial structure; we shall research and form the plan for new round of information technology development during the "Twelfth Five Year Plan" period and a longer period after that, emphasize the production, studying and research in using

---

[16]The concept of cloud computation was proposed by Google, this is a wonderful network application model. Cloud computer in narrow sense means the delivery and using model of IT infrastructure, meaning obtaining required resources through intent in on-demand and expandable manner; cloud computer in broad sense means service delivery and using model, meaning obtaining the required service through network in on-demand and expandable manner. Such service could be related to IT, software and internet, and could be any other service, it has the unique feature of super large scale, virtual, reliable and safe.

[17]Internet of things is another wave of information industry after computer, internet and mobile communication network. Internet of things means the kind of network that connect any item with internet based on agreed protocol with RFID sensors, GPS, laser scanner and other information sensing equipments, in order to make information exchange and communication, so as to realize intelligent recognition, positioning, follow up, monitoring and management. The US pays high attention to the strategic position of internet of things, its National Intelligence Committee listed, in the published report of *The Key Technologies that could Potentially Affect the Interests of the United States in 2025*, the internet of things as one of the six key technologies.

[18]Next-generation internet will be the network based on IP, from backbone network to terminal node, information will pass in form of IP. The next-generation internet and current internet would have three main differences: faster; bigger; and safer. The transmission speed of next-generation internet would be 1000–10,000 times higher than now.

[19]Next-generation mobile telecommunication technology means post-3G technology, normally meaning ITE technology and 4G technology.

[20]Chen Baoguo: *Influence of New Round of Information Technology Revolution Wave on China*, Scientific Decision Making, 11th Issue, 2010.

18.4 Countermeasures for Social Development Cost

coordinated innovation to build the standard system and independent innovation system of the internet of things industry, build new round of platform for the research and development, testing and experiment of information technology, actively participate in the research and formation of standards, and provide support for new round of information industry development and application in China.

Summarization of Chinese Social Development Cost

Over the thirty years since the reform and opening-up, China's economy and society have experienced in-depth transformation, the country has generally transformed from an agricultural society to an industrial society, transformed from a rural society to an urban society, and entered the middle stage of industrialization. However, the development method in the past that partially pursues GDP growth caused severe imbalance of Chinese economic and social development, China has made economic system reform and economic structure adjustment, but the social management system reform as well as the adjustment of urban-rural structure, regional structure, social hierarchy and other social structures are far from realized. It is indicated from the analysis of China's social development cost data that China's development level in science, education, culture, health service, social security and other social undertakes is 5–8 years behind the economic development level; China's overall social construction is 15 years behind the economic construction.

The international experiences have told us that rapid GDP growth doesn't necessarily represent excellent economic operation of a country, under more circumstances, the employment growth and social welfare improvement are more important. A trustworthy government policy, less corruption, more improved rule of law system and better education and welfare program could enable the government to achieve bigger success. Strengthening social construction, as the common rule of function transformation and modernization construction of different governments in the world, is a major trend of social advancement, and should put on important strategic position of national development; it is the urgent demand to adapt to the new characteristics and new challenges of social development of China in the new era, is an important guarantee to realize the strategic objective of comprehensively building a moderately well-off society, and is the necessary requirement to implement the Scientific Outlook on Development and the construction of Socialism Harmonious Society.

Part V
Chinese Cultural Development Cost

Chapter 19
China's Cost of Civilization Inheriting and National Customs Protection

19.1 Definition of Civilization and Recall of Chinese History of Civilization

19.1.1 Definition of Civilization and Connotation of Its Inheriting Cost

What is civilization[1]? From micro perspective, civilization can mean manners and knowledge of individuals, a "Civilized Person" should be a person who has received basic education, has good action capability and correct moral value; from general perspective, civilization can be interpreted as the entire material assets and spiritual assets created by the people of a nation. The social civilization discussed about in this book focuses more on social spiritual civilization. Spiritual civilization includes not only the entire scientific and cultural results created by human race,[2] but also the entire relationships and behavioral methods among people, between individual and society and between human being and the nature.[3]

---

[1] The word "Civilization" was quoted from *The Literary Mind and the Carving of Dragons*, "Minds to words, words to civilization". Ancient people believed that everything has its own principle, people could use their inner heart to sense the world. When you have something on your mind, you would want to express it, so language was invented, which led to text and article. So "Civilization" was originally meaning people's conception of the external world, and later also used to describe the state of human's social advancement.

[2] Such as the development scale and development level of education, science, culture, art, health, sports, etc., which should all be included in the scope of spiritual civilization.

[3] This aspect could also be specifically classified into moral cultivation, political appeal, folk custom, religious faith, etiquette standard, etc. The role of spiritual civilization is to provide ideological guarantee, spiritual power, political guarantee and intellectual support for the development of material civilization.

Since the formation of modern nation, people pay more attention to the social cost to be paid for advancement of civilization, because the development degree of social civilization is closely related to the rise, decline or development of nation as well as the living state of a people, if the cost couldn't be accurately controlled, the nation could take the wrong road, waste natural fortune and lag behind the forest of people in the world. Civilization inheriting of a nation mainly includes sorting-out of history and traditional culture, development and communication of traditional cultural and thought, preservation of historical relics and reinterpretation of national spirit. The efforts made in this process (such as legislation, policy modification, fiscal expenditure, etc.) as well as the mistakes made, windings occurred and damages caused, are all the construction cost paid for it (Chart 19.1).

19.1.2 Recall of the Process of Chinese Civilization

From formation to peaking, ancient Chinese society had created extremely complete, vast and developed feudal ruling system and made important contribution to human civilization. However, there were few dynasties in Chinese history that realized large-scale unification and stability,[4] the periods of the Spring and Autumn Period and the Warring States Period, the Wei, Jin, Northern and Southern States period, the Five Dynasties, the Dynasties of Liao, Song, Jin, Yuan, etc. were very turbulent.[5] The integration of ancient China was basically the process that the main Chinese civilization spreading to other ethnical people. The civilization of different ethnical people didn't have the capability of mutual balance and matching in historical development process. Therefore, local and ethnical declines often occurred, many ethnical people vanished in major conflicts and major integrations, and paid the price of ethnical perish or cultural disruption.[6]

---

[4]The dynasties in Chinese history that truly realized territorial unification only include Zhou, Qin, two Hans, Sui, Tang, Yuan, Ming and Qing.

[5]From the legendary Shen Nong Period since 26BC to the end of Qing Dynasty in 1911. During the about 4,500 years of long period, China had 3,791 wars recorded in texts. Source, Editing Group of Chinese Military History: *Chronology of Wars in History,* People's Liberation Army Press, 1986 Edition.

[6]The once strong races in history, such as Xiongnu, Tujue, Xianbei, Uyghur, Nanzhao, Tibet, Tangut, Qidan, Jurchen, have disappeared, they have either extinct or already integrated with other ethnic groups.

19.1 Definition of Civilization and Recall of Chinese History of Civilization

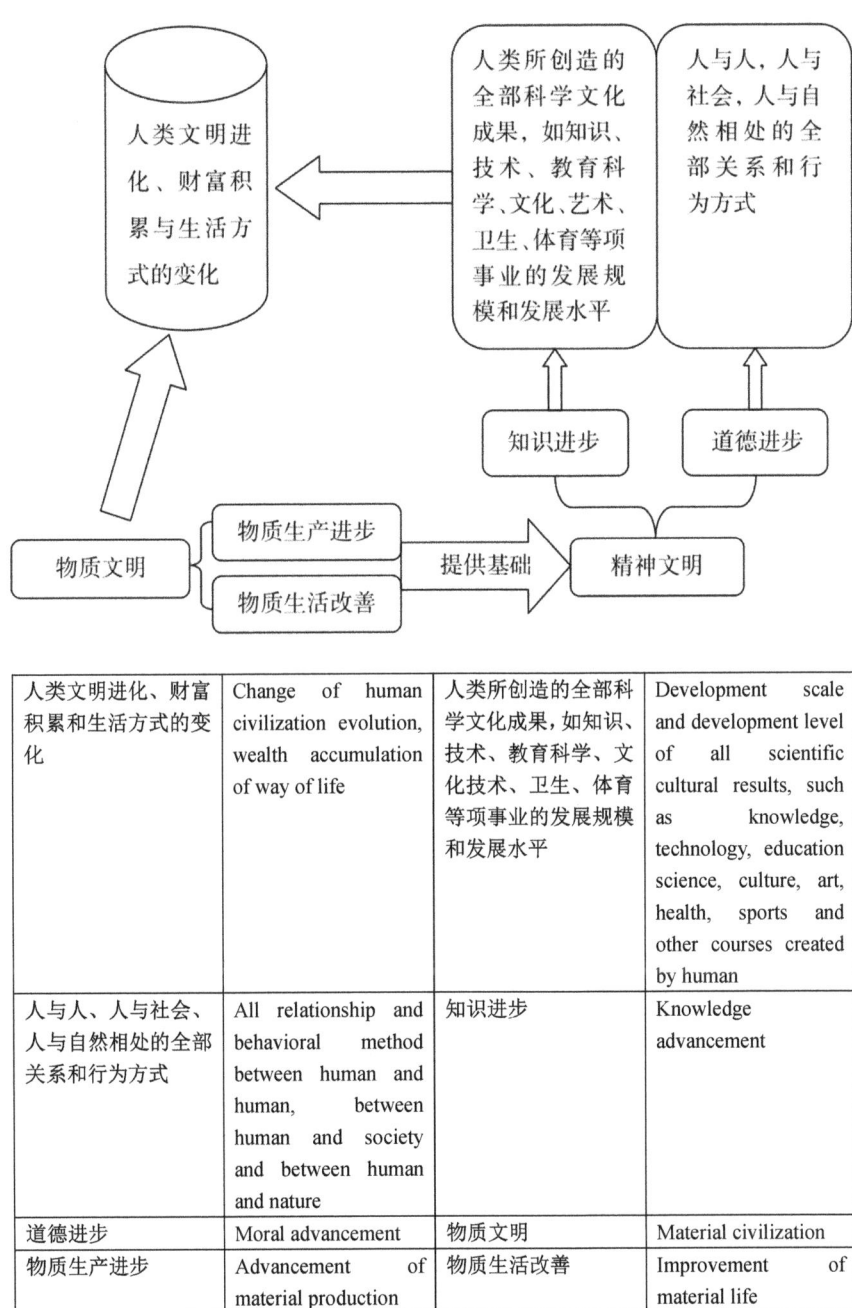

人类文明进化、财富积累和生活方式的变化	Change of human civilization evolution, wealth accumulation of way of life	人类所创造的全部科学文化成果，如知识、技术、教育科学、文化技术、卫生、体育等项事业的发展规模和发展水平	Development scale and development level of all scientific cultural results, such as knowledge, technology, education science, culture, art, health, sports and other courses created by human
人与人、人与社会、人与自然相处的全部关系和行为方式	All relationship and behavioral method between human and human, between human and society and between human and nature	知识进步	Knowledge advancement
道德进步	Moral advancement	物质文明	Material civilization
物质生产进步	Advancement of material production	物质生活改善	Improvement of material life
提供基础	Provide foundation	精神文明	Spiritual civilization

Chart 19.1 Complementary relationship between material civilization and spiritual civilization

Since the 20th Century, all ethnic people and nations in the world faced the process of transformation from traditional closed-type ethnical society to modern governing system and gradually entering the system of globalization. China is also facing the modern transformation in economic and political form. During transformation, many issues are inevitable. For example: how to measure advancement of a culture? How to maintain the balance between tradition and innovation? How to handle new cultural conflicts between different ethnic people? How to actively develop our own cultural system while avoiding "Cultural Penetration" by other nations? The map of civilizations of the world today has experienced countless wars and struggles, different civilizations have different origins, different forming eras and different styles, but they all have deep details and complete systems, they all have strong vitality, which decides that the conflict among different civilizations will be even more fierce, the eastern civilization and Chinese civilization are also dragged into the general environment of competition and mutual influence among civilizations, we not only need to pay participating cost, but also need to pay the competition and expansion cost.[7]

19.2 Forms of the Cost of Civilization Inheriting

From 1949 to 1978, China has experienced major up-and-down changes from material production field to spiritual civilization field. While from 1978 at the beginning of the reform and opening-up to today, China also experienced new historical development period, each aspect of social lives had second major change with modern significance. Behind the glorious achievements, when we investigate the inheriting and developing state of Chinese civilization and traditional culture, we can find that the economic cost, political cost and cultural cost that we have paid are far more than we imagined, and there are also phenomenon such as lack of partial effect or even imbalance due to lack or excessive cost inputting. For the convenience of readers, all costs referred to in this book will be classified into the following three forms based on their payment fields: civilization conflict and peace maintenance cost, civilization elimination and evolving cost, civilization inheriting and protection cost (Table 19.1).

---

[7]As an important part of international stage, China needs to bear more and more responsibilities as a power, and protect the rights, interests and status of its own country, so China will inevitably participate in international competition, so China should expand its own strength and fight for its own interests.

19.2 Forms of the Cost of Civilization Inheriting 269

Table 19.1 Forms of the cost of cultural inheriting/national customs protection cost

Classification of cost	Forms	Specific contents
Cost of civilization passing/ethnic custom protection	Cost of civilization conflict and peace maintenance	In history, it was reflected as conflict with herding civilization; conflict with Buddhism civilization; and the conflict with western civilization since modern times. Currently, it mainly evolved into the difference and struggle of ideology and way of thinking between China and the West
	Cost of civilization depletion and evolution	The negative factors that are left in civilization passing, and continuously obstruct social advancement
	Cost of civilization passing and protection	The social resources used, economic cost and labor cost consumed in the passing and promotion of traditional culture, protection of cultural relics and protection of intangible cultural heritage

19.2.1 Civilization Conflict and Peace Maintenance Cost

1. **External Civilization Conflict Cost**

Civilization conflicts with external side in Chinese history mainly included three types: firstly, conflict with nomadic civilization[8]; secondly, conflict with Buddhism civilization[9]; thirdly, conflict with western civilization in modern times.[10] The former two types of conflicts had long completed, while the third type of conflict still exists today. The direct conflict between Chinese and western culture is generally recognized as beginning since the preach of western Jesuit missionaries in China at the end of 16th Century, back then, the difference between China and the

---

[8] The "Difference between Chinese and Barbarian" during the Spring and Autumn Period was the earliest big discussion on ethnic issues in form of cultural debate, after the hundreds years of painful lessons from the Wu Hu Uprising of the Wei, Jin, Northern and Southern Dynasties, the political splitting, and introduction of Buddhism, Tang Dynasty's foreign policy was relatively open, the tendency of "Westernization" was relatively severe. Not only the court had group of Hu people, the Warlords and An Lushan Rebellion were also caused by Hu People. The two dynasties of Song and Ming learned the lesson from history, and began to enforce the policy of separation between Chinese and Barbarians, from Song to Qing, China gradually changed from a open country to a closed country.

[9] The conflicts and integration with Buddhism civilization was the severest during the period of Five Dynasties and Ten Kingdoms, but its final form was the localization of Buddhism.

[10] The early-stage missionaries were senior intellectuals from the West, received long-term training from the western Vatican, had high organizational and sacrifice sprit, they had political missions when dispatched to China, meaning, "Change Chinese into Barbarian", or in modern term "Peaceful Evolution".

west in etiquette culture caused doubts among people and encountered collective boycott by the literati and officialdom of Ming and Qing Dynasties,[11] they saw the huge political risks and uncontrollable trend that might be behind the civilization conflict between China and the West. However, with the development of industrial civilization, mercantilism entered its golden ages, the western nations eventually bullied China in form of invasion wars with their strong military force, China lost the opportunity of reform and was forced to start social transformation.

Since the reform and opening-up, it was when the US implements its neo-liberalism globalization movement, Chinese economy, politics, society and culture received the impact of the wave of wholesale westernization. Although China today already has strong economic strength and complete political system, the westernization thoughts and atmosphere of worshipping things from foreign and fawn on foreign countries are still flowing in unprecedented manner, which affect people's thinking method and living attitude, and change people's political ideas. In this sense, the traditional Chinese culture is in danger, and even partially degraded into the tool of profiting, the traditional value encounters great challenge, the development of social spiritual civilization prominently lags behind material civilization, chaotic phenomenon often occur in society, these are the huge social costs resulting from civilization conflict and development imbalance.

2. **Internal Civilization Conflict Cost**

In ancient history, internal civilization conflicts were mainly reflected as the running-in and coordination of religion, faith, cultural system, living customs, language, culture and other aspects with final result of national fusion.[12]

The period with the most fierce direct confrontation between feudal civilization and modern civilization was the period of the "May 4th Movement". The several political movements during Mao Era that involved cultural field caused unrecoverable trauma to traditional culture. Today's internal civilization conflicts are mainly national fusion issues in small scale and scope as well as the differences between different classes in cultural value, consumption value and core values, which are far from major conflict, but only interpreted as internal conflicts at the preliminary stage of socialism development.

19.2.2 Civilization Eliminating and Evolving Cost

Chinese civilization eliminating and evolving cost mainly means that traditional culture has long existence of negative factors, which cause large waste of social

---

[11]Lin Jinshui: *Literati and Officialdom during Ming and Qing Dynasty and Dispute of Etiquette between China and the West*, Historical Research, 2nd Issue in 1993.

[12]From this perspective, war is a way of running-in, political amnesty is also a way of running-in, but the most common way is cultural exchange and inter-ethnic mixed living through active economic activities.

resources, impede the advancement and enhancement of civilization, and eventually result in heavy reform price (Chart 19.2).

1. **Cost of Being Conservative on System Level**

The biggest characteristic of Chinese ancient laws and regulations is having complete organization, clear and compact chain of interest, focus on formality instead of reality.[13] Such systematic flaw caused the general mood of the society tend to be conservative.[14] To these days, the problems of conservative political system, relatively backward judicial system and corruption in the party, politics and military become increasingly severe, impede social development and advancement, and undermine social harmony. Therefore, the issue of how to break conservative thinking and enable innovation and reform to be spirit of the time needs correct guiding and active promotion by the party and government.[15]

2. **Cost of Being Conservative on Psychological Level**

The characteristic of Chinese culture is dominance by Confucianism with integration of Confucianism, Buddhism and Taoism, therefore, Chinese traditional culture's attitude has a style of its own in thinking, language and other behavioral ways, is not easy to be influenced by other foreign cultures, so as to reflect the self-reclusive phenomenon.[16] In the information era with rapid development of science, technology and culture, such conservative and closed psychological traits would doubtlessly restrain the development of scientific, technological and cultural industries. By comparing the creativity, operational ability and imagination of the

---

[13]The organization and management of the vast government institution cover every aspect of social, economic and cultural life, such as set of etiquette, bureaucratic establishment, military system, penalty system, education system, etc. Since the establishment date of national management organization, development was never stopped, but there was no major change either, any change would inevitably touch interest distribution, so major change could still be micro adjustment, the big speculation in formality and interest balance in actual work oppressed the real reform target to improve interest distribution system, which eventually caused the overall system to be strict but lack of internal vitality, and could never realize transcendence through self-criticism and denial.

[14]So China's feudal society lasted for more than two thousand years, and lost opportunities of reform for multiple times, at the end of Qing Dynasty, China even closed the nation, which caused the humiliation of China in modern times, like an egg, it was knocked open from outside, exposing its fragile internal, then began to have major change and major reform.

[15]The 6th Plenary Session of the 17th National Congress of CCP Central Committee passed the *Decision of the CCP Central Committee Regarding Several Major Issues Deepening the Cultural System Reform and Promoting the Major Development and Major Prosperity of Socialism*, which indicated that culture undertakings would become the key of social undertaking reform in the future.

[16]China's traditional philosophy, political theory, history, literature, etc. were all deeply influenced by the political idea of feudal unity, mutual support between king and people, and the ethic idea of patriarchal clan system as guideline, which is reflected as strong and fresh human affiliation, power above law, and people's way of thinking also tends to maintain tradition, boycott innovation, resist rational thinking and strict legal procedure, and pursue governance by doing nothing.

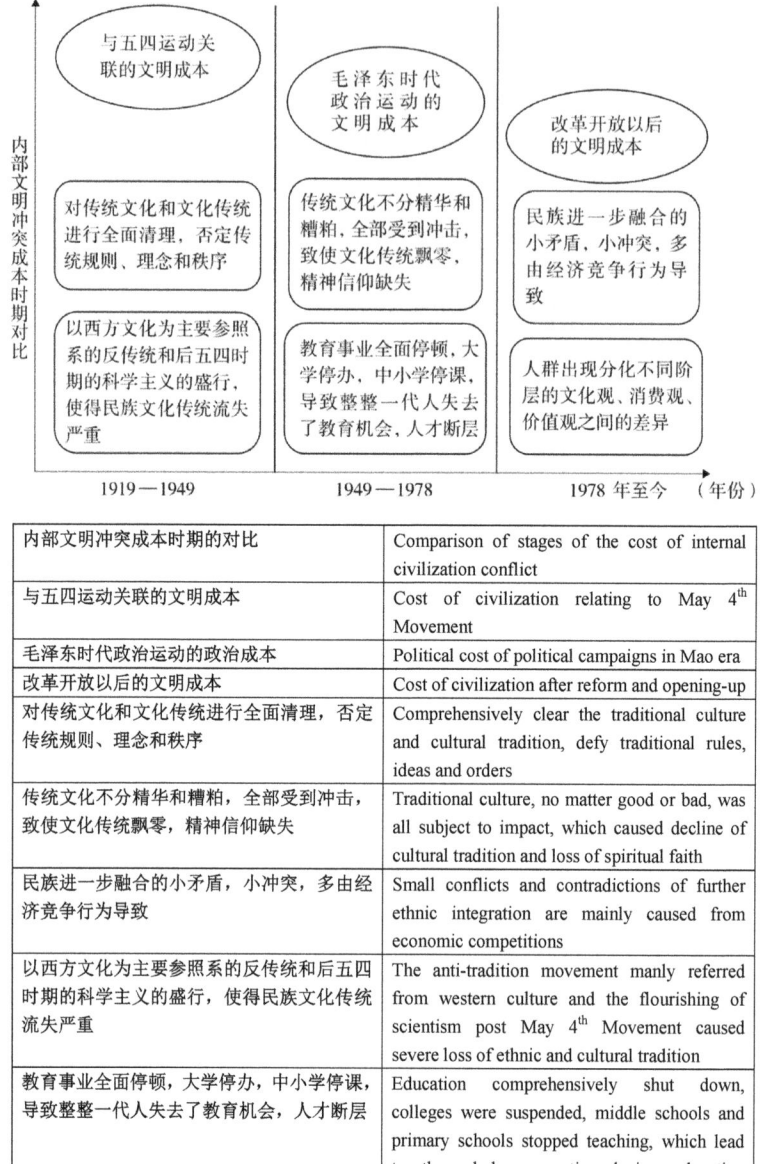

内部文明冲突成本时期的对比	Comparison of stages of the cost of internal civilization conflict
与五四运动关联的文明成本	Cost of civilization relating to May 4th Movement
毛泽东时代政治运动的政治成本	Political cost of political campaigns in Mao era
改革开放以后的文明成本	Cost of civilization after reform and opening-up
对传统文化和文化传统进行全面清理，否定传统规则、理念和秩序	Comprehensively clear the traditional culture and cultural tradition, defy traditional rules, ideas and orders
传统文化不分精华和糟粕，全部受到冲击，致使文化传统飘零，精神信仰缺失	Traditional culture, no matter good or bad, was all subject to impact, which caused decline of cultural tradition and loss of spiritual faith
民族进一步融合的小矛盾，小冲突，多由经济竞争行为导致	Small conflicts and contradictions of further ethnic integration are mainly caused from economic competitions
以西方文化为主要参照系的反传统和后五四时期的科学主义的盛行，使得民族文化传统流失严重	The anti-tradition movement manly referred from western culture and the flourishing of scientism post May 4th Movement caused severe loss of ethnic and cultural tradition
教育事业全面停顿，大学停办，中小学停课，导致整整一代人失去了教育机会，人才断层	Education comprehensively shut down, colleges were suspended, middle schools and primary schools stopped teaching, which lead to the whole generation losing education opportunity of interruption of talents
人群出现分化不同阶层的文化观、消费观、价值观之间的差异	People have difference among different classes in cultural outlook, consumption outlook and value
（年份）	(Year)

Chart 19.2 Description and comparison of the cost of Chinese internal civilization conflict since the "May 4th movement"

youth between China and foreign countries, it is easy to find out that although Chinese youth have good grades, but they don't know how to break thinking set, lack of imagination and have poor operational ability. In the competition of economic field, Chinese cultural creation industry just got started, the competitiveness in high-tech and innovative products is also lacking.

3. Cost of Restriction of Collective Thinking on Individual Thinking

Traditional Chinese culture requires individuals to actively devote to group, avoid individualism and don't need to excessively pursue individual perfection, "Wind will destroy the tree that is outstanding in forest". For individuals, to realize their own values, they must firstly obey and meet collective interest; for the society, there is no value of independence for the existence of individuals. The consequence of such cultural value is oppression to individual personality and denial to personal ability, which would restrain the various possibilities of cultural development, and impede the development of democracy and polyatomic ideologies.

4. Cost of Impact of Mass Culture to Elite Culture

The mass and the elite, as two types of different social groups, have differences in various aspects, including knowledge level, civilized degree, thinking method, etc. Their produced and created spiritual civilization and material civilization also have huge differences.[17] Although the spiritual assets created by elite culture are higher in quality, but also have narrower coverage scope and small communicative degree, its mass foundation is far weaken than that of mass culture, so it is easier to be resisted and impacted by mass culture, which would eventually affect the overall integration and harmony. Currently, the development process of cultural undertakings would inevitably have the phenomenon of mass culture gains the upper hand, and elite culture receives impacts. While the "Three Vulgar" culture that is most valueless, most easy to be pass away the sham as the genuine, and most easy to seduce people to degenerate in mass culture also have its own market.[18]

Special Column 19.1 The Damages of "Three Vulgar" Culture Brought to the Development of Spiritual Civilization in the Society

The "Vulgar" of "Three Vulgar" is greatly different from the "Custom" in "Folk Custom". The former is a vulgar culture that is lack of cultural common sense, moral quality and correct value, based on viewpoints of scholars: it is firstly

---

[17] As for China's ancient society, elite culture is relatively more reflected in spiritual life and ideology, such as compilation of etiquette and decorum, poetry, historical classics; while mass culture is relatively more reflected in material life and way of act, such as the common sense, skill and idea of folk custom, agriculture, handicraft industry and business production.

[18] On Jul 23, 2010, General Secretary Hu Jintao stressed at the 22nd Group Studying of the CCP Central Political Bureau that we need to guide the broad cultural workers and cultural units to actively practice the socialist core value system, adhere to the forward direction of socialist advanced culture, resolutely resist vulgar trend. This important speech by Comrade Hu Jintao started the new round of movement against vulgar in modern China, and drew high attention from the society.

reflected as the national nihilism that people follow the herd after the brainwashing by the hostile forces, they are indifferent to history, hostile to the government, irresponsibly and willingly criticize all current social undertakings; secondly reflected as the indifference to people's suffering and "Heartless" cultural attitude towards social injustice phenomenon, they totally ignore various social problems, present a false appearance of peace and prosperity, defile public opinions; and finally reflected as not understanding art, lack of basic art appreciation ability, but instead point fingers on art.

The vulgar style in the society has great damage to culture, to promote healthy, active, rich and elegant culture as well as the folk culture that people like to see, we have to firstly clear the unhealthy "Three Vulgar" culture, while clearing requires cost too, and it is a sum of cost that has to be paid for long term during the coming decades.

Source: Kong Qingdong: *How to Differentiate the Standard between "Elegant" and "Vulgar"*, People's Forum, the 24th Issue in 2010.

5. Cost of Corrosion of Cultural Dregs

Any nation and race inevitably have dregs while creating advanced civilization. Looking back to the development history of Chinese traditional culture, it is not hard to find out that dregs don't give up any opportunity, their threat to advanced culture is long existing. Such as the gangster logic,[19] gang's consciousness,[20] power-for-money deal, gender discrimination,[21] emperor's thick face and black heart theory,[22] corrupt living style, superstition, etc. which not only corrupt people's thinking, but also poison the society, undermines the rule of law, worsen morality, and would eventually jeopardize the peaceful living and working of the mass people, and affect people's normal living order.

---

[19] Mainly refer to the thinking method that doesn't make sense logically, resort to sophistry, and persist unreasonably. The "implication of the nine generations of a family" in ancient China was a penalty means that is against human nature and penalty means with gangster logic.

[20] Gang, as a civil association, was an ancient social phenomenon that was needed by people and society in history. It was canceled after the establishment of new China, and replaced by new form of organization, the Constitution regulated that citizens have the freedom of assembly and association.

[21] Although China promotes gender equality, but the reality is low birth rate of girls and severe imbalance of gender in population.

[22] The popularity of thick face and black heart theory in China's official circles caused severe decline of official quality. Many officials don't keep honesty, don't need consciousness, don't care about etiquette, don't have shame, don't have principle, consider seeking profit as the top priority and endless occurrence of corruption.

19.2.3 Cost of Civilization Inheriting and Protection

The cost of civilization inheriting and protection is mainly composed of three parts: protection and development cost of ancient cultural relics, ancient buildings, historical sites and other material cultural heritage; recognizing, protection and inheriting cost of non-material cultural heritage, such as traditional opera, music, handicraft, etc.; the cost of refining, promotion and studying of ancient cultural tradition, literature heritage, experience and knowledge.

1. **Inheriting and Protection Cost of Ancient Relics, Ancient Buildings, Historical Sites and Other Material Cultural Heritage**

    The part to be assumed by the government for the inheriting and protection cost of material cultural heritage mainly includes administrative legislation cost (establishment of policies, laws and regulations), fiscal expenditure (fiscal appropriation to cultural relics units and scientific research units, buy-back of cultural relics flew overseas, organization of scientific popularization, public-welfare and promotional activities in society, etc.).

    At the beginning since the establishment of New China, governments on each level set up the institutions specifically in charge of cultural relics, assumed the works and tasks of promoting cultural relics protection, supporting the training of talents, counting cultural relics resources, rescuing historical relics, etc., and started the first nationwide cultural relics consensus in 1957. Entering the reform and opening-up period, from 1981 to 1984, they undertook three-year long second nationwide cultural relics consensus, and determined the second batch of Major Historical Cultural Site Protected at National Level. In 1982, after the release of the *Cultural Relics Protection Law of the People's Republic of China*, Beijing was announced by the national government as one of the first batch of Famous Historical and Cultural Cities. Entering the 21st Century, the idea of cultural relics protection is increasingly recognized by people, in 2005, the State Council issued the *Notice by the State Council Regarding Strengthening of the Protection of Cultural Relics*, cultural relics such as industrial heritages, former residences of celebrities, vernacular architecture, cultural landscape, etc. gradually received broad attentions. The basic work of cultural relics protection is further strengthened, the effective of museum construction is prominent, and the cultural relics protection institutions are gradually improving. According to statistics of the Ministry of Culture, as of the end of 2010, there were 5207 cultural institutions nationwide, increasing by 1177 compared with that in 2005, there were 28.64 million cultural relics and objects, increasing by 24.3% compared with that in 2005. Besides, during the "Eleventh Five Year Plan" period, the total number of museums in China had already reached 3020 (Table 19.2).

Table 19.2 List of legislations for cultural relics protection since 1949

Date of issuing	Name	Main contents
1950	The law forbidding the export of precious cultural relics	The first management method by the new China's government to ban export of precious cultural relics
1961	Interim rule on cultural relics protection and management	Had basic instruction and regulation on each work of cultural relic protection
1982	Cultural relics protection law of the people's republic of China	The first law in cultural field, improved the legal system of cultural relics protection, symbolized the formation of China's historical and cultural relics protection system with cultural relics protection as the center.
1991	Modification of cultural relics protection law	Modified the two articles regarding criminal penalty, administrative penalty, pursuit of administrative responsibility and criminal responsibility
2002	Issued new cultural relics protection law	Key of modification: 1. Enforce the enlargement of cultural relics protection to each aspect of cultural relics protection; 2. Further standardized the laws and regulations regarding private collection of cultural relics; 3. Further determined the dominant position of cultural relics administrative department in law enforcement
Conclusion		After 60 years of effort, China has preliminarily formed the legal system of cultural relics protection law. It is composed of four parts: firstly, law, such as the *Cultural Relics Protection Law of the People's Republic of China*; secondly, administrative regulations, which was passed by executive meeting of the State Council, which issues rules or methods, such as the *Enforcement Rules of the Cultural Relics Protections Law of the People's Republic of China*; thirdly, local regulations, the cultural relics protection regulations appointed by provincial-level NPC Standing Committee; fourthly, regulations, the regulatory rules issued by ministries and commissions of the State Council, the cultural relics protection regulations issued by provincial people's government, such as the *Method of Beijing City on Protection of the Great Wall* issued by the People's Government of Beijing City

Source Concluded by the author

19.2 Forms of the Cost of Civilization Inheriting

Table 19.3 Main contents of non-material culture

Five major fields of intangible culture				
Oral legend and statement, including language as the media of intangible cultural heritage	Performance art, such as drama, etc.	Social custom, etiquette, ceremony	Knowledge and practice relating to the nature and universe	Traditional handicraft skill

2. Inheriting and Protection Cost of Ancient Non-material Cultural Heritage

The concept of non-material cultural heritage was originated from the *Proposal for Protection of Civil Creations* passed by the United Nations Educational Scientific and Cultural Organization at the 25th Convention held in Paris.[23] For the main contents, please refer to the following Table 19.3.

China's protection work of non-material cultural heritage is also led by the government and calling all walks of society to broadly participate, it started from the counting of the number of existing cultures, catalogue, application for the list of world heritage, exiting environment maintenance, scientific survey and research, academic research and other aspects, enabled many dying traditional items to have rebirth. According to statistics, China has nearly 870 thousand items of non-material cultural heritage, as many as 70 thousand of which entered the non-material cultural heritage list system on four levels of nation, province, municipality and county, 1028 items of which entered the "National List of Non-material Cultural Heritage". The Ministry of Culture released 3488 inheritors of national-level items, various provinces, autonomous regions and directly-governed municipalities announced 6332 inheritors of local items. Besides, China has also established 10 cultural and ecological protection experimental areas such as Southern Fujian, Huizhou, Qiang People of Sichuan, over 520 theme museums and 197 folk custom museums.[24] Meanwhile, the Ministry of Culture has also strengthened the protection of the cultural and ecological reserve area and the inheritors of non-material cultural heritage items. In 2007, the Ministry of Culture renamed the South Fujian Cultural and Ecological Protection Experiment Area and Huizhou Cultural and Ecological Protection Experiment Area. During this year's "Cultural Heritage Day", the Ministry of Culture announced the first batch of inheritors of non-material cultural heritage, including 226 inheritors of

---

[23] Such resolution defined civil creation as follows: "Civil creation (or traditional civil culture) means all the creations from certain cultural community, these creations use tradition as basis, expressed by certain group or some individuals, and recognized as the expression form that meet expectation of community and as their cultural or social characteristics; their codes and value are passed on through simulation or other oral forms, its forms include: language, literature, music, dance, game, myth, etiquette, custom, handicraft, architectural art and other arts". Since then, UNESCO used the concept of "Oral and intangible heritage" or "Intangible cultural heritage" to describe "Civil creation".

[24] Li Huilian: *Preliminary Discussion of the Current Situation and Countermeasures of China's Protection of Intangible Cultural Heritage,* Economist, 7th Issue in 2011.

national-level non-material cultural items in five categories, including folk literature, acrobatics and game, folk fine arts, traditional handicraft, traditional medicine, which have generated broad social influence.

3. **Screening, Studying and Developing Cost of Traditional Culture**

Since the establishment of New China, the traditional culture inheriting and developing cause has been mainly composed of two aspects of work. One aspect is to, in order to promote the traditional cultural industry to adapt to demand of the time, support private art troupes on fiscal level, establish special funds, meanwhile, concentrate private capitals to promote the development of local cultural and artistic undertakings; the other aspect is to promote the cultural quality education, strengthen the education of students of primary school, middle school, college and university in the aspects of literature, history, philosophy, art and other humanity and social science aspects, enhance the cultural taste, aesthetic taste, humanistic quality and scientific quality of the youth.

Special Column 19.2 China Energetically Promote the Education of Traditional Chinese Culture

In recent years, the Ministry of Education has listed the education of traditional Chinese culture into the items of the New Century Higher Education Teaching Reform Project, the related universities and publishing houses have formed and published books such as the *Summary of Traditional Chinese Culture, Generalization of Traditional Chinese Culture, Spirit of Traditional Chinese Culture, History of Chinese Culture*, etc., some of which have entered classrooms of colleges and universities as college curriculum of traditional culture.

Meanwhile, referring to the construction experiences of nationwide youth education base, various places are using museums, exhibition halls, parks and existing various education bases to provide the wide youth with facilities and activities of traditional Chinese culture, demonstrate the traditional culture parks with ethnical characteristics and regional features, enable them to be the cultural activity center for public good as the out-of-school classrooms of students and for people to receive education of traditional culture. Besides, news media also introduce traditional Chinese culture and develop traditional culture seminar activities by setting special columns, special issues, so as to enlarge publicity degree and form media environment; develop the creative performance activities with development of traditional culture as the theme, enable traditional culture to step on artistic stage and enter film or TV programs s well as literature pieces.

As of 2009, the central financial authority has input RMB 659 million in accumulation into the support of non-material cultural undertakings, in which the special expenditure of local transfer payment for support of non-material cultural heritage in 2009 was RMB 213 million, for the other years, please refer to Chart 19.3.

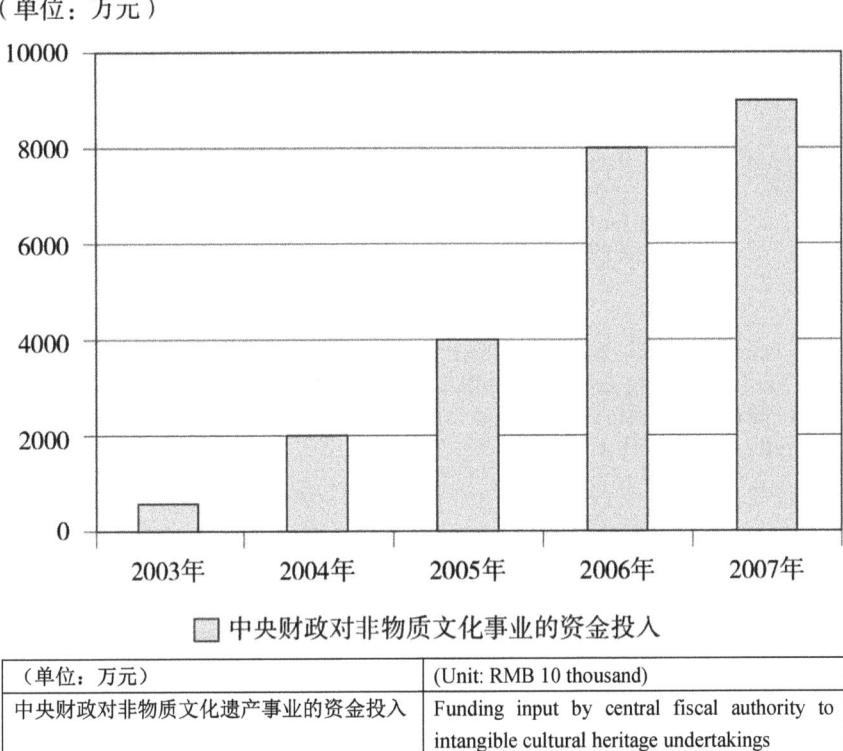

Chart 19.3 Funding input by central financial authority in non-material cultural undertakings.
Source Sorted out from the Statistical Yearbooks of Chinese Culture and Cultural Relics

19.3 Connotation, Definition of National Customs Protection and Forms of Cost

"Respecting the custom of minorities, each ethnic people have the freedom to maintain or reform their own custom" is the basic policy of the party and nation regarding the custom of minorities, and it protects the rights and freedom of each minority to live according to their own traditional custom and participate in social activities. However, due to long-term historical reasons, territorial differences, religion and other factors, it is not easy to maintain harmonious coexistence of different ethnical people and realize the advancement and development of various undertakings of ethnical regions. Protection of ethnical custom is mainly to solve ethnical conflicts, respect the natural law of national fusion, protect the freedom and rights of ethnical people in religion and faith, and implement preferential supportive policies to ethnical regions to help them realize rapid economic development.

Table 19.4 List of the forms of the cost of national customs protection

Classification of cost	Forms	Specific contents
Forms of the cost of national custom protection	Cost of handling ethnic conflicts	Various input caused from handling of ethnic conflicts
	Cost of ethnic integration and custom protection	Promote ethnic integration, protect the living habits of minorities from legislative and administrative perspectives
	Cost of religious faith protection	Protect and promote the development of religious undertakings and eliminate unharmonious factors

It is believed in this book that there are three forms of national customs protection cost, including ethnical conflict handling cost, national integration and customs protection cost, religion and faith protection cost, etc., for more details, please refer to Table 19.4.

19.3.1 Ethnical Conflict Handling Cost

At the stage of Chinese slavery society and feudal society, the establishment of nation with large-scale unification and the formation of Chinese nationality were always along with various conflicts. After the establishment of New China, the party and government's ethnic policy played important role in maintaining the national unification, strengthening the national unity, and guaranteeing social stability, the ethnical conflicts obviously decreased. But there were also major problems and painful prices paid (Table 19.5).

With the deepening of the reform and opening-up and the rapid development of market economy, some ethnic policies that were formed based on the planning economy thinking are no longer suitable for the ethnic relationship of current conditions, and no longer meet the demand of the economic and social development in ethnic regions. The lagging-behind of current ethnic policies is mainly reflected as the lagging-behind of the regional autonomy policy[25] and the lagging-behind of

---

[25]The minority regional autonomy is the basic policy of CCP to solve ethnic issue and a basic political system of the nation to guarantee the minorities to be the master of their own. But the ethnic regional autonomy system, during the process of political system evolution, should be gradually replaced by regular regional autonomy system, excessive outstanding of the independence of ethnic autonomy and strengthening the single national awareness and large-scale implementation of favorable ethnic policies could in turn cause the opportunity of splitting.

19.3 Connotation, Definition of National Customs Protection and Forms of Cost

Table 19.5 Ethnical conflicts and handling cost at the beginning since the establishment of New China

	Specific forms of conflicts	Handling method and handling cost
Preliminary stage after the establishment of New China (1949–1978)	Under the trend of massive development of cooperative across China, minority regions had hard time to adapt	Regional government didn't relied on the power of minorities themselves, didn't respect the right of minorities of being their own masters, adopted the method of Han cadres taking charge, forcing, ordering and copying the experiences in Han regions, which caused dissatisfaction from local minorities and even caused mass appeal from minorities
	In the movement of rectification and criticizing local nationalism, many contradictions among the people were considered as contradictions between ourselves and the enemy, wrongfully cracked down a batch of minority cadres, and hurt some leaders of ethnic and religious circle.	Due to the rash thinking of "Leftist" that occupied leading position in the party's economic work and ethnic work, different ethnic regions developed the criticism against the so-called "Special Theory", "Backward Theory", "Condition Theory" and "Gradual Theory", "Go all out and go fast" without consideration of conditions like Han regions, generally blew the tendency toward boasting and exaggeration and "Communism Tendency", which severely affected local economic and political situation
	Canceled favorable policies towards minorities in mixed region of ethnics, canceled and merged the autonomous regions of minorities, and forcefully changed the custom and habits of minorities	Caused huge waste of manpower, materials and funds as well as huge destruction of ecological environment in minority regions, severely frustrated the production initiative of people in minority regions, and resulted in huge economic losses. Greatly hurt the feeling of minorities, and damaged the gradually established mutual trust relations between Han people and minorities

economic preferential policy.[26] The conflicts occurred in the ethnic autonomous regions in recent years in Tibet, Xinjiang, Inner Mongolia, etc. were reflection of such conflicts. Although those events were completely handled, but their economic losses and social influences still need long time to eliminate.

Special Column 19.3 Imbalance in Economic Development Causes Increasing Political Risk Cost in the Regions Inhabited by Ethnic Minorities

Despite backward economy, Chinese minority regions have relatively abundant natural resources. China's large-scale rapid economic growth, while accelerates regional difference, greatly enhances the demand for natural resources. While main natural resources in China, such as energy and important ores, are all monopolized by state-owned enterprises, the inevitable consequence of it is that the central government expands its development of natural resources in minority regions. This would increase the minority's sense of exploitation. Meanwhile, economic development provides entrepreneurs with Han nationality considerable business opportunities. Because the local entrepreneurs in minority regions are lack of competitiveness and don't have developed commerce, the entrepreneurs in Han nationality and small vendors from outside are very easy to occupy the market. This would make minorities feel that the good things about economic development are fully occupied by the Han people from outside.

Targeting these rapidly economic and social changes, the ethnic policy of Chinese government didn't have major adjustment. Especially in religion policy, ethnic autonomy and other major and complex issues, the basic stance is "Maintaining stability is the top priority". Therefore the minority elite and mass who are dissatisfied with the current situation may lose confidence and hope in the government and finally resort to extreme. The large-scale bloodshed events occurred in Tibet and Xinjiang were exactly a type of reflection of sharpening ethnical conflicts in the regions. Before completely solving the three most sensitive issues in minority regions, namely political participation, cultural rights and religious freedom, the central government still needs to face considerable political risk cost of the minority region, and the economic support investment will also be very high.

Source: Pei Minxin: *China's Ethnic Problems will be Increasingly Severe*, Lianhe Zaobao, May 31, 2011.

---

[26]China's ethnic economic policy was established under the planning economy system, with the establishment of the socialist market economy system, the ethnic economic policy that was established depending on authority of central government, use of administrative orders, giving money and other "Blood Transfusion" supportive method is not suitable for the actual circumstance of minority regions and the requirements of market economy. Source: Luo Shujie and Xu Jieshun: *Thinking of Adjustment of China's Nationalities Policies at Transition of Centuries*, Journal of Guangxi Nationalities College (Philosophy and Social Science Edition), 2nd Issue in 1999.

19.3.2 National Fusion and National Customs Protection Cost

National fusion is an inevitable trend of historical development of multinational countries. The formation, change and development of national community in China and foreign countries and in ancient times and modern times are all closely connected with national fusion.[27] The national fusions in Chinese history mainly have the following three basic characteristics: firstly, based on cultural fusion; secondly, the scale of national fusion during discord times is larger; thirdly, the national fusions in the regions with multinational ruling are most common. The party and government considers the respecting and protection of the customs and different ethnics as a very important work in the process of national fusion. The ethnical work at the beginning since the establishment of New China had some detours, and accumulated many lessons. Currently, the ethnical work has walked on a path of healthy development (Table 19.6).

Strengthening the support of economic development in ethnic regions is one of the important means to promote national unity and eliminate ethnic conflicts. For many years, the national government has been implementing specific transfer payments to ethnic regions. Considering the factor of special expenditure in ethnic regions, the central fiscal authority also implemented preferential policies to ethnic regions through general transfer payments, such as enhance the payment coefficient of transfer payments for the economic development of ethnic regions, increase the amounts of general transfer payments, etc.[28] Meanwhile, the national government also implements preferential tax policies for the economic development in minority regions, for example, for start-up enterprises established in minority regions, as approved by governing tax authority, their income tax can be reduced or exempted for three years.[29]

---

[27] From historical perspective, national fusion means that two or more races in history, due to mixed living, marriage or other causes, have mutual penetration and mutual influence in society and culture, have difference reduced and common feature increased, and finally integrate into one and form one race.

[28] The main target of general transfer payment is to realize the equality of basic public service, it doesn't have regulation of specific uses, and independently arranged by local governments, which is a typical balance appropriation. From 2000 to 2007, the central fiscal authority arranged RMB 209.683 billion of general transfer payment in 8 minority provinces and regions. Source: Liu Haiyan: *Establishment of Service-type Government in Minority Regions,* Journal of Northwest University for Nationalities (philosophy and social science edition), 5th Issue in 2009.

[29] Besides, besides, for the minorities concentrated central and western regions whose local fiscal income decreased due to the exemption of agricultural tax and agricultural special duty except tobacco, the central fiscal authority gave proper subsidy.

Table 19.6 Comprehensive input in protection of national customs dominated by the government

Form of guarantee	Specific contents
Legal guarantee	The *Constitution of the People's Republic of China* regulates that each race shall have the freedom to maintain or reform its own ethnic custom and habit. The *Criminal Law of the People's Republic of China* regulates that "Any staff of national government institutions who illegally deprives citizens of their freedom of religious faith and violates any custom or habit of minority, if the circumstance is severe, shall be given term sentence or imprisonment for no more than two years"
Economic guarantee	National commercial departments shall pay attention to the arrangement of the production and supply of special items relating to custom and habit, and implement special economic support policy to the economic development in ethnic regions
Civil affairs guarantee	Civil affairs departments also issued the related regulations and required each place to respect the funeral habits of minorities, regularly or irregularly hold nation-wide minority art performance and traditional sports meeting, and regulated special holidays for the traditional holidays of minorities
Education guarantee	The nation and party central committee and the regions of multiple races have established publishing institutions for minority texts and the news agencies for minority language and text, published the journals and books of minority texts, as well as used the languages and texts of minorities to dub or translate films and TV series. Chinese government has also established ethnic language and text working institutions and research institutions in central government and each minority region, and set the related faculties or majors in some colleges and universities, as well as implemented double-language teaching in minority regions, cultivated large batch of talents engaging in minority language and text

Source Sorted out by the author

However, with the soaring of economy in ethnic regions, new problems also occurred accordingly. National traditions and customs were products of farming society, and suitable with the old production method and way of living, the impacts brought by market economy are difficult to absorb in short time, and could bring various problems demonstrated as follows Table 19.7.

19.3.3 Religion and Faith Protection Cost

China is a country with multiple religions, the believers mainly follow Buddhism, Taoism, Islamism, Catholicism and Christianity, etc. Chinese citizens may freely choose and express their own faith and express their religious identities. The "Cultural Revolution" occurred from 1966 to 1976 caused disastrous destruction on every aspect of the society, including religion, and resulted in severe social cost. Nowadays, the prosperity of Chinese religious undertakings fully learned the

Table 19.7 List of destructions of national traditional cultures and customs in market economy environment

Nature of destruction	Main forms	Example of actual cases
Economic interest in priority	Rough development of ethnic culture	Tourism developers, in order to appeal to the needs of tourists, couldn't differ the good and bad things of ethnic traditional culture, make sensational fake ethnic culture, but ignore the essence of culture
Artificial destruction	The image of ethnic history was changed	In urban construction and old town renovation, dismantled the historic blocks with ethnic cultural features, and destructed original urban image
Artificial denial	Burying of ethnic cultural techniques	In standard market behaviors, set market behavior access, which caused the folk doctors, handicraft masters and artiest lose employment qualification due to lack of related qualification certificates
Artificial ignorance	Fail to protect the ethnic and cultural property	China's intellectual property system requires that the protected works must have originality, specific entity of rights and limited period of protection. But there is no specific creator of ethnic traditional culture and knowledge as entity of rights, its creation is a constant, gradual and slow process, which requires its protection period to be permanent, therefore, it is very difficult for ethnic traditional culture to obtain protection from copyright law

Source Concluded by the author

detours and lessons made in the past, the protection of religion and faith is mainly composed of the following four parts: firstly, protection by law; secondly, judicial and administrative guarantee; thirdly, support to independently organized religious undertakings; fourthly, protection of the freedom and rights of minorities in religion and faith. In short, the policy is to adhere to the construction of rule of law in religion, increase fiscal input in religious undertakings and crack down illegal cult organizations.

Special Column 19.4 Several Important Legislations on Protection of Religious and Faith Freedom since the Establishment of New China
In the *Constitution of the People's Republic of China*, freedom of religious belief is a basic right of citizens. Article 36 of the *Constitution* regulated that "Citizens of the People's Republic of China has the freedom of religious belief." "No national authority, social organization or individual has the right to force citizens to believe in religion or not to believe in religion, nor discriminate against any citizen who has religious belief or the citizens who don't have religious belief." "The government shall protect normal religious activities." Meanwhile, it also stipulated that "No one is allowed to use religion to destruct social order, damage physical health of citizens

or impede activities of national education system." "Religious organizations and religious affairs shall not be subject to dominance by foreign forces."

It is also regulated in China's *Law of Regional National Autonomy, General Provisions of the Civil Law, Education Law, Labor Law, Compulsory Education Law, the Law on Election of People's Congress, Organic Law of Village Committees, Advertisement Law*, etc. that all citizens, despite their religious beliefs, shall have the right to vote and the right to be elected; the legal assets of religious organizations shall be protected by law; education and religion shall be separated, all citizens, despite their religious beliefs, shall have equal opportunity to receive education according to law; people in different ethnics shall respect each other's language, customs and religious beliefs; citizens shall not be discriminated in employment due to different religious beliefs; advertisements and trademarks shall not contain any content with national or religious discrimination.

Chinese government also issued the *Regulations Governing the Sites of Religious Activities* in order to protect the lawful rights and interests of religious activity sites. The Regulation stipulated that religious activity sites shall be managed by the management organization of such sites themselves, their lawful rights and interests as well as normal religious activities at such sites shall be protected by law, no organization or individual may violate or intervene. Anyone who violates lawful rights and interests of religious activity sites shall bear legal liabilities. Religious activities occurring at religious activity sites must also comply with laws and regulations.

19.4 Forming Causes of the Cost of Civilization Inheriting and National Customs Protection

The various forming causes of the cost of civilization inheriting and national customs protection are generated mainly from two points: "Respect of History" and "Respect of Human Rights". China is a ancient civilization with over five thousand years of long history, today's China has made radical changes, but the character and ideal of Chinese people are still "Richer and stronger China and its people", "Live and work in peace and contentment" and "Unity of heaven and man". For establishment of harmonious social value, we need to absorb nutrition from traditional culture, which contains the rule of governance and rule of people benefiting that has popular sense and could be learned today, this is the precious source for China to preserve and develop the spirit of our own people and advance along with time. The rise of market economy poses threat to traditional culture, how to handle the relationship between the two, we still need to think and explore. China has vast territory and many ethnics, the harmonious coexistence of multinational country is very important and related to national stability. China's development of different regions is imbalanced, and the national customs and religious beliefs are different, which pose great challenges to the government's management and guiding work, and also caused the relative heavy cost of ethnic affairs handling (Chart 19.4).

19.4 Forming Causes of the Cost of Civilization Inheriting ...

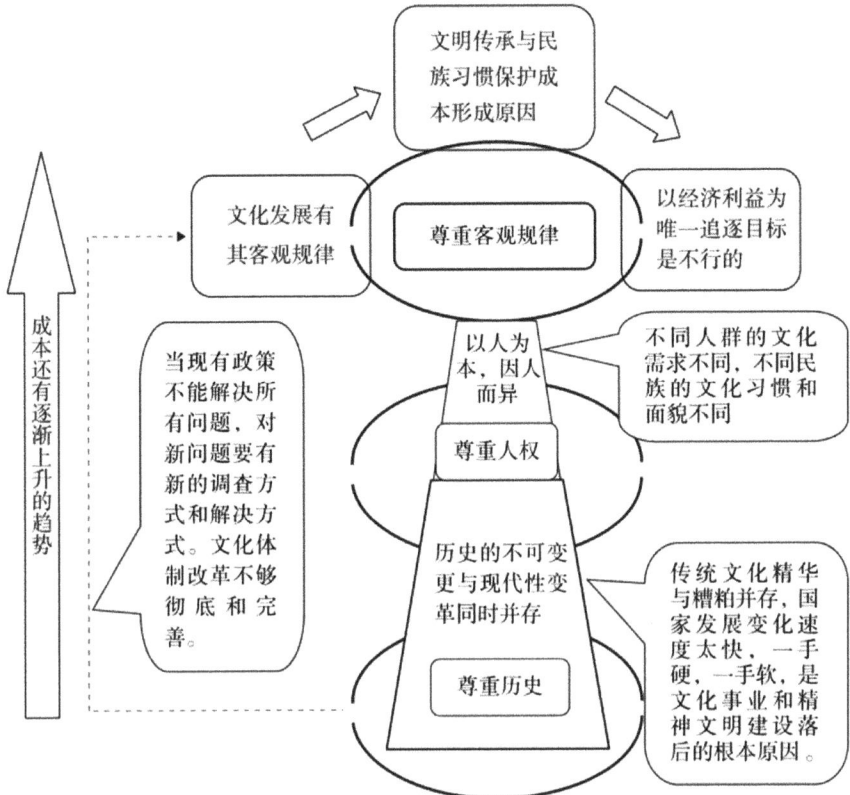

成本还有逐渐上升的趋势	Cost tends to gradually increase
文明传承与民族习惯保护成本形成原因	Causes of the cost of civilization transfer and ethnic habit protection
文化发展有其客观规律	Cultural development has its objective rule
总重客观规律	Respect objective rule
以经济利益为唯一追逐目标是不行的	Considering economic interest as the sole target is not plausible
当现有政策不能解决所有问题，对新问题要有新的调查方法和解决方式。文化体制改革不够彻底和完善	When the current policies are unable to solve all the problems, there should be new investigation methods and solutions for new problems. The cultural system reform is not thorough and improved enough.
以人为本，因人而异	People first, vary with each individual
尊重人权	Respect human rights

Chart 19.4 Forming causes of the cost of civilization inheriting and national customs protection

历史的不可变更与现代性变革同时并存	Unchangeable history and modern reform coexist
尊重历史	Respect the history
不同人群的文化需求不同，不同民族的文化习惯和面貌不同	Different groups of people have different cultural demands, different ethnic groups have different cultural habits and images
传统文化精华与糟粕并存，国家发展变化速度太快，一手硬，一手软，是文化事业和精神文明建设落后的根本原因	Coexistence of good things and bad things in traditional culture, excessively fast national development and change, one hand hard and one hand soft are the basic causes of the backward cultural undertakings and the construction of spiritual civilization.

Chart 19.4 (continued)

Table 19.8 Countermeasures for the cost of civilization inheriting and national customs protection

Countermeasure	Key points	Explanation of examples
Strengthen rule of law construction	Improve the minority habit law, cultural heritage protection law, etc.; strengthen the degree of related departments in law enforcement	For example: strengthen deeper research on the minorities habit law and local law, while guaranteeing the jurisdiction discretion of minority regions, reduce the conflicts between statute law and local law, and timely form effective legal rules to supplement flaws of current statute laws
Enhance administrative efficiency	Strengthen the rule of law awareness, cultural quality and law enforcement according to law awareness of local government and cadres	For example: strengthen the re-education and training on each level of government for ruling ideas. Enhance the ethnic rule of law awareness in government works, enable the cadres in ethnic regions to fully understand this region's ethnic culture, custom, habits and religion situation, guarantee the autonomy department in ethnic autonomous regions to exercise their autonomy

(continued)

Table 19.8 (continued)

Countermeasure	Key points	Explanation of examples
Reduce economic utilitarian ideas	"Developing Economy through Culture" is a wrong idea, we must respect the objective rule of cultural development	For example: some local governments take advantage of local ethnic cultural resources or cultural heritage to excessively introduce business investments, construct at will, destroy the urban image, or disrespect the special rule of traditional art or technique, for business interests, they even made fake or shoddy products. Such phenomenon must be avoided
Strengthen subject construction and input of scientific research	Eliminate the blind spots in cultural heritage protection and traditional culture learning. Enhance the technological content in operation process	For example: related subjects shall be established and increase scientific research input in the research on the protection and management mechanism of intangible cultural heritage, the research on the protection and legislation of intangible cultural heritage, protection of intangible cultural heritage and cultural diversity, intangible cultural resource and ecological environment protection, tourism industry and protection of intangible cultural heritage, protection of the successor of cultural heritage, etc.
Cultivate the active cultural awareness of young people	Establish the environment that is easy to contact traditional culture and different ethnic cultures	For example: build the quality education base for young people on traditional culture, enable the students from college, middle school and primary school to enter museums, visit more cultural relics, hold traditional poetry activities, so as to enable young people to be able to understand and love traditional culture

19.5 Countermeasures for the Cost of Civilization Inheriting and National Customs Protection

Summing up the afore-mentioned cost composition method and expenditure level, it is believed in this book that the countermeasures for further well distributing the cost of civilization inheriting and national customs protection are mainly composed of the following aspects (Table 19.8).

From ancient times continued to nowadays, Chinese nation not only created glorious history of civilization, but also paid heavy price and high cost and accumulated hard lessons. The reason why the inheriting of Chinese civilization and the protection of national customs could be continued to nowadays is because every Chinese person actively or passively bears the heavy duty of sustain civilization, protect the cultural tradition of our own ethnic and other brother ethnic groups.

Cultural development is closely related to economic development and social system development, the rise or fall of cultural undertakings would directly affect the speed of economic development, cause changes of social structure, and finally affect the entire social form. However, cultural development can't have the realization of political targets as the primary objective, nor use economic benefit as the sole measuring standard, this is in conflict with the culture's own characteristics of requiring independence, diversification and spiritualization.

Chapter 20
Cost of Chinese Ideological Evolution and Modern Media System Construction

20.1 Connotation and Forms of the Cost of Chinese Modern Ideological Evolution

The cost of Chinese modern ideological evolution mainly means the efforts made since the establishment of New China for ideological construction, the detours made as well as the conflicts and risks faced nowadays. China is a socialism nation that adheres to Marxism theories, which is essential different from the ideology of western countries. Therefore, every evolution and change in ideological field would cause major influence on the national construction causes and people's social lives, we have paid huge price of this. The cost of ideological construction is controllable, its orientation and regulatory method are also worthy of in-depth discussion, this is major guideline of nation.

It is believed in this book that based on the analysis in previous section, the cost of Chinese ideological evolution mainly includes four forms, meaning the cost of ideological evolution since the establishment of New China, the regular cost of ideological construction in current China, the risks and costs of Chinese ideological construction nowadays, the cost of ideological contradiction and conflict in international community. For more details, please refer to Table 20.1.

20.1.1 Cost of Ideological Evolution Since the Establishment of New China

1. **Cost of Generalization and Detours of Ideological Construction at the Beginning Since the Establishment of New China**

At the beginning since the establishment of New China, with the smooth development of various social undertakings, the main national leaders changed the original judgments and assumptions, the economic construction began to have

Table 20.1 List of the forms of the cost of Chinese ideological evolution

Classification of cost	Forms	Specific contents
Cost of China's ideological evolution	Cost of ideological evolution since the establishment of New China	Improper ideological construction, irrational political movements, and turbulent social situation caused loss of national construction and hurt feeling of the people
	Regular cost of China's current ideological construction	Formation, improvement and promotion of the party's theories. Meanwhile, when instructing the construction of other undertakings of society, it is inevitable to have conflicts and pay running-in cost
	Cost of risks in China's current ideological construction	Penetration of western thinking and culture; the domestic market economic construction has diversified values, the mainstream ideology is under impact. The proposition and advocacy of the socialist core value
	Cost of confrontation and conflict of international ideologies	The integration process into globalization brings new challenges to China, the penetration of western ideology strengthens. It requires innovation of thinking, adjustment of ideology, and participate in international competition in opener attitude

tendency of rushing things through and partial extravagance of subjective initiative.[1] At the beginning of the 1950s, relatively big mistakes were made and factors of social instability occurred. Since 1966, the "Leftist" thinking dominated the entire ideology, the ideology was in prevailing status,[2] the socialist ideology had severe detour. The final consequence was that ideology was severely generalized,

---

[1] In 1956, the international communism movement had severe problem, the influence of Poland and Hungary Incident and the 20th National Congress of CPSU affected China, which caused sudden tightening of Chinese political atmosphere. In 1957, Mao Zedong stressed class struggle again, and had the trend of expanded class struggle, the various policies of the party and country began to deviate from correct track, and had more and more severe mistakes.

[2] Back then, the dispute of ideological field was simply equivalent to confrontation of two social systems, requiring "Total Dictatorship" and "Continuous Revolution", with the increasingly stronger personality cult, the leader was praised by few people with certain agenda as "Genius" and "Peak", which stated unprecedented deity creating movement, continuously strengthening the superstition and blind following to the leader, any and all ideas that are inconsistent with the leader, classics and policies lost the reasonability to exist, the ideology has gradually changed into doctrinism, absolutism and sanctification, caused many correct viewpoints to be criticized. It should be denied that "cultural revolution" generated far-reaching influence on ideological work.

the combination of individual worshiping and mass campaign increased the irrational factors of ideology, which caused huge damage to the socialist ideology.[3] Behind the "Great Prosperity" of political movements from 1950s to 1970s, China paid dear economic cost and political cost, as well as severely damaged the image of the socialist ideology.

2. **Two Strategic Transformations of Ideological Construction**

Since 1980s, China's ideological construction had two strategic transformations, the various social undertakings returned to right track, and China entered the real prosperous development stage (refer to Table 20.2).

20.1.2 Regular Cost of Current Chinese Ideological Construction

1. **Cost of Creating Socialism Ideological Theories with Chinese Characteristics**

The scientific theoretical system of socialist ideology with Chinese characteristics is obtained through decades of discovery and from practice. In was clearly indicated in the party's 17th National Congress report that "The fundamental reason why we made all the achievements and progresses since the reform and opening-up can be concluded as: we have opened the socialism path with Chinese characteristics and formed the socialism theoretical system with Chinese characteristics."[4] The theoretical system of socialism with Chinese characteristics means the scientific theoretical system including the "Deng Xiaoping Theory", the important though of "Three Represents", the "Scientific Outlook on Development" and other major strategic thoughts.[5]

2. **Cost of Mutual Influence Between the Development of Chinese Ideology and Cultural Industry**

(1) Emerging stage of cultural industry and communicative dilemma of mainstream ideology

---

[3]After 1966, although the party's theory and policy had relatively big deviation and mistakes, people's living standard didn't have obvious improvement, there were economic stagnation and social turmoil, the ideology had already been disconnected or even conflicted with social development, people's lives. However, because the party had accumulated deep mass foundation in history, and the leader had great personal prestige, the socialist ideology didn't have obvious crisis during a very long period of time, but there was huge potential danger.

[4]*Guiding Textbook of the Report of the 17th National Congress of CCP*, People's Publishing House, 2007 Edition, pp. 10–11.

[5]*Guiding Textbook of the Report of the 17th National Congress of CCP*, People's Publishing House, 2007 Edition, p. 11.

Table 20.2 Two major strategic transformations of ideology since reform and opening-up

Leader	Time	Adjustment strategy	Great significance
Deng Xiaoping	1980s	Transformed from stressing the revolutionary function of ideology to stressing the service function and consolidating function of ideology	Correction of thinking route; criticized the doctrinarism and extremely "Leftist" thinking, ended the abnormal state in ideological field
Jiang Zemin	1990s	Ideological construction was mainly on innovation based on practice, preliminarily consolidated the basic ideological framework that is suitable to market economy	The new-round of top down "Reform" had obviously less obstacle, which enabled the overall administrative and management system to quickly re-unified based on utilitarian principle, and won warm response from the society
Hu Jintao	Beginning of the 21st century	The party's Sixth Plenary Session of the 16th Party Congress put the establishment of socialist core value system at prominent position, and clearly proposed the Marxism as the guidance, the socialist common faith with Chinese characteristics as the theme, the reform and innovation as the characteristics, the patriotism as the core, the national spirit as the essence and the socialist honor and disgrace outlook as the foundation, or the "Four in One" value system	It was another strategic transition after the "Cultural Revolution" that Deng Xiaoping included ideology work into the overall modernization construction with economic construction as the center. Such transition reflected the deep understanding of the characteristics of new era and historical conditions, as well as the attitude of carefully absorbing the experiences and lessons of domestic and foreign ideological construction

The period of 1978–1992 was the stage when people's thinking began to gradually activated after the completion of "Cultural Revolution", and also the stage when the socialism cultural undertakings returned to right track.[6] Cultural workers back then didn't know how to integrate ideology in cultural products, so they had to

---

[6]The big discussion regarding the standard of truth, the holding of the Third Plenary Session of the 11th National Congress of CCP, the emancipation of minds, and the re-establishment of the practical thinking route greatly liberated people's minds that had been restrained for long time, people broke through the jail of "Leftist" thinking, eager to understand new life and new knowledge, and continuously increased demands for cultural consumption. Cultural industry was started under such atmosphere. Because it was still the management model of planning economy, but the cultural market and relaxed cultural ecological environment that is critical to cultural industry hadn't been formed yet, which determined that the cultural industry at that stage could sparsely exist, but this couldn't prevent cultural industry from bringing certain impact on people's spiritual life and value.

partially simulate mass culture in the west, their works generated negative influence on China's mainstream ideology. Besides, due to the lack of management experiences in cultural industry development back then, plus the complex social environment and international background, there were several negative ideological trends in society, for which reason, the central government issued multiple documents and developed rectification works.[7] The cultural undertakings had always been the main channel for the party to promote ideology and undertake cultural construction. However, at that historical stage, the development of cultural industry didn't well play the role of promoting mainstream ideology, and even had differences in directions (Chart 20.1).

(2) Conflict between cultural industry with mainstream ideology at emerging stage

Since 1992, with the party and government's relaxation of cultural management policy, the development of cultural industry began to reflect the trend of diversification, and inevitably began to fight for ideological and cultural ground with mainstream ideology.[8] Although the development situation of mass culture has been good, but it always has two negative development trends: firstly, pursue economic benefits and ignore social benefits, equate vulgarism and shoddy manufacturing to popularization; secondly, partially separate cultural industry from political publicity, and believe that culture must get rid of the control and restriction of ideology. Facing the situation of diversification of ideology, the guiding methods adopted by the government included gradual proceeding of cultural system reform, strengthening the guiding and management of cultural market, transform from

---

[7]On Jun 6, 1983, the *Government Work Report* of the 1st Session of the 6th National Congress also indicated that "Most spiritual products are circulated in form of commodity, but we absolutely can't let the corrupted thinking of "everything is about money" to freely spread in our social lives". On Oct 12, 1983, Deng Xiaoping strictly indicated that such "Everything is about money" thinking and the trend of commercializing spiritual products are also reflected in other aspects of spiritual production. Some people who are mixed into art circle, publishing circle, cultural relics circle are becoming gainful businessmen". Source: *Selected Works of Deng Xiaoping*, Volume 3, People's Publishing House, 1993 Edition, p. 43.

[8]In 1992, the party's 14th National Congress determined the reform goal of establishing the socialist market economic system, which provided better environment for the cultural industry development that is closely correlated with market economy, China's cultural industrialization process accelerated apparently, and gradually expanded from circulation industry to manufacturing industry and service industry, and developed from relatively small enterprise units to relatively large cultural enterprise group. The cultural industry wave further expanded from private enterprise to large state-owned cultural units. Meanwhile, government cultural system reform was also started, the cultural department began to transform from "Direct Management" to "Indirect Management".

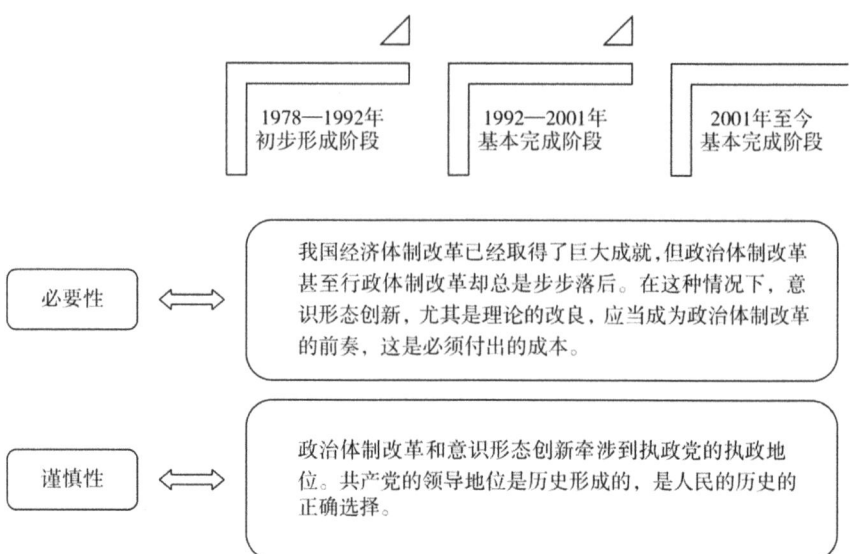

Chart 20.1 Formation and necessity of the scientific theoretical system of the socialism ideology with Chinese characteristics

administrative management to management according to law, adjusting the management degree of cultural policy on cultural industry, trying to reinforce the voice of mainstream ideology in the ideology publicity in cultural industry.[9] In the aspect of guiding and managing cultural market, the government management authority changed the means of purely using administrative documents to manage cultural market and actively developed crack-down actions.[10]

(3) Large-scale cultural industry and the normalization of the communication of mainstream ideology

Since the party's 16th National Congress, under the increasingly relaxed and harmonious cultural ecological environment, the development of cultural industry began to walk on the track of standardization, under the precondition of giving full scope to the theme of the times, obtaining bigger economic benefits have become the operating strategy and objective of most enterprises and public institutions of cultural industry. This was mainly resulted from the two aspects of efforts made by the Chinese Communist Party at this stage in handling cultural industry development and ideological communicative relationship: on one hand, implement macro-level management on cultural industry units through cultural system reform, more complete cultural industry policies and laws; on the other hand, continue to adhere to the cultural guideline of "Giving full scope to the theme of the times and promoting diversification", use the strengthening of mainstream ideological con-

---

[9]In 1992, the party's 14th National Congress Report clearly proposed to "improve cultural economic policy", in the same year, in the *Decision by CCP Central Committee and State Council Regarding Accelerated Development of Tertiary Industry* considered "cultural and health undertakings" as the key to accelerate the development of tertiary industry, the book *Major Strategic Decision-making—Accelerated Development of Tertiary Industry* that was published in the same year and edited by the General Office of the State Council, clearly used the concept of "Cultural Industry". The *Draft Report Regarding the National Economy and Social Development Plan Execution in 1998 and the National Economy and Social Development Plan in 1999* that was issued in 1999 clearly proposed to "promote the industrialization of culture", cultural industry was for the first time formally included in national development plan. In October, 2000, the *Suggestion by the CCP Central Committee Regarding the "Tenth Five Year Plan"* indicated to improve cultural industry policy, strengthen the construction and management of cultural market and promote the development of related cultural industry".

[10]1994 was the cultural market performance year, the key was the overall regulation of performance market; 1996 was the cultural market video year, which heavily cracked down illegal video products; 1998 was the cultural market legal year, which emphasized on the legal construction of cultural market.

struction to strengthen its appeal, so as to enable mainstream ideology to feasibly play leading role in the cultural development field, and guide diversified ideology to develop healthily.[11] China's cultural industry has since entered the industrialized development stage and scale operation.

20.1.3 Cost of Risks in Current Chinese Ideological Construction

Since the 1990s, under the background of economic globalization and new scientific and technological revolution, the rivalry of soft power among different countries has intensified, which poses huge challenge to China's mainstream ideology, therefore, in order to firmly improve mainstream ideological construction, China shall face the following risks, so as to realize "knowing the energy and knowing yourself and you can fight a hundred battles with no danger of defeat" (Table 20.3).

20.1.4 Cost of International Ideological Contradictions and Conflicts

1. **Cost of Chinese Socialism Ideology Responding to Globalization**

Ideology has always been deemed as the most difficult to reform and transform for a national power. However, since the reform and opening-up, Chinese Communist Party has successfully realized strategic transformation of ideology. The practices in the past have proved that communism system can be reformed, and can realize its own transformation through continuous adjustment in order to adapt to the globalization. But how the party could realize the legalization of such management and

---

[11]The party's 16th National Congress Report indicated to "actively develop cultural undertakings and cultural industry, continue to deepen the cultural system reform", and indicated that "developing cultural industry is an important way to boost the socialist culture under market economy condition and satisfy the spiritual and cultural demands of the people", which reflected that the party has unprecedented attention and strong support to the "Sunrise Industry" of culture. At the National Publicity Work Meeting in 2003, Hu Jintao stressed to "adhere to consider active development of cultural undertakings and cultural industry as an important task of publicity and cultural department". The *Resolution by the CCP Central Committee Regarding Several Issues on the Improvement of the Socialist Market Economy System* passed at the party's Third Plenary Session of the 16th National Congress indicated the transformation of government function in management of cultural undertakings, promote shareholding system in cultural field, implement diversified entities of investment, which made clear indication to China's cultural system reform. The *Decision by the CCP Central Committee Regarding the Strengthening of the Party's Ruling Ability Construction* passed at the party's Fourth Session of the 16th National Congress indicated again that "We need to deepen cultural system reform, liberate and develop cultural productivity …, further eliminate the system obstacles that restrain cultural development".

20.1 Connotation and Forms of the Cost of Chinese Modern …

Table 20.3 Detailed contents of the cost of risks in current Chinese ideological construction

Forms of risk	Possible consequence	Forms
Threat of cultural penetration by western hostile force on China's ideology	Peaceful evolution happened in socialist country	The west uses its primary advantages in economy, politics and military to promote the west's "Freedom and Democratic System", use the ethnic culture recognition crisis of developing country in modernization process, force or coerce them to recognize the western value
Social changes caused from new technology revolution weakens differences between the two major ideologies	All the countries under economic modernization may walk on the path of similar science and technology	The new technology revolution caused people to concentrate more attention on scientific and technological level, the instrumental rationality with science and technology as the main carrier have become the main ideology of modern society, and make people manipulated by instrumental rationality
The diversified value of socialist market economy brought negative influence on China's mainstream ideology	To certain extent constitutes the dispute and conflict between China's mainstream value and ideology	With the in-depth reform of China's economic system, under the complex background of intertwined and mutually affected various thinking and culture, different social classes and interest groups keep appearing, which provides realistic soil for the diversification of value and ideology
The communication method of internet age bring great challenge to China's ideological security	It not only provided new channel for the communication of socialist ideology, created new opportunities for strengthening the attractiveness and cohesion of China's mainstream ideology, and also imposed great challenge to China's ideological security	The development information network has changed traditional cultural communication method, network communication has the characteristics of real-time, massive, globalized, interactive, etc. which make the national measures to be completely exposed to the public, and provide good opportunity to enemy forces

further realize systemization of the nation is the major challenge that the party has to face directly.

Before the reform and opening-up, China had been deemed as a society established based on political ideology.[12] At the end of 1970s, the leadership with Deng Xiaoping as the core began to transform work focus to economy, the rapid expansion of private field extensively weakened the degree of political conflict, so as to reduce the party and government's burden of political reform, this had become a channel to national transformation.[13] However after integrating into the global system, the social order originally based on interest generated many new issues, the emerging social forces require reform of political order.[14] The current political order is continuously subjecting to internal impacts, and the pressure of political reform is also increasingly.[15]

2. Cost of Resisting Ideological Penetration from the West

With the end of the Cold War, under the joint efforts by the western governments and the capitalist class, they used their strong economic might to popularize and promote such idea that "Human history will end at the free and democratic value of the west" and the market economy model across the globe. Under such circumstance, China's special ideological model will inevitably be subject to impact from the west.

Nowadays, the western nations try to use the crossing and integration of many ways to promote their ideology, and integrate various social forces to form systematic controlling network. For example, using news papers, books, radios, films, television, internet and other media means to implement ideological penetration[16]; using various cultural exchanges and activities to strengthen the ideological influ-

---

[12]The society is organized based on highly unified political ideology, and use household registration system, unit system and other mass organization and other powerful organizational weapon to realize in-depth and complicated social projects and tasks, such as land reform, collectiveness, nationalization of industry and commerce, etc.

[13]In 1980s, CCP was not against the various economic form experiments of capitalism on actual policy level, but on ideological level, the legalization of capitalist economy was still not recognized.

[14]In 1990s, members of society were dedicated to make fortune, political passion waned down, this period reduced the party's political pressure, and the legitimacy of political system strengthened. But it also made the existing political order have no sufficient pressure to adapt to the interest-based social order.

[15]In recent years, advocacy to return the virtue and moral value has become the most important method for the leaders to enhance revitalize the party again. From 2000 to present, CCP launched large-scale cadre education training.

[16]Not only specially open media, such as the Voice of America, Free Europe Radio that affect the world to directly attack, but also assist and support the opposition party of object country to create media and brainwash the mass.

ence to the knowledge and cultural circle[17]; using the cultivation, support and funding of the "Dissidents" and "Opposition" in object countries to reach the purpose of ideological attack and subversion of national power of object countries.[18] Under such circumstance, China has to increase the corresponding prevention cost, build external promotion and exchange platform, adjust foreign strategy, implement cross using of economic means and cultural means, while consolidating our own stance, China must also strengthen the output the culture and value.

20.2 Connotation and Forms of the Construction Cost of Chinese Modern Media System

The construction cost of Chinese modern media system is correlated with the basic functions of media. Chinese media assumes the important "Mouthpiece" function and "Lookout" function, its construction cost is mainly reflected as the aspects including the party and government's operation and management of media undertakings (including fiscal expenditure and talent cultivation, etc.), media system reform at current stage and various possible crisis in responding to the new media age (Chart 20.2).

It is believed in this book that the construction cost of Chinese modern media system mainly includes the general management cost of the media system with Chinese characteristics and the reform cost of the modern media system, for more details, please refer to Table 20.4.

---

[17]In recent years, the western countries are paying more and more attention to using various foundations, research institutes, etc. to hold various names of international academic exchange activities, adopt the forms of scholar visits, lecture and senior education cross-border exchange, etc. so as to use the so-called cultural exchange to reach the political purpose of affecting intellectuals in thinking, and affect their public opinions in society. For example, the US Congress, the Pentagon, the National Security Commission, the US Academy of Navy and other government departments keep and sponsor a large batch of "research centers", foundations, university research institute, and have frequent activities in China in name of scholar exchange, sponsor project research, etc.

[18]China's Taiwan Issue, Xinjiang Issue and Tibet Issue are all the keys that the western countries pay attention to and take advantage of, especially the Tibet Issue. Source: Wei Mingliu: *Where is Dalai Group's Money Coming From*, Youth Reference, Apr 12, 2008.

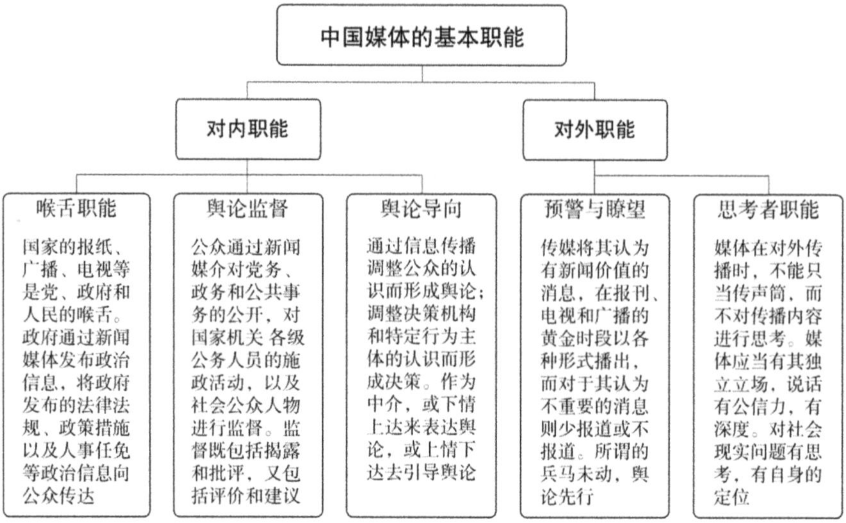

```
                    中国媒体的基本职能
                   /                \
              对内职能              对外职能
           /     |      \           /      \
     喉舌职能  舆论监督  舆论导向   预警与瞭望  思考者职能
```

中国媒体的基本职能	Basic functions of Chinese media
对内职能	Internal function
对外职能	External function
喉舌职能 国家的报纸、广播、电视等是党、政府和人民的喉舌。政府通过新闻媒体发布政治信息，将政府发布的法律法规、政策措施以及人事任免等政治信息向公众传达	Mouthpiece function The state government's newspaper, ratio, TV are the mouthpiece of the party, government and people. The government expresses political information through news media, deliver the laws, regulations, policies, measures and employment change to the public
舆论监督 公众通过新闻媒介对党务、政务和公共事务的公开，对国家机关各级公务人员的施政活动，以及社会公众人物进行监督。监督既包括揭露和批评，又包括评价和建议	Media supervision Through media, the public supervise the opening of party affairs, government affairs and public affairs, the governance by national departments and civil servants as well as public figures. Supervision includes not only exposure and criticism, but also evaluation and suggestion.
舆论导向 通过信息传播调整公众的认识而形成舆论；调整决策机构和特定行为主体的认识而形成决策。作为中介，或下情上达来表达舆论，或上情下达去引导舆论	Guidance of public opinion Form public opinion by adjusting the recognition of the public through information communication; form decision making by adjusting the recognition of decision-making body and certain entity of conducts. Express

Chart 20.2 Basic functions of Chinese media

		public opinion as intermediary to pass information top down or bottom up.
预警与瞭望 传媒将其认为有新闻价值的消息，在报刊、电视和广播的黄金时段以各种形式播出，而对于其认为不重要的消息则少报道或不报道。所谓的兵马未动，舆论先行		Early warning and lookout Media would release newsworthy information in various forms on newspaper, TV and radio at prime time, and rarely or not report the news that they deem not important. The so-called public opinion before action
思考者职能 媒体在对外传播时，不能只当传声筒，而不对传播内容进行思考。媒体应当有其独立立场，说话有公信力，有深度。对社会现实问题有思考，有自身的定位		Function of thinker When media broadcast externally, it should not only serve as a passer and not think about the broadcasted contents. Media should have its own stance, speak with creditability and depth. They should have thoughts about realistic issues in society and have their own positioning

Chart 20.2 (continued)

Table 20.4 Forms of the construction results of Chinese modern media system

Classification of contents	Forms	Specific contents
Cost of the construction of China's modern media system	General management cost of media system with Chinese characteristics	Original system of cultural institution is complex, redundant, large involvement, and consumes large unnecessary administrative cost; when reform approaches, it would change into the obstacles affecting and impeding reform
	Reform cost of modern media system	System reform means reconstruction of new management system and operation system, but under market economy environment, the sliding moral standard of media and appearance of new media both bring difficulty to reform. Besides, the problem of excessive administrative intervention still exists, which make media unable to fully release vitality

20.2.1 General Management Cost of the Media System with Chinese Characteristics

The cultural management institutions currently mainly include the Propaganda Department of the Central Committee of the CPC, the Ministry of Culture, the State

Administration of Radio, Film and Television, the General Administration of Press and Publication, the State Council Information Office (Foreign Propaganda Office) and the Xinhua News Agency.[19] The formation and operation of this management system has its deep historical reasons. However, with the deepening of the development of cultural undertakings, such system would obviously increase management cost,[20] and cause of increase of media operation cost.[21] In 2006, the CPC Central Committee and the State Council required in the *Several Opinions Regarding the Deepening of Cultural System Reform* that the government shall strengthen and improve the macro-level management on cultural field, accelerate the transformation of governmental functions, clear the duties of cultural

---

[19]If we say that the Propaganda Department of the Central Committee of the CPC is a comprehensive management department, then the rest five departments are specific management departments to certain sense, they respectively manage certain cultural fields. The Ministry of Culture manages culture and art, the State Administration of Radio, Film and Television manages radio and television, the General Administration of Press and Publication manages news publishing, the State Council Information Office is mainly responsible for external publicity, while the Xinhua News Agency approves the related operations of foreign news agencies in China.

[20]Compared with media in the past, the current media not only has diversified products, but also have various operation fields, abundant operation means, there are not only production of news products, but also non-news cultural and art activities; there are not only operation of certain industries and certain regions, but also the cross-industry and cross-region operation; not only traditional media operation, but also capital operation following market economy rules. Under such circumstance, although there is no change in traditional cultural management system, the things to do by different management departments are intertwined and difficult to differ the authorities.

[21]Under the background of the major development new technology, especially internet technology, the scope of authorities of different administrative departments are not clear any more, but intertwined. So the administrators of unclear authorities would hold their own grounds and argue, which made the cultural market and big media unable to adapt. Taking the internet for an example, the ones with administrative function includes the Ministry of Culture, the State Administration of Radio, Film and Television manages radio and television, the News Office of the State Council and the General Administration of Press and Publication manages news publishing, the Ministry of culture is "responsible for the prior approval of the online broadcasting of cultural and art products, responsible for implementing operation license management of internet café or other internet service sites, and regulate online game service (excluding prior approval of online game before release)"; the State Administration of Radio, Film and Television manages radio and television is responsible for the management of online video program service institution and business management and implement access and exit management, approving the contents and quality of information and online video program; the News Office of the State Council is responsible for "forming the development plan of internet news business, and instructing and coordinating the internet news reporting work"; the General Administration of Press and Publication manages news publishing is "responsible for the internet publishing activities and approval and regulating the internet publishing activities, cell phone magazine and literature businesses", its department of science, technology and digital publishing would not only "participate in the drafting of the regulations and rules of internet (including online game) publishing, cell phone publishing and digital publishing, forming the policies and important management measures of internet publishing, cell phone publishing and digital publishing, and supervise the implementation", but also "responsible for the regulation work of internet publishing, cell phone publishing and digital publishing activities, and implement access exit mechanism", "organize and punish the violation and lawbreaking behaviors of internet publishing and cell phone publishing".

administrative departments, smoothen the relationship between cultural administrative departments and their affiliated cultural enterprises and public institutions. In 2009, it was further indicated in the *National Layout Plan of Cultural Development during the "Eleventh Five Year Plan" Period* that the government shall actively promote the transformation of governmental functions, realize separation between government and enterprises, separation between government and public institutions, separation between government and capital and separation between management and handling, feasibly transform the governmental function from organizing culture to social management and public service." Although the document of central government was issued, the division of work among cultural management departments is still chaotic.[22]

20.2.2 Reform Cost of Modern Media System

1. Review of Reform History Since 1980s

China's media industrialization[23] began with the approval by the Ministry of Finance in 1978 for the attempt by 8 central news units including the *People's Daily* to promote enterprise management system, its development process generally experienced two stages. The first stage was 1978–1992, which was the period when China's media industry emerged. At that stage, news media gradually obtained proper operation autonomy, meanwhile, the forms of advertisements on media became increasingly diversified advertisement income continued to increase and accumulated large amount of capital for media enterprise. The second stage was from 1992 to present, which was the high-speed development period of Chinese media industry. At this stage, with the establishment of Chinese socialism market economy system, the market degree of media enterprises became higher, under the situation of facing internal and external competitions, China's media enterprise gradually realized conglomeration and scale operation through resource integration, restructuring and external expansion forms. In October, 2008, the Propaganda Department of the Central Committee of the CPC along with the Organizing

---

[22]For example, the tenth responsibility of the publicity management department of the State Administration of Radio, Film and Television is the "responsibility for the management of the publishing of radio, movie, book and magazine publishing of the current unit". There are two issues: firstly, State Administration of Radio, Film and Television is not only a cultural administrative department, but also a producer of cultural products, such method of serving both "Referee" and "Player" is obviously inconsistent with the market economy principle. Secondly, the responsibility of the Administration of News Publishing is to manage the news publishing industry across the country, but the books and magazines published by the State Administration of Radio, Film and Television are to be managed by the publicity management department of the State Administration of Radio, Film and Television, which is not reasonable.

[23]Media industrialization means that the production, sales and management of media products are operated according to the industrial rules formed by state industrial department, meaning using market means to distribute media resources.

Department of the Central Committee of the Communist Party of China, the Central Compilation Office, the National Development and Reform Commission, the Ministry of Finance, the Ministry of Human Resource and Social Security, the Ministry of Culture, the State Administration of Taxation, the State Administration of Industry and Commerce, the State Administration of Radio, Film and Television, the General Administration of Press and Publication and other related departments and units formed the *Regulation on the Transformation of Operational Cultural Public Institutions into Enterprises in Cultural System Reform* and the *Regulation on Supporting the Development of Operational Cultural Enterprises in Cultural System Reform*, which further expanded the scope of media's intangible assets usage and investment ratio, clearly proposed "targeting the characteristics of the cultural enterprises, research and form the evaluating and pledge method of copy rights, cultural brand and other intangible assets, guide commercial banks to grant loan support to cultural enterprises", "allow investors to invest and establish cultural enterprises with evaluation of intellectual property and other intangible assets, and the ratio of non-currency assets converted and invested in registered capital shall not exceed 70%". The new round of cultural system reform policy, while stressing the separation between the propaganda of party's news papers and journals and operational businesses, emphasized more on the restructuring of non-news journals and key news websites, recognized the specialty of the forms of intangible assets of media, enhanced the investment ratio of intangible assets, and explored a channel for media to have leap-forward development from communicative media to newspaper media and cultural media.

2. Additional Cost Brought from Media System Reform

(1) Sliding professional ethics of media, guidance of public opinion deviate from the right track

For government, one of the important functions of media is guidance of public opinion. However, under the current market environment, the media's role of guidance of public opinion tends to weaken, the mainstream ideology is facing great challenge. In order to absorb eyeball from the social public, operation of media even reflected vulgarization, repeating and false reporting.[24] On the other hand, under the seduction of interest, the professional ethics of media reflected degrading trend. The phenomenon of "Paid News" and "Paid No News" often occurred. What the public needs is objective and authentic report, while the "Paid News" of media to government or enterprises enables media to publish the unauthentic materials that are prepared by government or enterprises in advance and work as the effect of advertisement. When media obtains information that is negative to government or enterprise, they are supposed to publish according to truth

---

[24]For example, the current TV programs are filled with matchmaking programs, talent shows, but the programs with elegant arts are relatively few. Such programs and related news don't have too much goods to the general public, but occupy the time of elegant and civilized TV programs, so as to weaken the public opinion guidance function of media.

and realize the function of supervision by public opinions, but driven by interest, media would probably choose not to publish. Such phenomenon undermines the media's function of supervision by public opinions, the lack of supervision by media, high opinions of the public, untrue reports by media and the people being cheated indirectly caused the deepening of social conflicts.

(2) Traditional media couldn't adapt to market operation, the old system becomes the biggest shackle

Before the reform and opening-up, the top-down management method on media has strict hierarchy and poor liquidity. However, for the media that has already transformed or is about to transform into enterprise, what they need is a free flow including capital flow. The original system obviously set obstacle for such flow, so as to greatly restrain media enterprises.[25] On the contrary, the existing advantages of many media were obtained through monopoly, after public institutions transformed into enterprises, these media faced the issue of independent existence, they could no longer resort to fiscal support, because the market could only recognize those cultural products with real vitality. But if tossing their management and operation into the ocean of market economy, it is very worrisome whether they could survive or have the room of development.

(3) The issue of excessive administrative intervention and media freedom

With the stable and relatively fast development of the national economy, Chinese domestic consumer market has been active and the media advertisement market also reflected rising trend. In 2006, China's GDP reached over RMB 20 trillion, increasing by 10.7% compared with the same period in previous year, the total amount of advertisement market of Chinese media in that year was RMB 287.5 billion, increasing by 18% compared with the same period in previous year, which was greatly higher than that of many western developed nations.[26] But media professionals felt that the living environment is very cruel and difficult, they are

---

[25]The first thing is the four-level existence of media at "central—province—prefecture—county". In China, each level of administrative organization from the central government to county government has its own media. Although in the news paper and magazine governance in 2003, most county-level party newspapers were cut out, but in certain regions, county-level party newspaper still exists. These newspapers, along with county-level TV station, have many problems in news making, management and operation—their news are out of touch with the mass, the program quality is relatively low, or publishing and broadcasting irregular advertisements due to economic dilemma, etc.

[26]Cui Baoguo: 2007: *Chinese Media Industry Development Report,* Social Science Literature Press, 2007 Edition.

unable to develop the independence as a third party. The reason is that many local governments often adopt administrative means to arrogantly intervene the reporting rights or advertisement rights of media, their behaviors are not only meeting management regulations nor meeting market rules. For example, some local governments often, for the purpose of promoting political achievements or promoting the development of certain local undertaking, input advertisements in large scale on local media, although media make all efforts to coordinate, they could only receive symbolic compensation or any at all. For another example, the current policy has very strict restriction on new media with active thinking, the media don't have the rights to speak up, and they even don't have the right to truthfully report phenomenon in the society. Just taking the newspaper establishment in different place for an example, the previous policy had extremely strict control on newspaper establishment in different place, although slightly relaxed nowadays, its development is still very difficult. For example, from *Oriental Morning Post* to *Beijing News*, the several attempts by the Nanfang Media Group to expand outside Guangdong have been very difficult.[27] It can be said that with the deepening of media system reform, the government's management function will indeed be weakened, but how to weaken, how to handle the conflict between the freedom for media to survive and the rationality of government management, how to realize all flowers bloom together and contention of a hundred schools of thought while maintaining the unique ideology of China, it still needs time. The current situation is that the media don't have high degree of freedom, their independence is not strong, the quality of media professionals varies, and their level of work is not high. This indicates that China's modern media is still in development, the media development and government management are still in running-in phase, the necessary price and cost are to be paid.

(4) Cost of ideological security maintaining cost

It was indicated in the *Chinese New Media Development Report (2011)* jointly released by the News and Communication Research Institute of the Chinese Academy of Social Science and the Social Science Literature Press in Beijing on Jul 12, 2010 that the new media such as internet and cell phone continue to advance extensively and rapidly expand on global scale; the new media is penetrating deep into the society, and become the socialized media that promote the comprehensive

---

[27]Gan Xianfeng and Li Xu: *Promotion of System Reform of Traditional Media,* News Frontier, 5th Issue in 2009.

transformation of the nation and society. In this process, Chinese new media's characteristic of localized development is becoming increasingly clear, the development path of new media with Chinese characteristics gradually take shape.[28] New media has really developed, this year is a year with symbolic significance in the development history of new media. From the original media forms based on new information network technology such as micro-blog, social network website, instant message, forum, blog, vlog, search engine, cell phone short message, etc. to the new media innovated from traditional media, such as internet TV, online magazine, cell phone news, e-book, etc. emerged along with time, they have obvious development trend of integration, mobilization and socialization, which greatly enriched the forms of media, explored the living room of media, guided the brand-new method of news reporting and the orientation of social opinions, and will deeply affect Chinese political and cultural ecology, and will finally become an important engine of economic and industrial development, and deeply penetrate into the existing method of the society.

Meanwhile, the western media never stop their steps of using new media to implement ideological penetration, the internet has become the terminal tool to export western value to the whole world. It is not difficult to predict that the new media will become the main platform for the US to launch ideological war against China. The US, based on the consideration of its own interest and communicative effect, implements strategic transformation of communication against China, and smartly took advantage of the change of communicative situation from the rise of new media. New media will become the main carrier of the ideological hegemony of the US. Under such situation, ideological security has become the most important issue of current China. New media technology has generated extremely big influence on the establishment of ideology, its threshold of cultural and ideological penetration is being reduced. For China, the new media in globalized, networking and socialized development have explored the time and space for the existence of ideology, enabled the conflict and confrontation of different ideologies to become more direct, fierce and diversified. China must also directly face such radically changed form, win this battle of new media, make Chinese news to go out and pass on Chinese voice, instead of always passively responding to such battles.

---

[28] A set of statistic figures released in the foresaid report was inspiring. In 2010, the number of internet users across the globe had reached 2 billion, the number of cell phone users had reached 5 billion, the number of internet users in China had reached 450 and 900 million for cell phone users, China has become the real biggest nation of new media users across the globe. From 22.5 million in 2000 to 457 million people in 2010, the number of Chinese internet users increased by about 20 times within short 10 years.

20.3 Forming Causes of the Cost of Ideological Change and the Cost of Modern Media System Construction

Since the establishment of New China, Chinese ideology had certain reforms, the main causes of which is that during the process of Chinese social transformation in modern times, the interest relationships keep adjustments, reforms broke the original interest pattern, and transformations are made from the original single planning economy structure to the diversified structure with market contracting type. Meanwhile, social class and class structure had major changes, the subjective structure of interest in society reflect the trend of diversification. The transformation from unified society to diversified society changed the composition of social classes, different classes have higher requirements for political democracy and administration by rule of law, they hope that the political interests and rights of their own class or group could be satisfied under a democratic system, so as to generate more rights to participate in political life, this is a bottom-up demand. While from the top-down perspective, as the ruling party, the Chinese Communist Party shoulders the great mission of national prosperity and revival of nationhood, the party and government are the planners and promoters of development, in order to maintain the party's solid ruling status, it means they need to strengthen the legitimacy and authoritativeness of their ruling, build the harmonious political participation mechanism, expand the democracy within the party and democracy among people, coordinate the interest demands of different classes in society, therefore, ideological adjustment is imperative.

Modern media system reform is on one hand in order to coordinate the adjustment and publicity of ideology, and on the other hand in order to meet the media's own demand of survival and development. Different from western nations, China's media and government were originally the relationship of leading and being led. Later with the development of modern political civilization, people's democratic awareness continues to strengthen, the media's function of supervision by public opinions began to reflect, then the direct relationship between media and government generated the relationship of supervising and being supervised from the original relationship of leading and being led, the independence of media gradually reflected. The appearance of new media provided new means for Chinese ideological construction and cultural construction, and also brought great risks, this is a smokeless competition of value, national software and cultural attraction, in order to maintain the position of major world power and protect its ideological security, China has to respond to battle and can't back down (Chart 20.3).

20.3 Forming Causes of the Cost of Ideological Change ...

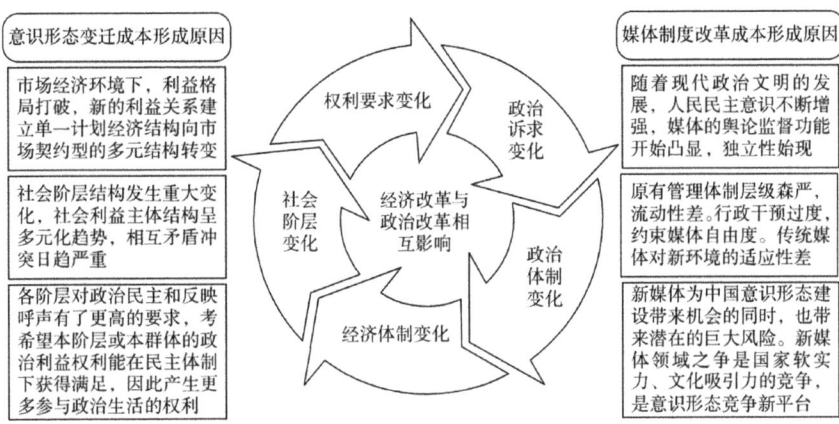

意识形态变迁成本形成原因	Forming causes of the cost of ideological change
媒体制度改革成本形成原因	Forming causes of the cost of media system reform
市场经济环境下，利益格局打破，新的利益关系建立单一计划经济结构向市场契约型的多元结构转变	Under the environment of market economy, the pattern of interest is broken, new interest relations establishes diversified structural transformation from single planning economic structure to market contracts
社会阶层结构发生重大变化，社会利益主体结构呈多元化趋势，相互矛盾冲突日趋严重	Structure of social class structure has major change, entity structure of social interest reflects the diversified trend, and mutual conflicts becomes increasingly severe
各阶层对政治民主和反映呼声有了更高的要求，考希望本阶层或本群体的政治利益权利能在民主体制下获得满足，因此产生更多参与政治生活的权利	Different classes have higher requirements on political democracy and appeals, they hope that the political rights and interests of their class or group could be satisfied under democracy system, so as to generate more rights of participating in political life
权利要求变化	Change of rights requirements
政治诉求变化	Change of political appeals
政治体制变化	Change of political system
经济体制变化	Change of economic system
社会阶层变化	Change of social class
经济改革与政治改革相互影响	Mutual influence between economic reform and political reform

Chart 20.3 Analysis of the forming causes of the cost of ideological change and the cost of Chinese modern media system construction

随着现代政治文明的发展，人民民主意识不断增强，媒体的舆论监督功能开始凸显，独立性始现	With the development of modern political civilization, people's democracy awareness continues to strengthen, the public opinion supervision function of media begin to show, and their independence began to reflect
原有管理体制层级森严，流动性差。行政干预过度，约束媒体自由度。传统媒体对新环境的适应性差	The original management system has strict hierarchy and poor flow. There is too much administrative intervention that restrain media freedom. The traditional media has poor adaptation to new environment
新媒体为中国意识形态建设带来机会的同时，也带来潜在的巨大风险。新媒体领域之争是国家软实力、文化吸引力的竞争，是意识形态竞争新平台	While new media brings opportunity to China's ideological construction, it also brings huge potential risks. The competition of new media field is the competition of national soft power and cultural attraction, and the new platform of ideological competition

Chart 20.3 (continued)

20.4 Countermeasures for the Cost of Ideological Change and the Cost of Modern Media System Construction

We believe that in order to actively respond to the cost of ideological change and reduce the cost of modern media system construction, we need to further liberate our thinking, effectively occupy the ideological and cultural battle field, actively respond to the challenge of increasing pressure of competition with foreign capital and cultural products after joining the WTO; deeply and constantly develop patriotism education; and actively prevent in foreign-related exchange and cooperation. We shall also deepen the reform of news publishing, radio, film and TV industries, adhere to and protect the party and government's control in this field, etc. Currently, the Chinese Communist Party has already gradually changed from being passive to being active in handling the relationship between the development of cultural industry to ideological communication, and allowing the development of diversified culture under the unified culture, which is a very good sign. Besides, we shall also correctly handle different advocates in ideological field, we shouldn't always forbid speeches and actions, we should give media more free room and let the media to actually develop their function of supervision by public opinions (Table 20.5).

China is powerfully responding to the globalization process, and carefully handling this new battle of ideological competition. In recent years, China has been continuously and gradually expending speaking right in international space of opinions, the ability of mainstream media to adopt new media for external communication is getting stronger, the private patriotism and nationalism have very

20.4 Countermeasures for the Cost of Ideological Change ...

Table 20.5 Countermeasures for the cost of ideological change and the cost of modern media system construction

Countermeasures	Key points	Explanation by examples
Allow diversified cultures under dominance by single culture	Respect the different cultural appeals and political appeals of different groups; make efforts to build the "modern China image" with characteristics of the age; make ideological saying innovation with patriotism as the core; and firmly adhere to the bottom line of socialist ideology	For example: comprehensively demonstrate the new image of modern China with "Chinese Characteristics" as the core idea, mainly reflecting three major images: firstly, the image of reform and opening-up that sets foot in China's development and closely following the trend of world. Secondly, the image of external peaceful development and internal stability, unity, cooperation and stabilization. Thirdly, the image of adhering to independence, responsibility and practicality
Emphasize on the media's role in public opinion supervision and guidance	The party committee and government emphasizes on the function of media, they should actively and closely connect with media, control the activeness and necessity of media supervision. They shouldn't simply oppress news report, respect the media workers, and improve the simple and rough style of media management work	For example: the mainstream media should establish public opinion monitoring mechanism, actively find public opinion supervision leads from forums, communities and micro-blogs. For the hot topics that may occur in the aspect of public opinion supervision, they should grab the active role of supervision, and correctly guide public opinion. Especially, they should control the guidance, not blindly follow trend, not make senseless sensation, and be good at guiding public opinions, and guide online supervision to correct direction
Be vigilant to the ideological trap of new technology revolution	While paying attention to the comprehensive national strength of different countries, and paying attention to the competition for talents and technology, more attentions should be paid to cultural and traditional ideological differences	For example, we should on one hand respect knowledge, respect talents, and energetically develop science and technology, on the other hand, we should also fully recognize the instrumental and reasonable nature of science and technology, we shall not be manipulated by foreign countries, and follow trend blindly. The homogenization of science and technology could lead to homogenization of science and technology, which needs special vigilance

(continued)

Table 20.5 (continued)

Countermeasures	Key points	Explanation by examples
Strengthen the communication method of internet scenario	Obtain the advantageous position in the international ideological struggle in internet scenario. Strengthen the authoritative recognition of socialist ideological form, easily guide and handle the diversification and multiple values caused from massive internet communications	For example: emphasizing the management of micro-blog, the government shall encourage the netizens to actively and healthily use the internet. Government officials of different levels shall emphasize on the playing of window role of micro-blog, strengthen communication with the mass, the party's publicity department shall inform the party's guidelines and policies, clear rumors, the basic-level publicity departments shall actively use the platform and window of micro-glob, actively "occupy" the battleground of micro-blog, develop healthy and wholesome internet culture. Only the development of healthy and wholesome internet culture could enable the internet to serve economic construction, and serve the enrichment of people's spiritual life

high interest in expressing on the internet. How to enable our socialism mainstream ideology not to deviate from practice and to be highly recognized and accepted by the main social groups through exploration of the communicative rule in the new age of media is an very important subject to be further and deeply studied for the maintenance of Chinese ideological security.

Chapter 21
Cost of Chinese Soft Power Construction and Response to International Cultural Invasion

21.1 Connotation of the Cost of Soft Power Construction

In 1990, US Scholar Joseph Nye published a series of thesis on the magazine of *Political Science Quarterly* and *Foreign Affairs*, and was the earliest to propose the concept of "Soft Power". He believed that soft power mainly includes three kinds of power and resource: attraction and emotional appeal in the aspect of culture, value and other ideology, the attraction and influence in social, political and economic system as well as development model, and the appetency of international image as well as the control over international rules and cultural development strategy and international mechanism in multilateral diplomacy.[1] Generally speaking, soft power has the following three characteristics, please refer to Chart 21.1.

In recent ten years, China has also noticed the importance of national soft power construction.[2] China's rise almost synchronized with the new round of globalization

---

[1] Joseph S Nye., "Foreign Policy Fall", *The Boston Glob,* 1990, (2): pp. 165–166.
[2] In 2006, Hu Jintao clearly proposed the concept of "Soft Power", and indicated "how to precisely find the direction of China's cultural development, create the new glory of ethnic culture, strengthen the international competitiveness of China's culture, and enhance the national cultural soft power is a major topic in front of us"; at the 17th National Congress of CCP, Hu Jintao clearly required the "enhancement of national cultural soft power", more actively promoted the big prosperity of cultural development". One of the biggest highlights of the 17th National Congress Report in the aspect of cultural construction was the clear proposal of the general strategic key and strategic thinking of current cultural construction. In the introduction of the cultural construction part of the 17th National Congress Report, the general task of cultural construction was clearly positioned: "adhere to the direction of advanced socialist culture, start new wave of socialist cultural construction, stimulate the cultural creation and vitality of all people, enhance the national cultural soft power, enable the people to have better guarantee of their basic cultural rights and interests, enable the social cultural lives to be more abundant and colorful, and enable the spiritual image of the people to be more active and wholesome".

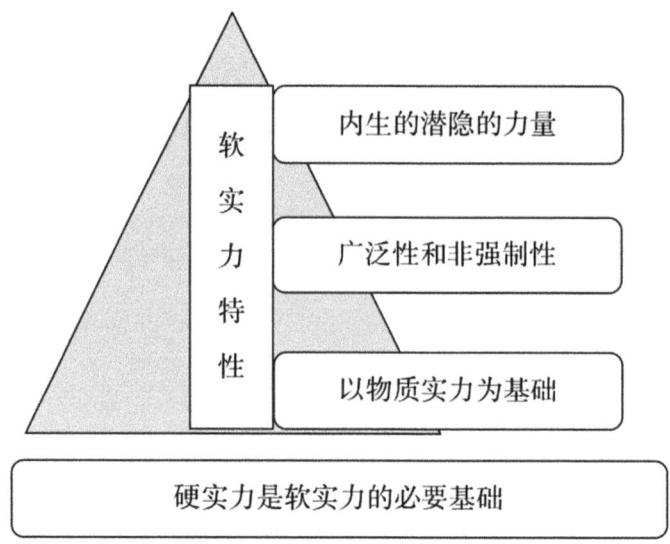

软实力特性	Characteristics of soft power
内生的潜影的力量	Endogenous and hidden power
广泛性和非强制性	Universality and optional nature
以物质实力为基础	Based on material strength
硬实力是软实力的必要基础	Hard power is the necessary foundation of soft power

Chart 21.1 Three characteristics of soft power

wave and transformation of international order.[3] China established the grand strategic framework with harmony idea as the core, upgraded soft power construction to the level of national strategy establishment, and the nation's cost of investments in various aspects of soft power construction also radically increased. Besides, China has been actively developing such construction on national level, regional level, organizational level and citizen level.

---

[3]The transfer of international strength not only brought China the advance that is beyond regular and beyond development stage, but also brought China severe challenges, the latter is mainly reflected as multiple problems in China, the tough stage of reform, the increase of international pressure, the deepening of doubts. This is a kind of "Rising Dilemma", which is completely different from the passive situation in mid 19th Century, China is also relatively easy to determine and promote the corresponding solutions.

21.2 Forms of the Cost of Soft Power Construction on Different Levels and the Cost of Response to Foreign Cultural Invasion

Based on the analysis above, it is believed in this book that the forms of the cost of soft power construction on different levels and the cost of response to foreign cultural invasion mainly include four aspects, for the corresponding national level, regional level, organizational (enterprise) level and citizen (individual) level, they are the cost of fighting for speaking right in international community, the cost of regional soft power construction, the cost of modern organization (enterprise) construction and soft power construction, the cost of the enhancement of scientific quality and cultural quality, for more details, please refer to Table 21.1.

Table 21.1 Forms of the cost of soft power construction on different levels and the cost of response to foreign cultural invasion

Classification of cost	Forms	Explanation of specific contents
Forms of soft power construction	National level: cost of international speaking right struggle	Mainly means that after the change of speaking right situation in the world, China wants to win active position, maintain China's national interest and play role at international stage. China must into large cost to strengthen the speaking right competition ability in the multiple fields of economy, diplomacy, media and culture
	Regional level: cost of regional soft power construction	Mainly means that the development of regional soft power and its input. The main realization means include policy engagement, economic connection and social atmosphere guidance. However, the excessive expansion of regional soft power would also bring negative effect, meaning the widening gap between regions, tight regional relations and increase of social conflicts

(continued)

Table 21.1 (continued)

Classification of cost	Forms	Explanation of specific contents
	Organization (enterprise) level: cost of the establishment of modern organization (enterprise) and soft power enhancement	Mainly means that the negative influence brought by weak development of Ngo on the society, and Chinese enterprises don't emphasize on soft power construction or don't understand soft power construction, which make enterprises difficult to go abroad, and the short life of enterprises could not effectively feedback to society
	Citizen level: cost of enhancing scientific quality and cultural quality	Mainly means the basic construction and spiritual civilization construction dominated by the government in order to enhance the scientific and cultural quality of citizens

21.2.1 National Level: Cost of Fighting for Speaking Right in International Community

1. Reform of the Situation of Speaking Rights in the World

"Speaking Right" is essentially a type of political and economic right.[4] Since the great geographical discovery in the 15th century, the European and American nations have been expanding backed by force and capital, so as to complete their primitive accumulation. In this process, the western developed nations, especially the US gradually obtained monopoly in international finance, and forced other civilizations to join the speaking right system with the framework of the so-called "Freedom and Democracy", in international exchanges, the west gradually monopolized the world in speaking right.

---

[4]Talking about the relations between eastern and western countries, people often use the concept of "Speaking Right", it could be sourced to the "Relation between speaking and social power" theory proposed by French sociologist Michael Foucault in 1970. His explanation was that speaking is not only a thinking symbol and communicative tool, but also a struggle means for people to express and fight for their own interest. So a word seemingly without ideological meaning was granted with the function of power and interest.

China has walked a difficult and struggling path from the loser and victim of international stage to gradual obtaining of speaking right.[5] After gradually strengthening, China will inevitably participate in more international competitions, such as competition in trade field, competition for financial dominance, without speaking right, it means lack of channels to maintain our own rights; besides, in the past, China has always wandered outside the international organizations, doesn't have sufficient understanding about the rules of international community, let alone the formation of international rules, which brought many difficulties for China to enter international community and participate in the adjustment of world situation. Therefore, in order to obtain speaking right and become maker of the rules, China still needs to input competition cost in many aspects.[6]

2. **Cost Paid by China for Obtaining International Speaking Right**

The competition for international speaking right is closely connected with the interests of different countries, such as the struggle in the issues like international code of human rights, humanitarian intervention, ideological code, climate change and standard of international greenroom gas emission reduction, exchange rate dispute, international financial system reform plan, etc., all of which are firstly reflected as the fight for international speaking right. Facing the international system currently dominated by the western nations, China shall maintain its own interest and must obtain more speaking rights. We believe that the cost of fighting for international speaking right is mainly composed of the following four types, meaning the cost of fighting for economic speaking right, the cost of fighting for media speaking right, the cost of fighting for diplomatic speaking right and the cost of fighting for cultural speaking right.

---

[5]After entering the 21st century, especially since this international financial crisis, different countries in the world suddenly found that China has become the second biggest economy of the world. With this international financial crisis as symbol and economic growth center transferring to developing countries, especially China, the world's relation of rights is having structural change. The growth of economic strength sets the material foundation of the enhancement of China's speaking right. After over 30 years of continuously rapid development since the reform and opening-up, China has become the country with the biggest foreign exchange reserve, the second biggest exporting country, and the second biggest economy. The correlation between China and the world has been unprecedentedly strengthened, China is widely participating in global and regional cooperation, actively participating in international and regional affairs, fulfilling its international duties, playing responsible and constructive role in international affairs such as anti-terrorism, nonproliferation, fight against pirates, etc., and becoming an important force to maintain world peace and promote common development.

[6]The rise of China doesn't mean that China's economic system, political structure and cultural value has replaced the hegemony status of the west. But the globally leading economic growth of China during the 32 years since the reform and opening-up, the economic might of New China during the 61 years since establishment, and the Chinese factors will increasingly become one of the most important influence in the global economic transformation, and will continue to deeply change the traditional world situation.

21.2.2 Regional Level: Cost of Regional Soft Power Construction

1. Definition of Regional Soft Power

Soft power is an important part of comprehensive competitiveness of a region, its composing parts could be classified into government management and service,[7] regional culture,[8] human resource,[9] regional image[10] and living

---

[7]It is mainly reflected in two aspects: the first is the quality and acceptance degree of government service, mainly including the citizens' recognition of government legitimacy, the citizens' acceptance degree of the public nature and reasonable nature of government system and public selection process, the degree of government operation, democracy and rule of law, the image of government staff in the eyes of the public, etc. Good government creditability could well guarantee the stability of system and predictability of policies, so as to attract large amount of funds and investors, strengthen local economic vitality, promote local economic growth, enhance local economic competitiveness. The second is the completion and execution of related system.

[8]Regional culture means the sum of the value, idea, thinking method, cultural attitude, ethnic art, custom, habit, moral standard, etc. with deep regional characteristics of people in certain region formed in material production, spiritual production and social lives. Regional culture, as an intangible force, quietly affects the regional development entity, and affects each link of regional economic development through the organization consolidation function of development entity. Regional culture, as the shared value, faith, attitude, habit and behavioral standard of a region, includes entrepreneurial spirit, innovation awareness, floating preference, credit idea, cooperation awareness, open thinking, etc., it has especially important significance for regional economic growth. Under the background of an innovative, creditable, cooperative and open regional culture, enterprises in the region could form a relation of mutual commitment and dependence, reduce uncertainty of information, and generate division of work and cooperation on the basis of competition. Meanwhile, cultural resources of the region are integrated through cultural transfer, cultural excavation and cultural construction to generate featured regional culture, form common value and idea of society, and strengthen the appeal and influence of regional culture through cultural industry construction and tourism industry construction, attract the inflow of external talents, capital and industry. Regional culture is an important foundation of regional soft power, good regional culture is the "Assistor" of regional economic development.

[9]Human resource mainly means human quality. Human quality requires people to adapt to the era in the aspects of attitude, idea, personality, professional skill, comprehensive quality, etc., continuously create, integrate and attract new spirit, personality, ability, and to output high-quality human resource for the social, economic and cultural development of the region.

[10]Mainly means the overall regional image, overall image of regional residents, image of regional natural environment and development, etc. As pointed by the Soft Power Topic Team of Peking University that regional image is highly correlated with various aspects of this region, including hard power, government service, human quality, etc., but we can't deny the importance of this indicator for this. A region's image is directly related to its internal coherence, and attractiveness to external resources. For example, if a region's image is not good, external investors would hesitate before entering. From this sense, regional image plays the role of promotion and develops brand effect.

21.2 Forms of the Cost of Soft Power Construction on Different Levels … 321

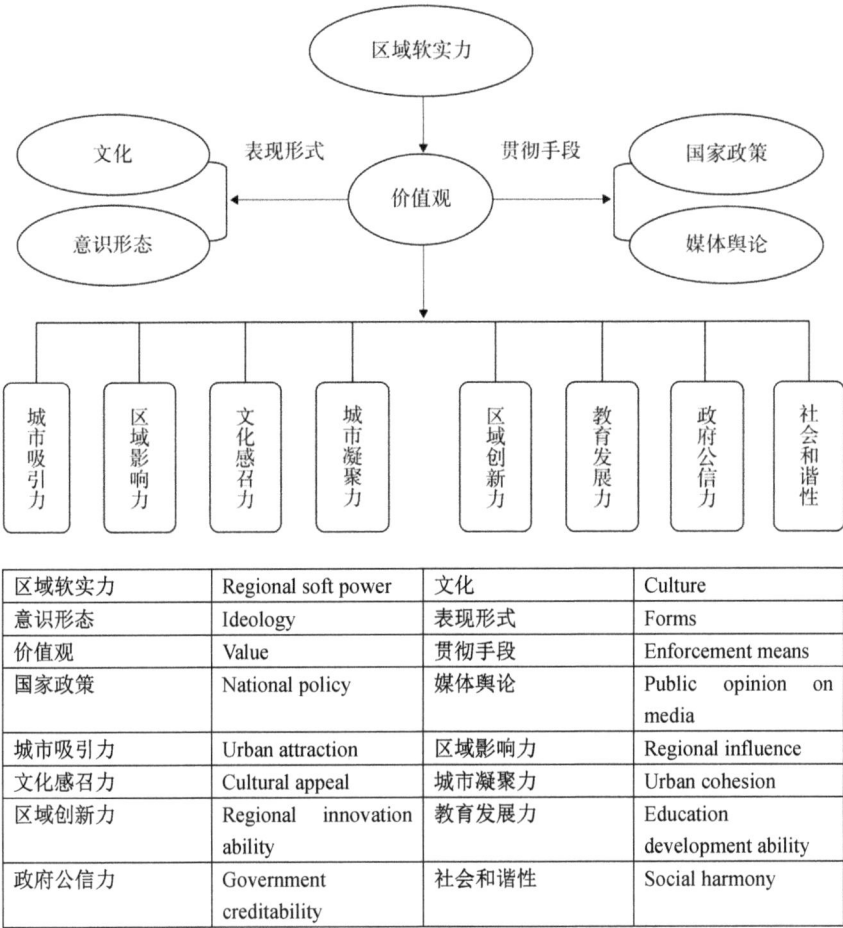

Chart 21.2 Composing parts of regional soft power

environment.[11] The soft power with unit of city is reflected in Chart 21.3, and mainly the combination of technological strength and cultural strength (Chart 21.2).

---

[11]Mainly means the hardware of living environment, because social environment is already reflected in several indicators above, for example, policy environment has been included in government management and service, cultural environment is also reflected in human quality and regional culture. Therefore, the living environment pointed out by us mainly includes two aspects: the first is natural environment, such as famous mountains, beautiful natural sceneries, which has considerable effect in concentrate internal coherence and attract external resources; the second is historical resources, although it has left obvious artificial trace in making, but such resource is also something that the outside couldn't copy and refer to.

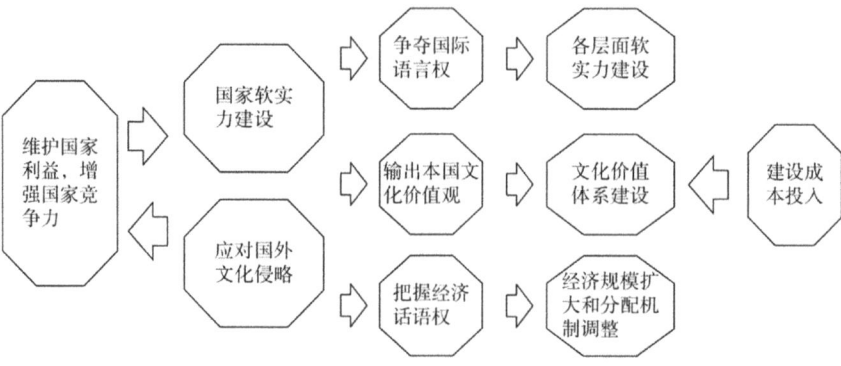

维护国家利益，增强国家竞争力	Protect national interest and strengthen national competitiveness
国家软实力建设	Construction of national soft power
应对国外文化侵略	Respond to invasion of foreign culture
争夺国际语言权	Compete for international speaking right
输出本国文化价值观	Export domestic culture and value
把握经济话语权	Control economic speaking right
各层面软实力建设	Construction of soft power on each level
文化价值体系建设	Construction of cultural value system
经济规模扩大和分配机制调整	Expansion of economic scale and adjustment of distribution system
建设成本投入	Input of construction cost

Chart 21.3 Forming causes of the cost of soft power construction and the cost of response to foreign cultural invasion

2. Result and Cost Input of the Soft Power Construction of Chinese Cities

The cost of Chinese regional soft power construction is mainly reflected as the prominent strengthening of regional economic advantages after the large-scale appearance of city clusters.[12] Such clustering of industries is the result of different industrial evolutions and also the result energetically guided by the government. Different industries, based on different comparison advantages, concentrate in different places, attract large number of enterprises to concentrate in this place, so as to further promote the city clustering of this region.

---

[12]Pearl River Delta, Yangtze River Delta, Bohai Ring District and other similar city clusters, the very important driving force of such city clusters is exactly special industries with certain advantages, the upstream, middle stream and downstream of these industries are highly concentrated, which would greatly reduce the development cost of different industries, and enable industries to have very strong competition advantage in the development of certain region.

Although the appearance of city clustering drove regional economic development, but also widened the gap with underdeveloped clustering, and formed regional conflicts, the gaps between the wide central and western regions and eastern regions in the aspects of urban and rural residents income and consumption level, marketization degree, especially in infrastructure, compulsory education, basic medical service, social security and other basic public service level, are also widening. Imbalanced regional development is the basic national situation of China, if the regional development gap keeps widening, and the regional development is uncoordinated for long term, it would generate severe influence on sustainable economic development and social harmony, and this is exactly the issue to be solved in the regional soft power construction in the future.

21.2.3 Organizational (Enterprise) Level: Cost of Modern Organization (Enterprise) Establishment and Soft Power Enhancement

In recent years, Chinese government has realized the position and role of non-government organizations in the aspect of foreign-related cultural exchange, and promoting outside world to recognize and understand China, so as to build the country's "Soft Power". But the discreet attitude of the government made non-governmental organizations unable to obtain strong support, and the independence of non-governmental organizations[13] is not recognized. As a result, the government is responsible for solving all sorts of social problems, consuming large quantity of funds, labors and materials. Meanwhile, the government can't take care of everything, which caused many social problems to be neglected, which led to more serious popular discontents and popular indignation.

While the development of non-governmental organization also faces various difficulties, for example, the legitimacy of identity,[14] the lack of policy support,[15]

---

[13]Non-government organization (NGO), or non-profit organization (NPO), means the social organization that is not government, not corporate, and not having the purpose of seeking profit. Such as various volunteer organizations, public-welfare organizations, charity institutions, environmental protection organizations, wild life protection organization, etc.

[14]Because there is no related policies, NGO currently is relatively difficult in registration and approval, especially the steps of approving international, Hong Kong and Taiwan organizations to enter Mainland China are not big, which make these organizations very difficult to operate with legal identity.

[15]Currently, China hasn't formed the legal system of supporting and standardizing NGO to legally operate and raise funds, which make many non-profit and public-welfare organizations difficult to develop even after obtain registration and approval.

the shortage of funding,[16] and the poor independence.[17] The lack of non-governmental organizations is one of the main causes of the vast operating cost of the government, as well as one of the reflections of Chinese society being lack of openness.

Over the 30 years since reform and opening-up, the development of Chinese enterprises mostly emphasized the construction of "Hard Power", which is restrained by the actual national situation. China's enterprises developed from small to big, during very long period, most enterprises have been rushing themselves from life or death, they couldn't care about soft power construction at all. It is a little harsh to require Chinese enterprises to have both soft power and hard power at their start-up stage. On the other hand, the construction of enterprise soft power is difficult to measure, having long cultivation period and slow effect, so it is naturally neglected by some growing enterprises that are oriented with the principle of seeking quick success and instant benefits. Therefore, the lack of soft power increasingly becomes the barrier of Chinese enterprises in expansion and development.

The construction of soft power of Chinese enterprises lies in not only the enterprises themselves, the preferential national policies caused private enterprises being in vulnerable position in production and competition process, while state-owned enterprises have the negative situation of being too big to fail in recent years, the incompletion of market mechanism, the excessive intervention of national policies and the thinking of seeking quick success and instant benefits in regulatory process have brought many obstacles for enterprise development.

21.2.4 Citizen Level: Cost of Enhancing Scientific Quality and Cultural Quality

1. Cost of Enhancing the Scientific Quality of Citizens

In 2010, China Association for Science and Technology released the result of the 8th Survey for the Scientific Quality of Chinese Citizens, the result indicated that during the "Eleventh Five Year Plan" period, the scientific quality of Chinese citizens had been steadily enhancing, the construction work of the scientific quality of Chinese citizens obtained prominent effect, particularly, the scientific quality

---

[16]Most local NGO in China are in semi-paralyzed state and unable to develop their original function due to lack of funding and stable funding sources.

[17]Currently there are few NGOs in China in real sense, most of them are semi-official organizations. In idea, organization, function, activity method and management system, they heavily depend on the government, and even play roles as a subsidiary institution of government.

Table 21.2 Cost of national input for the enhancement of scientific quality of Chinese citizens in recent years

Provide more opportunities and ways	Citizens enhance their own scientific quality mainly through scientific and technological education, communication, popularization and other means. Survey indicated that TV and news paper and other traditional media are still the main channels for Chinese citizens to obtain scientific and technological information, meanwhile, the internet provides more opportunities for citizens to use and obtain scientific and technological information
Provide more scientific popularization equipments	Mainly including construction of scientific and technological venues, such as zoo, aquarium, botanic garden, science and technology museum, natural museum, etc. as well as cultural & art venues such as public library, art gallery, exhibition hall, etc. Organize community and students from university, middle school and primary school to participate in various scientific popularization activities and visit exhibitions
Cultivate rational thinking of citizens	Promote and respect the ideas such as "natural rules, development and use of nature", "harmony between human and nature", etc. Teachers, scientists, doctors and other professions of scientific education and service are receiving respect from more people

level of urban labors and farmers has been enhancing in relatively fast speed.[18] The enhancement of scientific quality of citizens was mainly benefited from the construction inputs of the aspects in Table 21.2.

Currently, the scientific quality of Chinese citizens is only equivalent to the level of Japan, Canada, EU and other main developed countries and regions at the end of 1980s and the beginning of 1990s.[19] The main cause of low scientific quality level among citizens is that in our regular education, the process of scientific and technical education is too weak, for the enhancement of scientific quality level among citizens, the only thing that could have actual effect at current stage is the regular education in schools,[20] so we should energetically promote the reform of basic education.

---

[18]The ratio of urban labors having basic scientific quality increased from 2.37% in 2005 to 4.79% in 2010; the ratio of farmers having basic scientific quality increased from 0.72% in 2005 to 1.51% in 2010. The enhancement of scientific quality of urban labors and farmers played important role in the overall enhancement of the scientific quality of Chinese citizens. For the data, refer to the 8th survey of the scientific quality of Chinese citizens released by China Association for Science and Technology on Nov 25, 2011.

[19]China and the US released the ratio of citizens with basic scientific quality as of the end of 2007 respectively, back then, the ratio in the US was 25%, while the ratio in China was only 2.25%. It was found in the comparison with 15 EU countries, the US and Japan in 2001 that Chinese people's understanding of scientific knowledge ranked the last place. Same source with above.

[20]From the main measures of different countries, most also focus on the scientific and technological education in regular education. The first one is the issue in education reform. Just like the "2061 Program" of the US, the education reform of most western developed countries were in 1980s, especially the education reform of science and technology.

2. Cost of Enhancing the Cultural Quality of Citizens

The current situation of the ideological and cultural quality of Chinese citizens is the result of multiple factors in comprehensive effect. Although the traditional Chinese cultural is not lack of ethical and moral rules, the most developed is the personal morals that are reflected in the five cardinal relationships between father and son, between the emperor and subjects, between husband and wife, between the old and young, and between friends, but lack of public morality beyond the five cardinal relationships; when we transform from traditional subject society to modern society, the resources that traditional ethical and moral rules could contribute are limited, therefore, we need to develop new socialist ethics.

Besides, in the process of seeking modernization, we have walked many detours. In cultural field, we once denied all traditional culture without differentiation, damaged many active and excellent parts, especially during the ten years of "Cultural Revolution", China partially and extremely promoted the nihilism policy that destroyed everything, doubted everything and cracked down everything, which treated culture with anti-humanity method, which not only didn't bring real revolution of culture, but further destroyed the various existing cultural resources, and enabled many essentially ignorant ideological resources to come back. These historical reasons caused us to, while establishing new thinking and culture, face huge blankness to fill up. Over the thirty years since reform and opening-up, the development of Chinese society stepped on the right track again, but with the rapid development of economy, the negative influence of market economy also began to reflect in ideological and cultural field, which further intensified the confusion in ideology and culture, as well as bring obstacle to social stability and development.

Enhancing the citizens' cultural quality is an urgent and important work, requiring input of more labors and materials, as well as government leading and participation by all people.

21.3 Connotation of Social Civilization Cost in Modernization Construction Process and Forms of Cost

Social civilization cost in modernization construction process mainly means, with the acceleration of China's modernization process, the enlargement of the scale of market economy, the deepening of economic system reform, and the huge change in social relationship. Different classes have continuous frictions due to uneven distribution of interest, the mass at bottom level and emerging middle class have their power rights and requirements, while the social management level continuously has corruption of power, rent-seeking of public power, loss of integrity and other problems, long-standing malpractices become increasingly severe, which cause intensified conflicts between classes, increasing social conflict events and influence on social stability. On the other hand, the thinking of money as the first priority prevail, unscrupulous merchants ignore law and morality, and even challenge the bottom line of ethics, which brought severe damage to people's production and lives. Persistent occurrence of such phenomenon caused destructive influence on social order and national image.

Material civilization, spiritual civilization, political civilization and ecological civilization jointly constitute the system of socialist civilization, and have gradually enriched during the ruling process of the Chinese Communist Party. For spiritual civilization, the concept of social civilization need to be further enriched.

It is believed in this book that the main forms of the cost of social civilization change are as in Table 21.3.

Table 21.3 Forms of the cost of social civilization change

Classification of cost	Forms	Explanation of specific contents
Cost of civilization change	Cost of sliding morality of social groups	The problems of changing ethnic faith system, weakening of ethnic culture recognition, money first, power corruption and other social problems collectively reflect that the sliding morality of social groups have reached very severe level, which have caused very bad influence on social stability and social civilization, meanwhile, it also affects the steps of China's civilization advancement. In order to restrain morality sliding, the party, government and all walks of society need to carefully lear the socialist core value, make efforts to build the socialist core value
	Cost of changing social relations	The reform and opening-up released large quantity of free "social space" and "social resources", which caused polarization of social classes, new middle class emerges, the original interest distribution form needs to be broken again, there are a lot of conflicts. Power has become the main mechanism that shapes the operation of Chinese social class relations, the living situation of vulnerable groups is worrisome
	Cost of conflicts caused from formation of new social groups	The formation of middle class has affected the original combination form of other class of society. The interest appeal of middle class didn't receive sufficient attention, which made the entire class to be in the state of dissatisfaction and unease. Meanwhile, their formation and behavior also caused impact on the power class and China's democracy and rule of law, and urged the acceleration of democracy and rule of law

21.3.1 Cost of Sliding Collective Morality in Society

In recent years, the moral situation of Chinese society caused widespread worrying from the broad mass, such as bribery, corruption, abuse of power for personal gain, lack of integrity, food safety, fakery, infringement of rights, etc. All the astonishing behaviors of degraded morality exposed on media everyday have already become a social normal that we are getting used to. It is hard to imagine that a nation with excellent cultural tradition and upmost moral practice could have such trend where moral standards are unscrupulously challenged and the situation is getting worse. Such situation makes people very disappointed in the overall social morality, and caused great popular discontent. Cheng Zhihong proposed the cost of social dishonesty, and calculated that China has to pay RMB 585.5 billion of price (cost) for dishonesty.

A nation without enhancement of national quality and the power of ethics would never become a really strong power and a nation respected by people. Obviously China's citizen morality system has reached a key point that reconstruction is urgently needed. In summarization, China's morality problem and chaotic phenomenon in society are comprehensive reflection of the six types of social problems in Table 21.4.

Table 21.4 Classification of the morality problems and chaotic phenomenon in Chinese Society

Classification of issues	Nature of issues	Specific explanation
Change of ethnic faith system	After the breaking of old faith system, the new faith system has yet to be fully established, which makes the ethnical spirit lack of depending basis.	The faith system of Chinese nation is built on the traditional culture with Confucius system, but the social faith construction has for certain period completely deviated from the original faith, the new socialist core value has yet to be fully popularized, so the ethnic faith system is in unstable state
Weakened ethnic cultural recognition	On one hand, it is confusion over ethnic cultural, on the other hand, there is impact from western culture, which weaken the recognition of ethnic culture	The hundred years of humiliation history and the fact that although China now is rapidly developing economically, but the development is extremely unbalanced, and the spiritual civilization is lagging behind, which caused Chinese people to gradually lose psychological advantage and reduced cultural confidence, so the ethnic cohesion is also affected

(continued)

21.3 Connotation of Social Civilization Cost in Modernization … 329

Table 21.4 (continued)

Classification of issues	Nature of issues	Specific explanation
Sliding national morality	The morality sliding is currently already having no bottom, there are so many problems and happening so frequently, the difficulty in treatment is very big	The commodity exchange principle that is universal in market economy has eroded the social and political life as well as people's spiritual life, which caused forgetting friendship for profit, power-for-money deal, corruption and other negative phenomenon, the money worship, hedonism, and extreme individualism are spreading
Failing national education	The reform of education system is under a lot of disputes, the education industrialization indirectly makes economic benefit become the measuring standard of education success	The basic education considers examination as the purpose, which values more scores than virtue; higher education is oriented to instrument, only teaching instead of cultivation. Under the excuse of "Education Industrialization", education has even become a tool to make money. Economic benefit has become the standard to determine the success of education or not, the phenomenon of arbitrary charge persisted, many poor children can't afford education. Teachers are losing teacher's virtue, masters and doctors are produced in batches
Power corruption	National public servants abuse power, use power to seek profit, have power-for-money deals, and violate the basic interest of the mass	The typical privilege-type corruption is the "three public consumption issue", including eating and travelling with public funds (including travelling abroad with public funds), use of public cars. Some chiefs monopolize power, make excessive administrative intervention, abuse power and don't respect national laws
Interest above everything	Money worship becomes very popular, for money and interest, some people ignore morality and law, publically challenge public morality	Due to the spreading of money worship, the money-first conception has spread from economic field to every field. Driven by interest, food safety issue, commodity fraud issue, academic fraud issue, etc. all appeared

21.3.2 Cost of Change in Social Relations

China is under the age of major transformation, the biggest characteristic of which is to compact all the processes of pre-modern, modern and post-modern times into short ten to twenty years, which caused Chinese class relations to be extremely complex. To be specific, the current Chinese class interest relations are reflected in three relations: firstly, cooperation and mutual-benefiting relation. The reform and opening-up released large quantity of free "Social Space" and "Social Resource", which not only created conditions for the segmentation of social classes, but also provided certain opportunity for each class to obtain their own interests, and different classes also provided corresponding interest space to other classes during this process, this is the foundation of mutual cooperative relations between each class.[21] Secondly, collusive and exploitative relation, members of strong class jointed hands to exploit the interest of weak class through various means.[22] The characteristic of strong class is stressing their own interests and claims, extortion, and even to the point ignoring any morality and social justice, neglecting public welfare and the interest of weak class. Thirdly, competitive relation, market is also a kind of competitive mechanism that provides the platform for interest competition among different classes, meanwhile, other system reforms of the nation also introduced some competition mechanisms, which to certain extent intensified competition among different classes.[23]

Change of the three types of social relations above enabled power to become the main mechanism to shape the operation of social class relations in modern China. Although interest game is the main form of the class relations in current China,

---

[21]Without the large amount of foreign investments and private capital, the financial power of government alone is difficult to create so many employment opportunities. Many migrant workers enter cities to work, which is highly related to the input of various capitals. Similarly, the cooperation between government officials and enterprise bosses is even more obvious; government officials depend on economic development to create better political achievement, so they would favor the interest of businessmen, so different places issue various favorable investment attraction policies to increase local investment. Meanwhile, the central government's focus on stability also urge local government officials to have to look after livelihood besides forces, especially calming the mood of vulnerable groups, while many more common people are not willing to directly adopt the method of protesting against government to maintain their interests, because it would generate very high cost, they hope finding their interest through cooperation with the government.

[22]In order to resist exploitative behaviors of the strong class, the vulnerable groups whose interests are damaged would petition, sit for protest, more and more people would adopt extreme methods such as burning themselves or jumping off building to express their frustration and anger. Some vulnerable group would adopt violation against demolishers, or clash with local government.

[23]Currently, the vicious competition between classes in China is mainly reflected as that the strong classes use their force to exploit interest without limit, while the vulnerable classes also adopt the improper means that they can think of (fraud, cheating, etc.) to realize their interests, so as to result in the loss and lack of trust between different classes.

power becomes the shaping force of the class relations of modern China.[24] Most social members generally still recognize the ruling position of Chinese Communist Party, but there are indeed interest disputes and various conflicts between national and social managers and other classes, for example, the privilege of managers are completely not recognized by other classes; the corruption problem has severely damaged the legitimacy foundation of the national and social managers; certain national and social managers damage the interests of other classes in day-to-day management at will, and ignore appeals in society. Although managers have dominating ability over other classes, especially controlling ability, but their low public creditability, weak action capacity and incompetent handling of social events caused other classes to be increasingly discontent to such dominance relation, although they couldn't leave the support of the national and social manager class in their day-to-day lives,[25] so the entire society is under huge conflicts and frictions among different classes, the unstable factors in society have prominently increased.

21.3.3 Cost of Contradictions and Conflicts Caused from Formation of New Social Groups

At the stage of changing social relations and increasingly prominent segmentation of classes, a new group, meaning the middle class, begins to form.[26] From the perspective of dominance relation, the expanding of Chinese middle class has to great extent improved class relations in Chinese society, because the rise of middle class means that the economy has developed to new stage, people's income level has enhanced as well as improvement of civilization and quality. But China's middle class is not strong enough and hasn't formed common value, and they are

---

[24]In China's reform and opening-up over the past thirty years, the national government has to great extent delegated power to local governments, enterprises, individuals and society, which effectively strengthened the distribution ability of the latter. However, this didn't fundamentally change the original pattern of administrative governance, and didn't move the dominance position of the powerful classes.

[25]Any event that targets lawbreaking or violation of government officials would be actively participated by the mass, so there is so-called unrelated mass event. The recent class relationship began to spread from interest conflict to unrelated level, which all has such trend. This trend means that some class members are looking for venting their dissatisfaction and anger toward the ones with power, which further reduces the legitimacy of rulers.

[26]What role the middle class plays in class relations, it is a very controversial question. A viewpoint is that the middle class is the foundation or stabilizer of social stability. The middle class has certain dominance ability, but such ability is different from the dominance ability of higher class, and is relatively more acceptable by lower class. Many western scholars believe that middle class is the foundation of social stability and the symbol of mature society. However, another viewpoint is that with the expansion of middle class, they would have higher requirement for democracy, and have more distribution and mobilization abilities to launch democratic revolution.

subject to great influence from power and market, people within middle class are busy obtaining more interest for themselves, few people would put more energy and funds in bearing social responsibility, besides, to become a force with social dominance and become the society's management class, the middle class needs to break many obstacles. While the source of their own interest is not safe either, which is highly related to China's special economic development model and that situation that Chinese economy can't take care of its own under the trend of globalization, nevertheless, with the continuous expansion of the middle class, some people in this class begin to become an important social force, advocate social justice, fairness and equality, and obtained legitimate position to certain extent, which is good for promoting the democratic process of modern Chinese society.

But the rise of the middle class changed the constituting form of original social classes after all, which causes that different classes haven't formed the identity and position relations that are mutually recognized, the conflicts and contradictions are reflected in increasingly intensified manner. In society, the middle and upper class occupy considerable social resources and management power, but their abuse of such resources and power make other classes very dissatisfied, their flaunting of wealth and flaunting of connections by part of them are pure show-off and provocation, which severely offend other classes and cause dislike from the society, which would in turn weaken their social status. While the middle and lower classes of the society don't have enough income, lack of spiritual and cultural lives, and don't have social status, when they realize that their social status is directly linked to living quality, they would fall into the dilemma of "Populist", have intensive protesting movements, and encounter even more intensive identity conflicts and crack down. The excessively powerful management class and the excessively weak and young middle class constitute the new image of the composition of Chinese social classes. The identity and status position among classes are obviously affected or even decided by the interest relations and dominance relations, however, it would in turn affect the operation of the latter two. The biggest challenge of Chinese class relations is that the three relations trap into the state of inverse reinforcement, which is very bad for the maintenance of social order and construction of harmonious society.

21.4 Forming Causes of the Cost of Soft Power Construction and the Cost of Response to Foreign Cultural Invasion

Over the thirty years of reform and opening-up, China has made great economic achievements, the GDP statistics indicate that China has become the second biggest economy in the world and obtained the status of a major economic power. But China, as an economic power, is not a cultural power. On the international stage, China has always been manipulated by the west, always responding to western governments and media, and always defending itself. This is the case not only in politics, but also in economic and trading activities, in cultural exchange and visiting activities, China has also rarely had initiatives and leading power. This has brought damage to Chinese national interest. Facing the invasion of western value system and cultural wave, China, as a socialist nation, must have huge attraction and persuasive political ideas, political system, domestic policy and foreign policy, as well as have corresponding social core value and the cultural industry that serve and promote it. In the basic concepts involving the basic value of modern human being, such as democracy, freedom, human rights, rule of law, etc., China needs to fight for concept defining power, using power, topic setting power and story-telling power. While handling domestic issues, China needs to emphasize change of social relations and fair distribution of interest, so as to guarantee the stability and peace in domestic environment.

21.5 Countermeasures for the Cost of Soft Power Construction and the Cost of Response to Foreign Cultural Invasion

Analyzing based on the afore-mentioned forms and forming causes, we believe that in order to further strengthen soft power construction, the countermeasures for enabling Chinese culture to go abroad are mainly composed of the aspects in Table 21.5.

Table 21.5 Countermeasures for the cost of soft power construction and the cost of response to international cultural invasion

Countermeasures	Key points	Explanation by examples
Focus on the soft strength construction on each level	Avoid control all construction of soft power, treat differently on national level, regional level, organizational level and citizen level. Each level has key working projects	For example: the construction of soft power on national level is the competition for speaking right; the soft power on regional and organizational level mainly includes the construction of technological power and cultural strength; the construction of soft power on citizen level is the one people shouldn't ignore, cultivation of wise citizens who comply with disciplines and laws, and behave civilized is the foundation of strong soft power of a nation

(continued)

Table 21.5 (continued)

Countermeasures	Key points	Explanation by examples
Well handle domestic conflicts between classes	Research and adjust interest distribution, listen to the opinion of different classes in society, respect the appeal of each class, adjust the power mechanism of power structure, avoid excessive concentration and abuse of power	For example: pay special attention to the relations between basic-level township cadres in rural areas and farmers, the relations between urban municipal managers and citizens, the relations between state-owned enterprise managers and workers, and the relations between middle and senior party and political officials and the mass. Deepen the cadre system reform, strengthen the construction of party working style and integrity, enlarge anti-corruption movements, improve petition mechanism
Change the thinking of media construction	Promote media system reform, respect the independence of media, give certain space of freedom. Carefully cultivate media leaders and media elites, and cultivate the creditability of media	For example: the media also needs to promote morality construction, especially, building a batch of official media that have creditability, respect facts, speaking for the interest of common people, could feasibly reflect social problems instead of oppressing and smoothing things, or constantly speak bureaucratic tone. Meanwhile, allow different media to have different forms of exploration, protect the free rights of media coverage and respect the media professionals
Change the external communication method	Stop denying the reasonability of western value including democracy and freedom, but make argument that the western-style value of democracy and freedom is not suitable for the actual situation of China	For example, stop avoiding social transformation, political system reform and other such sensitive topics. Such open and tolerant attitude would speed up the effective communication between China and international community, and enable the world to understand the valuable nature of Chinese experiences. Besides, speaking with facts, recent years in some regions in Asia, Africa and Latin America, the "Beijing Consensus" of one-party ruling plus market economy has been pretty popular

(continued)

Table 21.5 (continued)

Countermeasures	Key points	Explanation by examples
Prudent use of government resources	Government resource of soft power includes political ideas, political system, domestic policy and foreign policy. Under the circumstance that non-government resources are available, hard output of government resources should be avoided	For example, in the fields that NGO could play good role, such as humanitarian aids, private cultural exchange, technological cooperation, etc., the power of NGO should be developed rather than the government taking care of everything. While the various overseas resources, including the resources of overseas Chinese people, should also be fully developed, we should mobilize their initiatives and avoid setting obstacles and labeling, instead we should give more tolerance and support

China has become the second biggest economy in the world, no matter in the aspects of global financial system, including sovereignty fund, investment market, or foreign investment and overseas trade, China obviously plays more and more important roles, which require China to play more important roles and input more costs and efforts in the forming process of the international speaking space as well as the new world rules in trading and non-trading fields.

The core of Chinese soft power construction is culture, but eventually would encounter the long-standing issues such as ideology, political appeals and conflicts, etc. Facing the penetration and strong position of the west in "Democracy", "Human Rights" and other value systems, How China could establish its own core value system, and guarantee the developing and stable cultural system is the key for China to enter the world and win over the world. Culture, as one of the factors of soft power, is naturally important, but what's more important is that China's modern culture, modern political thinking and system must be suitable to economic development, social harmony is naturally formed instead of forceful creation by power.

Summary of the Cost of Chinese Cultural Development

Chinese society is having in-depth and modernization-oriented changes, the modernization process is also the process of sorting out and reconstruction of social order and civilization. The role of culture in it is the rebuilding of the social advancement model and mass quality with the huge social influence of the widely recognized cultural value (core values, view of life, ideology and faith) and the cultural industry formed in the manner widely suitable for market economy operating method (news media industry, book publishing industry, radio, film and TV industry), reform is inevitable process of it. The great cause of Chinese reform is a process from increment to stock, and economics before politics, the steps of cultural and ideological reform always lag behind economic production. Therefore, during the past 60 years, the cost of cultural development is mainly reflected as lack of faith, sliding morality, severe threat faced by cultural diversity, severe insufficiency of creativity, poor market adaptability and other phenomenon that impact social stability and restraining national development. While the systematic and mechanism issues that restrain the development of cultural productivity haven't been fully solved, the freedom and openness required by cultural development are not implemented in place, the partial emphasis on economic nature, industrialization and utilitarianism of culture could only make the path of major cultural development even narrower. Currently the ratio of Chinese cultural industry in GDP is less than 5%, while the ratio of cultural industry of the US in GDP is over a quarter.[1] The cultural industry not only constitutes important economic support of a nation, but also important reflection of its soft power. Export of cultural products is the export of a nation's way of living and value, and is one of the fundamental means to expand its international influence. Only by directly facing China's cost of cultural development, deepening cultural system reform and freeing our minds, the major development and major prosperity of culture can be truly realized.

---

[1] The ratio of cultural industry in GDP is about 2.5%, while the ratio in western developed countries is normally over 10%, it can be seen that China's cultural industry still has great development potential and is a emerging industry. During the period from 2005 to 2008, the growth of added value of cultural industry was higher than that of the three major sectors and the GDP, and had obvious driving effect to the economic growth. Integrating the grey relation analysis, the influence of cultural industry on economic growth is obviously higher than the influence on the primary industry. In 2009, subject to the influence of financial crisis, the cultural manufacturing industry in cultural industry was subject to relatively big impact, and its added value also decreased along with the decrease of the added value of the secondary industry. Source: Li Zengfu and Liu Wanqi: *Empirical Research of the Influence of China's Cultural Industry on Economic Growth*, Industrial and Economic Review, 5th Issue in 2011.

Part VI
Cost of Opening to the Outside World and Development

Chapter 22
China's Opening Development Thoughts and Their Cost

22.1 Overview of Opening Development Thoughts and Their Cost

Just like the tough and winding process of China going global, the development of this opening-up thought that reflects this historical process also experienced detours and windings. In early 20th century, Mr. Sun Yat-sen was the first to clearly propose the "Theory of Opening" and appealed to the world to jointly develop China. At the same time, Comrade Mao Zedong at the stage of revolutionary base also clearly stated the thought of learning from other countries and opening-up, as well as put into practice in the economic construction at the beginning after the establishment of New China. However, with the establishment of the planning economy model and restriction of international environment back then, China's economic construction gradually returned to the closed or semi-closed state. After the Third Plenary Session of the 11th Central Committee of the Chinese Communist Party, based on the summarization of China's experiences and lesions in historical development, especially the experiences and lessons of economic construction since the establishment of New China, Comrade Deng Xiaoping restated the opening-up thought and enabled Chinese economy to go global again. Since the party's 14th National Congress, the third-generation leading group of the party with Comrade Jiang Zemin as the core led China to comprehensive opening-up. However, Chinese society also paid dear price in the forming process of development thoughts from no opening-up to comprehensive opening-up.

22.2 Forms of Opening Development Thoughts and Their Cost

The cost of Chinese opening development thought studied in this book means the price paid by Chinese society in the process from no opening-up to comprehensive opening-up, including the price of no opening-up, the cost of thinking game and the forming cost of comprehensive opening-up, please refer to Table 22.1. Cost of no opening-up means the huge opportunity cost caused to China's economic development and human resources since the establishment of New China due to the implementation of the high concentrated planning economic development model based on unitary public ownership and the partial focus on class struggle. The forming cost of opening-up thought involved certain negative influence from the "Cultural Revolution" and the practice of reform and opening-up in the forming and practicing process of the opening-up thought by the second-generation central leading group with Deng Xiaoping as the core. The cost of comprehensive opening-up means the price paid and risk faced by China in the aspects of economic independent right, ideology and value under the background of globalization and at the stage of comprehensive opening-up.

22.2.1 Price of No Opening-up and Cost of Thinking Game

1. Cost of No Opening-up: Taking Human Capital for an Example

Since the Opium War in 1840, Chinese traditional economy, under the oppression and provocation of foreign capitalism, began to gradually transform into modern economy, and Chinese society also slowly transformed into opening-up from closed state. Because Chinese modernization couldn't make primitive accumulation through colonial plunder like western countries did, and the funds provided by Chinese traditional society were very limited and far from meeting the demand of modern economic development. On the other hand, the declining and

Table 22.1 Forms of the cost of Chinese opening development thought

Classification of cost	Forms	Specific contents
Cost of China opening and developing thinking and ideas	Price of no opening and cost of thinking competition	Loss of human capital stock, cost of thinking competition, cost of route struggle, practical cost of preliminary formation of thinking
	Forming cost of comprehensive opening	Price paid and risk faced for economic independence, ideology and value

dissolving speed of Chinese traditional economy exceeded the development speed of modern economy. In this way, the slowly developing modern economy couldn't meet the most basic living guarantee for the broad population that became unemployed due to the dissolution of traditional economy, the social conflicts were extraordinarily complex and acute, the entire society was in turmoil.

Just at that time, Marxism as the reflection and salvation of the western modernization entered China. The Chinese Communist Party with Comrade Mao Zedong as the representative creatively developed the Marxism-Leninism, established revolutionary base in rural areas, implemented various effective economic policies, organically organized the increasingly failing traditional economy partially through planning means, and protected the basic livelihood of the broad farmers who relied on traditional economy. Since the establishment of New China, the economic development model of the revolutionary base quickly spread to the whole country, and developed into a type of highly concentrated planning economic development model based on unitary public ownership under the reference of the economic development model of the Soviet Union. With the establishment of such planning economic model, the opening-up thought of Mao Zedong was naturally based on planning economic system and gradually became closed state.

Since the end of 1950s, with the deterioration of the China-Soviet Union relations and the continuous blockage by the western capitalist nations, Comrade Mao Zedong began to stress on self-dependence, meanwhile, due to the lack of recognition of the situation of Chinese economy and society, and partial focus on class struggle, it made China lose the golden development opportunity period, and caused huge opportunity cost to Chinese economic development. The "Cultural Revolution" not only generated influence on economic growth, but also generated interruptive and long-term influence on the accumulating process of human capital, which was called by Deng Xiaoping as the biggest loss of the "Cultural Revolution[1]".

As shown in Table 22.2 as follows, during the ten years, the institutions of higher education and secondary schools missed the training of millions of professionals, the construction of Chinese intellectual teams had long-term blankness, according to the Census in 1982, the number of illiterate and semiliterate in China was over 230 million, accounting for nearly 1/4 of the total population back then.

---

[1]In July, 1978, Deng Xiaoping said that the "Cultural Revolution" had disturbance and destruction to every aspect, and had loss in each industry, but the biggest loss was in science and education, mainly in education, there were basically no talent during the 10 years, it held a generation back, and widened the gap between us and the advanced level in the world. The coverage of the loss in industrial and agricultural aspect is relatively easy, but the coverage of the loss in science, technology and education will need more time. Refer to Edited by CCP Central Committee Literature Research Office: *Chronicle of Deng Xiaoping (1975–1997)*, CCP Central Committee Literature Press, 2004 Edition, p. 347.

Table 22.2 Number of students recruited by institutions of higher learning and secondary schools

年份	高等学校 /万人	中等学校 /万人	中等专业学校 /万人	研究生 /人	出国留学生 /人
1965	—	673.0	20.8	1456	454
1966	—	298.0	4.6	—	—
1967	—	212.7	0.8	—	—
1968	—	713.3	1.8	—	—
1969	—	1128.3	1.3	—	—
1970	4.2	1420.7	5.4	—	—
1971	4.2	1577.5	21.3	—	—
1972	13.4	1752.9	26.8	—	36
1973	15.0	1620.4	20.4	—	259
1974	16.5	1918.9	32.7	—	180
1975	19.1	2478.0	34.4	—	245
1976	21.7	3240.2	34.8		277

Source Compiled by the State Statistics Bureau: *Fifty Years of New China: 1949-1999*, China Statistics Press, 1999 Edition, pp. 577, 581

年份	Year	高等学校/万人	Colleges and universities per 10 thousand people
中等学校/万人	Secondary schools per 10 thousand people	中等专业学校/万人	Trade schools per 10 thousand people
研究生/人	Graduate students /person	出国留学生/人	Overseas students/person

According to the calculation of the loss of human capital during the "Cultural Revolution" by Cai Fang and Du Yang, due to the change of human capital stock caused from shortening of the length of schooling (the length of schooling of primary schools and middle schools was reduced from 12 to 9 years or 10 years), suspension of universities and moderate special schools, change of teaching contents, cancellation of vocational education, as well as the influence on the human capital accumulation later, the average years of education for the population over

15 years old in 1982 was 4.8 years, if eliminating the influence of 'Cultural Revolution', it was estimated to be 5.6 years, which means that the 'Cultural Revolution' reduced the potential human capital stock by 14.3%".[2]

2. Cost of Thinking Game

After the decease of Comrade Mao Zedong, the second-generation central leading group of the Chinese Communist Party with Comrade Deng Xiaoping as the core summarized the historical experiences and lessons during the era of Mao Zedong in the victories and setbacks of socialism construction, and resolutely proposed and implemented the reform and opening-up policy, which enabled China to get rid of the closed state and integrated into the world again. However, the opening-up thought of the second-generation central leading group with Deng Xiaoping as the core was not formed in one day, but continuously adjusted in the struggle of the two political lines as well as continuously enriched and developed in social practices, huge price was also paid in its forming and developing process.

As early as April, 1957, Comrade Deng Xiaoping clearly pointed out in the report made at the Xi'an Cadres Meeting that our country is a poor country, to build such as poor and backward country into an advanced socialist industrialized country, we are still primary students. Therefore, we shall learn all advanced experiences in the world, from all countries in the world, including the US, we have to learn their advanced things.[3] In 1962, when talking about recovery of agricultural production, Comrade Deng Xiaoping proposed again the importation of the raw materials and equipments from foreign countries for production of vinylon in order to reduce the pressure of cotton supply in China. That was essentially the opening-up policy that was energetically implemented later.

Special Column 22.1 Effect of Studying in Soviet Union to the Formation of Comrade Deng Xiaoping's Reform and Opening-up Thought

When Deng Xiaoping studied in Soviet Union, Soviet Union was at the state of opening to the capitalism world. After experiencing a series failures at the preliminary stage of political power establishment, the national leaders of Soviet Union all recognized that Soviet Union couldn't independently complete socialism construction under the closed condition, Lenin clearly pointed out that "Socialist republic couldn't survive without having connection with the world, under current circumstance, we should connect our own survival with the relationship with capitalism", we must obtain the funds, technologies, equipments and talents of the

---

[2]Cai Fang and Du Yang: *Destruction of Physical Capital and Human Capital by "Cultural Revolution"*, Economics, 4th Issue, Volume 2, 2003.

[3]Deng Xiaoping: *Congratulations at the Third National Congress of the Chinese New Democratic Youth League, Selected Works of Deng Xiaoping*, 1994 Edition, p. 326.

capitalism world through opening-up, so as to promote the development of socialism construction. In November, 1920, the Soviet People's Committee issued lease and transfer decree to lease enterprises, mines and forests to foreign capitalists for operation and development, the latter, besides inputting funds and technologies, had to contribute partial products to the nation of Soviet Union. Since then, Soviet Union opened its gate of opening-up, the Soviet Union and foreign capitalists established joint-venture companies, introduced foreign funds, technologies, machinery and equipments, hired foreign experts and technicians, etc. Well-known capitalists in the capitalism world, such as Henry Ford, Armand Hammer, etc., invested and established plants in Soviet Union, while the former was a resolute anti-communist person. The period when Deng Xiaoping studied in Soviet Union was the period of full-blown opening-up of Soviet Union, which would generate certain influence on Deng Xiaoping. By comparing the opening-up thoughts of Lenin and Deng Xiaoping, it is not hard to see that Deng Xiaoping not only inherited the theory of Lenin, but also made exploration and innovation on its basis, and promoted Lenin's opening-up thought a big step forward, which was mainly reflected as: firstly, exploring the economic opening-up to broader fields. Subject to the objective conditions back then and limitation of practical experiences, Lenin's opening-up thought was still limited to economic significance, and he believed that only when "the nation is still extremely vulnerable in economy", "the use of capital of the capitalist class" is necessary. While Deng Xiaoping pointed out that the implementation of opening-up "is a strategic issue" and has "political and strategic significance", is a "long-term policy", it can be seen that Deng Xiaoping's opening-up thought was not only limited to economic field, but explored to political field and cultural field. Secondly, extending from partial opening-up to comprehensive opening-up, although Lenin also had high regards in opening-up, but in terms of object of opening-up, it was limited to capitalist nations, in the region of opening-up, it was still partial regions and didn't extend to the whole nation. Deng Xiaoping's opening-up thought was more comprehensive than that of Lenin, in terms of the object of opening-up, "opening-up means opening-up to all countries, and opening-up to various types of nations"; in terms of the scope of opening-up, it formed the comprehensive opening-up situation of "special economic zone—coastal opening-up cities—coastal economic opening-up regions—inlands", and finally realized opening-up on national scale. Thirdly, the form of opening-up was developed from the promotion of lease system to establishment of special economic zone. It was pointed out above that the Soviet Union's opening-up adopted lease system as the main form, Deng Xiaoping, based on the actual circumstance of China's socialism construction, boldly proposed the thought of establishing special economic zone, which can be said as essential leap-forward compared with the lease system of Lenin.

Source: Cui Yanhong: Influence of Studying in Soviet Union on Deng Xiaoping's Reform and Opening-up Thought, Journal of Guangdong Foreign Language and Foreign Trade University, the 5th Issue in 2009.

At the initial stage of the "Cultural Revolution", Deng Xiaoping was stroke down as the No. 2 "Capitalist Roader" of the "Liu-Deng Capitalist Class Headquarter", received wrongful criticism and struggle, was deprived of all positions, and his whole family was involved, then he was demoted to reform through labor in the Tractor Repairs and Production Factory in Xinjian County of Jiangxi. After recovery in 1973, targeting the severely difficult situation caused by the "Cultural Revolution", Comrade Deng Xiaoping decisively abandoned the blockage and isolation policy adopted back then, clearly proposed the opening-up policy and reflected extraordinary courage and boldness of vision. In 1975 when Comrade Zhou Enlai was in serious disease, Comrade Deng Xiaoping received support from Comrade Mao Zedong to preside over the works of Central Government, proposed a set of important rectifying thoughts and developed comprehensive rectification. Later, he tried to put opening-up into practice as policy. In August, 1975, in the *Several Opinions Regarding Industrial Development,* he clearly proposed the "Grand Policy" of "introducing new technologies and new equipments, and expanding imports and exports".[4] If the "Cultural Revolution" was to combat and prevent revisionism, and the rehearsal to prevent the comeback of capitalism, then the "Comprehensive Rectification" was the rehearsal of China's reform and opening-up.[5] However, the situation didn't last long, Comrade Deng Xiaoping's thoughts and practices of opening-up had to come to a halt because he was stroke down again. During this process, Chen Yun, Hu Yaobang and Deng Xiaoping were all stroke down and lost the leading position as well as the opportunity to fight against the "Gang of Four".

The Third Plenary Session of the 11th Central Committee of the Chinese Communist Party held in December, 1978 was a great turning point with far-reaching significance in the history of the party and nation since the establishment of New China. At the Central Working Conference prior to the plenary meeting, Comrade Deng Xiaoping made the speech with topic of "emancipating the mind, seeking truth from the facts, unite as one looking into the future", which relatively systematically explained the basic thought of reform and opening-up. To that point, the reform and opening-up were formally determined as the party's fundamental guidelines and policies, Deng Xiaoping's opening-up thought was basically formed, and put into large-scale practice.

---

[4]Deng Xiaoping: *Several Opinions Regarding the Development of Industry,* Selected Works of Deng Xiaoping, Volume 2, People's Publishing House, 1994 Edition, p. 29.
[5]Gong Yuzhi: *From Mao Zedong to Deng Xiaoping*, CPC Party History Press, 1994 Edition, pp. 285–287.

As everyone knows, China indeed paid certain price in the reform and opening-up as well as the socialist modernization construction. Firstly, China paid certain price in reforming the traditional planning economic system and political system, brought many new conflicts and problems, including "(1) caused from benefiting oneself at the expense of public interests by few people resulted from incomplete and irregular policies; (2) caused from spreading corruption of power-for-money deals resulted from rule by man management and incomplete rule of law; (3) caused from rising evil and sliding morality brought by low ideological quality of partial party members and cadres due to the lagging construction of political thinking theories".[6]

Secondly, China inevitably took certain risks and paid certain prices for establishing new system and implementing new policies, China implemented opening-up policy, attracted investments, established special economic zones, and made adjustments and reforms on state-owned large and medium-sized enterprises, so as to pay certain prices. For example, since the opening-up, China's economy has developed, but the corruptive culture of the western nations had great influence on us.

22.2.2 Forming Cost of Comprehensive Opening-up

In the process of active participation in economic globalization, China's economic development will be subject to the restriction and influence by the world market and global trade, as well as the pressure of unreasonable economic order formed by developed nations. As for the current situation, the economic globalization benefits the developed nations more, while the developing nations, due to relatively poor competitiveness, not only have fewer benefits than that of developed nations, but also will receive even bigger impact at the time of international economic fluctuations.

At the same time of economic globalization, the western nations are also seeking the convergence and integration of political thought, ideology and value selection. While promoting economic hegemony to developing nations, the western developed nations have never stopped political subversion and ideological penetration and corrosion. While promoting economic globalization, they energetically advocate the "National Weakening Theory" and "Neo-interventionism", etc. which have caused national states (weak countries to be precise) to gradually lose their rights.[7]

---

[6]Fan Yanning: *Historical Observation and Realistic Analysis of Social Development Cost Issue,* Journal of Wuhan University, 4th Issue in 2001.

[7]Chen Bao: *Strategy of Economic Globalization and Chinese Social Development,* Journal of Anhui Education College, Jul, 2002.

The cost of comprehensive opening-up is actually reflected as China's all preparations and challenge responses from gradual opening-up to comprehensive opening-up, because it already has related representations and studied in the cost of economic transformation (transformation cost of globalization), it is not explained here.

22.3 Forming Causes of Opening Development Thoughts and Their Cost

22.3.1 Long-term No Opening-up Due to Decision-Making Mistakes of Leader

From the institutional arrangement of the party and nation, from the party's 8th National Congress in 1956 to the party's 10th National Congress in 1973, the *Constitution of the Chinese Communist Party* that has been modified for many times didn't regulate the term of party leaders. Since the establishment of New China, Comrade Mao Zedong has been the supreme leader of Chinese ruling party for as long as 27 years, during the later stage of his ruling, he made the "Great Leap-forward", the "Cultural Revolution" and other major, long-term and comprehensive historical decision-making mistakes. From the implementation of system, the democratic concentration system within the party, the collective leading and individual accountability system wasn't fully implemented, as a result, when Comrade Mao Zedong made major decision-making mistakes, the party was already lack of the capability to restrain him, limit him, correct him and replace him, so the consequences became more and more severe. Besides, the method of class struggle for strategic disputes was also an important reflection back then, Chinese leaders had different viewpoints and selections on China's industrialization path and economic development strategy, but such differences in opinions were handled in the method of class struggle, which caused leaders within the party unable to express different opinions, and caused the "Twice Striking Down" of Comrade Deng Xiaoping during the "Cultural Revolution", and caused Comrade Deng Xiaoping's Opening-up Thought unable to be put into practice, or suspension after short-term practice, which resulted in the confusion in thinking and practice, and also delayed China's opening-up over and over again.

22.3.2 Necessary Price for Exploration and Development of New System

On one hand, due to the limitation of resources, a reform must break the original social order and interest pattern, enable certain region or industry to firstly develop, "cause a series of deep transformation in economic life, social life, working method and spiritual state", cause gap of living state and people's psychological gap. On the other hand, because China was at the transformation stage from planning economy to market economy, the establishment of socialist market economy system would inevitably generate frictions and conflicts among different systems, and encounter disturbance from "Leftists" and "Rightists", which would inevitably cause change of social structure and certain social irregularity.

22.3.3 Associated Effect of Globalization Trend

Many rules in international economic operation were made not by sovereignty nations but the international economic organizations that are controlled by western developed nations. The developed nations take advantage of their economic strength and advantages to establish unfair and unequal international economic rules or order, so as to make economic exploitation on the developing nations, and make efforts to enable them to accept unfair trading rules in international economic activities.

The western developed nations use modern information technology and media means to promote the values in the west, such as democracy, freedom, human rights, etc., seeming to enable the "US Civilization" or "Western Civilization" to become the destiny of the modern civilization of human being, the "Democratic Market and Capitalism" have become an eternal model and have threatened the traditional culture and value of many ancient civilizations. This requires us to be alert about the political plot of western nations when participating in global activities.

The forming causes of the cost of Chinese opening-up thoughts are concluded as follows, please refer to Chart 22.1.

22.4 Countermeasures for Opening Development Thoughts and Their Cost

22.4.1 Implement Democratic Centralism Through System Reform

The "Infinite Term System" and "Lifelong Term System"[8] of national leaders were abolished, in 1982, the *Constitution* clearly regulated that national leaders shall have a "System with Two Terms", in 1987, the party's 13th National Congress ended the "Lifelong Term System" of leaders, so as to restrain the rights of leaders

---

[8]Hu Angang: *Historical Review of China's Politics and Economy (1949–1976)*, Tsinghua University Press, 2008 Edition, p. 495.

22.4 Countermeasures for Opening Development Thoughts and Their Cost

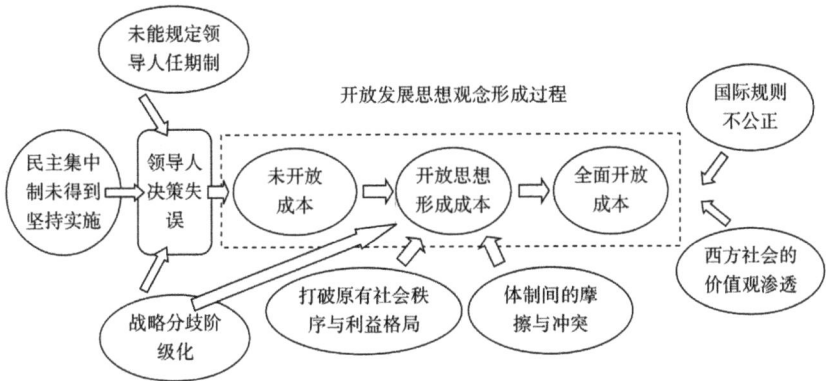

未能规定领导人任期制	Failed to regulate the leader tenure system
开放发展思想观念形成过程	Forming process of the open development thinking and idea
国际规则不公正	Unjust international rules
民主集中制未得到坚持实施	Democratic centralism didn't receive persistent implementation
领导人决策失误	Decision-making mistakes by leaders
未开放成本	Cost of no opening
开放思想形成成本	Cost of forming open thinking
全面开放成本	Cost of comprehensive opening
战略分歧阶级化	Strategic dispute became class struggle
打破原有社会秩序与利益格局	Broke original social order and interest pattern
体制间的摩擦与冲突	Friction and conflict between systems
西方社会的价值观渗透	Value penetration by western society

Chart 22.1 Forming causes of Chinese opening-up thoughts

and implement the democratic centralism, prevented personal dictatorship from the level of constitution, and ensured collective decision-making among leaders from systematic perspective.

22.4.2 Correctly Treat and Handle the Necessary Cost of the Formation of Reform and Opening-up Idea

There is no existing model to copy for China's socialism market economy system reform, and there is no existing experience to follow. As indicated by Comrade Deng Xiaoping: "What we are doing is a new undertaking, Marx didn't talk about it, our predecessors hadn't done it, and other socialist nations hadn't done it, so we don't have existing experience to learn. We can only learn in practice and explore in practice.⁹" From the basic rule of paying certain price in social development, we shall correctly treat the problems occurred in China's reform and opening-up, we shouldn't lose confidence in reform and opening-up or deny the achievements in reform and opening-up just because of the problems occurred. When the 1989 political disturbance was just quiet down, Comrade Deng Xiaoping clearly answered in the talk on June 9th that "The basic point of reform and opening-up is wrong or not? It is not wrong. Without the reform and opening-up, how could we achieve what we have today? People's lives had big advancement over the past ten years, it should be said that we have taken a higher level, despite the problems such as inflation, we have to fully estimate the achievements of reform and opening-up in the past ten years.¹⁰"

Adhering to rule of law means that we shall, according to the provisions of the *Constitution* and other laws, manage various national affairs through various means and forms, so as to gradually realize the socialist democratic rule of law. Meanwhile, we shall feasibly enforce the guideline of ruling the country by virtue. Marx and Engels once said that "The thought of the ruling class is the dominating thought in every era.¹¹" The ethical level of leaders and cadres is the indicator of the construction of social morality system, therefore, we should focus on the morality construction of leaders and cadres, enable them to be restrained and improved on institutional level. Deng Xiaoping indicated that "Good systems in these aspects could make bad people unable to do whatever they want, while bad systems could make good people unable to do good things, or even take to the other side.¹²"

22.4.3 Countermeasures for Associated Risks of Globalization

First of all, we shall enhance the ideological, moral, scientific and cultural quality of all people, adhere to the socialist orientation, and abandon all the thoughts and ideas that are in conflict with the cause of socialist modernization. Secondly, in the process of globalization, we shall adhere to the principle of independence, avoid trust in the western society, draw on advantages and avoid disadvantages, in the struggle and cooperation with western nations, make efforts to establish the truly reasonable and fair new order for international politics and economy. Finally, while alerting the western enemy forces, we shall actively participate in the development

---

⁹*Selected Works of Deng Xiaoping,* Volume 3, People's Publishing House, 1993 Edition, pp. 258–259.

¹⁰CCP Central Committee Literature Research Office: *Compilation of Selected Important Literature since the Party's 13th National Congress,* People's Publishing House, 1991 Edition, p. 540.

¹¹*Karl Marx and Frederick Engels,* Vol. 1, People's Publishing House, 1995 Edition, p. 98.

¹²*Selected Works of Deng Xiaoping,* Volume 2, People's Publishing House, 1994 Edition, p. 333.

22.4 Countermeasures for Opening Development Thoughts and Their Cost

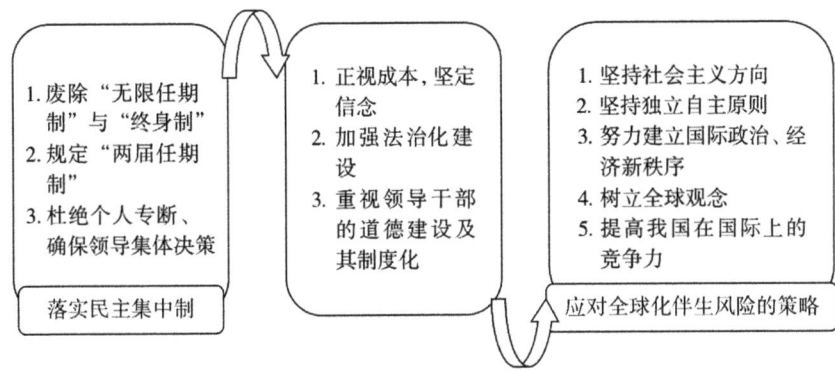

1. 废除"无限任期制"与"终身制"	1. Cancel the "Infinite Tenure System" and "Lifelong Tenure System"
2. 规定"两届任期制"	2. Regulate the "Two-term Tenure System"
3. 杜绝个人专断、确保领导集体决策	3. Avoid personal arbitrary, ensure leadership group decision making
落实民主集中制	Implement democratic centralism
1. 正视成本，坚定信念	1. Recognize cost and strengthen the conviction
2. 加强法治化建设	2. Strengthen the construction of rule of law
3. 重视领导干部的道德建设及其制度化	3. Emphasize the morality construction of leaders and its institutionalization
1. 坚持社会主义方向	1. Adhere to socialist orientation
2. 坚持独立自主原则	2. Adhere to the principle of independence
3. 努力建立国际政治、经济新秩序	3. Make efforts to build new order of international politics and economy
4. 树立全球观念	4. Establish global awareness
5. 提高我国在国际上的竞争力	5. Enhance China's competitiveness in international community
应对全球化伴生风险的策略	Strategy to respond to the risks brought by globalization

Chart 22.2 Countermeasures for the cost of opening development idea

of economic globalization, improve the comprehensive, multi-level and wide-range opening-up pattern, so as to improve China's competitiveness in international community. For the countermeasures for the cost of opening development idea, please refer to Chart 22.2.

Chapter 23
China's Cost of International Exchange and Consensus

23.1 Overview and Forms of the Cost of International Exchange and Consensus

Since the establishment of New China, China and other nations and international organizations have formed wide range of international consensus regarding international exchange, experienced long, tough and complex process, and China has made extremely hard and bitter efforts. As a large developing country that is rising, with the enhancement of economic strength, China has actively developed multiple channels and multiple fields of international exchange, joined and learnt to negotiate and cooperate with different international organizations, participated in the formation of international systems and obtained certain speaking rights, as well as actively enabled the world to understand a real China, which has great significance for China to further develop itself and play more active role in global and regional affairs.

Generally speaking, at the preliminary stage of New China, China's international exchanges were relatively few, which had relatively strong political purposes and at relatively closed state. With the recovery of China's lawful seat in the United Nations and the gradual development of the reform and opening-up, China's international exchange reflected the trend of comprehensive development, which not only reflected as participating in international organizations, conferences and conventions with more active attitude, but also reflected as actively planning the strategies of regional and international organizations based the development of international situation, and has preliminarily established a regional organization network surrounding China; meanwhile, China has also proposed its own claims facing major international public affairs and crisis, as a responsible major power, China actively bear responsibilities, actively promotes and maintain the interests of the wide developing countries. For more details about the forms of the cost of international exchange and consensus, please refer to Table 23.1.

Table 23.1 Forms of the cost of international exchange and consensus

Classification of cost	Forms	Specific contents
Cost of international exchange and consensus	Cost of international exchange and participation	Efforts made to participate in international organizations and form international conventions at each stage after the establishment of New China
	Organizing cost of international exchange	Efforts made to initialize or establish forum mechanism, build China's international organization network, organize and establish large international exchange
	Fulfilling cost of international consensus	When facing major international affairs and crisis, actively bear certain responsibilities and promote the formation of new international order

23.1.1 Cost of International Exchange and Participation

During the period from the establishment of New China and 1970, China's international exchanges were still in relatively closed state. The international exchanges developed by China at that stage were centered at obtaining international recognition, China made limited contacts and cooperation with multiple nations and international organizations. Surrounding the center of international recognition, China established the international exchange with purpose of recovering China's lawful seat in the United Nations, and implemented struggle against the US' conducts of obstructing China's recovery of its lawful seat in the United Nations.[1]

China tried to make efforts in recovering its lawful seat in the United Nations, and establishing contacts with other international organizations, for example, during 1950s, China proposed joining applications to multiple global organizations including the World Health Organization, the World Meteorological Organization, the World Federation of Trade Unions, IMF, the World Bank, etc. respectively, but all the applications were obstructed by the US, back then, China only joined some international organizations and institutions of the Socialism Camp led by the Soviet Union. Subject to the lockage and isolation by the West led by the US, China had long been excluded from the system of the United Nations. China considered the UN as a voting machine of the US.

---

[1] On Nov 5, 1949, Foreign Minister Zhou Enlai contacted the UN Secretary General Trygve Halvdan Lie and President of the 4th UN General Assembly Carlos P. Romulo and solemnly stated that "The People's Republic of China is the only legitimate government that represent Chinese people, the KMT government has lost any and all legal and factual basis to represent Chinese people, and required immediate cancellation of all the rights of the KMT Government Delegation to continuously represent Chinese people in participating the UN".

The People's Republic of China recovered its lawful seat in the United Nations,[2] which completely solved China's voting right issue in the United Nations, and also meant that China's international exchanges had entered to a new stage. However, China at that time was still wandering between insider and outsider of international organizations, the international image of passive and negative participant was very prominent. China at that stage focused on developing foreign political relations, China established contacts with auxiliary institutions of the United Nations one after another, and participated in their activities. In 1972, China participated in the formal activities of the United Nations Development Program, the United Nations Environment Program, the United Nations Industrial Development Organization, the United Nations Conference on Trade and Development, etc. In November, 1973, Chinese Delegation attended the 17th General Conference of the United Nations Food Agriculture Organization, and was elected as a council member country. Besides, China at that stage also recovered and developed friendly relations with many international organizations, including the Nuclear Weapon Ban Organization of Latin America, the International Commission on Large Dam, the International Union of Geodesy and Geophysics, the International Organization for Standardization, the International Olympic Committee, the Asian Games Federation, the Organization of African Trade Union Unity, etc. At that stage, the number of international organizations that China joined developed from 1 in 1966 to 21 in 1977, and the number of non-governmental organizations that China joined developed from 58 to 71.[3]

With the transfer of working focus from class struggle to economic construction in 1979, reform and opening-up became the guideline of the party and national government, and China's international exchange also had unprecedented promotion, China began to comprehensively participate in international exchange. At that stage, the quantity, activeness and quality of international organizations that China joined surpassed that in previous two stages (Table 23.2).

The number of international organizations that China joined and the number of treaties China signed had significant increase, but from overall perspective, in the process of China's international exchange and participation at the early stage since the reform and opening-up, China was more often expressing statements and stance in principle, rarely participated in formation of procedure, and lack of the awareness of forming procedures, therefore, China was relatively passive in actions.

---

[2]On Oct 25, 1971, after fierce debate of the 26th UN General Assembly, the lawful seat of the People's Republic of China at the UN was recovered through majority votes.
[3]Gerald Chan, *China's Compliance in Global Affairs,* New Jersey: World Scientific Publishing Company, 2006, p. 51.

Table 23.2 Abstract of International Organizations/Treaties China Joined since the reform and opening-up

Field	Year	Name of International Organization/Treaty Joined
Economy	1980	Became member of IMF and World Bank, and joined the International Agricultural Development Fund
	1986	Joined Asian Development Bank
	2001	Became member of World Trade Organization
	2006	Joined the *Treaty of World Intellectual Property Rights Organization*
Disarmament	1988	Became member of UN Peace Keeping Action Committee
	1989	Dispatched five military observers to UNTSO
Environment	1992	Signed the *UN Framework Convention on Climate Change*
	1998	Signed the *Kyoto Protocol*
Nuclear nonproliferation	1983	Signed the *Treaty on Antarctica* and *Outer Space Treaty*
	1987	Signed the *South Pacific Non-nuclear Region Treaty*
	1992	Signed the *Africa Non-nuclear Region Treaty*
	1996	Signed the *Comprehensive Nuclear Test Ban Treaty*
Human rights	1997	Signed the *International Convention on Economic, Social and Cultural Rights*
	1998	Signed the *International Convention on Civil and Political Rights*

Source Concluded from Wang Yizhou's *Sixty Years of Chinese Diplomacy*, Chinese Social Science Press, 2009 Edition, p. 58

23.1.2 Cost of Organizing International Exchange

With the further deepening of international exchange, China clearly realized that the participation in the formation and implementation of international systems are a very important condition to win reputation on international stage. In major global issues in international community, China has been increasingly integrating in international system and becoming more cooperative.

On one hand, China not only participates in exchange of international forum in form of talks, participates in international organizations and comprehensively develop international exchange, but also actively initiates and establishes forum mechanism in order to establish China's network of international organizations.

In 1993, China and some countries jointly developed the Informal APEC Economic Leaders' Meetings, and established a systematic platform to hold meeting once a year, as of 2011, 19 such meetings were held. On Sep 25, 1999, the Group 20 was formally established. China was one of its founding nations, and actively participated in its activities as a representative of the emerging economies, as well as played important role. In October, 2000, the 1st Ministerial Conference of the China-Africa Cooperation Forum was held in Beijing. Over 80 national or government leaders, ministers from 44 countries in Africa as well as representatives from 17 regions and sub-regions participated in the Conference, which determined the group dialogue mechanism for China and Africa to regularly discuss with each other. In 2004, China's Ministry of Foreign Affairs and the Secretariat of the Arab League jointly established the China-Arab Cooperation Forum, which held the first Ministerial Conference in Cairo in September, 2004. In April, 2006, China-Pacific Islands Economic Development and Cooperation Forum was established under the initiative by Chinese Ministry of Foreign Affairs, and the first Ministerial Conference was held in Fiji. Wen Jiabao and the leaders and related ministers from 8 Pacific Islands participated in the forum, the 9 countries jointly signed the Program of Action for China-Pacific Islands Countries Economic Development and Cooperation. China-Caribbean Region Economic & Trade Cooperation Forum was established under the initiative by China in 2004, which was designed to promote the economic and trade exchange and cooperation between China and Caribbean Region and realize joint development.

China puts the key and benchmark of the network of organizations on the surrounding regions, because the stability and development of surrounding regions are directly correlated with China's peace and development. Different from some big powers' handling of peripheral relations in the past, China conforms to the developing trend of globalization and regional integration, and implements it in form of regional cooperation.

What is worth to be mentioned more that on Apr 26, 1996 in Shanghai, the five nations, China, Russia, Kazakhstan, Kyrgyzstan, and Tajikistan formally signed the agreement regarding strengthening trust in military field at border areas, and established the "Shanghai Five Mechanism". In 2001, the "Shanghai Five Mechanism" was developed into the Shanghai Cooperation Organization. As the first intergovernmental international organization that is named after a Chinese city, Shanghai Cooperation Organization has symbolic significance in the history of Chinese and international organization's development of relations. With Shanghai Cooperation Organization as the platform, China not only found a path of cooperation to solve border issues and disputes, but also established a stage for multi-field cooperation among China, Russia and Central Asia. Meanwhile, China also established the diversified East Asian Cooperation Mechanism, such as the ASEAN + China ("10 + 1"), ASEAN + China, Japan and Korea ("10 + 3"), etc., and had positive actions in Korean Nuclear Issue, developed multilateral diplomacy and promoted the Six Party Talks to solve Korean Nuclear Issue. In recent years, China has been actively participating in the dialogue and meetings between the G8 and developing nations.

On the other hand, Chinese government has also been energetically supporting the application, planning and holding of large international activities such as the Olympic Games and EXPO, so as to establish "Invitational" exchanging platform for the international community to understand China more thoroughly, enable China to quickly integrate into the international "Big Family", win over the "Attention" of the world, as well as spread the Chinese civilization enhance the national image. While being proud for this, we should also notice that behind various international effects, China has also paid extremely large efforts and prices. The analysis of its cost is as shown in Chart 23.1.

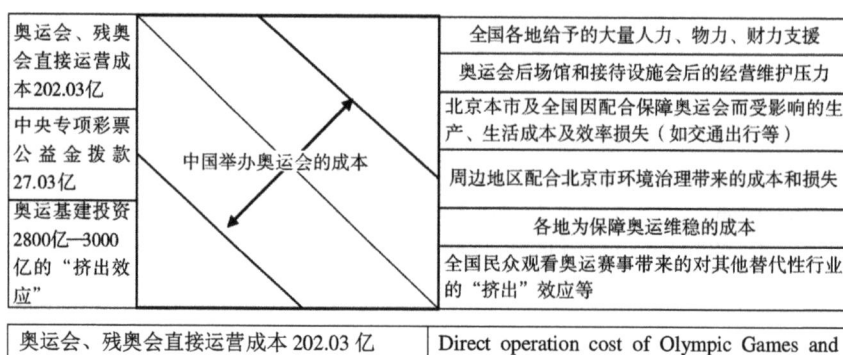

奥运会、残奥会直接运营成本 202.03 亿	Direct operation cost of Olympic Games and Paralympic Games at RMB 20.203 billion
中央专项彩票公益金拨款 27.03 亿	Special lottery public welfare appropriation from central government at RMB 2.703 billion
奥运基建投资 2800 亿-3000 亿的"挤出效应"	Crowding-out effect of the RMB 280-300 billion of Olympics infrastructure investment
中国举办奥运会的成本	The cost of China holding the Olympic Games
全国各地给予的大量人力、物力、财力支援	Large amount of manpower, material and financial support given by all other places in China
奥运会后场馆和接待设施会后的经营维护压力	Operation and maintenance pressure of the venues and reception facilities after the Olympic Games
北京本市及全国因配合保障奥运会而受影响的生产、生活成本及效率损失(如交通出行等)	Beijing and China's cost of affected production and living as well as efficiency loss (such as traffic and travelling) due to coordination and guarantee of Olympic Games
周边地区配合北京市环境治理带来的成本和损失	Cost and loss of surrounding regions from environmental treatment in order to coordinate with Beijing City
各地为保障奥运维稳的成本	Cost of stability maintenance in different places in order to support the Olympic Games
全国民众观看奥运赛事带来的对其他替代性行业的"挤出"效应等	Crowing-out effect of national audiences to other alternative industries in order to watch Olympic Games, etc.

Chart 23.1 Cost of organizing international exchanges, taking the cost of chinese hosting of the olympic games for an example. *Source* Concluded by the author

23.1.3 Cost of Fulfilling International Consensus

With the acceleration of Chinese procedure of economic development and political democratization, China's ability to shoulder international responsibilities and perform international consensus has been continuously increasing. After the beginning of the Asian Financial Crisis in 1997, Chinese Government promised to the world not to devalue the RMB, under the condition of not having too much foreign exchange reserve, China provided Thailand and other Southeast Asian countries with over USD 4 billion of aid in total, and formally proposed the strategic objective of "being a responsible major power in international community". The proposition of such strategic objective meant that China's recognition of itself had a huge change, the core recognition of sovereignty-centered independent large country was integrated with the new recognition of being a responsible major power, the latter has direct correlation with the international systems, China's international behaviors are increasingly subject to regulation of the international systems, the objective of having a constructive and responsible international image has thus been established in the interaction between China and international systems.

As for the issue of response to global climate change alone, China, as a responsible large nation, made its supposed contribution in the aspect of fulfilling the international consensus of protecting the global climate, so China is a very important and active force. In September, 2002, even without the quota and obligation of reducing greenhouse gas emission, China announced to approve the *Kyoto Protocol*. During the period from 1990 to 2005, the energy consumption per RMB 10 thousand of GDP decreased by 47%, equivalent to saving of 800 million tons of standard coal and equivalent to reduction of the emission of 1.8 billion tons of carbon dioxide. During the same period, China absorbed another 5 billion tons of carbon dioxide through plantation and forest protection. Especially since the 20th century, China's implementation of the family planning policy prevented 300 million of births. Based on the standard in 2004, every person in the world releases 4 tons of carbon dioxide in each year, the number of 300 million people means that China emitted 1.2 billion tons less carbon dioxide in last year.[4] In June, 2007, China formally released the *National Program of China's Response to Climate Change*, which was China's first comprehensive policy document in response to climate change and also the first national program issued by a developing country in response to climate change. The issuing and enforcement of such program reflected the attitude of Chinese Government as a responsible power, and will play active role in China's work of responding climate change, as well as make new contribution to the world response to climate change.

China's actions in nuclear test ban, environment and other aspects are also increasingly active, China has obtained huge reputation and begins to reflect the characteristics of a active participant, its procedure establishment capability has been enhanced. Especially since the 21st century, in the aspect involving UN reform framework, China has been emphasizing very much on the interest of developing nations. In the documents regarding China's stance on UN reform, it

---

[4]Li Huiming: *Responsible Power in International Society—Identity Resort and Practical Construction of Modern China,* Journal of the University of International Relations, 1st Issue in 2008.

was clearly indicated that the reform should to most extent satisfy the requirements and concerns of most member nations, especially the developing nations. Poverty, disease and environmental deterioration also pose serious challenge to the international community. We shall emphasize the needs of developing nations, realize the coordinated, balanced and common development across the globe. We shall guide the globalized and balanced development, strengthen the developing nations' rights of equal participation and decision making in international economic affairs. Different countries shall develop sustainable international cooperation based on the principle of "common but differentiated responsibilities", the key is to help developing nations to effectively respond to environmental challenges, especially help them solve the urgent problems such as water resource shortage, urban air pollution, ecological deterioration, desertification, etc. The developed nations should implement their commitments, provide developing nations with the related technical transfers and funding support, and help the developing nations to have capacity building. China supports the reinforcement of the functions of the Anti-terrorism Committee of UN Security Council, expand the authority limit of the execution bureau, especially help the developing nations to strengthen the building of anti-terrorism capacity. China supports proper streamlining of conference agenda, optimization of schedule, in each year, some major and substantial issues of different parties, especially the issues that developing nations care about, shall be discussed. In priority, we should increase the representation of developing countries. Developing nations already account for over 2/3 of the total members of the UN, but their representation at the UN Security Council is severely lacking. This situation must be corrected. More nations, especially medium and small nations, should have more opportunities to enter the UN Security Council and participate in its decision making.

In 2008 when the international community encountered the most severe economic crisis since the World War II, China didn't keep out of the affair, but actively and jointly responded to this crisis with international organizations as a responsible member of the international community, and shouldered the responsibilities as a major developing power. At the two G20 Summits at the end of 2008 and the beginning of March, 2009, China was not only an active participant, but also a main role at the summits. In the G20 Summit, China clearly proposed its own advocates, required the promotion of the international financial order to develop towards the direction of equality, fairness, tolerance and order based on the principles of comprehensiveness, balance, gradualness and effectiveness; international financial institutions shall strengthen the aids to developing nations, improve the governance structure of the International Monetary Fund and the World Bank, and enhance the representation and speaking right of developing nations. The international community should pay high attention to and greatly reduce the damage of international financial crisis caused to the developing nations, especially the most underdeveloped nations; the international community, especially the developed nations, should bear their supposed responsibilities and obligations, continuously fulfill their commitments in aids and deleveraging, feasibly maintain and increase the aids to developing nations, feasibly help the developing nations to maintain financial stability, promote economic growth, feasibly help the developing nations, especially the African nations to overcome difficulties, and continuously improve the external environment for the development of these nations.

23.2 Forming Causes of the Cost of International Exchange and Consensus

It can be seen from the afore-mentioned efforts of China in participating in international exchanges, organizing international exchanges and fulfilling international consensus that prior to 1970s, China's limited international exchange was mainly in order to obtain acknowledgement of international community, of course primarily to obtain acknowledge from socialist nations, or the acknowledgement of some friendly nations or third-world nations. Since 1970s, especially after the reform and opening-up, China has been increasingly active in the participation of international exchange, which was mainly to create a peaceful international environment, fully use the international rules and system to serve the domestic economic construction, China understood the international rules and system in form of tool theory, and the steps and methods of participation were closely correlated with China's reform and opening-up. Since the 1990s, China's comprehensive national strength already had great development, China began to comprehensively connect with the international community, both the width and depth of participation in international communities were unprecedented. With the acceleration of the global steps of standardization and systemization, and the sovereignty nations increasingly involving in the process of international integration, the role of non-national entities is increasingly stronger, the interdependence in international relations deepens, the main theme changed from international conflict to international cooperation, win-win and common win principle gradually become the mainstream of international relations. International mechanism becomes an important leverage to regulate the relations among nations and increasingly plays the role of standardizing party of international behaviors. This provided dynamic and opportunity for China to comprehensively participate in international mechanism.[5] Since entering the 21st century, China deeply recognizes that its own development constitute an inseparable part of the world development, China's destiny is closely connected with the world's destiny, in international organizations, China actively participates in the reform and improvement of international organizations and systems with the responsibility and awareness of a major nation for peaceful development, and promotes the international organizations to develop towards the more fair and reasonable direction.

---

[5]Li Huiming: *Responsible Power in International Society—Identity Resort and Practical Construction of Modern China,* Journal of the University of International Relations, 1st Issue in 2008.

China is at the key stage of peaceful development, the building of positive and active image of responsible power and creation of an excellent international atmosphere are correlated to the big picture of national modernization and development. The establishment of national image and participation in international mechanism are a dialectical and unified process. Active participation in international mechanism could provide broader strategic space for the exploration of China's national interest, and could help improving the international image of China; the enhancement of China's international image is good for reduce the obstacle in participation in international mechanism, and good for playing important role in the regional and international mechanism.

23.3 Countermeasures for the Cost of International Exchange and Consensus

Firstly, focus on maintaining of national interest, and bear the responsibilities granted by the international community. First of all, China should set foot in Asia, play constructive role in the affairs of Asia, especially the East Asia, participate in the establishment, improvement and maintenance of the international mechanism in East Asia and Asia.

Secondly, expand the space of international activities, try to participate in all global international mechanism, be a genuine "responsible major power in international community", unite with most nations in the world to jointly restrain hegemony and power politics, make supposed contribution to the establishment of a fair, reasonable and democratic international order.

Thirdly, the current world situation is an international system that has always been dominated by the Western Nations led by China, while complying with the rules in international system, China should also actively fight for speaking right with strength.

Fourthly, under the circumstance of participating more in the innovation of international mechanism, China should not only actively provide public goods to the international community, establish its own image of major power, but also clearly face its own nation and do things according to its abilities.

Chapter 24
China's Cost of Investment and Construction of Investment Environment

24.1 Overview and Forms of the Cost of Investment and Construction of Investment Environment

The large-scale use of foreign investments raised large amount of urgently needed funds for the construction of China's modernization and played great promoting role in China's infrastructure construction. The inflow of foreign investments also brought advanced technical and management experiences, promoted the development of emerging industries and the transformation of traditional industry into modern industry. Meanwhile, the pouring-in foreign investments also increased the fluctuation of China's economy, imbalance of industrial structure and regional difference in economic development, and to certain extent restrained the development of national industry and growth of young industry, as well as increased China's burden of paying off debts plus interests.[1] Based on scholars' research and analysis, in order to attract foreign investments and improve investment environment, besides maintaining social stability, strengthening supervision, building financial infrastructure, cultivating human capital, implementing related laws and regulations to protect the rights and interests of investors and intellectual property rights, providing investors with most-favored-nation treatment and super-national treatment, etc., China also bears various prices brought by foreign investment attraction and economic development.

In this book, the cost of investment and investment environment is classified into the cost of providing investment environment, which is mainly reflected as great increase of dependence on global economy, certain risks faced by financial security, and impact of unfair competition of foreign enterprises on national industry; and China's cost of foreign investment, which is mainly reflected as the continuous increase of direct foreign investment, as well as the loss of investment efficiency,

---

[1]Zhang Daoquan: *Establishment of Price Awareness in Foreign Economic Relations,* Economic Issue Exploration, 11th Issue in 2004, pp. 76–78.

loss of time and loss of assets in the overseas acquisitions by Chinese enterprises, for more details, please refer to Table 24.1.

24.1.1 Cost of Providing Investment Environment

Firstly, while creating an open investment environment for foreign capital, China's domestic economy becomes increasingly easy to fluctuate along with the global economy. The energetic development of foreign economic relations enables China's economy to be increasingly close with the global economy, the mutual dependency is strengthened to unprecedented level, the world economy, especially the economy of developed nations, increasingly influences China, and China's economy also influences the world economy. According to the related materials, China's rate of foreign trade dependence was 9.8% in 1978, which increased to 45.5% in 1994, then decreased to 36.1% in 1997. Since 2000, this data ranged between 43 and 47%, which is far higher than the developed capitalism nations such as the US and Japan and the major developing nations such as India and Brazil, and far exceeding the safety line of 25%.[2] At such rate of foreign trade dependence, we could on one hand share the opportunity of global economic prosperity, and on the other hand must bear the negative influences brought by external economic turmoil. Once the world economy, especially the economy of China's main trade partners, has turmoil or recession, China's import/export trade and the entire national economy would be inevitably subject to impact, the negative influence of the Asian financial crisis on China's economy was an example.

Secondly, it makes China's financial security to be subject to certain degree of influence. In recent years, affected by China's high-speed economic growth and the expectation of appreciating RMB, "Hot Money" poured into China in large scale (refer to Table 24.2). The large-scale entrance of hot money affected the effect of monetary policy, disturbed the normal operation of financial system, and intensified the domestic inflation. Most of the "Hot Money" flowing into China were invested in real estate, stock and other virtual economic markets, which caused the hot situation of virtual economic market and affected the development of the real economy.[3] When the economic fundamentals have fluctuation, "Hot Money" would become "Bird on a Wire" and quickly evacuate. Once the "Hot Money" evacuate in large scale suddenly, it would generate huge impact on this nation's capital market or even the entire country's financial system and financial order.

Thirdly, the unfair competition has to certain extent restrained and impacted the development of China's national industry. One of the main purposes of foreign

---

[2]Zhang Daoquan: *Establishment of Price Awareness in Foreign Economic Relations*, Economic Issue Exploration, 11th Issue in 2004.

[3]Li Hairong and Li Zhengdan: *Influence of "Hot Money" on Chinese Economy and Response Measures*, Fujian Finance, 4th Issue in 2011.

24.1 Overview and Forms of the Cost of Investment and Construction ...

Table 24.1 Forms of the cost of China's investment and construction of investment environment

Classification of cost	Forms	Specific contents
Constructing cost of China's investment and investment environment	Cost to provide investment environment	Dependence on global economy, impact on financial market, and loss of national industry
	Cost of China's foreign investment	Continuously increasing foreign direct investment, loss and cost of investment efficiency, cost of time loss, and cost of capital loss

Table 24.2 Estimation of net flow of "Hot Money" in China during 2001–2010

(Unit: USD 100 million)

	外贸顺差①	直接投资净流入②	境外投资收益③	境外上市融资④	前四项合计⑤	外汇储备增量⑥	"热钱"流动净额⑦=⑥−⑤
2001 年	225	398	91	9	723	466	−257
2002 年	304	500	77	23	904	742	−162
2003 年	255	507	148	65	975	1377	402
2004 年	321	551	185	78	1135	1904	769
2005 年	1021	481	356	206	2064	2526	462
2006 年	1775	454	503	394	3126	2853	−273
2007 年	2643	499	762	127	4031	4609	578
2008 年	2981	505	925	46	4457	4783	326
2009 年	1957	422	994	157	3530	3821	291
2010 年	1831	467	1289	354	3941	4696	755
合计	13313	4785	5330	1459	24887	27777	2890

Source International Payments Analysis Team of the State Administration of Foreign Exchange: *Monitoring Report of China's Cross-border Fund Flow in 2010*, China Finance Press, 2011 Edition

外贸顺差①	Foreign trade surplus①
直接投资净流入②	Net inflow from direct investment②
境外投资收益③	Overseas investment return③
境外上市融资④	Overseas IPO financing④
前四项合计⑤	Sum of the previous four items⑤
外汇储备增量⑥	Increment of foreign exchange reserve⑥
"热钱"流动净额⑦=⑥−⑤	Net flow of "hot money" ⑦=⑥−⑤
合计	Total

investment in China is to occupy the vast market of China, the "Market for Technology" policy adopted by China provided opportunity for this. In the competing process with foreign enterprises, the market shares of domestic enterprises declined, some industries already reflect the situation of control or even monopoly by foreign investors, some Chinese enterprises therefore became subsidiary or auxiliary organs of multinationals, and become simple workshops that engage in processing for others. China's machinery industry, oil industry, daily-use chemical industry, electronic telecommunication industry are to certain extent subject to impact from foreign enterprises.

24.1.2 Cost of China's Foreign Investment

Since the 1990s, the FDI outflow on China's international balance sheet increased from less than USD 1 billion to over RMB 2 billion in each year, which draw wide attentions. After the East Asia Financial Crisis in 1997–1998, Chinese Government proposed the "Go-out" strategy and encouraged enterprises to go out in the world to undertake direct foreign investment activities. Under this strategy, the direct foreign investments by Chinese enterprises had extensive growth and rapid development. In 2009, the FDI across the globe sharply declined by 43%, while China's FDI flow created the new historical record of USD 56.53 billion, which increased by 1.1% compared with the same period in previous year, in which the non-financial FDI was USD 47.8 billion, which increased by 14.2% compared with the same period in previous year.

The *2011 World Investment Report* of UNCTAD indicated that the outflow of foreign direct investment across the globe in 2010 was USD 1.32 trillion, and the year-end stock was USD 20.4 trillion, calculating on this basis, China's foreign direct investment in 2010 accounted for 5.2 and 1.6% respectively in the global flow and stock in that year, in 2010, China's foreign direct investment flow ranked the 5th among all countries (regions) in the world, and the stock ranked the 17th place. This indicates that China is gradually becoming one of the major foreign investment nations (Chart 24.1).

The *Statistical Bulletin of China's Foreign Direct Investment in 2010* released by State Administration of Statistics indicates that in recent years, the ratio of international direct investments made in form of acquisition and merge becomes higher and higher, cross-border acquisition and merge increasingly become a prominent characteristic of cross-border direct investment. The ratio of overseas M&A amount by Chinese enterprises in FDI has been in the state of great fluctuation, which is specifically reflected as increasing transaction in overseas M&A, and its ratio in China's total direct foreign investment has been continuously expanding. Table 24.3 is specific reflection of this situation.

Since 1997, the overseas M&A by Chinese enterprises generally experienced two M&A waves under different backgrounds. The main region of overseas M&A by Chinese enterprises are mainly neighboring nations, and the targets concentrate on oil, telecommunication, traffic and other similar national resource and infrastructure industry. Since China joined WTO in 2001, the second wave of overseas

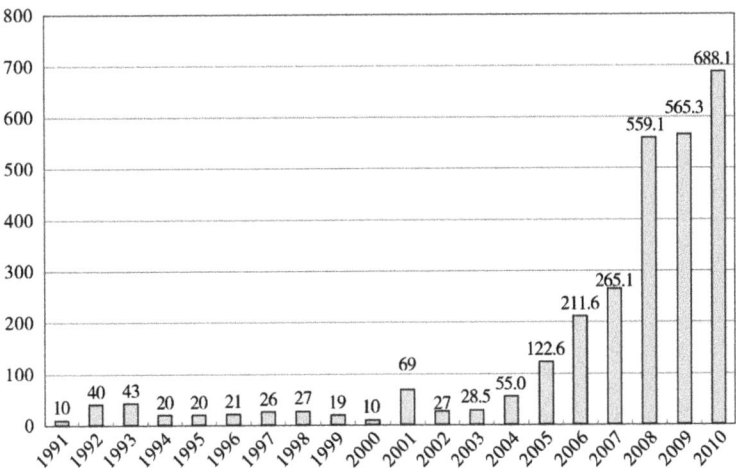

Chart 24.1 Growth of China's foreign direct investment during 1991–2010. *Source* The data of China's foreign direct investment during 1991–2001 and the data of the overseas M&A amount during 1991–2008 were cited from the *World Investment Reports* of the UNCTAD, the data of China's foreign direct investment during 2002–2008 was cited from the statistical data of Chinese Ministry of Commerce

M&A emerged, during which stage, there were a series of overseas M&A events with major influences (for more details, please refer to Table 24.4).

However, the "going-out" of Chinese enterprises was all smooth. On one hand, going out in the world and realize overseas M&A would encounter layers of obstacles. In 1990, China Aviation Technology Import/Export Corporation acquired the Seattle Manko Company, in 2008, China's Huawei Company united with the US Bain Capital to start competitive acquisition on 3Com Company, in 2010, the company acquired the road network of Motorola's mobile network business, in 2011, the company was forced to quit the acquisition of partial assets of the US server technology company, the 3 Leaf Systems, due to the layers of obstacles by the US Government, the acquisition didn't make it.

On the other hand, there were also many cases that the companies completed acquisition but failed to reach expected objectives.

Special Column 24.1 TCL's Mistake in Acquiring Thomson and Alcatel
In 2004, TCL completed the acquisition of the French Thomson Colored TV business and the telephone business of Alcatel. But one year later, the performance report in the third quarter of 2005 reflected the fact of sharply sliding performance. The failure of overseas acquisition caused the comprehensive decline of economic performance of the entire enterprise, and severely affected the enterprise's development. Based on the explanation by Li Dongsheng, Board Chairman of TCL, the reason for having such huge loss was that firstly, the enterprise had mistake in judging the technical advantage and brand value of Thomson colored television business; secondly, the enterprise had insufficient estimation on the consolidating

Table 24.3 Ratio of overseas M&A by Chinese Enterprises in China's FDI during 1991–2008

年份	海外并购交易额（百万美元）	FDI（百万美元）	海外并购占FDI的比例（%）	年份	海外并购交易额（百万美元）	FDI（百万美元）	海外并购占FDI的比例（%）
1991	3	1000	0.30	2000	470	1000	47.00
1992	573	4000	14.33	2001	452	6900	6.55
1993	485	4300	11.28	2002	1047	2700	38.78
1994	307	2000	15.35	2003	1647	2850	57.79
1995	249	2000	12.45	2004	1125	5500	20.45
1996	451	2100	21.48	2005	5279	12260	43.06
1997	799	2600	30.73	2006	14904	21160	70.43
1998	1276	2700	47.26	2007	−2388	26510	−9.01
1999	101	1900	5.32	2008	36861	55910	65.93

Source the data of China's foreign direct investment during 1991-2001 and the data of the overseas M&A amount during 1991-2008 were cited from the *World Investment Reports* of the UNCTAD, the data of China's foreign direct investment during 2002-2008 was cited from the statistical data of Chinese Ministry of Commerce

年份	Year
海外并购交易额（百万美元）	Overseas acquisition amount (USD million)
FDI（百万美元）	FDI (USD million)
海外并购占FDI的比例（%）	Ratio of overseas acquisition in FDI (%)

cost of overseas M&A. In other words, the enterprise bought the colored TV technology and brand that didn't have advanced technology, underestimated the consolidating difficulty and cost after M&A, which caused the enterprise to face severe test.

Source Liao Yunfeng: *Overseas M&A by Chinese Enterprises*, China Economic Press, 2006 Edition, pp. 123–124.

While making overseas investment, some insiders colluded with foreign enterprises in transferring state-owned assets or appropriating and transferring state-owned assets in overseas operation, which caused large-scale loss of state-owned assets in acquisitions. In other words, because foreign capital understood the governance structure problem of Chinese state-owned enterprises, so they colluded with acquirers in the M&A in order to cut state-owned assets. The specific method could be falsely high acquiring price; or huge consolidating cost, or

24.1 Overview and Forms of the Cost of Investment and Construction … 371

Table 24.4 List of major overseas M&A events since 2001

No.	Date	Event	Amount
1	July, 2011	Shanghai Construction Group acquired HKC (holdings) Limited	HKD 200 million
2	September, 2011	Holley Group acquired the Royal Philips' CDMA Mobile Telecommunication Department in the US	USD 180 million
3	October, 2001	Guangdong Midea Group acquired the Magnetrons Busienss of Japanese Sanyo Electric	JPY 2.35 billion
4	January, 2002	CNOOC acquired partial equities of the five major oil fields of Indonesia held by Spanish Raposo Company	USD 585 million
5	April, 2002	CNPC acquired oil and gas assets of US Devon Energy Group in Indonesia	USD 216 million
6	July, 2002	CNOOC acquired equities of Indonesian Tangguh gas field held by BP	HKD 7.8 billion
7	September, 2002	CNC formally acquired US Asia Global Crossing Ltd.	USD 80 million
8	October, 2002	Shanghai Automotive Industry Corporation acquired 10% equities of GM Daewoo	USD 59.7 million
9	February	BOE acquired the TFT-LCD business from Korean HYDIS	USD 380 million
10	March, 2003	CNOOC and Sinopec both acquired 1/12 equities of the Northern Caspian Sea Project of Kazakhstan held by BP for USD 615 million	USD 1.23 billion
11	April, 2003	CNPC International Company along with Malaysian National Oil Company Petromas to acquire AHIH	USD 82 million
12	April, 2004	TCL acquired cell phone business from Alcatel	Euro 55 million
13	July, 2004	SAIC acquired Korean Ssangyong Motor	About USD 500 million
14	August, 2004	China Aviation Oil acquired 20.6% equities of Singapore National Oil Company	SGD 543 million
15	September, 2004	China Minmetals Corp attempted to acquire Canadian Noranda Mineral Co., Ltd. (failed)	Offer of USD 6 billion
16	November, 2004	SHANGGONG CO acquired equities of a subsidiary of German FAG	USD 24.3 million
17	December, 2004	Lenovo Group acquired PC business of IBM	USD 650 million of cash and USD 600 million of stock

(continued)

Table 24.4 (continued)

No.	Date	Event	Amount
18	December, 2004	SNDA acquired Korean Aetoz Soft Company	USD 91.7 million
19	January, 2005	CNC acquired 20% equities of PCCW Hong Kong	USD 1.016 billion
20	April, 2005	CNOOC acquired equities of Canadian MEG Company	USD 123 million
21	June, 2005	Haier Electronics attempted to acquire US Maytag Company (quit)	Offer of 1.28 billion
22	July, 2005	CNOOC tried to acquire the US UNOCAL Corp. (quit)	Offer of 18.5 billion
23	July, 2005	MG successfully bid the UK Rover Automobile (Hundred years old factory) and its engine production department	USD 87 million
24	October, 2005	Wholly-owned subsidiary of CNPC, CNPC International acquired 100% of Kazakhstan Oil Company	USD 4.18 billion

Source Liao Yunfeng: *Overseas M&A by Chinese Enterprises*, China Economic Press, 2006 Edition, pp. 7–10

avoidance of acquisition for core technology. These methods extravagated the value of acquired enterprise, enabled the state-owned enterprises as acquirers to pay the prices that were far exceeding the value of acquired enterprises, while the enterprise insiders and government officials who made decisions and participated in acquisition obtained extra benefits.

24.2 Forming Causes of the Cost of Investment and Construction of Investment Environment

24.2.1 Forming Causes of the Cost of Providing Investment Environment

There is no relatively broad regional policy based on individual countries in China's foreign trade development, which caused unbalance of trade partners, large amount of trades concentrates in few nations and regions. Since the reform and opening-up, China gradually established the export-oriented foreign trade policy, the national government sustained the favorable policies back then, such as foreign exchange retention, tax rebate of export, export subsidy, etc., which encouraged enterprises to expand export volume, and enabled export amount to grow at annual average rate of 12.4%. The excessive attraction of foreign investment and inappropriate use of foreign investment caused the total import/export amount realized by foreign investment enterprises in China to far exceed the import/export amount of China's own enterprises.

One of the important parts of economic globalization is financial globalization, the core content of which is the free flow and reasonable allocation of financial capital worldwide. This opens the gate of convenience for short-term capital (or Idle Fund) in international scope, while the biggest characteristic of short-term capital is its very strong speculative nature, which decides that it would get in and out of the capital markets in different countries at very fast flowing speed, while bringing capital supply to different countries, it would also bring huge impact on the financial markets of different countries.[4] The Asian Financial Crisis, the US Subprime Mortgage Crisis and other international financial crisis in history proved that no open economy could avoid impact of international capital.

At the beginning since the reform and opening-up, in order to improve investment environment and accelerate the introduction of foreign investment, China implemented a series of special favorable policies on foreign capital, plus some local governments implemented excessively favorable policies on foreign investments, which enabled foreign investment enterprises to enjoy a series of "Super-national Treatment", as a result, it to certain extent restrained the development of other enterprises in the same industry, especially made state-owned enterprises to be in unequal competitive status for long term. In the era of economic shortage, due to the huge market development space, reflection of such circumstance was not obvious, but after the seller's market was transformed into the buyer's market, such circumstance became very prominent.

24.2.2 Forming Causes of the Cost of Foreign Investment

The success or failure of foreign direct investment and cross-border M&A depend on multiple factors, it not only depends on the various conditions of enterprises themselves, but also depends on the external environment of the two countries involved. For China, due to its special national situation and economic system, the factors that affect the success or failure of overseas M&A of enterprises are even more complex and having even higher risks.

Mutual independence and fair trade of property are the basic conditions for smooth and effective M&A, while overseas M&A is enterprise M&A between different countries, so its requirements for clear property right would be more specific. However, the main entities in China that participate in overseas M&A are state-owned enterprises or the companies with shares controlled by the state. Because laws and regulations in some nations regulate that those enterprises of important industries shall not be acquired by state-owned enterprises of other

---

[4]Zhang Daoquan: *Establishment of Price Awareness in Foreign Economic Relations,* Economic Issue Exploration, 11th Issue in 2004.

countries, or domestic enterprises have priority in acquisition of important enterprises, even sales to foreign state-owned enterprises would go though complex and fussy approving procedures, when selling important enterprises, some countries often have multiple layers of reviews. Chinese enterprises encountered many such issues in the practice of overseas M&A, which caused some major M&A with considerable economic and social benefits to fail eventually.

In overseas M&A by state-owned enterprises, there are wide problem of being crave for greatness and success and blind M&A. Many major enterprises in China consider overseas M&A as a "Political Achievement" project, or pursue overseas M&A as ideal, some even operate huge amount of overseas M&A as "Monument" of the entrepreneur's glorious retirement. Such blindness was the result of "Cheep Voting Right" of officials at governmental institutions, was blind behavior due to the lack of effective property restraint, and caused efficiency loss in M&A deals.

On the other hand, from international perspective, many host nations have relatively strict market access requirements for Chinese investment, there is natural necessity. As a new player of the international direct investment market, no matter for individual, the mass, institution or the government, there should be a process of adaption and getting used to it. However, what's worse is that as the defending power, considerable forces in western nations consider the rising China as an emerging and powerful challenger, which makes things even more complex, this point is most outstanding in the US. The direct investment projects of Chinese manufacturing enterprises in the capital and technology intensive industries in the US were impeded over and over again due to the cause of "National Security" on the US side.

The forming causes of the cost of China's investment and the construction of investment environment are concluded as follows, please refer to Chart 24.2.

24.3 Countermeasures for the Cost of Investment and Construction of Investment Environment

24.3.1 Countermeasures for the Cost of Providing Investment Environment

China shall expand domestic market demand, enhance consumption ratio, as well as support and encourage enterprises to develop the emerging market, reduce the risks and potential hazards brought by the high foreign trade dependence.[5] China shall enlarge the efforts in controlling the illegal inflow of foreign exchange by letting the regulatory departments to control the total volume, structure, source and direction

---

[5]Jing Yuemei: *Research on the Influence and Response Strategy of High Foreign Trade Dependence on China's Economy,* Journal of Anhui Electronic Information Occupational College, 2nd Issue in 2011.

24.3 Countermeasures for the Cost of Investment and Construction …

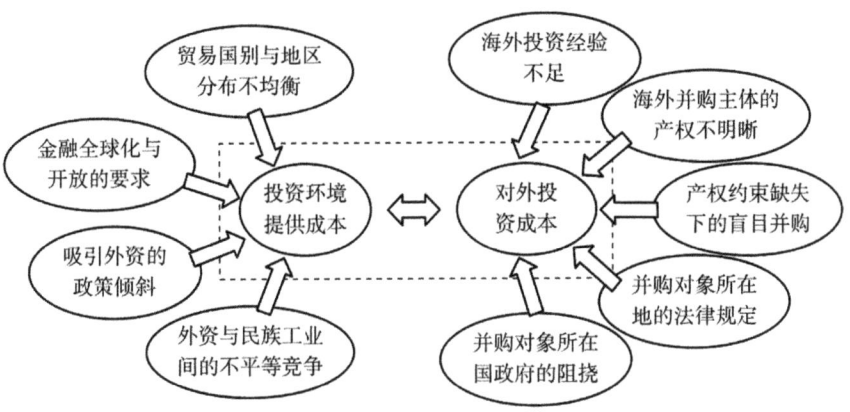

贸易国别与地区分布不均衡	Unbalanced distribution of trade countries and regions
金融全球化与开放的要求	Requirement of financial globalization and opening-up
吸引外资的政策倾斜	Favorable policies to attract foreign investment
外资与民族工业间的不平等竞争	Unequal competition between foreign investment and national industry
投资环境提供成本	Cost to provide investment environment
对外投资成本	Cost of foreign investment
海外投资经验不足	Lack of foreign investment experiences
海外并购主体的产权不明晰	Unclear properties of the overseas acquiring entities
产权约束缺失下的盲目并购	Blind acquisition under the lack of ownership restraint
并购对象所在地的法律规定	Local law and regulation of acquiring object
并购对象所在国政府的阻挠	Obstacle by local government of acquiring object

Chart 24.2 Forming causes of the cost of China's investment and construction of investment environment

of funds, improve macroeconomic regulation, actively maintain the economic stability and financial health.[6] China shall coordinate the relationship between the development of Chinese economy and protection of national industry, the market opening should be gradual and orderly, the protection of national industry should follow the following principles: (1) there must be limit in protection. (2) the purpose of protection is to enhance the international competitiveness of national industry. (3) there must be selection of the objects of protection. (4) the degree of protection must have a process from big to small.

---

[6]Guan Tao: *Chinese Countermeasure in Currency War: Actively Respond to the Impact of International Capital Flow,* International Economic Review, 2nd Issue in 2011.

24.3.2 Countermeasures for the Cost of Foreign Investment

Firstly, the government should speed up in forming the overall development plan for the implementation of the going-out strategy, regularly select and release the key industries for foreign investment, develop the comparative advantage of China, implement policy guiding for the enterprises that implement overseas investment. Secondly, also consider the private enterprises that have development potential and competitive advantage as the objects receiving favorable policies, so as to form the diversified situation of the entities of overseas investments. Thirdly, strengthen the

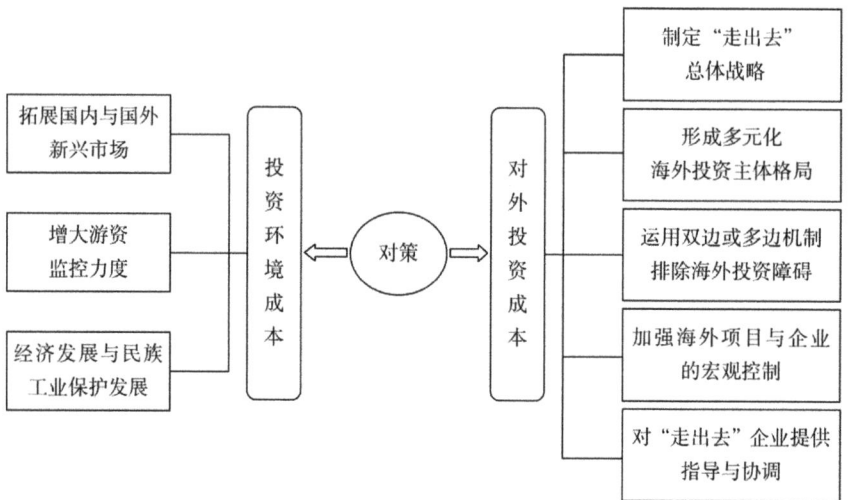

拓展国内与国外新兴市场	Explore emerging market in China and foreign countries
增大游资监控力度	Increase regulation on hot money
经济发展与民族工业保护发展	Economic development and the protection and development of national industry
投资环境成本	Cost of investment environment
对策	Countermeasure
对外投资成本	Cost of foreign investment
制定"走出去"总体战略	Form general strategy of "going-out"
形成多元化海外投资主体格局	Form diversified main arrangement of overseas investment
运用双边或多边机制排除海外投资障碍	Use bilateral or multiple mechanism to eliminate overseas investment obstacle
加强海外项目与企业的宏观控制	Strengthen the general control of overseas projects and enterprises
对"走出去"企业提供指导与协调	Provide "going-out" enterprises with instruction and coordination

Chart 24.3 Countermeasures for the cost of China's investment and construction of investment environment

government management functions, use multilateral and bilateral mechanisms to solve the barriers in foreign investment, and escort the going-out way of Chinese enterprises. Fourthly, strengthen the government's supervision function, it is suggested that the national government should establish a special supervisory organ for overseas investments, which should make dynamic management supervision on the operating situation of overseas investment projects, and strengthen the macroeconomic adjustment of the going-out enterprises. Fifthly, develop the role of the related industrial organizations and intermediary institutions, provide instruction and coordination to the enterprises that implement going-out strategy (Chart 24.3).

Chapter 25
Cost of International Trade Development

25.1 Overview of the Cost of International Trade Development

The thirty years of reform and opening-up was also the thirty years when China's commodity import/export had rapid development. China's commodity import/export amount increased from USD 20.64 billion in 1978 to USD 2.17383 trillion in 2007, and the trade balance also changed from USD 1.14 billion of trade deficit in 1978 to USD 262.2 billion of trade surplus in 2007. The annual growth rate of commodity trade in China was higher than the GDP growth rate during the same period in previous year, the trade dependence increased from 23% in 1985 to 67% in 2007.[1] The practices have proved that the development of China's foreign trade had significant promoting effect on China's economic growth. China so far is already deep in the process of participation in the economic globalization.

For the thirty years of reform and opening-up, China's auxiliary reforms on foreign trade, foreign capital, foreign exchange and other systems as well as the bold promotion of the opening-up policy greatly mobilized the initiatives of the foreign trade enterprises, operating departments, and foreign enterprises in their investments in China, provided effective incentive mechanism and system protection for China to accelerate the process of actively participating in the economic globalization and promote the extensive development of foreign trade. However, some policies implemented at this stage, while promoted the radical development of foreign trade, also made China's economy to pay extremely high price.

---

[1]Jin Zhesong and Li Jun: *China's Foreign Trade Growth and Economic Development: Review and Prospect of the Thirty-year Anniversary of the Reform and Opening-up,* China Renmin University Press, 2008 Edition, pp. 1–2.

Table 25.1 Forms of international trade cost

Classification of cost	Forms	Specific contents
Cost of international trade	Burden of tax rebate	Multiple decline of tax rebate rate promoted continuous and repaid growth of foreign trade and export, but also continuously increased the burden of tax rebate on the central fiscal authority, there was the phenomenon of owed tax rebate, which formed severe conflict with the burdening ability of fiscal authority
	Opportunity cost and risk of high foreign exchange reserve	Increase of foreign exchange reserve would cause the increase of monetary supply, form long-term inflation pressure, increase the opportunity cost of foreign exchange reserve and enlarge exchange rate risks
	Cost of handling trade frictions	Export commodities were often subject to technical barriers and anti-dumping restrictions set by foreign countries, which severely impeded further growth of export

25.2 Forms of the Cost of International Trade Development

The forms of China's trade cost include tax rebate burden, opportunity cost and risk of high foreign exchange reserve, and cost of response to trade frictions. For more details, please refer to Table 25.1.

25.2.1 Tax Rebate Burden

In 1985, China classified the industrial and commercial tax into product tax, value adding tax, business tax and salt tax. In march of the same year, the State Council approved the *Regulation on the Levy and Rebate of Product Tax and VAT on Exported Products*. This regulation began to enforce since April 1 of the same year, which symbolized the reestablishment of China's export tax rebate system. The export tax rebate system, especially the zero tax rate policy implemented on exporting goods since 1994, greatly mobilized the initiative of export enterprises, and enabled China's export trade to have high-speed growth for continuous years.[2]

---

[2]Jin Zhesong and Li Jun: *China's Foreign Trade Growth and Economic Development: Review and Prospect of the Thirty-year Anniversary of the Reform and Opening-up*, China Renmin University Press, 2008 Edition, pp. 1–2.

Meanwhile, it also brought continuous tax rebate burden on the national finance, and becomes unsustainable.[3]

In 1994, China implemented the zero tax rate policy on exporting goods according to international practice, the tax rebate rate for exporting goods was 17 and 13% respectively; for the exporting goods purchased by small-scale taxpayers for specially approved tax rebate, the tax rebate rate was 6%, calculating on this basis, the average export tax rebate rate was 16.13%, however, since Jul 1, 1995, the State Council regulated that the tax rebate rate of exporting goods shall be formed based on the actual tax burden of goods, so the tax rebate rate was reduced by 3.7%, and the average tax rebate rate was reduced to 12.90%; since Jan 1, 1996, it was further reduced by 46%, the export tax rebate rate of 3 main classes was eventually reduced to 9, 6 and 3% respectively, the average tax rate was reduced by 8.29%; it was regulated that those with annual tax rebate amount exceeding the quota shall transfer to the next year for tax rebate. Since 1998–1999, in order to reduce the influence of the Asian Financial Crisis on foreign trade and export, the national government had small-range hikes of export tax rebate rate for 8 times, those recovering the 17% of tax rebate rate in the first batch were the textile products and textile machinery, in 1998, the average export tax rebate rate reached 9.32%. After multiple times of adjustment, the comprehensive export tax rebate rate reached around 15.11%. The increase of export tax rebate rate promoted the continuously rapid growth of foreign trade and export, but also made the export tax rebate burden of the central finance authority continuously increase. Due to the limitation of financial strength, there was big gap between the export tax rebate quota determined by the Ministry of Finance in each year and the actual amount of payable tax rebates. According to statistics, during 1997–2002, the annual growth rate of tax rebate quota was 17.8%, while the annual average growth rate of payable tax rebate reached 37.4%, the gap between the two was nearly 20%. Since 2002, the extensive growth of foreign trade and export as well as the huge amount of export tax rebate brought along became unbearable to the Central Finance Authority, in order to reduce the pressure of export tax rebate to the Central Finance Authority, on Oct 13, 2003, the Ministry of Finance and the State Administration of Taxation jointly issued document to implement structural adjustment on the existing tax rebate rate of VAT on exporting goods. Based on difference of exporting products, the exporting tax rebate rate was adjusted to five classes, 17, 13, 11, 8 and 5%, which was enforced since Jan 1, 2004. Based on the current export structure, the average level of export tax rebate rate was reduced by around 3% from 15.11 to 12.16%, such adjustment actually transformed the neutral rebate principle of "rebating all those supposed rebates" to the non-neutral principle of "Differentiated Tax Rebate", since then, export tax rebate has to great extent changed from a system to a policy tool for the government to regulate export and industrial structure. In 2005 and 2007, the government reduced the export tax rebate rate

---

[3]Pei Changhong and Gao Peiyong: *Export Tax Rebates and China's Foreign Trade*, Social Science Literature Press, 2008 Edition, pp. 5–12.

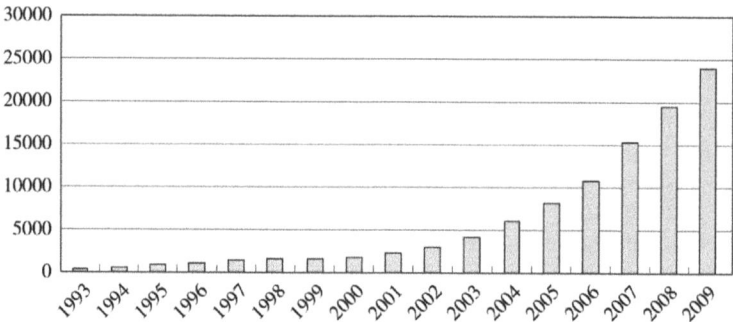

Chart 25.1 Growth of China's foreign exchange reserve since 1993. *Source* Concluded from the related data in the *China Statistical Yearbook in 2011*

again, although the two adjustments had no direct correlation with financial pressure, but the scale of tax rebate borne by the finance authority had reached RMB 337.2 billion and RMB 428.5 billion, the problem of the sustainability of export tax rebate policy actually occurred.

25.2.2 Opportunity Cost and Risk of High Foreign Exchange Reserve

Since the implementation of the reform and opening-up policy in 1978, China's commodity import/export trade had rapid development. It can be seen from Chart 25.1 that since 1978, China's total amount of commodity import/export had rapid growth. In 1978, the total amount of China's commodity import/export trade was only USD 20.64 billion; in 2010, the total amount of China's import/export trade reached USD 2.97276 trillion, which was 144 times of that in 1978. Since 1990, the development of China's foreign trade reflected 21 years of trade surplus (except in 1993), and the trade surplus kept expanding, and reached the highest level of USD 295.46 billion in 2008.

Correspondingly, China's foreign exchange reserve also had very rapid growth (refer to Chart 25.1). Based on statistics of the State Administration of Foreign Exchange, in 2010, China's foreign exchange reserve already reached USD 2.847338 trillion, which increased by USD 448.1 billion compared with that at the end of previous year, and China became the country with the largest foreign exchange reserve in the world.[4]

The increase of foreign exchange reserve led to the increase of the funds outstanding for foreign exchange, which further causes increase of monetary base, while the increase of monetary base results in increase of monetary supply through

---

[4]Sorted from the related data from *2011 China Statistic Yearbook*.

25.2 Forms of the Cost of International Trade Development

the monetary multiplier. In theory, in short term, foreign exchange reserve directly constitutes a part of a nation's monetary base, with other conditions remain unchanged, the increase of foreign exchange reserve would cause increase of monetary supply, from long-term perspective, foreign exchange reserve would indirectly affect consumer price index through its influence on monetary issuance, and therefore form the pressure on rising prices.

The growth of foreign exchange reserve would cause stronger expectation of RMB appreciation, which results in the increase of hot money inflow, which causes further increase of foreign exchange reserve, and then the expectation of RMB appreciation further strengthens, then the hot money inflow further increases, so as to form a vicious circle and causes increasing pressure for RMB appreciation. Under the expectation of appreciating local currency, large quantity of speculative capital entered China, which supported the formation of economic bubble and generated extremely high damage to the healthy and excellent operation of economy.

We can also see that the continuous increase of foreign exchange reserve brings negative influence on the economy and enables the cost of holding reserve to increase.

There is opportunity cost for holding foreign exchange reserve. The opportunity cost of holding foreign exchange reserve $\approx$ domestic investment income—interest yield of holding foreign exchange reserve. For a developing country, its domestic investment return rate or the growth rate of national economy brought by import of goods far exceeds the interest rate from holding foreign exchange reserve, such difference constitutes the opportunity cost of holding foreign exchange reserve.

Two scholars in the US, Dean Beker and Karl Walentin, once systematically analyzed the huge opportunity cost paid by developing nations for the increase of foreign exchange reserve, the price paid by East Asian nations in the past ten years for accumulation of foreign exchange reserve was equivalent to over 20% of the GDP.[5] Besides, According to the statistics by the World Bank, the cost of holding foreign exchange reserve by the developing nations is as follows in Table 25.2.

Since the 1970s, with the continuous deterioration of the US' international payments state, the status of US dollar as international reserve currency began to continuously decline, the reserve currency developed towards the direction of polarization, the ratios of Deutsche Mark, Japanese Yen, Swiss Franc, etc. gradually increased, since the birth of Euro, its ratio in reserve currency gradually enhanced. However, due to the historical reason and the aspect of international currency system, the US dollar still has absolute advantage among different international reserve currencies.

However, since 2002, the US dollar began to continuous devalue. Economists widely believe that the Bush Administration has already firmly implemented the dollar policy of "Orderly Devaluation", in the future, US dollar is likely to devalue

---

[5]Wang Zhen: *Research on the Management of China's Foreign Exchange Reserve*, China Finance Press, 2007 Edition, p. 132.

Table 25.2 Cost of holding foreign exchange reserve by developing nations (ratio in GDP)

Region	Cost in each year (%)		Cost in each 10 years (%)	
	Lowest	Highest	Lowest	Highest
East Asia and Pacific Region	1.2	2.5	12.6	25.2
South Asia	0.4	0.9	4.5	9.0
Latin America and Caribbean Sea Region	0.8	1.6	9.4	18.8
Sub-saharan Africa	0.8	1.6	9.5	18.2
Middle East and North Africa	0.9	1.8	10.5	21.1

Source Wang Zhen: *Research on the Management of China's Foreign Exchange Reserve*, China Finance Press, 2007 Edition, page 132

by 15–20%. The continuous and extensive devaluation of US dollar caused the dollar-denominated strategic resources across the global to extensively appreciate, which resulted in the increase of China's payments for imports. Besides, as estimated by IMF, if the US dollar devalues by 25%, the loss of the holders of dollar-denominated assets will reach 10% of US GDP, in which the loss of East Asian economies due to the holding of US dollar reserve will account for 1.5% of the US GDP.

Well-known US Economist Mckinnon indicated that East Asian nations widely have the "Virtue of Conflict". On one hand, because the saving rates among East Asian countries are higher than their investment rate, so the East Asian nations widely use their current account surplus to finance the Americans who spend money like water; on the other hand, exactly because East Asian countries have accumulated large amount of foreign exchange reserve through fixed exchange rate system, which would restrain the East Asian countries' ability to change exchange rate system. Because once East Asian countries adopt floating exchange rate system, the currencies of East Asian countries would widely and extensively appreciate against US dollar, while the foreign exchange reserve of East Asian countries are mostly held in form of dollar-denominated assets, then letting the exchange rate float means extensive shrinking of foreign exchange reserve. Taking China as an example, based on the *2010 Financial Statistics Data Report* issued by the Central Bank, at the end of 2010, the surplus of the national foreign reserve was USD 2.8473 trillion, assuming that about 50% of the foreign exchange reserve was held in form of dollar-denominated assets, then 10% of US dollar devaluation against RMB means that China's foreign exchange reserve would suffer the loss of USD 142.365 billion. This puts the East Asian economies in an awkward situation: on one hand, the more foreign exchange reserve they hold, the easier they become the "Hostage" of the dollar standard system. Because, as long as the dollar collapses comprehensively, the biggest loser won't be the US, but the foreign governments and investors who hold dollar-denominated assets, what's more, the US economists Roberta & Gough and Oberst Field believed that in order to eliminate the vast current account deficit, the US needs to devaluate by at least 20–30%. Therefore, if the Chinese Government lets RMB exchange rate float, not to mention possible losses in other aspects, the loss of foreign exchange reserve alone could reach as high as USD 300–400 billon. Therefore, China faces extremely high exchange rate risks for holding such a large scale of foreign exchange reserve.

25.2.3 Cost of Handling Trade Frictions

It is believed in this book that the handling cost of trade frictions is mainly reflected as encountering trade protectionism, influence by the hidden rules of international community, the loss in antidumping lawsuits, etc.

In recent years, with the trend of the trade liberation and the stronger role of multilateral trade system, the traditional custom duties, quota, license and other measures that restrain trade have been greatly retrained and weakened due to the increasing transparency. Under such trend, the new trade protectionism emerges. The technical barriers and antidumping that are highly invisible gradually become the main means for some countries to protect their domestic market. For China as a major exporting country, its exported commodities have been frequently subject to restriction of the technical barriers and antidumping measures set by foreign countries, which severely impeded the further growth of export.

China's exported commodities, especially textile products, have often been accused in international market for antidumping. From 1995 when WTO was established to the end of 2006, there had been 3044 cases of investigations established for antidumping, 1941 of which finally adopted antidumping measures, accounting for 64% of the total number. The number of cases of antidumping investigation encountered by China ranked the first in the world, reaching 536 cases, accounting for 28% of the world total. From the number of cases adopting antidumping measures, China also ranked the first place with 375 cases, both numbers far exceeded the later Korea and the US.[6] The cause of China encountering antidumping frictions was on one hand from the competition by developing countries, due to the isomorphic nature of the low-end products and narrow market, trade frictions occurred from some developing countries. For example, India was the nation that filed the most antidumping investigations in the world, ranking number one at 457 cases, 196 of which were against China, accounting for 42% of the total. Taking the competition between China and Turkey in textile products export for another example, in the European and US market, due to the expansion of the export of Chinese textile products, trade frictions occurred from Turkey, during 2001–2006, Turkey was also a developing country that filed relatively many antidumping investigations against China, and 1/6 of which involved textile product trade. On the other hand, it was due to the low price of export products from developing countries. For example, the price of chemical fabric that China exports to EU was only USD $0.9/m^2$, which was far lower than the USD 3.29 of Japan and USD 1.94 of Korea; the price of silks and satins that China exported to India was 25% lower than the price of that exported to other countries; in 2005, the number of shoes that China exported accounted for 60% of the total in the world market, the average price for a pair of shoes was only USD 2.7, which was only 1/5 of that in Spain and 1/12 of that in the Italy.

Besides, technical trade barriers also severely affected the development of China's exports. In each year, China has about USD 20 billion of exported products subject to influence of technical trade barriers set by foreign countries. Taking electromechanical products for an example, the technical barriers encountered by China in recent years are as follows in Table 25.3.

25.3 Forming Causes of the Cost of International Trade Development

Except 1998, China has been maintaining the "Double Surplus" of current account and capital account since 1994.[7] As we know, based on the analysis above, the continuous increase of China's trade surplus was actually the result of the transfer of international industrial structure. Basically, the current situation of China's foreign trade is decided by the industrial structure of the entire world, which can't be changed overnight.

---

[6]Li Guifang: *Analysis Report of the Foreign Direct Investment by Chinese Enterprises*, China Economic Press, 2007 Edition, pp. 53–54.

[7]Wang Zhen: *Research on the Management of China's Foreign Exchange Reserve*, China Finance Press, 2007 Edition, p. 190.

25.3 Forming Causes of the Cost of International Trade Development

Table 25.3 Technical barriers encountered by Chinese electromechanical products in recent years

Cause	Imposing country	Chinese enterprises or products subject to TBT	Specific year
Incompliance with standard regulation	UK	Ningbo Toaster and Grass Lamps	2003
	Japan and EU	Information telecommunication, technical remote communication equipments (copy machine, printing machine, computer)	2001
	EU CR Regulation	Lighters from Wenzhou	2001
	Gambia	Donglin Diesel Engine	1991
	Brazil	Steel wire rope	2000
Incompliance with certification	US	Jiangsu Little Swan	1999
	US (GS)	Electronic appliance switches	2000
	US (CE)	Hefei and Baoji Forklift Factory	2001
	US, EU, Canada, Australia	High and medium voltage valves of Nantong Valve Factory	2000
	EU	Radio	1999
	US, EU	Chunlan Air Conditioning	2000
	Italy (IMQ)	Festive lighting	2003
Incompliance in packaging or labeling	US, EU	Wooden package of machinery or equipment products	1999
	EU (labeling)	Small electronic appliances and lighting from Wenzhou	2003
	US (UL labeling)	Transformers from Shenzhen Longgang	1999
Incompliance with environmental protection requirements (green barrier)	EU	Small-flux motorcycles under 1500CC	2003
	US, EU	Xinfei and Haier refrigerators	1999
	US (EPA)	Changzhou Diesel Factory	1999
	US, EU	Fluorescent powder for color TV from Shanghai Yuelong Chemical Factory	1999

Source Li Guifang: *Analysis Report of the Foreign Direct Investment by Chinese Enterprises*, China Economic Press, 2007 Edition, pp. 53–54

25.3.1 Conflict Between "Rebating All Rebates" and Finance Support Capacity

The frequent adjustment of export tax rebate rate reflects the conflict between the export tax rebate principle of "Rebating All Rebates" and the finance support capacity. Based on the theoretical design of VAT and universal practice, export tax rebates are supposed to be "Rebating All Rebates". However, the result of China's implementation of the "Rebating All Rebates" principle was the severe "Tax Owing" from government to enterprises, the conflict between "Rebating All Rebates" and finance support capacity has been very prominent. The reduction degree of tax rebate rate is restrained by export situation, excessively low tax rebate rate would bring negative effect on exports, so as to affect the overall economic growth.

The management of export tax rebate quota was the direct cause of problems, the scale of export tax rebate was concluded based on the arrangement and estimation on the development plan. In recent years, due to the extensive and excessive growth of foreign trade and export beyond plan for continuous years, there is relatively big gap between the export tax rebate quota arranged by budget and the amount of payable tax rebate that actually occurred, which resulted in large tax owing. The current VAT is a type of shared tax between the central government and local governments, the distribution system is the source of such problem. Since 1994, VAT was defined as a type of shared tax between the central government and local governments.[8] Such tax rebate system resulted in the fact that after the central finance authority bears export tax rebates, the actually received VAT income is less than 75%, while the actually received VAT income of local governments is higher than 25%; after local governments obtain 25% of income based on the gross income of VAT, the central finance authority distributes the export tax rebates back to the enterprises in different places, and the local governments would also obtain income tax from it, which means transfer payment from the central finance authority to local governments, which increases the pressure of financial expenditure.

25.3.2 Necessity of Foreign Exchange Reserve for Economic Development

It causes the large-scale increase of foreign exchange reserve. With the advancement and development of China's economy, in coming future, China's demand for foreign exchange reserve would be huge, which determines that China must hold large amount of foreign exchange reserve. Sufficient foreign exchange reserve is also very necessary for the macroeconomic stability of China, which is the huge invisible income of holding foreign exchange reserve.

---

[8]Meaning that 75% of the VAT generated from export sales shall be owned by the central government, while 25% shall be owned by local governments, and the entire export tax rebate shall be paid by the central government. Besides, there is a reward mechanism, meaning that for the part of additional VAT revenue compared with that of previous year, besides the 25% share, local governments shall receive reward share of 1:03.

25.3 Forming Causes of the Cost of International Trade Development

25.3.3 *Cause of Intensifying Trade Frictions*

Since the 911 Terrorist Attack in 2001, the economies in the world widely had recessions, and various trade protectionism emerged. Firstly, the trade frictions encountered by China has something to do with the downturn of world economy. Secondly, China's continuously high-speed growth of export indeed caused impact to the related industries of related countries to certain extent. Thirdly, China's economic structure is relatively backward, the exported products are mostly labor intensive, therefore, the product organizations are usually dispersed, the exporting order is relatively chaotic, which caused price competitions to become the main means for exporting enterprises, so it is difficult to avoid having the situation of low-price dumping. Fourthly, enterprises don't have sufficient understanding of international rules and international standards, which led to frequent impact on Chinese enterprises from technical barriers and other similar measures.

For the forming causes of China's trade cost, please refer to Chart 25.2.

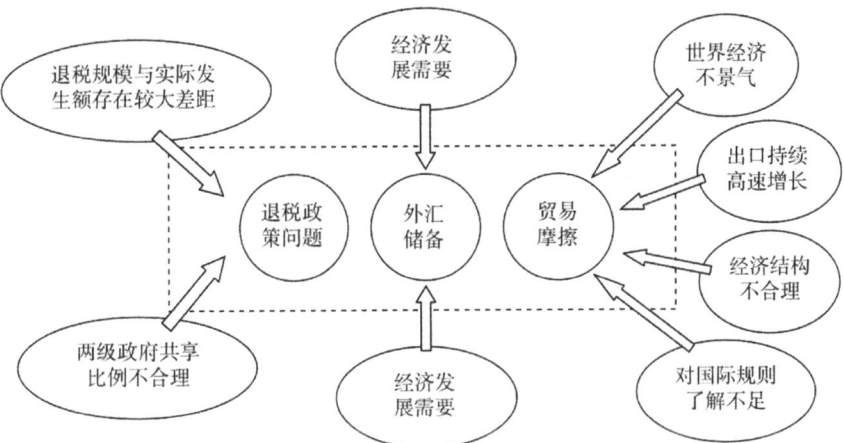

退税规模与实际发生额存在较大差异	Relatively big difference between tax rebate and actual occurred amount
经济发展需要	Demand of economic development
世界经济不景气	World economic downturn
退税政策问题	Tax rebate policy issue
外汇储备	Foreign exchange reserve
贸易摩擦	Trade friction
出口持续高速增长	Continuous high-speed growth of export
经济结构不合理	Irrational economic structure
两级政府共享比例不合理	Irrational sharing ratio between two levels of government
经济发展需要	Demand of economic development
对国际规则了解不足	Insufficient understanding of international rules

Chart 25.2 Analysis of the forming causes of international trade cost

25.4 Countermeasures for International Trade Cost

25.4.1 Reform the Export Tax Rebate System

Firstly, China shall fundamentally improve the joint burden mechanism of export tax rebate, further clarify the responsibilities of different governments in financial expenditure; improve the local tax system structure, strengthen the financial strength of local governments. Secondly, reform the management method of export tax rebate quota, cancel the planned quota management of export tax rebate amount in budget, undertake tax rebate management based on the actual amount of tax rebate, and establish the scientific and dynamic monitoring system of export tax rebate. Thirdly, seek the self-balancing mechanism between export tax rebate and import/export tax, using import tax to solve the tax rebate in export is feasible and rational, the self-balance of import and export tax could be realized through "Using import tax to support export tax rebate". Fourthly, from long-term strategy, China could consider classify VAT into the tax for Central Government, and the central finance authority shall bear the entire export tax rebates. From international experiences, VAT in most countries are central government's tax, so export tax rebates should also be borne by the central finance authority, this is the matures and most stable export tax rebate system, it is good for the development of the "Neutral" role of VAT, and could promote economic growth to most extent.

25.4.2 Improve Foreign Exchange Reserve Management

China shall strengthen the construction of foreign exchange market, promote the reform of foreign exchange rate system, gradually expand the floating range of exchange rate, and strengthen the flexibility of exchange rate. The ratio of China's foreign direct investment and the attracted FDI is far lower than the level of developed nations, therefore at current stage where foreign exchange reserve is sufficient and continuously growing, China should energetically promote China's foreign direct investment, so as to relieve the pressure caused from trade surplus and large-scale inflow of foreign investment.

25.4.3 Actively and Strategically Handle Trade Frictions

Firstly, we shall adjust the economic development strategy, system and policy to eliminate our own factors that could cause trade frictions, fade the effect of export but value the effect of import. Secondly, China shall actively participate in the formation and improvement of international economic rules, and promote the trade liberation in other countries. Thirdly, China shall establish the enterprise handling system for international economic frictions, normalize the behaviors of enterprises themselves to actively respond to occurrence of trade frictions. Fourthly, China

25.4 Countermeasures for International Trade Cost

Chart 25.3 Countermeasures for international trade cost

shall be good at using WTO dispute solving mechanism to maintain the lawful rights and interests of Chinese enterprises.

The strategies for handling of Chinese international trade cost could be concluded as follows in Chart 25.3.

Chapter 26
Construction Cost of China's Three-Dimensional Transportation

An open economy needs a globalized and networked traffic & transportation system, whether having a complete traffic & transportation system for internal and external connection is a symbol to measure the economic openness degree and development degree of a country.[1] Since 1980s, economic globalization has become the trend of current economic development, it requires providing convenient and fast service system of traffic and transportation across the globe. Besides cheap labor and preferential tax policy in invested countries, whether the traffic & transportation are convenient and whether its public facilities are complete have become an important factor in investment decision making and consideration of international capital.

So far, China has preliminarily established a comprehensive transportation system that is composed of five modernized transportation means, namely railway transportation, road transportation, waterway transportation, airway transportation and pipeline transportation, as well as has certain scale and transporting capacity.[2] China's foreign opening-up pattern has been expanded from the coastal region to the inland regions that are along rivers, along borders and along roads, traffic & transportation play the key role. The preliminary establishment of this three-dimensional transportation network is concentrated with huge social and economic costs.

---

[1]Bai Xuejie and Wang Yan: *Reform and Development of Chinese Traffic & Transportation Industry*, Economic Management Press, 2009 Edition, p. 18.
[2]Bai Xuejie and Wang Yan: *Reform and Development of Chinese Traffic & Transportation Industry*, Economic Management Press, 2009 Edition, pp. 18–19.

26.1 Overview of the Construction Cost of Three-Dimensional Transportation

From general perspective, China's traffic & transportation infrastructure have always been the "Bottleneck" of economic growth and to great extent unable to meet the demand of economic development. Since the reform and opening-up, Chinese Government gradually valued the strategic status of infrastructure industry in national economic development, enlarged the investments in infrastructure and accelerated the development steps of traffic and transportation. Since 1985, the fixed assets investment of Chinese traffic & transportation industry has been basically maintaining at about 6% of GDP.[3] In the late 1990s, with the implementation of China's active financial policy, further acceleration of the traffic & transportation infrastructure development and continuous expansion in scale, the traffic & transportation structure had prominent improvement, which promoted the development of national economy. Especially in highway, high-speed railway, etc., China's transportation infrastructure witnessed a period from nothing to something and rapid development, as well as quickly leaped forward to world first class in many technologies and indicators.

The preliminary forming process of the comprehensive three-dimensional transportation network was concentrated with the funding support from many parties, including the central and local governments, finance authority and private capital, and made huge achievement. However, the entire Chinese society and economy also paid great price due to lack of experience, incomplete management system and many other reasons.

26.2 Forms of the Construction Cost of Three-Dimensional Transportation

The development of Chinese transportation industry over the last sixty years was also formed based on paying many prices. Generally speaking, on one hand, it can be said that the preliminary formation of China's modern three-dimensional transportation network was based on relatively high level of fixed assets investments. On the other hand, the general efficiency of transportation infrastructure construction investments is still not ideal, there are hidden investment efficiency

---

[3]Gao Feng: *Investment of Transportation Infrastructure and Economic Growth*, Chinese Fiscal Economy Press, 2005 Edition, pp. 57–58.

26.2 Forms of the Construction Cost of Three-Dimensional Transportation

Table 26.1 Forms of the construction cost of China's three-dimensional transportation

Classification of cost	Forms	Specific contents
Construction cost of three-dimensional transportation	Cost of infrastructure investment	Construction input of traffic infrastructure
	Opportunity cost of infrastructure investment	Low investment efficiency caused from repeated construction and overcapacity[a]
	Loss of management efficiency	Operation efficiency loss caused from incomplete traffic network mechanism

[a]Meaning the opportunity cost brought by failure to invest in the fields with relatively high efficiency

loss caused from repeated construction and overcapacity,[4] and operating loss caused from setting of management system and supervising departments. The forms of the cost of three-dimensional transportation development are shown in Table 26.1.

26.2.1 Input of Transportation Infrastructure Construction

The development of traffic & transportation must be established on certain scale of transportation infrastructure. While the construction of transportation infrastructure requires huge amount of funding input, in the development process of China's traffic & transportation, the long-term funding shortage was the biggest barrier of the accelerated development of China's transportation industry.[5]

China's traffic & transportation was developed under the background that the foundation was relatively weak and its leading role had been ignored for long term. During the period from the establishment of New China to 1980s, the national development strategy was put more on the revitalization of industry and agriculture, the traffic & transportation, as the leading industry of economic development, didn't receive its deserved status and emphasis, meanwhile, the national economy that was established on the impoverished foundation had very limited resources to input in

---

[4]Potential investment efficiency loss means the efficiency loss in investment activities that are difficult to judge or not easy to reflect. Many investment projects are lack of rational and objective decision making before commencement, in overall perspective, they are lack of long-term and general scientific development plan, as a result, the seemingly successful projects couldn't fully reflect their roles during operation process, their contribution rate to economic growth is lower than predicted level, so as to cause potential economic loss.
[5]Bai Xuejie and Wang Yan: *Reform and Development of Chinese Traffic & Transportation Industry*, Economic Management Press, 2009 Edition, p. 31.

the development of traffic & transportation, such state accumulated for long term and restrained the development of national economy for long time.

1990s was the key period when traffic & transportation began to enter rapid development. With Comrade Deng Xiaoping's South Talks as the symbol, different local governments began to have rethinking and breakthrough from the original thinking method in the process of reform and construction. They explored new methods in raising the funds for transportation construction, expanded the funding sources for transportation construction, and accelerated the steps of transportation facility development.[6] The ratio of China's investment in traffic & transportation in national infrastructure investment also had rapid growth in each year (refer to Table 26.2). The rapid growth of transportation investment at this stage was mainly benefited from the issuance of transportation investment and financing reform policy after the transportation investment system reform. The state issued a series of new regulations and new policies. The central and local governments gradually formed the new financing guidelines of widely opening channels, raising funds from multiple parties, focusing on real effects and having flexible methods, especially in the field of roads and ports, the new pattern has been formed that there are multiple funding sources, including "national investment, local fundraising, social fundraising, domestic loan, use of foreign investment, etc."

Over the 60 years since the establishment of New China, especially since the reform and opening-up, China has made great achievements in the construction of transportation infrastructure, and the comprehensive transportation capacity has realized great enhancement. According to the data from *2010 Chinese Transportation Yearbook*, to the end of 2009, the total length of comprehensive transportation routes had reached 6.4842 million kilometers; the operating length of railway had reached 85.5 thousand kilometers, in which the length of double tracking railways had reached 29 thousand kilometers, electric length had reached 30 thousand kilometers; the length of completed roads had reached 3.8608 million kilometers, including 65.1 thousand kilometers of highways; the length of inland navigation had reached 123.7 thousand kilometers; the length of civil aviation had reached 2.3451 million kilometers; the length of oil and gas pipelines had reached 69.1 thousand kilometers. Please refer to Chart 26.1.

26.2.2 *Opportunity Cost of Transportation Infrastructure Construction*

The construction of railways, ports, airports and other traffic & transportation infrastructure was lack of overall consideration, which caused the air-sea-land

---

[6] Bai Xuejie and Wang Yan: *Reform and Development of Chinese Traffic & Transportation Industry*, Economic Management Press, 2009 Edition, p. 32.

26.2 Forms of the Construction Cost of Three-Dimensional Transportation

Table 26.2 Ratio of China's investment in traffic & transportation industry in national infrastructure investment

年份/时期	全国	交通运输业	比重%	年份	全国	交通运输业	比重%
"一五"	588.47	90.15	15.30	1989	1551.74	166.51	10.70
"二五"	1206.09	163.30	13.50	1990	1703.90	207.16	12.20
1963—1965	421.89	53.78	12.70	1991	2115.80	340.18	16.10
"三五"	976.03	150.01	15.40	1992	3012.65	457.58	15.20
"四五"	1763.95	317.59	18.00	1993	4615.50	901.24	19.50
"五五"	2342.17	302.45	12.90	1994	6436.74	1372.94	21.30
1980	558.89	62.34	11.20	1995	7403.62	1587.53	21.40
1981	442.91	40.47	9.10	1996	8570.79	1844.62	21.50
1982	555.53	57.21	10.30	1997	9917.02	2197.45	22.20
1983	564.13	78.40	13.90	1998	11916.42	3252.19	27.30
1984	743.15	108.46	14.60	1999	12455.28	3429.28	27.50
1985	1074.37	178.10	16.60	2000	13427.27	3641.94	27.10
1986	1176.11	180.81	15.40	2001	14820.10	4116.43	27.80
1987	1343.10	189.73	14.10	2002	17666.62	4393.98	24.90
1988	1574.30	212.17	13.50	2003	22908.60	4892.71	21.40

Source summarized and calculated based on the *Compilation of the Statistical Materials of the Fifty Years of New China*, and the *China Statistical Yearbook* in previous years.

年份/时期	Year/period	全国	National
交通运输业	Transport industry	比重%	Ratio %
"一五"	"First Five Year Plan" period	"二五"	"Second Five Year Plan" period
"三五"	"Third Five Year Plan" period	"四五"	"Fourth Five Year Plan" period
"五五"	"Fifth Five Year Plan" period		

network and port stations to be not rational enough in arrangement,[7] and resulted in the problems of repeated construction and overcapacity, led to huge waste in lands and funds.

---

[7] Editing Department of *World Shipping*: *Current Situation of Chinese Port*, World Shipping, 4th Issue in 2010.

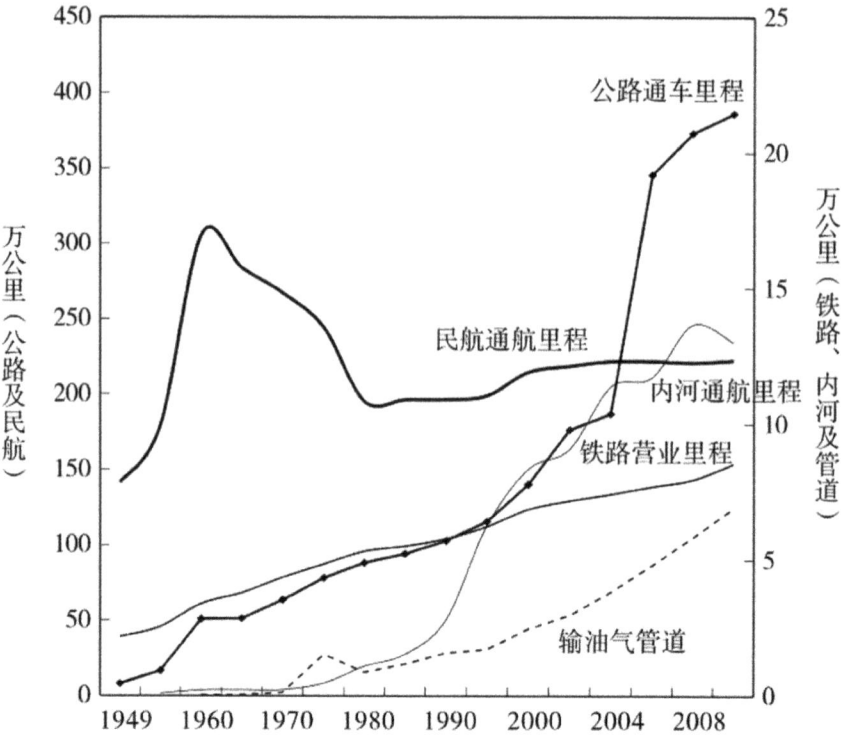

Source sorted out, calculated and drawn from the data of *2010 Chinese Transportation Yearbook*.

万公里（公路及民航）	10 thousand km (road and civil aviation)
公路通车里程	Mileage in highway open to traffic
民航通航里程	Aviation range of civil aviation
内河通航里程	Cruising range of inland river
铁路营业里程	Length of railroad lines in service
输油气管道	Oil and gas pipeline
万公里（铁路、内河及管道）	10 thousand km (railway, inland river and pipeline)

Chart 26.1 Development of transportation since the establishment of New China (1949–2009). *Source* Sorted out, calculated and drawn from the data of *2010 Chinese Transportation Yearbook*

Taking the high-speed train for an example, "The rise of Chinese high-speed train… was a direct challenge to China's aviation transportation industry, when the Beijing-Shanghai high-speed train was opened, the price of air tickets between

Beijing and Shanghai had comprehensive fall".[8] The opening of high-speed railway on one hand provides travelers with another choice to arrive at destination in faster manner under the precondition of affordability, but on the other hand, involves certain degree of repeated construction, please refer to Special Column 26.1.

Special Column 26.1 Observation of Transportation Resource Optimization through Beijing-Shanghai High-speed Railway
It should be pointed out that there are also flaws in the construction of Beijing-Shanghai high-speed railway, such as the problems of repeated construction and construction sequence. Before the Beijing-Shanghai high-speed railway, the 120 km Beijing-Tianjin Intercity High-speed Railway and the 300 km Shanghai-Ningbo Intercity High-speed Railway had already been completed. Due to the separation of operating entity, the company of Beijing-Shanghai High-speed Railway ignored the doubt about repeated construction, and completed the construction of the 1318 km Beijing-Shanghai High-speed Railway at once, as a result, there are two high-speed railways between Beijing and Tianjin as well as between Shanghai and Nanjing, this is huge waste of transportation resource allocation. There is no precedent in the world history of high-speed railway construction—building two high-speed railways between two cities at the same time. Based on the experiences of the Japanese Shinkansen, the annual passenger capacity of one high-speed railway could reach 100 million person times, while the Beijing-Tianjin Intercity High-speed Railway and Shanghai-Ningbo Intercity High-speed Railway haven't reached the designed passenger capacity, the opening of the Beijing-Shanghai High-speed Railway would no doubt intensify the idling passenger capacity of the two intercity high-speed railways, so as to cause vicious competition within the high-speed railway market, and eventually cause lose-lose situation.

From the perspective of the optimization of transportation resource allocation, when the decisions were made about Beijing-Shanghai High-speed Railway, if the resources were concentrated in firstly constructing the 900 km line between Tianjin and Nanjing, then using the already completed Beijing-Tianjin Intercity High-speed Railway and Shanghai-Ningbo Intercity High-speed Railway to open the full line of Beijing-Shanghai High-speed Railway, it could not only build 420 km less of high-speed railway, and save nearly RMB 70 billion of investment, and the construction could be completed 1 year earlier to (catch up the Shanghai EXPO) develop the benefits of high-speed railway.

Source Shen Peijun: *Observation of Transportation Resource Optimization through Beijing-Shanghai High-speed Railway*, Comprehensive Transportation, the 7th Issue in 2011

As early as 2005, an official in the Ministry of Land and Resources pointed out that the repeated construction of ports is very prominent, in some costal and

---

[8]Shen Peijun: *Observation of Traffic Resource Optimization from Beijing-Shanghai High Speed Rail*, Comprehensive Transportation, 7th Issue in 2011.

riverside regions, there is one wharf in every one kilometer, their handling capacities are all very big, there is phenomenon of idling capacity.

In the aspect of road and bridge construction, there are 39 completed and opened bridges on Yangtze River between Yibing and Shanghai, plus 11 bridges under construction and 17 bridges under planning, in the coming 10 years, there will be 100 bridges on Yangtze River. In the main stream of nearly 3000 km of Yangtze River, there will be one cross-river bridge in less than 30 km on average.

26.2.3 Loss of Management Efficiency

China's traffic & transportation have been basically developing in various separated transportation types and having facility complement based on requirements of respective transportation organizations, in the network construction, station construction and computerization construction, they rarely considered mutual coordination, common use or interconnection from the perspective of comprehensive transportation system, so the efficiency and effect level of mutual connection and coordination is very low. This is typical thinking of considering how to meet production requirements from the perspective of its own system producer, and has deep thinking of closed system; and it is also related to the different marketization degrees of various transportation types.

Railway (meaning National Railway) is a national monopolized operation that is a sole operation in China, different railway bureaus and railway branches are only production workshops, they don't really have legal person property rights, and they can't independently make decisions in market activities, independently operate and be responsible for its own profit and loss, besides, they are not only unable to independently provide transportation service, but also unable to independently obtain income from the market, and the railway system itself adopts a "Big Pot" financial settlement system that is based on costs, therefore, different railway bureaus and railway branches not only don't have decision-making power, but also don't have the initiative to have construction cooperation with other transportation means, all cooperation must be decided by the Ministry of Railway. While roads and water transports are basically in local regions and constructed and operated by companies, such extremely imbalance of entities in negotiations of construction cooperation caused difficulty in realization of cooperation.

Station hub is a link connecting different transportation means, and one of the most basic conditions to realize "Zero-distance transfer" and "Seamless connection" of comprehensive transportation, however, due to the difference of China's various transportation means in operation and management system,[9] in station construction,

---

[9]For example, railway hubs and railway routes are an entirety, which is under the unified construction operation by the Ministry of Railways, while road station hubs are mainly built by enterprises or jointly built by local government and enterprises, and operated by enterprise.

the two are in extremely unbalanced status in cooperated construction of station hub, besides, because of the difference in development objective and the operation and management system, it is usually very difficult to reach consensus, this also causes that fact that the stations of various transportation means are separately planned based on respective transportation and production requirements, separately constructed, and form system separately, so they couldn't form comprehensive transportation hubs, the connection between different transportation means must be transferred between different stations, and the efficiency of system is low.

Special Column 26.2 Example of Low Efficiency of Transportation System
None of the 45 main road hubs and 100 secondary road hubs planned in China is constructed jointly with railway hub. Taking Beijing for an example, Beijing is one of the city in China with the largest passenger and cargo flow volume, it currently has over 50 railway cargo stations, 1.67 million square meter of total goods yards, 3 railway marshaling yards, and 26 road passenger stations. However, these stations and yards are basically in gross-style dispersed arrangement, there is not one large comprehensive hub that closely integrate different transportation means, besides, the railway passenger stations are in downtown area, road passenger stations are basically in edge of city, railway cargo stations are even closer to city, the main hubs of roads are basically outside the railway circle, the relations between stations are not close, which is basically in detached state, combined transport between the two is only a kind of simply section-by-section integration, connection in the middle must rely on automobile to transfer again before completion.

Source Compiling Committee of "Chinese Comprehensive Transportation Development Strategy": *Chinese Comprehensive Transportation Development Strategy*, Xi'an Jiaotong University Press, 2004 Edition, pp. 21–22.

So far as concerned, there is not one city in China that has built a large concentrated transfer center that is composed of multiple transportation means and connecting intercity transportation and urban traffic. The lagging of the planning and construction of comprehensive transportation hub stations has severely restrained the enhancement of the efficiency and effect level of China's comprehensive transportation system.

26.3 Forming Causes of the Construction Cost of Three-Dimensional Transportation

26.3.1 *Promote the Demand of National Economic Development*

China's construction of transportation infrastructure has achieved huge development since the establishment of New China, especially the thirty years since the reform and opening-up. Traffic & transportation industry is a basic industry of national economy and an important material base of the advancement of human society, the development of traffic & transportation industry has very important influence on the social and economic development of a nation. The contribution of traffic & transportation industry to national economy is mainly reflected in two aspects: firstly, the contribution of traffic & transportation to GDP. It is not only reflected as the increase of the economic output of the operation of transportation infrastructure and traffic & transportation industry, but also reflected as the increase of economic output of other industries relating to traffic & transportation industry, such as processing and manufacturing industry, metallurgy industry, foreign trade, financial industry, etc., meaning the ripple effect. Secondly, the contribution of traffic & transportation industry to employment, employment level is also an important macroeconomic indicator of a nation, the contribution of traffic & transportation industry to employment is mainly reflected as the increase of unit output in transportation industry, and the increase of employment in transportation industry and other industries.

The scale of transportation investment is mainly decided by the investment system and funding channel. Development of transportation industry must be supported by high investment policy. This has been proved by development experiences in many countries in the world. Currently in economically developed countries, the investment in traffic & transportation normally accounts for 10–14% of total investment, while the ratio in developing countries is normally above 20%, the World Bank suggested that the minimum limit of transportation investment in developing countries is 20–28%, in the period of rapid economic development in developed countries, transportation investment even accounted for over 30%.[10] Enlargement of the investment in the construction of transportation infrastructure is a requirement to promote the development of national economy.

26.3.2 Lack of General Vision in Planning

If construction is made completely in accordance with national transportation development plan, in theory, there will be no problem of repeated construction and overcapacity. But in fact, China's construction of transportation infrastructure and facility is lack of unified planning, different provincial and municipal government form their own plans, basically, there are many additional projects besides those under national planning.

Scientific long-term development plan is the foundation of success in investment projects, if we can't establish the arrangement of investment project on the base of

---

[10]Editing Committee of the *China Comprehensive Transport Development Strategy*: *China Comprehensive Transport Development Strategy*, Xi'an Jiaotong University Press, 2004 Edition, pp. 21–22.

long-term development plan, not only the function of project is difficult to develop, but also it may sometimes affect the economic development of certain region or sector, or even lead to loss of overall economic interest. In China, decision-making of many investment projects is lack of exactly such support of scientific long-term development planning. Sometimes, there is no long-term development planning, even there is, they are lack of scientific practice and operability; sometimes, there is long-term development plan, but there is no thorough consideration in project decision-making. Over the years, local governments all have formed long-term development plans, which all passed review by NPC, but they are lack of enforceability. As a result, long-term development plans are put to "sleep", investors are on their own, which causes influence to the overall optimization of resource allocation.

26.3.3 Incomplete Management System

China's current traffic & transportation system is in separated state under management by multiple ends. For long term, the Ministry of Transportation, the Ministry of Railway, Civil Aviation Administration of China, the Ministry of Construction and the Ministry of Public Security each performs its own functions.[11] Such situation remained until Mar 23, 2008.[12] Generally speaking, the functions of ministries and commissions can directly penetrate to local provincial departments and bureaus. The function scope of administrative organs is divided based on administrative regions, which makes the connection between different transportation means relatively poor, and causes low transportation efficiency, waste of resources and high social costs. Each transportation means stresses its role in certain aspect, they all build their own transportation hubs, but never generally consider rational division of work and coordinated development.

The forming causes of the construction cost of Chinese three-dimensional transportation could be comprehensively reflected in Chart 26.2.

---

[11] The Ministry of Transport, Ministry of Railways and Civil Aviation Administration are mainly responsible for the planning, construction and management of out-of-city national road network, waterway transport facilities, railway facilities, civil aviation transport facilities respectively. The Ministry of Construction and Ministry of Public Security are responsible for the planning and construction of transport facilities within cities, and the security of public transport and urban road safety, and maintenance of transport.

[12] Based on the institutional reform plan of the State Council in 2008, the Ministry of Transport of the People's Republic of China was built based on the original Ministry of Transport, the CAAC, State Post Bureau and other departments were transferred to the Ministry of Transport for management, but the Ministry of Railways is still on the same level with the Ministry of Transport.

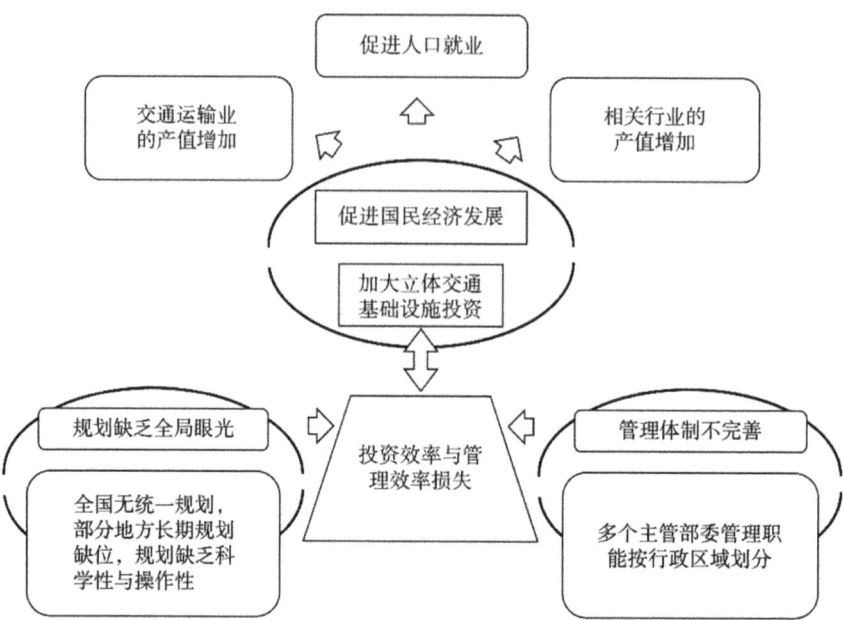

促进人口就业	Promote employment of population
交通运输业的产值增加	Increase of output of transport industry
相关行业的产值增加	Increase of output of related industry
促进国民经济发展	Promote the development of national economy
加大立体交通基础设施投资	Enlarge three-dimensional traffic infrastructure investment
规划缺乏全局眼光	Lack of overall perspective in planning
全国无统一规划，部分地方长期规划缺位，规划缺乏科学性与操作性	Lack of nationally unified planning, some places are lack of long-term planning, and lack of scientific and operable nature in planning
投资效率与管理效率损失	Loss of investment efficiency and management efficiency
管理体制不完善	Incomplete management system
多个主管部委管理职能按行政区域划分	Many competent ministries and commissions have their management functions divided by administrative regions

Chart 26.2 Forming causes of the construction of Chinese three-dimensional transportation

26.4 Countermeasures for the Construction Cost of Three-Dimensional Transportation

26.4.1 Improve the Investment and Financing System of the Construction of Chinese Transportation Infrastructure

Because the current investment and financing entities and investment means are still relatively simple, while the investment decision-making entities are diversified, China shall build a rational and highly efficient investment and financing system as soon as possible, reform the investment and financing system, explore the investment and financing channels, introduce venture capital mechanism, further open the traffic & transportation capital market to the entire society, besides, the state should form the related policies in fiscal and taxation aspects to support the construction of transportation infrastructure.

26.4.2 Realize the Comprehensive Management of Traffic & Transportation Industry

China shall build and arrange rational comprehensive traffic & transportation network based on the demands for transportation capacity and transportation quality, such as the comprehensive construction of the well-off society, strategic adjustment of economic structure, coordinated development of regional economy, comprehensive development of rural economy, construction of urbanization, economic globalization, guarantee of national security and stability, and implementation of sustainable development strategy. It is necessary to organically integrate the construction of transportation channel and fast traffic network in urban economic zones, realize the effective connection and coordinated development between different transportation means, strengthen the scientific management of comprehensive transportation system, and promote the reform of traffic & transportation management system and comprehensive traffic & transportation network operation mechanism.

Meanwhile, China should change the state of the long-term separated planning, construction and management of different transportation means as soon as possible, realize the comprehensive planning and management of traffic & transportation industry, accelerate the reform of market orientation of various transportation means, accelerate the construction of modern enterprise system, and open up the transportation infrastructure construction and operation market.

26.4.3 Strengthen the Coordinated Development Planning and Layout

To realize rational layout, the basic work is to coordinate the thinking of development and strengthen the planning. On one hand, the development of traffic & transportation should be coordinated with the economic and social development, in productivity layout and planning, the state should fully consider the layout and planning of transportation, and the layout and planning of transportation should also be coordinated with the layout and planning of productivity, we should make all efforts to reduce repeating, detour in transportation, reduce the transportation intensity, reduce the excessive occupation of transportation resources, so as to reduce the damage of the traffic & transportation on external environment in society; on the other hand, we should strengthen the unified planning between different means of traffic & transportation, enforce the idea of sustainable development throughout transportation development planning, based on the characteristics of different transportation means, enhance advantages and avoid disadvantages, form an organic and rational comprehensive transportation system, enhance the efficiency and service level of transportation, reduce the influence of traffic & transportation on external environment to minimum, so as to realize the sustainable development of transportation.

The establishment of a modernized traffic & transportation system is the urgent demand of the social and economic development in the 21st century. In order to timely seize the opportunity brought by integration of world economy, China should fully absorb the experiences and lessons of developed nations in their establishment of modernized traffic & transportation system, fully develop the late-mover advantage, enhance the level of network and modernization of China's traffic & transportation infrastructure, strengthen the management and operation level of China's traffic & transportation organizations, realize the modernization and internationalization of traffic & transportation, and strengthen China's advantage in the competition of economic globalization.

The countermeasures for the construction cost of China's three-dimensional transportation are concluded as follows in Chart 26.3.

26.4 Countermeasures for the Construction Cost of Three-Dimensional ... 407

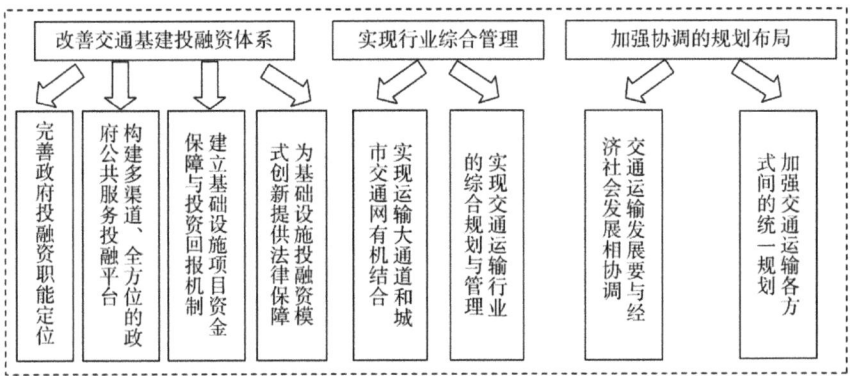

改善交通基建投融资体系	Improve transport infrastructure financing and investment system
实现行业综合管理	Realize comprehensive management of industry
加强协调的规划布局	Strengthen the coordinated planning layout
完善政府投融资职能定位	Improve the government's function positioning in investment and financing
构建多渠道、全方位的政府公共服务投融平台	Establish multi-channel and comprehensive platform of government public service investment and financing
建立基础设施项目资金保障与投资回报机制	Establish the funding guarantee and investment return mechanism for infrastructure projects
为基础设施投融资模式创新提供法律保障	Provide legal guarantee for the innovation of infrastructure investment and financing model
实现运输大通道和城市交通网有机结合	Realize organic integration of big transport channel and urban transport network
实现交通运输行业的综合规划与管理	Realize the comprehensive planning and management of transport industry
交通运输发展要与经济社会发展相协调	Transport development shall be coordinated with economic and social development
加强交通运输各方式间的统一规划	Strengthen the unified planning of different forms of transport

Chart 26.3 Countermeasures for the construction of China's three-dimensional transportation

Chapter 27
The Cost of China's Creditability Construction and Handling of Threats

27.1 Overview and Forms of the Cost of China's Creditability Construction and Handling of Threats

Creditability is interpreted in the *Modern Chinese Dictionary* as the power the make the public trust. The world creditability is derived from the English word "Accountability", meaning the responsibility to report, explain and defend for certain matter; taking responsibility for the behaviors of oneself, and accept questioning. Creditability means the trust of fairness, justice, efficiency, humanity, democracy and responsibility reflected by public power in social public life when facing time difference, public communication and exchange of interests. Creditability is not only a kind of trust of social system, but also authentic expression of public authority, which is in the scope of political ethic. Creditability is a major issue for a nation and a group of people.

The cost of Chinese creditability construction and handling of threats means the cost of China building its international creditability and the cost of handling international and domestic threats, including the maintaining cost of Chinese territory and sovereignty integrity, the cost of national unity, the cost of foreign aids, the cost of bearing international responsibilities, and the cost of communicating national image, please refer to Table 27.1.

27.1.1 Maintaining Cost of China's Territory and Sovereignty Integrity

Sovereignty means "A country's domestic high authority and external right to independence. In *International Law*, it is the fundamental attribute of a nation and the foundation of the basic national rights".[1] National unity and territory integrity

---
[1] Ci Hai Editing Committee: *Ci Hai* (Middle Volume), Shanghai Cishu Press, 1999 Edition, p. 3049.

Table 27.1 Forms of the cost of China's creditability construction and handling of threats

Classification of cost	Forms	Specific contents
Cost of creditability construction and threat response in China	Cost of maintaining the territory and sovereign integrity of China	Cost paid for maintaining the territory and sovereign integrity of China
	Cost of national unity	Policy cost of maintaining national unity and development
	Cost of foreign aids	Input into foreign aids to the international community and countries subject to disasters
	Cost of bearing international responsibilities	Cost paid for bearing international responsibilities
	Cost of promoting big power image	Input and efforts made to promoting active and responsible big power image

are the concentrated expression of national sovereignty, and also the main support of a nation's international creditability. In order to maintain national unity and territory integrity, the Chinese Government under the central leadership has made unremitting efforts.

After the establishment of the People's Republic of China, as the interim constitution of New China, the *Common Guiding Principle* announced that "The People's Republic of China must completely cancel all privileges of imperialist nations in China".[2] "For the various treaties and agreements entered into by and between the Nationalist Government and foreign governments, the Central Government of the People's Republic of China shall review, upon their contents, they will be acknowledged, or canceled, or modified, or reformed respectively".[3] The Party Central Committee with Comrade Mao as the core made all efforts in realizing national equality and unity, had strong vigilance of the plots of domestic and foreign reactionary forces to divide China, smoothly completed the peaceful liberation of Tibet, the suppression of rebellion in Xinjiang, achieved the great victory of the war to resist the US aggression and aid Korea and the war to aid Vietnam and resist America, so as to maintain the national sovereignty and national dignity.

The party's second-generation leadership with Deng Xiaoping as the core integrated the characteristics of time and China's situation, creatively proposed the new concepts, new strategies and new viewpoints in maintaining national sovereignty. Firstly, when involving the basic standpoint of national sovereignty,

---

[2]CCP Central Committee Archives: *Selected Documents of CCP Central Committee,* Volume 14, CCP Central Committee Party School Press, 1987 Edition, p. 732.

[3]CCP Central Committee Archives: *Selected Documents of CCP Central Committee,* Volume 14, CCP Central Committee Party School Press, 1987 Edition, p. 734.

Deng Xiaoping stressed that "The sovereignty issue is not an issue to be discussed",[4] and resolutely protect the fundamental interest of Chinese people in national unity and territory integrity. In order to solve the issue of Hong Kong, Macao and Taiwan, so as to realize national unification, Deng Xiaoping creatively proposed the great concept of "One Country Two Systems". Secondly, in the strategy of maintaining national sovereignty, on one hand, Deng Xiaoping didn't give up using revolutionary war to stop the violation of Chinese sovereignty from antirevolutionary war. For example, in China's Taiwan issue, we will never promise giving up using force. On the other hand, he also believed that the main factors damaging Chinese sovereignty would be non-war, meaning hegemonism and power politics. Therefore, he stressed that "Whether China could resist the pressure of hegemonism and power politics, and adhere to our socialist system is precisely depending on whether we could make relatively fast growth speed and realize our development strategy".[5]

The party's third-generation central leadership with Jiang Zemin as the core successfully practiced the great concept of Deng Xiaoping's "One Country Two Systems" concept, Hong Kong and Macao smoothly returned. Meanwhile, the leadership prudently observed the words and deeds of the leaders in Chinese Taiwan District, once Chinese Taiwan side is found out having any words or deeds that intend separation, the leadership would give strong denounce and crack down, and also didn't promise giving up using force. On the other hand, the party's third-generation leadership provided feasible solution of improving the cross-strait relations. In January, 1995, Jiang Zemin made the "Eight-Point Proposal" to develop cross-strait relations and promote national unification.[6] Besides, the third-generation leadership of the party actively responded the challenges of economic globalization on national sovereignty, well handled the dialectical relationship between the absoluteness and relativity of sovereignty. On the basis of adhering to the principle of the supremacy of national interests, the leadership took the opportunity to use national power to prevent and avoid the negative effect of economic globalization.

---

[4]*Selected Works of Deng Xiaoping,* Volume 3, People's Publishing House, 1993 Edition, p. 12.
[5]*Selected Works of Deng Xiaoping,* Volume 3, People's Publishing House, 1993 Edition, p. 356.
[6]On Jan 30, 1995, General Secretary Jiang made a speech of *Keep Fighting for Promoting the Completion of the Motherland Unification Course* at the Spring Tea Party. In this speech, General Secretary Jiang proposed eight viewpoints and claims regarding developing the cross-strait relations at current stage, and promoting the peaceful unification process of motherland, which was later referred to as "Eight Viewpoints of Jiang". Its main contents include: adhere to the one China policy; holding no dispute over Taiwan's development of civil economic and cultural relations with foreign countries; implement negotiation for the peaceful unification across the strait; make efforts to realize peaceful unification; energetically develop cross-strait economic exchange and cooperation; jointly pass and develop the excellent tradition of Chinese culture; Taiwan compatriots are also our brothers; welcome the leader of Taiwan to visit Mainland in proper identity, we are also willing to accept invitation from Taiwan to visit Taiwan. This was the succession and development of the "One China Two Systems" conception.

27.1.2 Cost of National Unity

For the minorities, Chinese Government not only implements the policies such as national equality and unity, regional autonomy, respect for customs, freedom of religious belief, development of language, etc., but also actively adopts a series of favorable policies for the economy of minorities, makes efforts to reduce the gap of economic and cultural development between different people and regions, promote the common prosperity and advancement of different people, comprehensively make efforts in the construction of the development capacity of minority regions, so as to feasibly accelerate the development progress of the modernization in minority regions.

For a long term, Chinese Government has been specially regarding the support the minorities to develop economy, based on the regional characteristics of different minorities, the characteristics of different industries and sectors, and the social and economic development stages, the Chinese Government has adopted different favorable tax policies (refer to Table 27.2), makes efforts to reduce the economic burden of minorities, so as to maintain the normal operation of husbandry, industry, transportation and other construction courses in minority regions.

Targeting the characteristics in minority region of low economic development starting point, backward cultural education level and extremely lack of talents, the government not only gave special favors in minority regions in the aspects of fiscal and tax policies, but also gave proper support to minorities in the aspects of population development, cultural education, talent cultivation that are closely related to economic development (refer to Table 27.3), so as to honestly and sincerely help the minority to make economic and cultural development.

According to statistics, since the establishment of New China, the Party's Central Committee issued and implemented 123 articles of minority economic policies regarding finance, taxation, cultural, education, poverty relief, industry, etc., which greatly promoted the economic development and livelihood improvement in minority regions, and energetically maintained China's national unity.

27.1.3 Cost of Foreign Aids

Providing foreign aids[7] is an important means for sovereignty nations to reflect national power, maintain and enhance international creditability. The quantity of aids provided has become an important indicator to measure national strength and international contribution.

---

[7]Foreign aid is an important content of modern international relations. It means that a country or group of nations provides free or preferential paid goods or funds to solve the political and economic difficulties or problems faced by the aided country, or certain means to reach the special purpose of the aiding country.

Table 27.2 Favorable tax policies in minority regions

No.	Name of policy	Period (year)
1	The state exempted or reduced industrial and commercial tax in border and minority regions	1950–1993
2	The state implemented light tax favor to the agriculture and husbandry industry of minority regions	1953–present
3	The state reduced agricultural tax to the minority regions with difficulty in living	1958–present
4	The state exempted the township enterprises of border countries and minority counties of industrial and commercial tax for five years	1979–1985
5	The state implemented 3:7 sharing of the reduced cost of the infrastructure enterprise in eight minority provinces and regions	1979–1985
6	The state exempted or reduced the income tax in the "old revolutionary, minorities, border and poor" regions	1985–present
7	The state implemented favorable tax policy on border trade	1991–1994
8	The state regulated the reduction or exemption of the fixed assets investment adjustable tax in minority regions	1992–present
9	The state exempted or reduced tax for border trade imported commodities in 12 categories and 162 items	1992–1995
10	The state regulated reduction or exemption of income tax for new enterprises in "old revolutionary, minorities, border and poor" regions	1994–1997
11	The state imposed 10% of special agricultural product tax on the enterprises that purchasing border-traded tea materials	1994–present
12	The state reduced enterprise income tax for the foreign investment enterprises that are set in central and western regions at the tax rate of 15% for three years	2000–2002
13	The state may regularly reduce or exempt enterprise income tax for the local enterprises in western regions and autonomous regions	2001–2010
14	The state imposed the two-year exemption and three-year half levy of income tax for the newly established enterprises in transport, electricity, water power, postal, radio and TV industries in western regions	2001–2010
15	The state exempted VAT for the fixed production, dealership and distribution of tea	2001–2005
16	The state exempted special agricultural product tax in western regions for ten years on the special agricultural products generated from returning gain plots to ecological forestry and grassland in order to protect the ecological environment	2001–2005

Source Wen Jun: *Evaluation of Chinese Economic Policies for Minorities* (1949–2001), National Situation Report, the 42nd Issue in 2003

Since the early 1950s, China started providing economic and technical aids under the circumstance that China was very tight in finance and very lacking in materials. At the beginning, foreign aids was an important means for China to develop relations with third-world nations, and also played important role in recovering the lawful seat at UN and opposing "Taiwan Independence". With the

Table 27.3 Favorable policies for minorities in population, education and employment

No.		Name of policy	Period (year)
Population birth policy	1	The state implements "Prosperous Population" policy on minorities	1951–1980
	2	The state promotes family planning on minorities	1982–present
	3	The state implements family planning in the concentrated regions of minorities	1992–present
	4	The state implements preferential birth policies on the population of minorities	1984–present
Cultural and education policy	5	The state determined the establishment of national colleges	1950–present
	6	The state set up special subsidies for minorities education	1952–present
	7	The state established administrative institution for minority education	1952–present
	8	The state regulated the consideration of minority characteristics in development of minority education	1951–present
	9	The state regulated the expenditure of minority colleges	1963–present
	10	The state implemented preferential college entrance examination policies on minority students	1977–present
	11	The state implemented preferential policy on the minorities living in mixed regions	1979–present
	12	The state implemented preferential policy on the private teaches at border areas	1979–1987
	13	The state determined the guideline and policy on development of minority education	1981–present
	14	The state adopted special policy on the occupational education in minority regions	1992–present
	15	The state implemented key support policy on the compulsory education in minority poor regions	1995–2010
	16	The state formed special policy for the education of poor female students from poor families in minority regions	1997–present
	17	The state implemented special policy in western regions for attracting talents and development technological education	2001–2010
Labor and employment policy	18	The state's main policies for minorities in the aspects of employment and occupation	1984–present

Source Wen Jun: *Evaluation of Chinese Economic Policies for Minorities* (1949–2001), National Situation Report, the 42nd Issue in 2003

strengthening of comprehensive national strength, the quantity of foreign aids provided by China becomes more and more, and its scope is also continuously expanding.

Besides, Chinese Government has been actively responding to advocates of the international community and related nations, after major disasters such as the major earthquakes in Algeria, Iran and Pakistan, the Indian Ocean Tsunami, the American Hurricane, the Chinese Government provided urgent humanitarian aids in shortest time, and effectively supported the disaster relief works in related national governments and people.[8]

Although China still faces a series of serious development issues itself, China's steps of foreign aids are still not slowed down, which reflects the spirit of sincere cooperation in international relations and obtained high praises from the aided governments and people. In December, 2004, BBC made a survey on 23 thousand people in 22 countries across the globe, which indicated that China has won respect of the world. Generally speaking, nearly half (48%) of the people believed that "China's international influence" has positive significance, which was 10% higher than that of the US.[9] Through sincere and selfless aids, China provided great assistance to the economic development in other developing nations, and also enabled the development model with Chinese characteristics to be promoted and recognized by the international community, the wide developing nations hope to learn and refer to the successful experiences of Chinese development (Chart 27.1).

27.1.4 Cost of Bearing International Responsibilities

In recent years, Chinese Government and people have used feasible actions that benefiting the people and the world to enable China's international image as a responsible major power to be gradually established and won praises of the world. China has been actively participating peacekeeping actions of the UN, bearing the responsibilities of maintaining and promoting world peace and development, and establishing the image of a responsible political power. Since April, 1990 when China dispatched five military observers to the United Nations Truce Supervision Organization, China has dispatched over 14 thousand of military, police and civil officials to participate 24 peacekeeping actions, China has become the country among the five permanent member states of UN Security Council having sent the most peacekeeping personnel. Currently, there are still over 2100 Chinese peacekeeping personnel developing UN peacekeeping actions in 10 task regions. Chinese peacekeeping personnel, no matter they are military personnel or police, are lack of

---

[8]Dai Chunxia: *Discussion of the Role of Foreign Aids in the Enhancement of Chinese Soft Power,* Rule of Law and Society, 9th Issue in 2008.

[9]Zhao Lei: *Three Dimensions to Understand China's Soft Power: Cultural Diplomacy, Multilateral Diplomacy and Foreign Aids Policy,* Social Science Forum (Volume of Academic Review), 5th Issue in 2007.

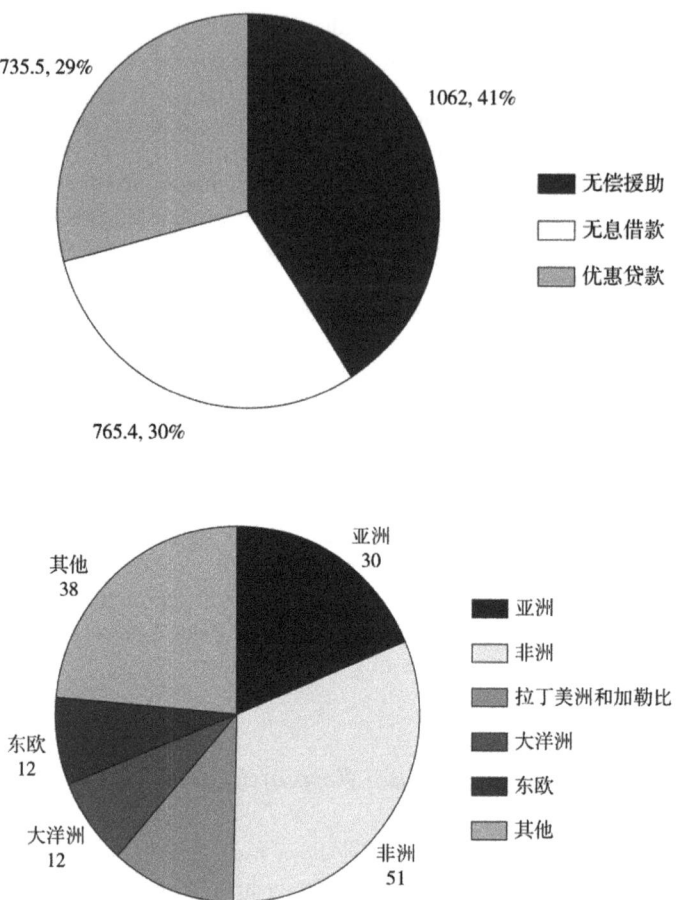

Chart. 27.1 Forms of Chinese foreign aids and regional distribution of aided countries. *Source* News Office of the State Council of the People's Republic of China: *China's Foreign Aids*, People's Publishing House, 2011 Edition, p. 6

active self-defense force, and at dangerous state of being attacked at any time. The peacekeeping task regions deployed by UN are usually poor and small countries in Asia, Africa and Latin America, the painful sufferings they experience now have been experienced by China before. When peacekeeping in these countries and regions, China can feel their pain and empathy. China knows that what these countries urgently need is peace and what they hope for is reconstruction rather than foreign military oppression and intervention. Therefore, China's peacekeeping force is shown in the image of peaceful construction, and winning the respect from the world and the people in task regions through real actions.

China has been actively bearing the responsibilities of promoting the world's economic stability and healthy development, and increasingly establishing the image of responsible economic power. In 1997, facing the sudden Asian financial storm, out of surprise of the world, Chinese Government resisted different opinions, resisted the pressure of shrinking export, solemnly announced to the world that "China will adhere to the standpoint of not devaluating RMB, and bear the historical responsibility of stabilizing the financial environment of Asia". That was a milestone event of China establishing the image of responsible major power. In 2009, facing the financial crisis that swept the world, Chinese Government released a basket of programs to stimulate the economy, adopted active and responsible attitude, explored domestic demands, adjusted the structure, appealed to promote cooperation in international community, jointly overcome the difficulties, reform the existing international financial order, make all efforts to promote economic recovery of the world, so as to become the locomotive and dynamic of world economic development and win respect from the world.[10]

27.1.5 Communication Cost of Major Power Image

In recent years, it is not unusual to see unfair reports by international media about Chinese economic and political issues, the open China has achieved huge advancements in the aspects of economy, society, democracy and rule of law, and also plays increasingly important role in international community, but the mainstream western media still undermine China's image at all costs, which can also enable us to see the complex feelings of the western world has when facing the rising China. China is not seeking hegemony, contrarily, the strategy of "Keeping a low profile"[11] since the era of Deng Xiaoping is still working and will be adhered to

---

[10]Long Xiaonong and Zhang Yuqiang: *Observation of the Construction of China's Image as Responsible Major Power from Aids to Haiti,* Foreign Communication, March, 2010.

[11]At the critical point at the end of 1980s and beginning of 1990s when there was change to East Europe and the crossing of domestic reform and opening-up, Deng Xiaoping, with high political wisdom as a strategist and boldness, thoroughly analyzed the situation, and proposed the guideline to "calmly observe, stabilize the foothold, calmly respond, be good at being conservative, never stand out, keep a low profile, and do something". He indicated that the "international situation could be summarized as three sentences: first sentence, calmly observe; second sentence, stabilize foothold; third sentence, calmly respond. Never feel anxious, which is useless anyway. We must keep calm, calm and calm again, do practical works, take care of one thing, our own thing". "Some countries in the third world hope that China should lead. But we would never lead, this is a

for a very long time. In 2009, Premier Wen Jiabao particularly expressed the Chinese Government's attitude towards the sensational expression of "G2", and said that it was "groundless and wrong".[12]

Apparently, there is huge difference between the real China and the China that the westerners recognizes through various media. For the rising China, how to eliminate all sorts of doubts in different countries towards China, seize proper opportunity to build and communicate the image of active, positive and responsible major power has become an important task of China's external communication.

The "SARS" epidemic situation in 2003 directly promoted the rapid development of China's news speaker system and the enhancement of officials' ability in responding to media. On Apr 5, 2006, the *Regulation of Government Information Disclosure of the People's Republic of China* was released, and formally enforced since May 1, 2008. The approach of Beijing Olympics further promoted this work. Besides, the crisis handling ability of Chinese Government is also continuously enhancing.

Seeing is believing. The Olympics have proved that enabling more foreigners to come to China and personally experience China's development and culture is apparently the best communication means of national image; the Shanghai EXPO in 2010 also enabled more people to understand China. Besides, the world was deeply impressed with China's stable performance in the world financial crisis, its rational, energetic and decent attitude in the China-US trade friction, its confidence reflected at the 60th National Anniversary, its independence in climate issue and performances in a series of major events.

On Dec 7, 2009, which was the first day after the opening of the Copenhagen Conference, the website of the Foreign Ministry released the news that Chinese Delegation established the "Chinese News and Communication Center" at Copenhagen Center". China became the only developing country that separately set up news and exchange center at the conference, becoming one of the three with EU and the US' news centers. Airing the voice of China to the world from Copenhagen, besides establishing the image of open mind, brave bearing of responsibilities, defending its interests with reasons and standing up against others, there were also full of wisdom and humor. It proves that the public relations ability of Chinese Government has made great improvement.

In January, 2011, the Chinese National Image Video represented by *Figure Chapter* and *Perspective Chapter* was formally broadcasted on the media of the US and UK, which provided a set of sensational experiences to viewers about China, and since then opened the "Era of Public Relations" of China's national

---

(Footnote 11 continued)

fundamental national policy. We can't afford to lead, and we don't have enough power to do it anyway. There will be no benefit from it, many countries that lead have disappeared. China will always stand on the side of third world, China will never claim hegemony, and China will never lead" and other important statements. Refer to *Selected Works of Deng Xiaoping,* Volume 3, People's Publishing House, 1993 Edition, pp. 321 and 363.

[12]Wang Bo: *Tell "Chinese Story" to the World,* World Knowledge, 1st Issue in 2010.

27.1 Overview and Forms of the Cost of China's Creditability Construction ... 419

image. With Hu Jintao's visit to the US, the 30-s long Chinese national image video's *Figure Chapter* were repeatedly broadcasted on the big screens of the busiest Time Square in New York, which broadcasted for over 8000 times in accumulation. Such major promotion video was active exploration for the establishment of China's national image and seeking of world's recognition towards China.

27.2 Forming Causes of the Cost of China's Creditability Construction and Handling of Threats

27.2.1 The Demand for China to Enhance Comprehensive National Strength and Build National Image

Sovereignty, territory integrity and national unity are important parts of comprehensive national strength and the foundation for China to form hard power and soft power. In today when the globalization process keeps accelerating and national sovereignty is affected and weakened by cross-border organizations, only by adhering to the independence of national sovereignty, China could, while obtain mutual benefit and win-win situation with the world, maintain its own security, and create a peaceful external environment for China's development. Only by strengthening national unity, China could form strong internal force, promote China to continuously develop, further enhance comprehensive national strength and realize the bright vision of more prosperity and fortune.

The flow of foreign aids funds will inevitably lead to flow of culture, knowledge, technology and way of living. The flow of foreign aid funds will help us establish and maintain close relationship with the aided countries, affect (but not direct control and intervene) the development direction of the political and economic development of the aided countries, enhance China's international image and win the recognition and cooperation of the international community.

27.2.2 The Ulterior "Chinese Threat Theory"

But we should also see that in the international community, some politicians in the West expressed suspicion or distortion towards China's resolute adherence to the Socialist Path since the reform and opening-up, and the huge achievements made, plus the separation brought by ideology, etc., they believed that the large foreign aids made by China have "Agenda", or complained that China's foreign aids were too small to meet the responsibility of big power. These would all generate negative influence on China's international image. For example, Joseph Nai believed that China's culture and soft power in image cohesion is more worrying than China's military force. He emphasized that the rise of China's soft power will threaten the interest of the US, and proposed to contain the development of China's soft power. It can be seen that the soft power threat has become another version of the "Chinese Threat Theory".[13] China's rapid development and increasingly expanding political influence have affected and weakened the dominant position of the West in international affairs, strengthened the speaking right of developing countries and increased the collective influence of developing countries. The enhancement of the international status of developing countries makes the strategic interest of western nations to be balanced, which is the thing that western countries don't want to see, so they slander China and distort the viewpoint of international opinions on China. Economically, China's rapid development has driven the overall development of developing countries and objectively weakened the western countries' control over underdeveloped countries, so the western countries spare no effort in criticizing the intension of China's foreign aids, the purpose of which is to disturb the normal friendly cooperation between China and other underdeveloped or developing countries, so as to smear China's international image.[14]

27.3 Countermeasures for the Cost of China's Creditability Construction and Handling of Threats

27.3.1 Enhance Communication Ability of National Image

Judging from the current international communication order, the situation of strong west and weak east is still very obvious. Whenever there is any dispute, the international opinion would form the encircling trend from West to East. First of all, China needs to strengthen the communicating ability of its own image. Currently, Chinese Government has made efforts in promoting the establishment of its own major international media, which was a good start.[15] Secondly, when building national image, communication contents and means should be as close to the international mainstream political, cultural and value norms system as possible. To put it simply, meaning having exchange and communication in the language that others can understand, so as to reach the purpose of influencing others' recognition. In the communication of government image, we need to get rid of the relatively rigid and empty traditional lecturing model in "Political Propaganda", but use specific and fresh facts to reflect the down-to-earth, honest and low-key side.

---

[13]Dai Chunxia: *Discussion of the Role of Foreign Aids in the Enhancement of Chinese Soft Power*, Rule of Law and Society, 9th Issue in 2008.

[14]Wang Hongyi: *Preliminary Discussion of 'Chinese Threat Theories'*, Western Asia and Africa, August, 2006.

[15]Long Xiaonong and Zhang Yuqiang: *Observation of the Construction of China's Image as Responsible Major Power from Aids to Haiti,* Foreign Communication, March, 2010.

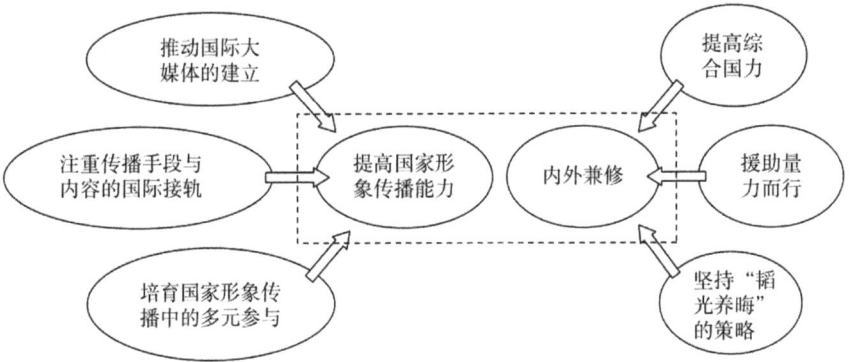

推动国际大媒体的建立	Promote the establishment of international big media
注重传播手段与内容的国际接轨	Focus on the connection of communication means and contents to international standard
培育国家形象传播中的多元参加	Nurture the diversified participation in communication of national image
提高国家形象传播能力	Enhance the communication ability of national image
内外兼修	Get refined internally and externally
提高综合国力	Enhance comprehensive national strength
援助量力而行	Grant foreign aids according to our own abilities
坚持"韬光养晦"的策略	Adhere to the strategy of "Keeping a low profile"

Chart. 27.2 Cost of creditability construction and handling of threats

Finally, in the process of building national image, China needs to actively cultivate diversified entities to participate, reduce direct operation by government, explore the channels for Chinese private organizations and individuals to contact foreign private society.[16]

27.3.2 Enhance China's Comprehensive National Strength

Nakae Sakamoto, Former Ambassador to China, said that "if a country with great population and nuclear weapon is a threat, then the US also has great population and nuclear weapon, why don't we say the US also constitutes threat?[17]" There is only one reason: the US is too powerful, people feel that there is no need any more

---

[16] Li Geqing: *Growth of Big Nation and Building of China's National Image,* Modern International Relations, 10th Issue in 2008.

[17] Chang Gong: *China shouldn't have Self-inflicted Setbacks: China's Attitude, Global Role and No Self-inflicted Setbacks,* Jiuzhou Press, 2009 Edition, p. 60.

to call the US threatening and dangerous. So the fundamental solution is that China should be strong and stronger, which is the fundamental method and most effective way to get rid of the "China Threat Theory".[18] Besides, while providing foreign aids and bearing international responsibilities, China should still adhere to the policy of "Keeping a low profile", get refined internally and externally according its own abilities, so as to avoid generate unrealistic expectation for China by some developing countries and the international community or the lack of understanding and support within China.[19]

The countermeasures for the cost of China's trades can be concluded as follows in Chart 27.2.

---

[18]Huang Xinghua: *Origin and Development of "China Threat Theory"*, Jiangxi Higher Education Press, 2010 Edition, p. 233.

[19]Long Xiaonong and Zhang Yuqiang: *Observation of the Construction of China's Image as Responsible Major Power from Aids to Haiti,* Foreign Communication, March, 2010.

Summary of the Cost of Foreign Opening-up Development

Since the reform and opening-up, China's opening-up and development thinking has gradually formed and developed and guided China to actively expand international exchange, energetically develop foreign trades and actively attract the technologies and investments by foreign investors. Chinese enterprises are no longer content with introducing in, but beginning to boldly enter the international market, and beginning to make foreign investments and M&A; meanwhile, from the central government to local governments, China has enlarge various investments, created the policy environment, legal environment and transportation network that promoted the economic and social development, so as to set foundation for the deepening of reform, further opening-up and promoted high-speed economic growth.

With the enhancement of China's economic strength, China plays increasingly important role on the international stage. As a major developing country, China didn't choose only taking care of its own, but actively bearing international responsibilities, having provided treat assistance to other developing countries in their economic development, promoted the international order to continuously develop to the direction of fairness, justice, tolerance and order.

Since the opening-up, China's economic development has long been under the model of investment driving and export driving, large amount of fiscal funds flew to transportation and other infrastructure field in order to promote regional opening-up, but there was also phenomenon of severe blind investment and concentrated investment (taking cross-river bridge for an example, in the coming 10 years, there will be over 100 Yangtze River Bridges, above the mainstream of Yangtze River with the length of 3000 km, there will be one cross-river bridge in less than every 30 km on average); the excessive dependence on foreign investment and export has caused the irrational industrial structure and caused the continuous widening of trade surplus, which peaked in 2008 at USD 295.46 billion; in 2010, China's foreign exchange reserve has reached USD 2.847338 trillion, which contains extremely high exchange rate risks (assuming about 50% of foreign exchange reserve is held in form of dollar-denominated assets, then 10% of US dollar devaluation against RMB means that China's foreign exchange reserve will lose USD 142.365 billion); the economic foreign dependence has increased from 12.54% in 1980 to 50.28% in 2010, which far exceeded the internationally recognized safety line of 25%; the development of national industry has been restrained; the cost of international frictions continuously rise, from 1995 when WTO was established to the end of 2006, the number of antidumping cases and cases of antidumping measures encountered by China both ranked number one in the world, and reached 536 cases and 375 cases respectively, while the various forms of "Soft Barriers" and trade suits have caused serious economic and social influences. During this process, different countries, out of their different political, economic or other purposes, proposed various versions of "China Threat Theories", and attempted to influence and control the international community's judgment on China's peaceful rise and their relations with China, and demonized China's international image. In order to promote the foreign communications in different fields, including the political, economic, diplomatic, cultural, official and private fields, establish and maintain China's image and creditability as a responsible major power, enable the international community to correctly recognize China, and create an excellent external environment for China's development, China has paid dear costs, no matter it's foreign aids or international exchange, and the cots to be paid are still rising.

In the process of opening-up and participation in globalization, the afore-mentioned cost and prices are inevitable for China to transform from planning economy to market economy. However in today, thirty years after the reform and opening-up, we need to correctly treat and rethink about these occurred and occurring costs, under the instruction of new development viewpoint, we should use overall and long-term perspective, under the precondition of promoting the sustainable development of China's economy and society, enhance the deepening of reform, expand opening-up, further enhance the cost efficiency of the integration into globalization.

Part VII
Cost of China's Natural Development

Chapter 28
Cost of China's Resource and Energy

28.1 Overview of the Cost of China's Resource and Energy

Resource means "all the material and non-material factors needed for the human being's survival, development and enjoyment".[1] Resource in narrow sense only means natural resource. Natural resource means the generic term of the natural material or natural environment that have social function and relative scarcity. Resource in narrow sense means that no matter it is overall environment or other certain part, if it is able to be utilized by human being, it can be concluded as natural resource.[2] In other words, we can interpret natural resource as the generic term of the materials that can be utilized in the human being's production and life and could generate certain social effect.[3] Natural resource could also be classified into land resource, climate resource, water resource, biological resource, mineral resource, ocean resource, energy resource, tourism resource, etc. In this chapter, we will mainly research land resource and water resource.

Energy could be interpreted as source of power, meaning all the resources that could provide the power for human being for production and living. Therefore, energy could be classified based on different classifying standards, for more details, please refer to Table 28.1.

Through primary understanding of resource and energy above, we can consider resource and energy as the "Gift" of the Mother Nature to human being. In the advancement and development of human society as well as modernized production,

---

[1] Wang Jingping: *Discussion on the Meaning and Characteristics of Natural Resources*, Journal of Dezhou College, 2nd Issue, Volume 7, 2011.
[2] Li Jinchang: *Several Issues Regarding Natural Resources*, Journal of Natural Resources, 7th Issue in 1992.
[3] The UN explained the concept of natural resource in a document in 1970 in this way: any and all components that human finds in natural environment, as long as they could provide welfare to human being in any way, is a natural resource.

Table 28.1 Classification of energy from different perspectives

Classification basis	Type of energy	Specific types of energy
Renewable or not	Renewable energy	Solar power, wind power, biomass energy, ocean energy, water energy, hydrogen energy, fuel cell, etc.
	Non-renewable energy	Coal, oil, natural gas, etc.
Source of energy	From the earth itself	Nuclear power, geothermal energy
	From extraterritorial objects	cosmic radiation, solar power
	From interaction between the earth and other stars	Wind power, water energy, wave energy, ocean temperature difference energy, biomass energy, photosynthesis, fossil fuel, tidal energy that are generated from solar power
Classification on degree of utilization	Regular energy	Long utilizing period, relatively mature technology, mass production, wide use, such as nuclear fission
	New energy	Few utilization, under research, such as nuclear fusion
Based on obtaining method	Primary energy	The energy that exists in nature and could be directly utilized, such as coal, oil, natural gas, wind power, water power, etc.
	Secondary energy	The energy that is generated from direct or indirect processing or transformation, such as electricity, steam, coal, coal gas, hydrogen, etc.

Source The materials are concluded by the author of this book

there is not one thing that doesn't depend on resource and energy. Therefore, resource plays immeasurable role in the development of human society. But the resources and energy that the Earth grants to human being are limited, most resources and energy are non-renewable resources. In the continuous development process of human society, more and more resources and energy are developed and utilized. In the process of resources being utilized by human being, it must generate the using cost. The understanding of cost price in Economics can be divided into three parts: the value of means of production consumed in commodity production, the value of labor[4] (remuneration paid to labors) and the additional surplus value.[5] The former two are the basic components of cost. Observing the cost of resource and energy from the perspective of economics, we can understand the cost of resource and energy as all the prices paid by human being in order to meet the demands of human being in the process of developing and utilizing resources and energy. Its meaning has the following aspects: firstly, the capital cost and labor cost

---
[4]The former two are the basic composing factors of the cost.
[5]Liu Zhaohong: *Discussion Regarding Resource Cost Awareness Theory,* China's Economics of Land and Resources, 5th Issue in 2007.

28.1 Overview of the Cost of China's Resource and Energy   429

Table 28.2 Composition of the cost of resources and energy

	Name of cost	Specific reflection
Part I	Capital input and labor cost	Such as exploitation, processing and smelting of mineral resources, input of facilities for development and using and energy, compensation paid to labors in the process of development and utilization of resources and energy, etc.
Part II	Practice cost of scientific research	Such as the research and development of the processing and reutilization of sewage water, sea water desalination and new energy

input in order to guarantee the resources and energy needed for the exploitation for actual production and living; the second is the cost of scientific research and practice input by human being for enhancing the utilization rate of resources and energy in order to avoid expected pressure in the future, for more details, please refer to Table 28.2.

The quantity of the cost of resource and energy depends on the human being's demands for resources and energy in production and living activities. If the demands continuously grow, the limited resources and energy existing in real nature would be unable to meet human's demands, which would inevitably enhance the using cost of resources and energy. If with the advancement and development of human's science and technology, the development and utilization of new energy, the enhancement of the using efficiency of alternative and recyclable resources, enabling the demands of resources and energy to meet human's production and living activities, thus, the cost of resources and energy would have the room of further decrease.

In modernized production, the use of resources and energy is the main source of social advancement. Observing from the development experiences of some developed countries, these countries were unable to become world leading powers within very short period of time, the deciding factor was their large-scale development and utilization of resources and energy. Statistic experiences indicate that a country's total national production is positively correlated with its consumption of resources and energy. The higher the consumption of resources and energy, the higher the social development level, and the higher the people's living standard, the sum of population of developed countries such as the US, Russia, Japan, Germany, the UK, France, Italy, etc. only accounts for 20% of world's total population, but their energy consumption accounts for about 67% of the world's total energy consumption.[6]

China at current stage has relatively abundant total volume of energy resources. In 2006, the per capita occupying volume of coal resource was about 50% of world average, while the per capita occupying volume of oil and natural gas resource was about 1/15 of the world average.[7] What China's rapid economic development

---

[6]Gui Xiue: *Problems and Countermeasures Faced by China's Sustainable Use of Farmland Resource*, Theoretical Guide, 7th Issue in 2008.

[7]Li Yanming, Zhang Qing and Zhang Chuanping: *Analysis of the Value Loss of Chinese Oil and Gas Resources based on User Cost Method*, Financial and Economics, 6th Issue in 2010.

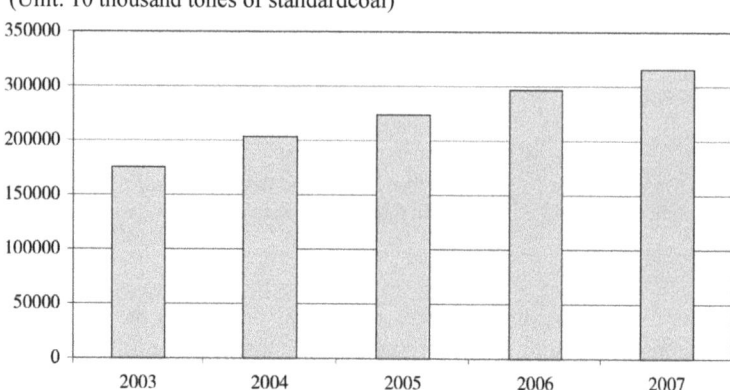

Chart 28.1 Growth of China's energy consumption (2003–2007). *Source* Concluded from sorting and drawing of the *China statistics yearbooks* during 2004–2008

depended on after the reform and opening-up was a gross-type economic growth method, high energy consumption, high investment and low added value are the development characteristics at this stage. Such economic growth method caused rapid enhancement of demand for energy.[8] Observing the total energy consumption during the 5 years from 2003 to 2007, the demand of Chinese social development for energy continuously increased, for more details,[9] please refer to Chart 28.1.

Currently, China has become the second biggest energy production and consumption country in the world. During the process of continuous social advancement and the continuously accelerated development of urbanization and industrialization, the conflict between people's demand for resources and energy and the sustainable development of economic growth and the maintain of a environment of harmonious coexistence between human and nature in the future becomes increasingly intensified. At each stage of production, transportation and consumption, energy would cause certain degree of pollution on the environment, while most energies are non-renewable energy, if this generation uses more, the next generation would have to use less, so in the process of continuous developing society, and in the process of developing and using limited resources and energy, the consideration of the cost of resources and energy is very necessary and significant.

---

[8] Lin Boqiang, Wei Weixian and Li Pidong: *China's Long-term Coal Demand: Influence and Policy Selection*, Economic Research, 2nd Issue in 2007.

[9] Energy Statistics Department of State Statistics Bureau: *China Energy Statistics Yearbook 2006*, China Statistics Press, 2006 Edition, p. 125.

28.2 Forms of the Cost of Resources and Energy

The cost of resources and energy could have many forms. It could be classified into R&D cost, exploitation cost, using cost, consumption cost, equipment inputting cost, labor cost, and late-stage maintenance cost based on the use of resource and energy as well as the production process. It could also be classified into resource cost[10] and energy cost[11] based on the different variety of resource and energy. This book will focus on researching the quantitative issue of the farmland cost, forestland cost and mineral cost in resource cost; as well as the quantitative issue of the cost of oil, natural gas and other energy. In China, it is reflected as some unrecoverable and non-renewable "Resource", including the decrease of farmland, decrease of forestland, consumption of mineral resource and energy resource (coal, oil, natural gas), etc.

28.2.1 Farmland Cost

The measurement method of farmland cost is obtained from the evaluation method of the depletion value[12] of farmland resource and risk cost. Considering the farmland redline of 0.8 mu per person set by the UN, when per capita farmland exceeds 0.8 mu, the cost of farmland is the depletion value of farmland; when per capita farmland resource is lower than 0.8 mu, it means some people will face lack of basic living security at any time, and the cost of farmland should also include the living cost of excessive population. Meaning:

$$\text{Cost of farmland} = (\text{Farmland area of current year } - \text{ Farmland area in 1978})$$
$$\times \text{ Annual yield price of farmland} + \text{Excessive population} \times \text{Living cost} \tag{28.1}$$

In which:[13]

---

[10]Resource cost could be further classified into land resource cost, climate resource cost, water resource cost, biological resource cost, mineral resource cost, oceanic resource cost, energy resource cost, tourism resource cost, etc.

[11]Energy cost could be further classified into the cost of coal, oil, natural gas, wind power, water energy, etc.

[12]Depletion volume of farmland adopts the number in 1978 as the standard, the farmland decreased compared with 1978 is the depletion cost of farmland, the increased farmland compared with that in 1978 is the negative increase of farmland cost.

[13]Living cost means the total amount of equivalent fortune that a natural person could obtain through labor in his/her entire life on average.

Excessive population = Total population in current year − Farmland area in current year ÷ 0.8 mu, excessive population is always a positive value, when it is negative, assuming it is zero.

Because the per capita farmland area of China during 1978–2009 exceeded 0.8 mu, if it was 1.53 mu/person in 1979 and 1.37 mu/person in 2009, so there was no excessive population. In 2009, the farmland area was 121.72 million hectares, increasing by 2233.1 ha from that in 1978. The farmland in 1995 was the smallest, only 9510.13 ha, the loss of farmland depletion reached over RMB 76 billion (refer to Table 28.1).

28.2.2 Cost of Forest Land

The measurement of forest land is relatively simpler compared with that of farmland, this is because forest land, compared with farmland, is not something people depend on for living, so the risk cost brought by them is not considered. For the measurement of forest land depletion, the value in 1978 was still adopted as standard, the relatively decreasing resource is positive cost, and the relatively increasing resource is negative cost. Then the cost calculation formula of forest is:

$$\text{Forest land cost} = (\text{Forest land area in 1978} - \text{Forest land area in current year}) \times \text{Annual yield price of forest land} \tag{28.2}$$

Through the formula above, substitute the statistic data[14] of forest land area of China from 1978 to 2009, as well as the statistic data of annual yield price[15] of forest land, the farmland area in 1978 was 115 million hectares, over the last 300 years, the forest land area has been increasing, the number in 2009 was 195.4522 million hectares. The highest value of national per capita forest land area during 1978–2009 was only 0.146 hectare, while the world average was 0.6 hectare, less than a quarter, although there is no direct cost, it also reflects the situation of insufficient ecological construction and loss of forest land resource.

---

[14]Calculating the cost of Chinese forest land from 1978 to 2009, for more details about the statistic data of various variables, please refer to the Appendix.

[15]Due to the incompletion of forest land price in historical years, the annual yield price of forest land is on the benchmark of 1991, the rest years could all be calculated in comparable price.

28.2 Forms of the Cost of Resources and Energy

28.2.3 Cost of Mineral Resource

Mineral resource is non-renewable, every part used is lot for ever, all the use quantity should be calculated into cost, while the general calculation of forest land and farmland resources is unrecoverable. This chapter mainly studies the quantity of iron ore cost, the calculation of cost is the output in current year multiplying the market value of related mineral in current year. The calculation formula of the cost of mineral resource is:

$$\text{Iron ore cost} = \text{Iron ore output} \times \text{Market price of iron ore} \quad (28.3)$$

Through the formula above, substitute the statistic data of iron ore output of China from 1980 to 2000, and the market price of iron ore, the iron core cost of China from 1980 to 2000 could be calculated (for more details, please refer to Chart 28.2), we can see from the chart that the iron ore cost of China since 1980 reflected rising trend, and reached the highest value of about RMB 2.25 billion in 2000. Prior to 1992, the iron ore cost of China has always been growing at a relatively even speed. From 1993 to 1998, the increase of iron ore cost in China was very massive, equivalent to twice of the growth rate of that from 1980 to 1992.

28.2.4 Cost of Coal

The coast of coal could be measured through the coal exploitation volume in historical years (excluding the coal imported from other regions), and the market value of coal. The calculating formula of the cost of coal is:

$$\begin{aligned}\text{Cost of coal} = &\ \text{Exploitation volume of coal in current year} \\ &\times \text{Market price of coal}\end{aligned} \quad (28.4)$$

Through the formula above, substitute the statistic data of coal exploitation volume of China from 1978 to 2009, and the statistic data[16] of market price[17] of coal, the cost of coal in China from 1978 to 2009 could be calculated (for more details, please refer to Chart 28.3), we can see from the chart that the cost of coal in China since 1978 reflected rising trend, as of 2009, it was already as high as RMB 70 billion, in which the growth of coal cost in China from 1997 to 2003 was slow.

---

[16] Calculating the cost of Chinese coal from 1978 to 2009, for more details about the statistic data of various variables, please refer to the Appendix.

[17] Due to the incompletion of coal price in historical years, the annual yield price of coal is on the benchmark of 1991, the rest years could all be calculated in comparable price.

(Unit: RMB 10 thousand)

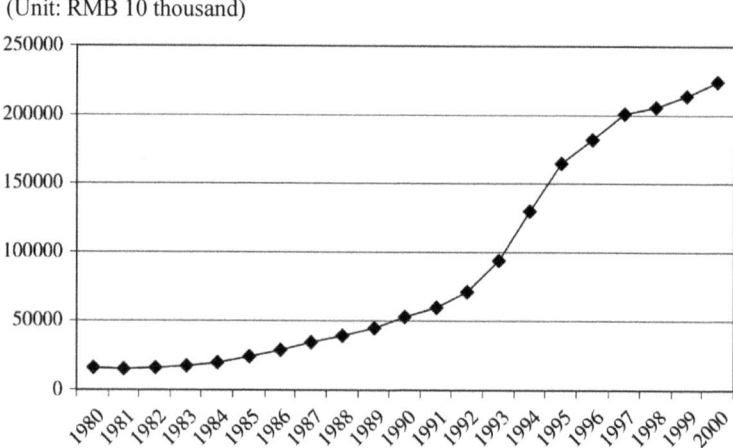

Chart 28.2 Iron ore cost of China from 1980 to 2000. *Source* Calculated based on the data in Table 28.3

(Unit: RMB 10 thousand)

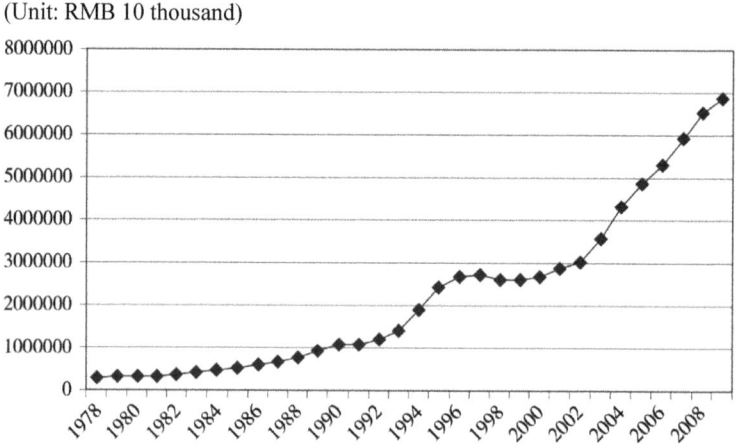

Chart 28.3 Coal cost of China from 1978 to 2009. *Source* Calculated based on the data in Table 28.3

28.2.5 Cost of Oil

The cost of oil could be measured based on the substitution of China's crude oil output in historical years (excluding the oil imported from external regions) and the market price of crude oil in historical years. The calculation formula of oil cost is:

28.2 Forms of the Cost of Resources and Energy

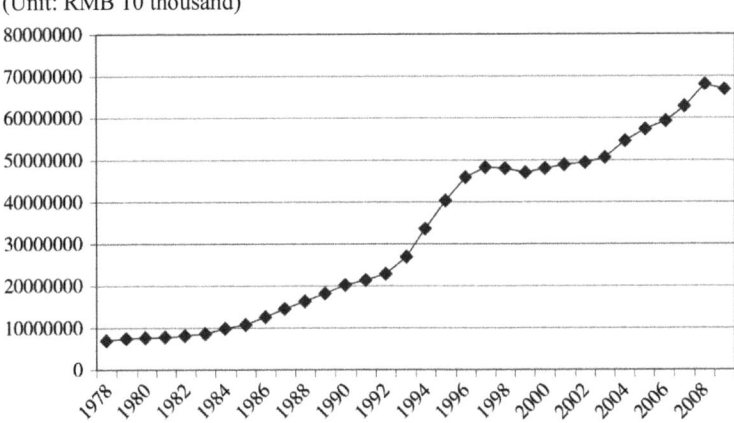

Chart 28.4 Oil cost of China from 1978 to 2009. *Source* Calculated based on the data in Table 28.5

$$\text{Cost of oil} = \text{Crude oil output in current year} \times \text{Market price of crude oil} \tag{28.5}$$

Through the formula above, substitute the statistic data of crude oil of China from 1978 to 2009, and the statistic data[18] of the market price[19] of crude oil, the oil cost of China from 1978 to 2009 could be calculated (for more details, please refer to Chart 28.4), we can see from the chart that the oil cost of China since 1978 reflected the increasing trend, as of 2008, it was RMB 690 billion, in which the oil cost of China from 1993 to 1997 entered fast increase period, and reflected flat increase from 1997 to 2002.

28.2.6 Cost of Natural Gas

The cost of natural gas is measured based on the exploitation volume of natural gas of China in historical years and the market price of natural gas. The specific calculation formula is:

---

[18] Calculating the cost of Chinese oil from 1978 to 2009, for more details about the statistic data of various variables, please refer to the Appendix.

[19] Due to the incompletion of oil price in historical years, the annual yield price of oil is on the benchmark of 1991, the rest years could all be calculated in comparable price.

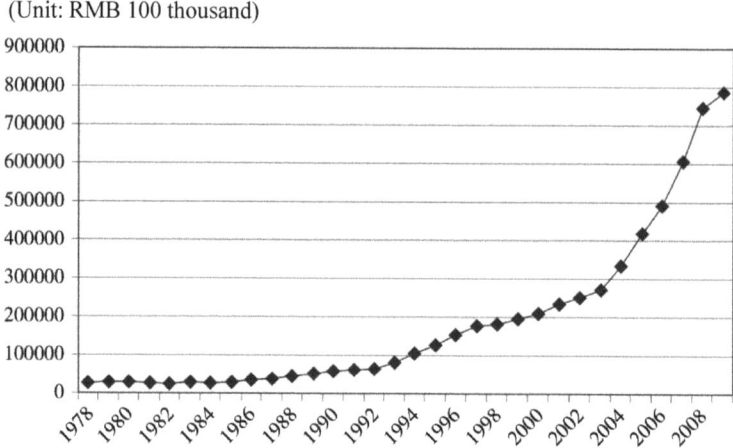

Chart 28.5 Cost of natural gas of China from 1978 to 2009. *Source* Calculated based on the data in Table 28.6

$$\text{Cost of natural gas} = \text{Exploitation volume of natural gas in current year} \\ \times \text{ Market price of natural gas}$$

(28.6)

Through the formula above, substitute the statistic data of natural gas output of China from 1978 to 2009 and the statistic data[20] of the market price[21] of natural gas, the cost of natural gas of China from 1978 to 2009 could be calculated (for more details, please refer to Chart 28.7), we can see from the chart that the cost of natural gas in China since 1978 reflected increasing trend, from 1978 to 2002, the cost of natural gas of China increased slowly, since 2003, the cost of natural gas in China had sharp increase, as of 2009, it reached the highest value of about RMB 80 billion (Chart 28.5).

28.3 Forming Causes of the Cost of Resource and Energy

In the production and living of human being, the development and utilization of different resources and energies will generate different costs, the common cause of the generation of the cost of resource and energy could be concluded as the following aspects.

---

[20] Calculating the cost of Chinese natural gas from 1978 to 2009, for more details about the statistic data of various variables, please refer to the Appendix.

[21] Due to the incompletion of natural gas price in historical years, the annual yield price of natural gas is on the benchmark of 1991, the rest years could all be calculated in comparable price.

28.3 Forming Causes of the Cost of Resource and Energy

Firstly, because most resources and energies are non-renewable, limited resource and energy have been consumed on large scale in the production and living of human being, which causes continuous decrease of the usable quantity of remaining resource and energy, which to certain extent would increase the using cost of the remaining usable resource and energy.

Secondly, under the influence of general economic environment, China's inflation rate has been continuously increasing (for more details, please refer to Table 28.4). We can see from Table 28.3 that the CPI changing rates of China during the 6 years from 2003 to 2008 were all positive, meaning that China's CPI during the six years was in increasing state, which also led to the rising prices in different fields, so the exploitation equipments of resource and energy as well as labor cost would also increase.

However, due to the difference and specialty of these resources and energies, the generating causes of these costs in the process of human being developing and utilizing these resources and energies are also different. As for the formation of the cost of resource, taking the forming cause of farmland cost as an example, the unclear ownership and property right of rural collective land,[22] the continuous increase of construction land and decreasing quality of farmland are the important influencing factors of the formation of farmland cost.

First of all, the unclear ownership and property right of rural collective lands will increase the difficulty in farmland protection in China. It is stipulated in Article 8 of the *Land Management Law of the People's Republic of China* that for the lands in rural area and urban outskirts, expect those belonging to the state according to laws and regulations, the rest shall be collectively owned by the farmers. The *Constitution* hasn't made specific interpretation of "Collective Ownership" so far. In the *General Provisions of the Civil Law*, "Collective Ownership" is defined as ownership by two levels, township and village; in the *Agriculture Law* and *Land Management Law*, "Collective Ownership" is defined as the collective economic organization of township, village or within village.[23] Such situation of unclear land ownership would cause instability of farmland use right, cause occupation of large-scale farmland resources, so as to result in unnecessary loss.

Secondly, the continuous growth of construction land is also the main factor of the formation and growth of farmland cost. With the continuous development and advancement of Chinese society as well as the continuous promotion of industrialization and urbanization process, the demands for industrial, residential and construction lands are ever increasing, so as to cause occupation and waste of large quantity of farmlands. According to statistics, during the "Eleventh Five Year Plan" period alone, there were 32.85 million mu of additional construction lands, including about 16.41 million mu of occupied farmlands. As of 2006, the total

---

[22] Rural collective land ownership system means the system arrangement that constitutes the collective land property structure and property relations, and the specific reflection of the collective land economic relations in rural areas.

[23] Pan Mingcai: *Issues to be Solved in Implementation of Strict Farmland Protection System*, Theoretical Vision, 4th Issue in 2006.

Table 28.3 Inflation changing rate of China from 2003 to 2008

年份	CPI 年变化率%
2008	5.86
2007	4.75
2006	1.46
2005	1.82
2004	3.88
2003	1.16

Source International Financial Statistics, IMF Members' Financial Data by Country, 14 May, 2010

年份	Year	CPI 年变化率%	Annual change rate of CPI %

construction lands in China were 485 million mu, the annual additional construction lands were 7.709 million mu. In 2006, the additional construction lands in China were 404.3 thousand hectares, increased by 15.3% compared with the same period in previous year.[24]

Besides, the increase of farmland cost caused from decrease of farmland quality also can't be ignored. According to statistics, 30% of Chinese farmlands faces the crisis of water and soil loss, the annual losses of farmland organic matter and the nutrient elements such as nitrogen, phosphorus and potassium were 2.7×10^7t, 5.5×10^6t, 6.0×10^3t and 5.0×10^6t.[25] The decrease of farmland quality would lead to decrease of farmland utilization rate, so as to cause the relative increase of farmland cost.

Similar to the forming causes of resource cost, the forming causes of energy cost are also closely correlated with its characteristic of limitation. When pricing energy, the current resource pricing theoretical framework didn't make the assumption of renewable natural resources.[26] While due to the special nature of energy, meaning

---

[24]Liu Zhengshan: *China's Grain Security and Farmland Protection*, Science of Finance and Economics, 7th Issue in 2006.

[25]Zhang Qizhen and Yang Ming: *Farmland Protection and Sustainable Development*, Journal of Zhoukou Normal College, 5th Issue in 2004.

[26]When pricing energies, these theoretical frameworks would only consider the related factors in the rocess of regular energy exploitation, such as the energy exploration cost, energy exploitation cost, energy exploiting equipment cost and labor cost, but didn't consider its non-renewable nature and the damage to the environment.

28.3 Forming Causes of the Cost of Resource and Energy

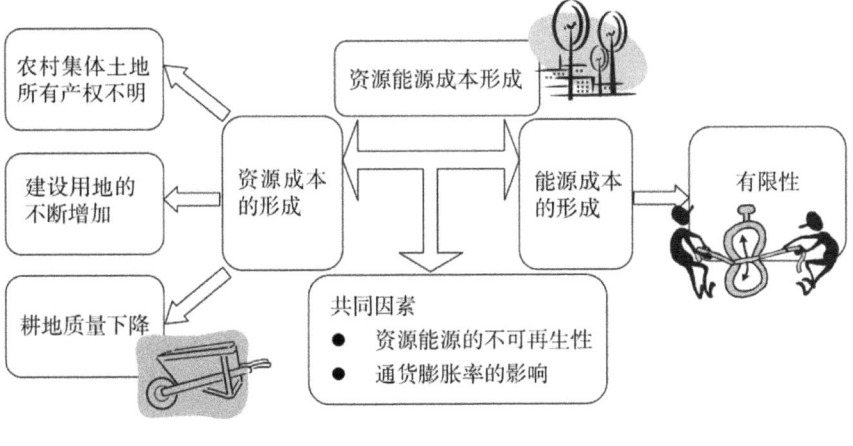

农村集体土地所有产权不明	Unclear ownership of rural collective lands
资源能源成本形成	Formation of the cost of resource and energy
建设用地的不断增加	Continuous increase of construction lands
资源成本的形成	Formation of resource cost
能源成本的形成	Formation of energy cost
有限性	Limitation
耕地质量下降	Decline of farmland quality
共同因素 ●资源能源的不可再生性 ●通货膨胀率的影响	Common factors ●Non-renewable nature of resources and energies ●Influence of inflation rate

Chart 28.6 Forming causes of the cost of resource and energy

its non-renewability, it is also one of the forming causes of energy cost. Under the circumstance of not finding new reserve or alternative energy, the current exploitation of energy includes the opportunity cost of the future, meaning the depletion cost of energy. With the increasing scarcity caused from continuous depletion of energy, the using cost of energy in the future will also continuously increase.

It can be seen from above that the forming costs of resource and energy are diversified, the formation of cost not only has the effect of common factor, but also have differentiated forming factors of resource and energy cost. In order to facilitate the readers to understand, the forming factors of the cost of resource and energy could be concluded as follows in Chart 28.6.

28.4 Countermeasures for the Cost of Resource and Energy

In order to correctly handle the relationship between the use of resource and energy and economic benefit, and between social benefit and environmental benefit, we should begin establishing a scientific ecological idea, fully consider the carrying capacity of resource, energy and the nature, strictly control the irrational utilization method of resource and energy, form the rational handling measures and solutions, pay attention to the utilization of resource and energy, and the common development of the comprehensive benefit in the three aspects, meaning the use of resource and energy, the society and the ecological environment. Targeting the formation of the cost of resource and energy, the specific countermeasures could be concluded as follows in Chart 28.7.

28.4.1 Countermeasures for the Control of Resource Cost

In the aspect of resource cost, targeting the farmland cost issue generated from unclear ownership and rural and collective lands, the state could actively promote the land acquisition system reform, strictly manage land resources, rationalize and standardize the use of lands as soon as possible, explore and establish the farmland acquisition hearing system, determine the use of farmlands, deny the phenomenon of farmland abuse, so as to further reduce the depletion cost of lands.

Targeting the farmland cost caused from continuously increasing construction lands, we could constantly improve the legal construction in the aspect of lands, and clarify the rural lands ownership system. Besides, China should strengthen the construction of related laws and regulations, strengthen the degree of supervision and inspection by the law enforcement departments. Once any farmland-related law violation is found out, the related authorities on different levels shall immediately transfer it to the related judicial department, strictly punish the criminal activities of farmland occupation, so as to reduce the cost of farmland occupation.[27]

Besides, proper reclamation of land resources and enhancement of farmland quality could also be deemed as measures to handle the farmland cost resulted from decreasing farmland quality. Proper development of land resources could increase effective farmland area, so as to reduce farmland cost. Only the enhancement of productivity per area of farmland could, under limited farmland resource conditions, enhance the using rate of farmland resources, so as to reduce the cost of farmland resource in other form.

---

[27]Zhao Qiguo and et al.: *Change of China's Farmland Resources and Its Sustainable Use and Protection Countermeasures,* Journal of Soil, 4th Issue in 2006.

28.4 Countermeasures for the Cost of Resource and Energy

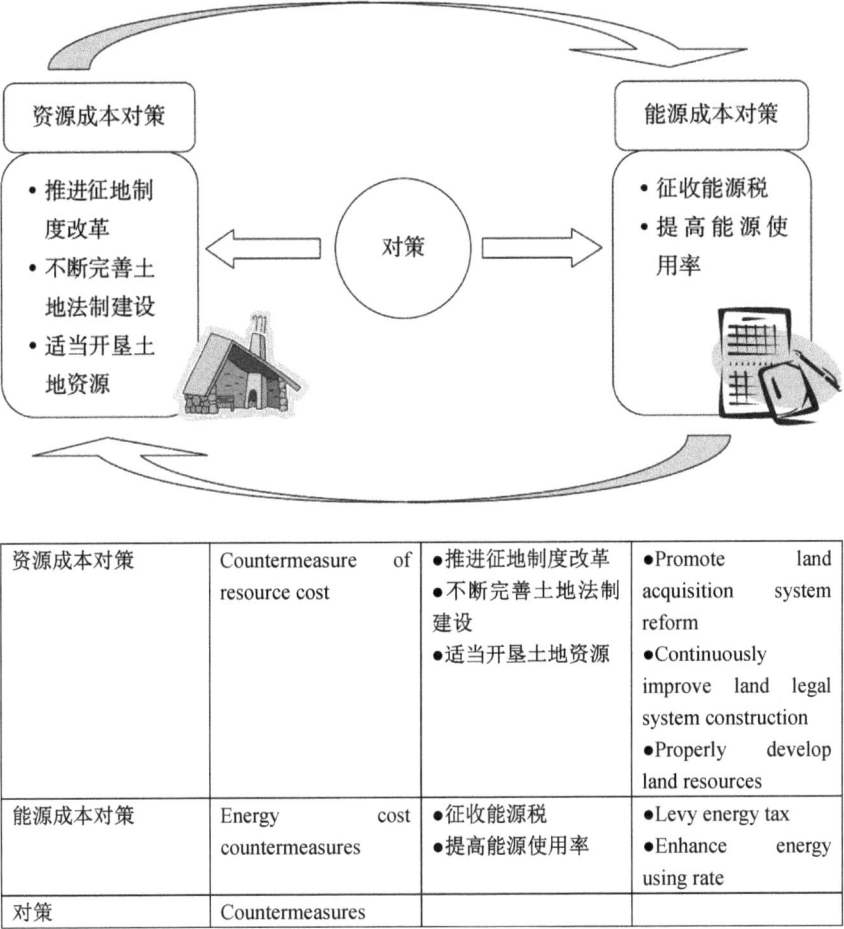

资源成本对策	Countermeasure of resource cost	●推进征地制度改革 ●不断完善土地法制建设 ●适当开垦土地资源	●Promote land acquisition system reform ●Continuously improve land legal system construction ●Properly develop land resources
能源成本对策	Energy cost countermeasures	●征收能源税 ●提高能源使用率	●Levy energy tax ●Enhance energy using rate
对策	Countermeasures		

Chart 28.7 Countermeasures for the cost of resource and energy

28.4.2 Countermeasures for the Control of Energy Cost

In the aspect of energy cost control, the cost generated from depletion of energy could be adjusted and controlled through collection of energy tax. From economic perspective, although the collection of energy tax could directly enhance the production cost of energy exploitation enterprises, from another perspective, the collection of energy tax could to certain extent increase the utilization rate of energy, so as to further restrain energy consumption, and would somewhat relieve the phenomenon of monopolized operation in the field of energy. Based on the survey and research of some scholars, the collection of energy tax could reduce the monopoly rents and surplus profit of state-owned oil enterprises, enabling the related enterprises to actively improve exploitation model and enhance the exploitation rate of the scarce oil and gas resources.[28] Through such method of stimulating and enhancing productivity, we could to certain extent control the consumption cost of energy. From some international experiences, we could find that although Japan and Europe have relatively heavy energy tax, while the US, Australia and other nations have relatively light energy tax,[29] the influence of heavy tax policy didn't reduce the competitiveness of Japan and other nations, on the contrary, it simulated the related enterprises' attention to the efficiency of energy utilization,[30] so as to enhance energy use rate and in other form reduce the depletion cost of energy.

Appendix

See Tables 28.4, 28.5, 28.6, 28.7, 28.8 and 28.9.

---

[28] Lin Boqiang and He Xiaoping: *Cost of Depletion of Chinese Oil and Gas Resources and the Macroeconomic Influence of Policy Selection,* Economic Research, 5th Issue in 2008.

[29] Energy tax rate of different countries: 120% for Japan, 73% for UK, 260% for Germany, 300% for France and 30% for the US.

[30] Meng Li and Shang Kewen: *Comparison Analysis of Oil Tax Systems in China and Foreign Countries and Suggestions for Policy Reform,* Taxation and Economics, 6th Issue in 2007.

Appendix

Table 28.4 China's farmland cost during 1978–2009

年份	耕地面积 （万公顷）	耕地价格 （元/万公顷）	当年人口 （人）	人均耕地面积 （亩/人）	耕地成本 （亿元）
1978	9938.9	446912421.8	962590000	1.54	—
1979	9949.8	467470393.2	975420000	1.53	—
1980	9930.5	489369101.9	987050000	1.51	—
1981	9903.5	507692511.2	1000720000	1.48	—
1982	9860.6	526462832.9	1016540000	1.46	—
1983	9835.9	545680067	1030080000	1.43	—
1984	9785.3	565791126	1043570000	1.41	—
1985	9684.6	585902185	1058510000	1.37	149.0
1986	9622.9	647576099.2	1075070000	1.34	204.6
1987	9588.8	715506787.3	1093000000	1.32	250.5
1988	9572.1	790588074.2	1110260000	1.29	289.9
1989	9565.6	873713784.6	1127040000	1.27	326.2
1990	9567.2	967118480.8	1143330000	1.26	359.5
1991	9565.3	1000190000	1158230000	1.24	373.7
1992	9542.5	1064098476	1171710000	1.22	421.8
1993	9510.1	1220517824	1185170000	1.20	523.4
1994	9510.1	1515033110	1198500000	1.19	649.6
1995	9510.1	1773795402	1211210000	1.18	760.6
1996	13003.9	1921276501	1223890000	1.59	—
1997	12990.3	1974905992	1236260000	1.58	—
1998	12964.2	1959264057	1247610000	1.56	—
1999	12920.5	1931555487	1257860000	1.54	—
2000	12824.3	1939599911	1267430000	1.52	—
2001	12761.5	1953007283	1276270000	1.50	—
2002	12592.9	1937365349	1284530000	1.47	—
2003	12339.2	1960604794	1292270000	1.43	—
2004	12244.4	2037026819	1299880000	1.41	—
2005	12206.6	2073673637	1307560000	1.40	—
2006	12177.5	2104957507	1314480000	1.39	—
2007	12173.5	2205959714	1321290000	1.38	—
2008	12171.6	2336011229	1328020000	1.37	—
2009	12172	2319475469	1334740000	1.37	—

Source the farmland data were sorted out and calculated based on the *Agricultural Data Net*, and the *China Statistic Yearbooks*. The farmland prices in 1991 were cited from Wang Xianfeng, etc.; *Evaluation of the Resource and Environmental Price of Economic Growth in Hunan Province, Resource and Environment in Yangtze River Basin*, 2nd Issue in 2007; the prices in other years were calculated based on the comparable prices in previous years.

年份	Year
耕地面积（万公顷）	Farmland area (10 thousand hectares)
耕地价格（元/万公顷）	Farmland price (RMB/10 thousand hectares)
当年人口（人）	Population in current year (person)
人均耕地面积（亩/人）	Per capita farmland area (Mu/person)
耕地成本（亿元）	Farmland cost (RMB 100 million)

Table 28.5 Area and price of China's forest lands during 1978–2009

年份	林地面积 (万公顷)	林地价格 (亿元/万公顷)
1978	11500	4946.8275
1979	11500	5174.3816
1980	11500	5416.7761
1981	11500	5619.5961
1982	12000	5827.3628
1983	12000	6040.0764
1984	12500	6262.6836
1985	12500	6485.2909
1986	12500	7167.9531
1987	12500	7919.8709
1988	12500	8750.9379
1989	13400	9671.0478
1990	13400	10704.935
1991	13400	11071
1992	13400	11778.396
1993	13400	13509.786
1994	15900	16769.745
1995	15900	19633.958
1996	12863	21266.412
1997	12863	21860.031
1998	13370	21686.892
1999	13370	21380.189
2000	15894	21469.231
2001	15894	21617.636
2002	15894	21444.497
2003	15894.09	21701.732
2004	17490.92	22547.64
2005	17490.92	22953.28
2006	17490.92	23299.558
2007	17490.92	24417.541
2008	17490.92	25857.067
2009	19545.22	25674.035

Source the data were sorted out and calculated based on the the *China Statistic Yearbooks, China Papermaking Yearbooks, etc.* The forest land prices in 1991 were cited from Wang Kaifeng, etc.; *Evaluation of the Resource and Environmental Price of Economic Growth in Hunan Province, Resource and Environment in Yangtze River Basin*, 2nd Issue in 2007; the prices in other years were calculated based on the comparable prices in previous years.

年份	Year
林地面积（万公顷）	Area of forest land (10 thousand hectares)
林地价格（亿元/万公顷）	Forest land price (RMB 100 million/10 thousand hectares)

Table 28.6 Cost of China's iron ores during 1980–2000

年份	铁矿产量 (万吨)	铁矿市场价格 (元/万吨)	1978—2009 价格指数	铁矿成本 (万元)
1980	3802	43252.01072	109.5	16444.41
1981	3417	44871.4924	113.6	15332.59
1982	3550	46530.47364	117.8	16518.32
1983	3738	48228.95442	122.1	18027.98
1984	4001	50006.43432	126.6	20007.57
1985	4679	51783.91421	131.1	24229.69
1986	5064	57234.85255	144.9	28983.73
1987	5503	63238.78463	160.1	34800.3
1988	5704	69874.70956	176.9	39856.53
1989	5820	77221.62645	195.5	44942.99
1990	6237	85477.03307	216.4	53312.03
1991	6765	88400	223.8	59802.6
1992	7589	94048.4361	238.1	71373.36
1993	8738	107873.2797	273.1	94259.67
1994	9741	133903.4853	339	130435.4
1995	10529	156773.7265	396.9	165067.1
1996	10721	169808.5791	429.9	182051.8
1997	11511	174548.5255	441.9	200922.8
1998	11852	173166.0411	438.4	205236.4
1999	12533	170717.0688	432.2	213959.7
2000	13101	171428.0608	434	224587.9

Source the iron ore production data were cited from the Yearbook Editing Committee of the Ministry of Foreign Trade and Economic Cooperation: *China Economic Trade Yearbook 2001*, China Economic Press 2001 Edition, Page 45, the iron ore market price of 1991 was cited from Wang Kaifeng, etc. *Evaluation of the Resource and Environmental Price of Economic Growth in Hunan Province*, Resource and Environment of Yangtze River Basin, 2nd Issue in 2007; the prices in other years were calculated based on the comparable prices in previous years.

年份	Year
铁矿产量（万吨）	Iron ore production (10 thousand tons)
铁矿市场价格（元/万吨）	Market price of iron ore (RMB/10 thousand tons)
1978-2009 价格指数	Price index during 1978-2009
铁矿成本（万元）	Cost of iron ore (RMB 10 thousand)

Appendix

Table 28.7 Cost of China's coal during 1978–2009

年份	煤炭市场价格 （万元/万吨）	煤炭开采量 （万吨）	1978—2009 价格指数	煤炭成本 （万元）
1978	4.463806971	61800	100	275863.2708
1979	4.669142091	63500	104.6	296490.5228
1980	4.887868633	62000	109.5	303047.8552
1981	5.070884718	62200	113.6	315409.0295
1982	5.258364611	66600	117.8	350207.0831
1983	5.450308311	71500	122.1	389697.0442
1984	5.651179625	78900	126.6	445878.0724
1985	5.852050938	87200	131.1	510298.8418
1986	6.4680563	89400	144.9	578244.2332
1987	7.14655496	92800	160.1	663200.3003
1988	7.896474531	98000	176.9	773854.504
1989	8.726742627	105400	195.5	919798.6729
1990	9.659678284	108000	216.4	1043245.255
1991	9.99	108000	223.8	1078920
1992	10.6283244	111600	238.1	1186121.003
1993	12.19065684	115100	273.1	1403144.602
1994	15.13230563	124200	339	1879432.359
1995	17.71684987	136100	396.9	2411263.267
1996	19.18990617	139700	429.9	2680829.891
1997	19.725563	138800	441.9	2737908.145
1998	19.56932976	133200	438.4	2606634.724
1999	19.29257373	136400	432.2	2631507.056
2000	19.37292225	138400	434	2681212.44
2001	19.50683646	147200	437	2871406.327
2002	19.35060322	155000	433.5	2999343.499
2003	19.58272118	183500	438.7	3593429.336
2004	20.34603217	212300	455.8	4319462.63
2005	20.71206434	235000	464	4867335.121
2006	21.02453083	252900	471	5317103.847
2007	22.03335121	269200	493.6	5931378.145
2008	23.33231903	280200	522.7	6537715.794
2009	23.16715818	297300	519	6887596.126

Source the data of coal production were sorted out and calculated based on the *China Statistic Year Books*. the coal market price of 1991 was cited from Wang Kaifeng, etc. *Evaluation of the Resource and Environmental Price of Economic Growth in Hunan Province*, Resource and Environment of Yangtze River Basin, 2nd Issue in 2007; the prices in other years were calculated based on the comparable prices in previous years.

年份	Year
煤炭市场价格（万元/万吨）	Market price of coal (RMB 10 thousand/10 thousand tons)
煤炭开采量（万吨）	Exploiting quantity of coal (10 thousand tons)
1978-2009 价格指数	Price index during 1978-2009
煤炭成本（万元）	Cost of coal (RMB 10 thousand)

Appendix

Table 28.8 Cost of China's oil during 1978–2009

年份	原油市场价 (万元/万吨)	1978—2009 价格指数	原油产量 (万吨)	原油成本 (万元)
1978	681.7533719	100	10405	7093643.834
1979	713.114027	104.6	10615	7569705.396
1980	746.5199422	109.5	10595	7909378.788
1981	774.4718304	113.6	10122	7839203.868
1982	803.1054721	117.8	10212	8201313.081
1983	832.4208671	122.1	10607	8829488.137
1984	863.0997688	126.6	11461	9891986.45
1985	893.7786705	131.1	12490	11163295.59
1986	987.8606358	144.9	13069	12910350.65
1987	1091.487148	160.1	13414	14641208.61
1988	1206.021715	176.9	13705	16528527.6
1989	1332.827842	195.5	13764	18345042.42
1990	1475.314297	216.4	13831	20405072.04
1991	1525.764046	223.8	14099	21511747.29
1992	1623.254778	238.1	14210	23066450.4
1993	1861.868459	273.1	14524	27041777.49
1994	2311.143931	339	14608	33761190.54
1995	2705.879133	396.9	15005	40601716.39
1996	2930.857746	429.9	15733	46111184.91
1997	3012.66815	441.9	16074	48425627.85
1998	2988.806782	438.4	16100	48119789.19
1999	2946.538073	432.2	16000	47144609.17
2000	2958.809634	434	16300	48228597.03
2001	2979.262235	437	16396	48847983.61
2002	2955.400867	433.5	16700	49355194.48
2003	2990.852042	438.7	16960	50724850.64
2004	3107.431869	455.8	17587	54650404.28
2005	3163.335645	464	18135	57367091.93
2006	3211.058382	471	18477	59330725.72
2007	3365.134644	493.6	18631	62695823.54
2008	3563.524875	522.7	19043	67860204.19
2009	3538.3	519	18949	67047246.7

Source the data of oil production were sorted out based on the *China Statistic Year Books*. the crude oil market price of 1991 was cited from Wang Kaifeng, etc. *Evaluation of the Resource and Environmental Price of Economic Growth in Hunan Province*, Resource and Environment of Yangtze River Basin, 2nd Issue in 2007; the prices in other years were calculated based on the comparable prices in previous years.

年份	Year
原油市场价（万元/万吨）	Market price of crude oil (RMB 10 thousand/10 thousand tons)
1978-2009 价格指数	Price index during 1978-2009
原油产量（万吨）	Production of crude oil (10 thousand tons)
原油成本（万元）	Cost of crude oil (RMB 10 thousand)

Table 28.9 Cost of China's natural gas during 1978–2009

年份	天然气市场价 (十万元/亿立方米)	1978—2009 价格指数	天然气产量 (亿立方米)	天然气成本 (十万元)
1978	178.2273603	100	137.3	24470.61657
1979	186.4258189	104.6	145.1	27050.38632
1980	195.1589595	109.5	142.7	27849.18353
1981	202.4662813	113.6	127.4	25794.20424
1982	209.9518304	117.8	119.3	25047.25337
1983	217.6156069	122.1	122.1	26570.86561
1984	225.6358382	126.6	124.3	28046.53468
1985	233.6560694	131.1	129.3	30211.72977
1986	258.2514451	144.9	137.6	35535.39884
1987	285.3420039	160.1	138.9	39634.00434
1988	315.2842004	176.9	142.6	44959.52697
1989	348.4344894	195.5	150.5	52439.39066
1990	385.6840077	216.4	153	59009.65318
1991	398.8728324	223.8	160.7	64098.86416
1992	424.3593449	238.1	157.9	67006.34056
1993	486.738921	273.1	167.7	81626.11705
1994	604.1907514	339	175.6	106095.896
1995	707.3843931	396.9	179.5	126975.4986
1996	766.199422	429.9	201.1	154082.7038
1997	787.5867052	441.9	227	178782.1821
1998	781.3487476	438.4	232.79	181890.175
1999	770.2986513	432.2	251.98	194099.8541
2000	773.5067437	434	272	210393.8343
2001	778.8535645	437	303.29	236218.4976
2002	772.6156069	433.5	326.6	252336.2572
2003	781.8834297	438.7	350.15	273776.4829
2004	812.3603083	455.8	414.6	336804.5838
2005	826.9749518	464	509.44	421294.1195
2006	839.4508671	471	585.53	491523.6662
2007	879.7302505	493.6	692.4	609125.2254
2008	931.5944123	522.7	802.99	748060.9972
2009	925	519	852.69	788738.25

Source the data of natural production were sorted out based on the *China Statistic Year Books*. the natural gas market price of 1991 was cited from Wang Kaifeng, etc. *Evaluation of the Resource and Environmental Price of Economic Growth in Hunan Province*, Resource and Environment of Yangtze River Basin, 2nd Issue in 2007; the prices in other years were calculated based on the comparable prices in previous years.

年份	Year
天然气市场价（十万元/亿立方米）	Market price of natural gas (RMB 100 thousand/100 million m^3)
1978-2009 价格指数	Price index during 1978-2009
天然气产量（亿立方米）	Production of natural gas (100 million m^3)
天然气成本（十万元）	Cost of natural gas (RMB 100 thousand)

Chapter 29
Cost of China's Ecological Environment

29.1 Overview of the Cost of the China's Ecological Environment

In recent years, with the rapid economic development, people have seriously neglected the environmental problems brought by the development of human society, and made excessive development and utilization of the ecological environment, which have caused resource depletion and continuous deterioration of ecological environment, as a result, the living environment of human being in the future has been severely threatened. According to statistics, in 1990s, 2/3 of China's economic growth was based on environmental overdraft. The economic loss caused from atmospheric pollution accounted for about 8% of the GDP.[1] China's waste discharge has exceeded the threshold of environmental capacity, so as to bring a lot of pressure on the ecological environment.[2] Besides, based on the National Class II Air Quality Standard, the environmental capacity of sulfur dioxide in China's urban air quality should be about 12 million tons, but if the situation goes on in this way, China's emission of sulfur dioxide would reach 29 million tons in 2020,[3] which is multiple times of the national standard, and it would extensively exceed the capacity of ecological environment. Therefore, Liu (2004) indicated that "the Chinese economy is a heavily loss economic system with the cost of ecological environment exceeding the GDP, the operation of such economic system is maintained depending on 'Environmental Overdraft' and 'Ecological Deficit', the excessive resource consumption, severe environmental pollution and radical

---

[1]Dai Li: *Beyond Growth—Economics of Sustainable Development,* Shanghai Translation Press, 2001 Edition, p. 18.
[2]The *2010 World Bank Development Report* also indicated that the content of organics in China's water environment at current stage is 70% more than the environmental capacity, in which the annual emission of sulfur dioxide exceeds the environmental capacity by over 60%.
[3]World Bank: *2010 World Development Report—Development and Climate Change,* Tsinghua University Press, 2010 Edition, p. 105.

ecological destruction have made the ecological capacity of China's ecological system reach the threshold, and the capacity of some resources and environment has reached the supporting limit".[4] Therefore, it is very necessary and important to integrate the consideration of natural environment cost into the production and living process of human being.

The so-called ecological environment cost could also be referred to as environmental cost, meaning the price generated from the influence of human being on the environment during living and production process. But from different perspectives, we could have different interpretation of environmental cost. In the System of Integrated Environmental and Economic Accounting issued by the United Nations Statistics Division and the US Environmental Management Committee in 1993, the cost of environment was defined as (for more details, please refer to Table 29.1).

The Japanese Ministry of Environment defined the cost of environment as "the costs and related expenses paid for the purpose of reducing the environmental load generated from the activities of enterprise and public institutions, including the investment amounts and current expenses of environmental preservations".[5] The State Statistics Bureau of Holland defined the cost of environment as "the cost incurred from the environmental protection activities that are implemented in order to prevent a facility from causing negative influence on surrounding environment".[6]

There are also many interpretations in China on the cost of environment, Guo (1992) established the cost of environment on the foundation of "Ecological Environment Cost", and defined the cost of environment; while Luo (1997) believed that "the cost of environment means the value of ecological factors to be consumed in the production and operational activities of enterprises and the various expenditures incurred for the recovery of ecological environment quality" (for more details, please refer to Table 29.2).

29.2 Forms of Environmental Cost

Based on the analysis in previous section, it is proposed in this book that the cost of ecological environment has the following four main forms, including the cost of water pollution, the cost of sulfur dioxide, smoke and dust emission, the cost of industrial dust emission, the cost of solid waste, etc., for more details, please refer to Table 29.3.

---

[4]Liu Sihua: *Several Issues Regarding the Outlook of Scientific Development,* Journal of Inner Mongolian College of Finance and Economics, 6th Issue in 2004.

[5]Liang Lihui: *Environment—Economic-oriented Analysis of Enterprise Environment Income and Cost,* Financial and Accounting Communication, 6th Issue in 2007.

[6]Wang Zheli: *Preliminary Discussion of Environmental Cost and Its Measurement,* Financial and Accounting Research, 24th Issue in 2010.

29.2 Forms of Environmental Cost

Table 29.1 Definition of the cost of environment by the United Nations statistics division and the US environmental management committee

United Nations statistics division	US environmental management committee
• Economic loss caused from consumption of natural resources and quality degradation	• Cost of environmental loss, meaning the cost or expenditure caused from environmental pollution itself
• Actual expenditure in the aspect of environment, meaning the various expenses incurred to prevent environmental pollution and various expenditures incurred in order to improvement environment, or recover the quantity or quality of natural resources	• Cost of environmental protection, meaning the expenses incurred to separate itself from pollution
	• Cost of environmental affairs, meaning various expenses incurred from the cell-phone environmental pollution intelligence, calculation of pollution degree and implementation of pollution prevention and control policy in order to manage the environment
	• Environmental pollution elimination expense, meaning the expenses incurred to eliminate current environmental pollution

Source Richard, Smith. "Accounting for the Environment: the Role of Strategic Management Accounting", *Management Accounting,* Vol. 4, No. 2. 1997

Table 29.2 Definitions of environmental cost by Chinese scholars

No.	Definition of environment cost by Professor Guo Daoyang[a]	Definition of environment cost by Professor Luo Guomin[b]
1	Input added to treat ecological environment due to environmental deterioration	Expenditure from maintaining environment, meaning the expenditure made before environmental pollution in order to maintain environmental quality and prevent hazard from happening
2	Loss caused from deterioration of ecological environment caused from major responsibility accident and the environment treatment expense and fine incurred hereof	Expenditure to prevent pollution
3	Fine caused from project investment without permission from environmental protection department	Expenditure to treat environment
4	Investment loss and waste from inefficient environmental treatment	Loss caused from artificial damage of ecological environment

Source Concluded by the author
[a]Guo Daoyang: *Preliminary Discussion on the Control of Green Cost,* Monthly Journal of Accounting, 5th Issue in 1997
[b]Luo Guomin and others: *Green Marketing,* Economic Science Press, 1997 Edition, p. 229

Table 29.3 Forms of the cost of ecological environment

Classification of cost	Forms	Specific contents
Cost of ecological environment	Cost of water pollution	Cost generated from treatment of waste water produced from people's production and living
	Cost of emission of sulfur dioxide and smoke	Cost generated from treatment of sulfur dioxide and smoke generated from people's production and living
	Cost of emission of industrial dust	Cost generated from treatment of industrial dust produced from people's production and living
	Cost of solid waste	Cost generated from solid waste generated from people's production and living

29.2.1 Cost of Water Pollution

Water pollution mainly involves discharge of industrial waste water, therefore, the cost of water pollution mainly involves the discharge cost of waste water, which is calculated through the following formula:

$$\text{Discharging cost of waste water} = \text{Discharging volume of waste water} \times \text{Treatment price of waste water} \quad (29.1)$$

With the rapid growth of economy as well as the development and advancement of human society, China's water pollution problem becomes increasingly prominent, partial water resource pollution is also intensifying. Therefore, the treatment of water pollution should draw more attentions from people. Enlarging the degree of water pollution treatment is one of the effective means to maintain the natural environment of human being and also the necessary trend in the aspect of China's natural environmental protection.

By substituting China's statistics[7] during 1980–2009 into the formula above, we could calculate the industrial water discharging cost of China from 1980 to 2009 (for more details, please refer to Chart 29.1). We can see from the chart that China's discharging cost of industrial waste water since 1980 reflected increasing trend, as of 2008, it already reached the highest value of RMB 1.9 billion.

---

[7]Calculating the discharge cost of industrial waste water of China from 1980 to 2009, for more details about statistic data of each variable, please refer to the Appendix.

29.2 Forms of Environmental Cost

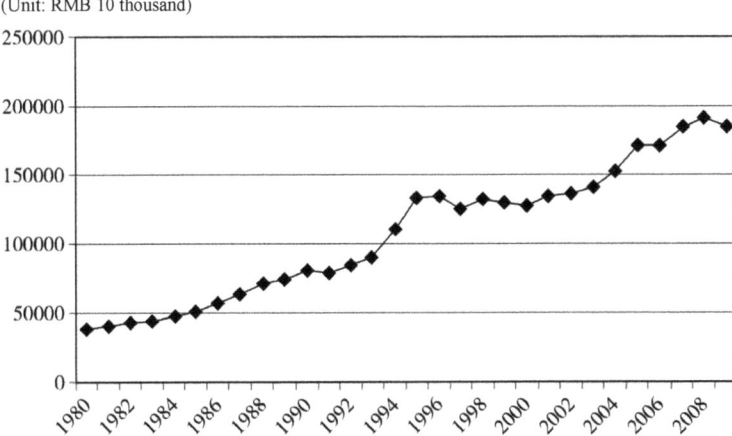

Chart 29.1 China's discharging cost of industrial waste water from 1980 to 2009. *Source* Calculated based on the data in Annexed Table 29.1

29.2.2 Emitting Cost Sulfur Dioxide and Dust

For calculation of the emitting cost of sulfur dioxide and dust, the following calculation formula is proposed in this book:

$$\text{Emitting cost of sulfur dioxide and dust} = \text{Total emitting volume} \times \text{Treatment price} \quad (29.2)$$

Because China's emitting volume of sulfur dioxide has far exceeded the self-purification capacity of the environment in recent years, nearly one third of national lands are subject to severe pollution of acid rain, therefore, different regions in China have enlarged investments in the emission of sulfur dioxide and dust. By substituting China's statistics[8] during 1992-2008 into the formula above, we could calculate the emitting cost of sulfur dioxide and dust of China from 1992 to 2008 (for more details, please refer to Chart 29.2), we can see from the chart that during the period from 1992 to 2005, China's emitting cost of sulfur dioxide and dust reflected the trend of escalated increase, as of 2005, it has reached the highest level of about RMB 140 billion. During the period from 2005 to 2008, China's emitting cost of sulfur dioxide and dust declined year by year.

---

[8] Calculating the emission cost of sulfur dioxide and smoke of China from 1992 to 2008, for more details about statistic data of each variable, please refer to the Appendix.

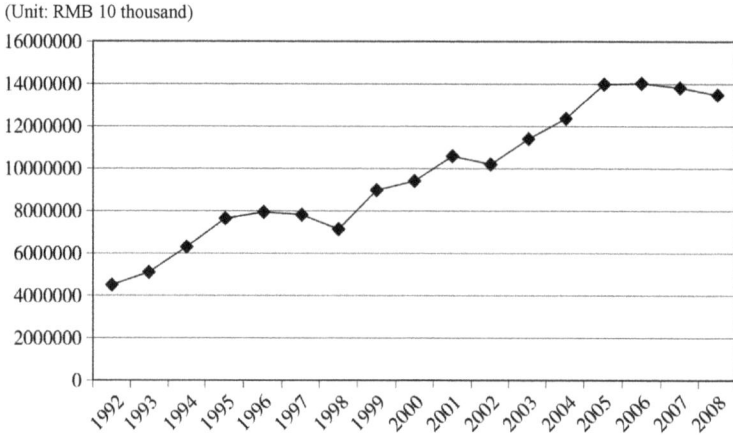

Chart 29.2 China's emitting cost of sulfur dioxide and dust from 1992 to 2008. *Source* Calculated from the data in Annexed Table 29.2

29.2.3 Discharging Cost of Industrial Dust

Discharging cost of industrial dust means the cost paid by people in the treatment process of particulate matters[9] discharged by enterprises in production process. Discharging cost of industrial dust could be calculated through the following formula:

$$\text{Discharging cost of industrial dust} = \text{Total discharging volume of industrial dust} \times \text{Treatment cost} \quad (29.3)$$

The large-scale discharge of industrial dust would not only cause pollution to the natural environment of human being, but also result in damage to people's physical health.[10] Therefore, the enlargement of the treatment of industrial dust discharged has active role in not only the natural environment that people live in but also people's health. By substituting China's statistic data[11] from 1992 to 2009 into the formula above, we can calculate China's discharging cost of industrial dust from

---

[9] Industrial particulate matter mainly means the particulate matter emitted during production process, such as the fireproofing material dust in steel enterprises, dust of coke screening system in coke making enterprise, dust of sintering machine, dust of lime kiln, cement dust of construction material enterprise, etc. excluding the smoke and dust emitted to the atmosphere by power plants.

[10] For the dusts entering respiratory tract and lung, although most of them could be discharged from body, but there will be few that stay in lung organization. If the total stock of dusts accumulated in lung exceeds 4 g, it could form different degrees of dust lung.

[11] Calculating the emission cost of industrial dust and smoke of China from 1992 to 2009, for more details about statistic data of each variable, please refer to the Appendix.

29.2 Forms of Environmental Cost

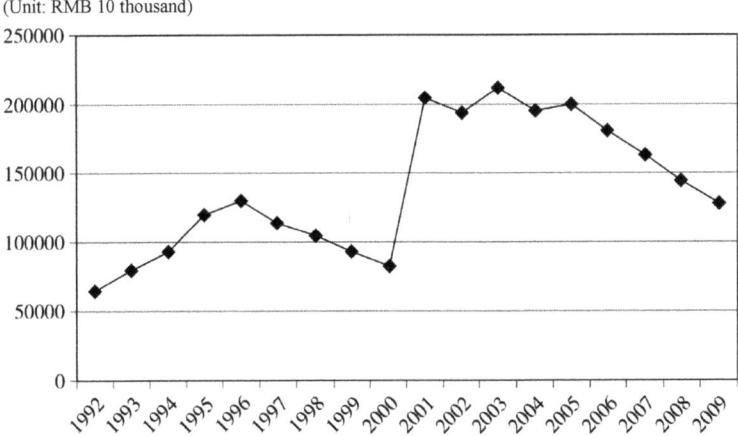

Chart 29.3 China's discharging cost of industrial dust from 1992 to 2009. *Source* Calculated based on the data in Annexed Table 29.3

1992 to 2009 (for more details, please refer to Chart 29.3), we can see through the chart that China's discharging cost of industrial dust formed a watershed in 2000. Before 2000, China's discharging cost of industrial dust was basically at a relatively low level, while during the year from 2000 to 2001, China's discharging cost of industrial dust soared, and maintained a relatively high level from 2000 to 2005. After 2005, China's discharging cost of industrial dust gradually declined. As of 2009, China's discharging cost of industrial dust has declined to about RMB 13 billion.

29.2.4 Cost of Solid Waste

With the economic development, increased population and radical acceleration of urbanization, the output of solid waste has been increasing and becoming increasingly complex in variety. Therefore, the expense of handling solid waste becomes a kind of direct cost, while the importance of measuring such variety of cost also becomes increasingly prominent. The cost of solid waste in the text is measured[12] through the variable of the market value of solid waste, discharging volume of solid waste and utilization volume of solid waste, the calculation formula is:

---

[12]Due to the lack of data, China's collection quantity of solid wastes from 1980 to 2009 is set to be zero, the discharging cost of solid wastes is the cost paid to handle the total discharging quantity of solid wastes in previous years.

(Unit: RMB 10 thousand)

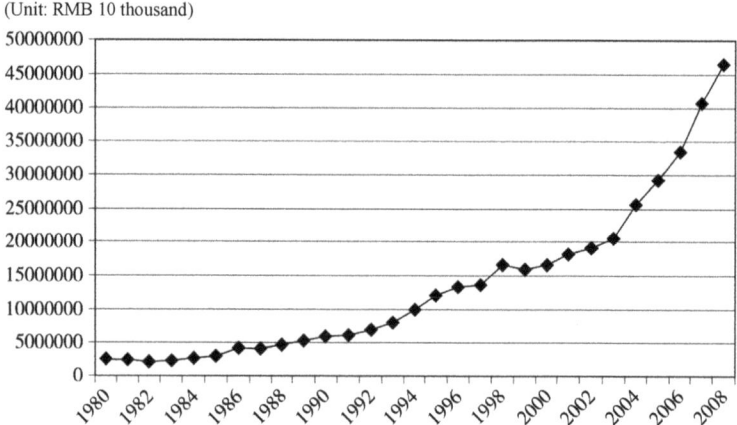

Chart 29.4 China's treatment cost of solid waste from 1980 to 2008. *Source* Calculated based on the data in Annexed Table 29.4

$$\text{Cost of solid waste} = \text{Discharging volume of solid waste} \\ - \text{Utilization volume of solid waste} \times \text{handling price} \tag{29.4}$$

By substituting China's statistic data[13] from 1980 to 2008 into the formula above, we can calculate China's treatment cost of solid waste from 1980 to 2008 (for more details, please refer to Chart 29.4), we can see from the chart that the treatment cost of China's solid waste before 1994 has reflected a even growth trend. The growth rate maintained at same level from 1980 to 1995. Since 2003, China's treatment cost of solid waste has been growing rapidly, which increased from RMB 256.681 billion in 2004 to the highest value of RMB 466.257 billion in 2008.

29.3 Forming Causes of the Cost of Ecological Environment

The forming causes of the cost of ecological environment are diversified, and could be specifically concluded as Chart 29.5.

Firstly, the historical issue left since the establishment of New China is one of the important factors of generating the cost of environmental cost. The repeated construction, irrational economic structure and industrial layout have caused increasingly sharp conflict between the economic activities in social development

---

[13]Calculating the handling cost of solid wastes of China from 1980 to 2008, for more details about statistic data of each variable, please refer to the Appendix.

29.3 Forming Causes of the Cost of Ecological Environment

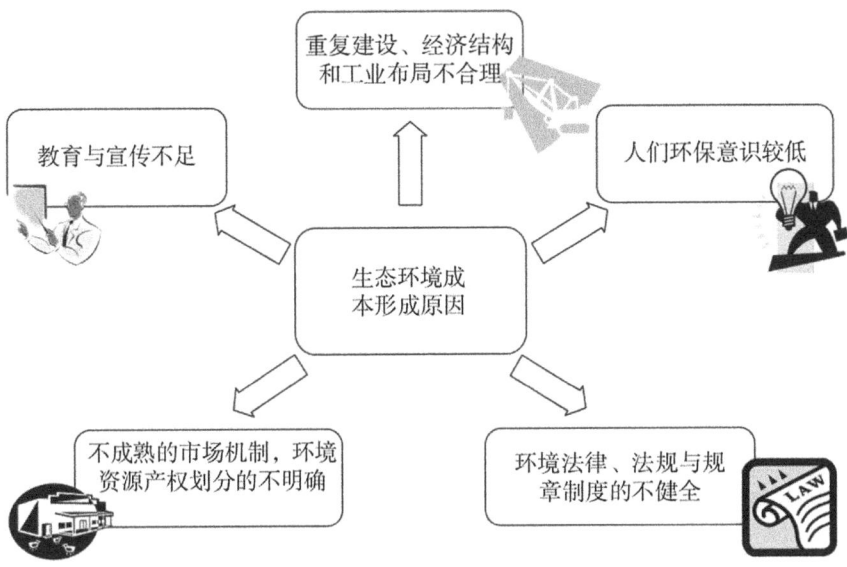

重复建设、经济结构和工业布局不合理	Repeated construction, irrational economic structure and industrial layout
教育与宣传不足	Insufficient education and publicity
人们环保意识较低	Relative low environment-protecting awareness among people
生态环境成本形成原因	Forming causes of the ecological environment cost
不成熟的市场机制，环境资源产权划分的不明确	Immature market mechanism, unclear classification of environmental resource property
环境法律、法规与规章制度的不健全	Incomplete laws, regulations and rules regarding environment

Chart 29.5 Forming causes of the cost of ecological environment

process and the environment. In the development process of some regions, due to poor economic foundation and low technical level, the competitiveness of output products is also relatively low, in order to pursue economic benefits, these regions blindly expanded production scale. This would cause large-scale wastes of minerals and other resources. Due to the low production technology level, large volume of dust and pollutant from energy consumption were emitted into the atmosphere without any treatment or only through simple processing, the cost of ecological environment generated from such operation model would be multiple times higher than the benefits obtained from production. Considering China's situation, because China didn't have general planning on cities in China's urban construction process during 1950s–1970s, 80% of industrial enterprises are basically concentrated in cities, and many enterprises were established in residential area and water source area.[14] Besides, in China's development of industrialization, the ratio of heavy-pollution industry is relatively high. Such situation would no doubt intensify the pollution degree of China's cities, which to certain extent increased China's pollution cost of ecological environment.

Secondly, China's recognition of resource cost began relatively late, so China's education and promotion on environmental resource protection is insufficient. Such drawback and flaw are mainly reflected as: (1) the degree of education of school students about environment and resource protection is low and the contents involved are few. (2) The degree of training on the labors participating in production and living activities about environmental protection awareness is low, and the depth is not sufficient. The government's inputs on environmental protection promoting work and the education on establishing people's awareness of environmental cost are widely insufficient, and the forms are too simple, as a result, people widely have weak awareness of environmental cost, which has greatly increased the possibility of enabling people to increase environmental cost in the production and living activities in the future.

Thirdly, the social development level and residents living standard is also one of the important influencing factors on the formation of the cost of ecological environment. The awareness of environmental cost is closely correlated with the social development level and residents living standard. When a country's social development level and residents living standard are at a relatively low level, people's surviving awareness will be in dominant position rather than paying too much attention on the cost issue brought by environmental pollution. After solving the problems of food and clothing, people would start the diversification of individual and household consumption, but people at this stage would still not have deeper understanding of environmental cost; only when the social level develops to a relatively high stage, people would gradually change consumption awareness and step into spiritual civilization. China is currently at the second stage mentioned above, the social development level and residents living standard are still have gap with that of developed nations, but China is also actively adopting the related preventive and treatment measures to reduce the excessive environmental cost generated in China's development process to minimum.

Fourthly, due to China's immature market mechanism, and the unclear classification of environmental resource ownership (such as exploitation, use, income and supervision), some enterprises had speculative conducts in production activities. Most enterprises are on the purpose of pursing maximized interests, they usually won't think too much about if they have caused environmental pollution and increased the cost of ecological environment, in addition, China's pollution

---

[14]Yang Mingying: *Economic Benefit and Resource Price,* Research of Quantitative Economy and Technical Economy, 7th Issue in 1995.

discharging cost is widely lower than treatment expense, and the ecological compensation expense is widely lower than recovery expense,[15] so most enterprises would choose less evil, and are more willing to pay pollution discharging expense or ecological compensation expense rather than paying treatment expense and recovery expense, this is also one of the important factors causing China's cost of environmental pollution.

Fifthly, the incompletion of laws, regulations and rules on environment is also a factor causing environmental cost. Although China's environment and resource laws have formed a relatively complete system, but laws usually have the characteristics of lagging and insufficient predictability, so it may not timely solve some new circumstances and new problems only depending on the original laws. For example, owners of large area of forest let plant disease spread and caused the degradation of the ecological system of forest, or rural contracting operators overuse forests but pay little attention to maintenance, etc.

29.4 Countermeasures for the Cost of Ecological Environment

Firstly, establishing scientific model of coordinated development between ecology and economy is one of the effective ways of controlling environmental cost. Due to the development of society and economy, the ecological environment of human being is increasingly deteriorated, so such irrational economic development model of high consumption, high investment and high pollution should be transformed. Especially after the proposal of China's Outlook on Scientific Development, the transformation from traditional irrational economic model to the "Green" economic model has become inevitable. American scholar Daley mentioned in *Beyond Growth—Economics of Sustainable Development* that economy is sub-system of environment. He believed that with the advancement of human society, the potential of economic development would gradually decrease, while ecological capital is replacing artificial capital to become scarcity cost of economic development.[16] Similarly, American scholar Brown also proposed the necessity of the transformation of economic development model, transforming from model A to model B, meaning the model with coordinated development between ecology and economy, only this could reach the sustainable development of economy and society.[17] Therefore, China should promote the cleaning of production, green

---

[15]Chang Jiwen: *Discussion on the Generating Causes of China's Existing Secondary Environmental Resource Problems,* Environment and Development, 2nd Issue, Volume 14, 1999.

[16]Dai Li, Zhu Dajian and others: *Overloaded Growth—Economics of Sustainable Development,* Shanghai Translation Press, 2011 Edition, p. 63.

[17]Lin Zixin and others: *Model B—Saving the Earth and Sustaining the Civilization,* Oriental Press, 2003 Edition, p. 39.

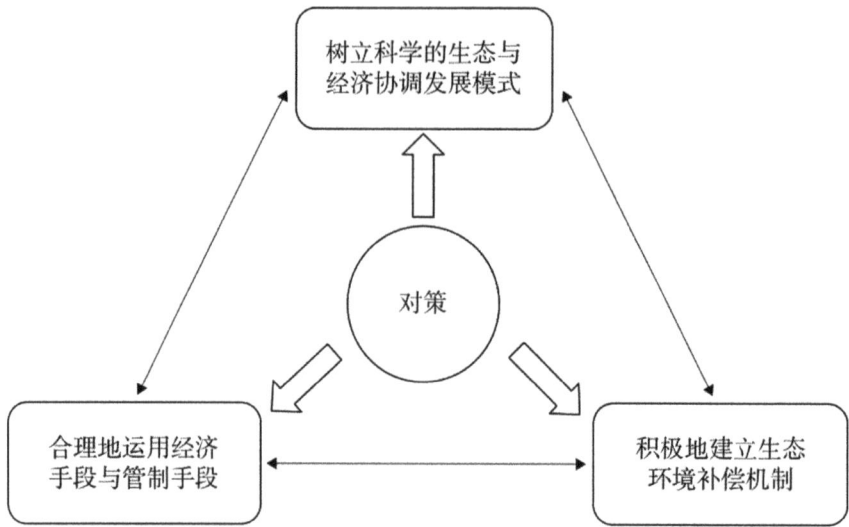

树立科学的生态与经济协调发展模式	Establish scientific model with coordinated development between ecology and economy
对策	Countermeasures
合理地运用经济手段与管制手段	Rationally use economic means and regulatory means
积极地建立生态环境补偿机制	Actively establish ecological compensation mechanism

Chart 29.6 Countermeasures for ecological environment treatment cost

consumption and development of recycling economy,[18] promote scientific environmental awareness in production and living process, so as to fundamentally reduce the possibility of the occurrence of environmental cost.

Secondly, reasonably adopt economic means and regulatory means to effectively reduce environmental cost. Proper increase of pollutant discharging fee is a very effective means to reduce environmental cost. China's pollutant discharging fee system has relatively early implementation and wide adoption, and has accumulated precious experiences in practice. Taking the emission fee of sulfur dioxide of power plants for an example, the pilot of sulfur dioxide emission fee began since 1992, expanded in 1995, as of 1997, the pollutant discharging fee in electricity industry reached RMB 315 million; with the emission fee standard of sulfur dioxide was

---

[18]Recycling economy is essentially a type of ecological economy, it requires using ecological rules rather than mechanical rules to instruct economic activities of human society, recycling economy reduces resource consumption and discharge of wastes through repeated utilization of materials and energy.

increased to RMB 0.6, the regulation became stricter, the emission volume of sulfur dioxide of different power plants obviously declined, the capacity of sulfur-removal power plants built in 2006 exceeded 100 million KW.[19] Besides, China should properly collect environmental tax to promote enterprises to increase the awareness of environmental cost. Due to the relatively heavy burden and low technical level of Chinese enterprise, the collection of environmental tax on enterprises has never been actually implemented (Chart 29.6).

Thirdly, actively establish the ecological environment compensation mechanism,[20] which is also one of the means to reduce the occurring possibility of environmental cost. From some international experiences, the establishment of ecological environment compensation mechanism is very necessary. For example, Japanese enterprises have paid heavy price for environmental pollution, so as to compensate the destruction of ecological environment.

Appendix

See Tables 29.4, 29.5, 29.6 and 29.7.

---

[19]Ma Jiantang: *Theoretical Analysis of the Internalization of the External Cost of Environment—Theoretical Framework of Pollution Emission Reduction and Policy Suggestions*, Qinghai Environment, 3rd Issue, Volume 17, 2007.

[20]Economic compensation for environment means a policy that enterprises compensate victims for their environmental pollution.

Table 29.4 China's data of waste water discharging cost during 1980–2009

年份	工业废水排放总量（万吨）	工业废水治理价格（元/万吨）	1980—2009年可比价格	工业废水治理成本（万元）
1980	2335512	166.3538874	109.5	38852.15003
1981	2379272	172.5826631	113.6	41062.1098
1982	2394396	178.9633601	117.8	42850.91537
1983	2387744	185.4959786	122.1	44291.69098
1984	2513637	192.3324397	126.6	48345.39367
1985	2574009	199.1689008	131.1	51266.25432
1986	2602380	220.1340483	144.9	57287.24445
1987	2637531	243.2260947	160.1	64151.63649
1988	2683886	268.7488829	176.9	72129.13644
1989	2520945	297.0062556	195.5	74873.6435
1990	2486861	328.7578195	216.4	81757.49997
1991	2358687	340	223.8	80195.358
1992	2338534	361.7247542	238.1	84590.56364
1993	2194919	414.8972297	273.1	91066.58124
1994	2155111	515.0134048	339	110991.1054
1995	2218943	602.9758713	396.9	133796.9089
1996	2058881	653.1099196	429.9	134467.5604
1997	1883296	671.3404826	441.9	126433.2845
1998	2004658	666.023235	438.4	133514.8806
1999	1973036	656.6041108	432.2	129550.3548
2000	1942405	659.3386953	434	128070.2778
2001	2026282	663.896336	437	134524.1196
2002	2071885	658.5790885	433.5	136450.0135
2003	2122527	666.4789991	438.7	141461.9671
2004	2211424	692.4575514	455.8	153131.7248
2005	2431121	704.9151028	464	171373.391
2006	2401946	715.5495979	471	171871.1494
2007	2466493	749.8838248	493.6	184958.3205
2008	2416511	794.0929401	522.7	191893.4325
2009	2343857	788.4718499	519	184806.5265

Source : the data of total discharging volume of industrial waste water were based on statistics of previous years; the market prices of waste water treatment market in 1991 were cited from Wang Kaifeng, etc.; *Evaluation of the Resource and Environmental Price of Economic Growth in Hunan Province*, Resource and Environment of Yangtze River Basin, 2nd Issue in 2007; the prices in other years were calculated based on the comparable prices in previous years.

Table 29.4 (continued)

年份	Year
工业废水排放总量（万吨）	Total discharged volume of industrial waste water (10 thousand tons)
工业废水治理价格（元/万吨）	Treatment price of industrial waste water (RMB/10 thousand tons)
1980-2009 年可比价格	Comparable price during 1980-2009
工业废水治理成本（万元）	Treatment cost of industrial waste water (RMB 10 thousand)

Table 29.5 China's data of sulfur dioxide and dust emission cost during 1992–2008

年份	二氧化硫与烟尘总排放量（万吨）	二氧化硫与烟尘治理价格（万元/万吨）	1992—2008年可比价格	二氧化硫与烟尘排放成本（万元）
1992	2192.4	2063	238.1	4522921.2
1993	2172.4	2366.254935	273.1	5140452.221
1994	2147.8	2937.240655	339	6308605.479
1995	2243	3438.910962	396.9	7713477.287
1996	2146.9	3724.837043	429.9	7996852.648
1997	2048	3828.810164	441.9	7841403.215
1998	1890	3798.48467	438.4	7179136.027
1999	2414	3744.765225	432.2	9039863.252
2000	2512.1	3760.361193	434	9446403.352
2001	2799.9	3786.354473	437	10601413.89
2002	2730.8	3756.028979	433.5	10256963.94
2003	3004.57	3801.083998	438.7	11420622.95
2004	3141.4	3949.245695	455.8	12406160.43
2005	3498.3	4020.293994	464	14064194.48
2006	3453	4080.944981	471	14091503.02
2007	3239.24	4276.761025	493.6	13853455.38
2008	2991.95	4528.895842	522.7	13550229.91

Source the data of total emission volume of sulfur dioxide and dust were sorted out based on the *China Statistics Yearbooks*; the market prices of treatment market in 1991 were cited from Wang Kaifeng, etc.; *Evaluation of the Resource and Environmental Price of Economic Growth in Hunan Province*, Resource and Environment of Yangtze River Basin, 2nd Issue in 2007; the prices in other years were calculated based on the comparable prices in previous years.

年份	Year
二氧化硫与烟尘总排放量（万吨）	Total emission volume of sulfur dioxide and smoke & dust (10 thousand tons)
二氧化硫与烟尘治理价格（万元/万吨）	Treatment price of sulfur dioxide and smoke & dust (RMB 10 thousand/10 thousand tons)
1992-2008年可比价格	Comparable price during 1992-2008
二氧化硫与烟尘排放成本（万元）	Emission cost of sulfur dioxide and smoke & dust (RMB 10 thousand)

Table 29.6 China's data of industrial dust emission cost during 1978–2009

年份	工业粉尘排放总量（万吨）	治理价格（万元/万吨）	1992—2008年可比价格	工业粉尘排放成本（万元）
1992	576	1130	238.1	650880
1993	617	1296.106678	273.1	799697.8202
1994	583	1608.861823	339	937966.4427
1995	639	1883.649727	396.9	1203652.176
1996	639	2040.264595	429.9	1303729.076
1997	548	2097.215456	441.9	1149274.07
1998	506	2080.604788	438.4	1052786.023
1999	458	2051.180176	432.2	939440.5208
2000	404	2059.722806	434	832128.0134
2001	990.6	2073.960521	437	2054465.292
2002	941	2057.349853	433.5	1935966.212
2003	1021.31	2082.028559	438.7	2126396.588
2004	904.8	2163.183536	455.8	1957248.464
2005	911.2	2202.099958	464	2006553.482
2006	808.4	2235.321294	471	1807033.734
2007	698.7435	2342.578748	493.6	1636861.674
2008	584.9478	2480.684586	522.7	1451070.991
2009	523.6472	2463.124738	519	1289808.372

Source the data of total emission volume of industrial dust were sorted out based on the *China Statistics Yearbooks*; the market prices of treatment market in 1991 were cited from Wang Kaifeng, etc.; *Evaluation of the Resource and Environmental Price of Economic Growth in Hunan Province*, Resource and Environment of Yangtze River Basin, 2nd Issue in 2007; the prices in other years were calculated based on the comparable prices in previous years.

年份	Year
工业粉尘排放总量（万吨）	Total emission volume of industrial dust (10 thousand tons)
治理价格（万元/万吨）	Treatment price (RMB 10 thousand/10 thousand tons)
1992-2008年可比价格	Comparable price during 1992-2008
工业粉尘排放成本（万元）	Emission cost of industrial dust (RMB 10 thousand)

Table 29.7 China's data of solid waste treatment during 1980–2008

年份	固体废物排放量（万吨）	固体废弃物治理价格（万元/万吨）	1980—2008年可比价格	固体废弃物治理成本
1980	48725	51.37399464	109.5	2503197.889
1981	43055	53.29758713	113.6	2294727.614
1982	40501	55.26809651	117.8	2238413.177
1983	41185	57.28552279	122.1	2359304.256
1984	45211	59.39678284	126.6	2685387.949
1985	48409	61.5080429	131.1	2977542.849
1986	60364	67.98257373	144.9	4103700.08
1987	53541	75.11394102	160.1	4021675.516
1988	56132	82.99597855	176.9	4658730.268
1989	57173	91.72252011	195.5	5244051.642
1990	57797	101.5281501	216.4	5868022.493
1991	58759	105	223.8	6169695
1992	61884	111.7091153	238.1	6913006.89
1993	61708	128.1300268	273.1	7906647.694
1994	61704	159.0482574	339	9813913.673
1995	64474	186.2131367	396.9	12005905.78
1996	65897	201.6957105	429.9	13291142.23
1997	65750	207.3257373	441.9	13631667.23
1998	80068	205.6836461	438.4	16468678.18
1999	78442	202.7747989	432.2	15906060.78
2000	81608	203.6193029	434	16616964.08
2001	88840	205.0268097	437	18214581.77
2002	94509	203.3847185	433.5	19221686.36
2003	100428	205.8243968	438.7	20670532.52
2004	120030	213.847185	455.8	25668077.61
2005	134449	217.69437	464	29268790.35
2006	151541	220.9785523	471	33487310.79
2007	175632	231.5817694	493.6	40673169.33
2008	190127	245.2345845	522.7	46625715.84

Source the data of solid waste treatment were sorted out based on the *China Statistics Yearbooks*; the market prices of treatment market in 1991 were cited from Wang Kaifeng, etc.; *Evaluation of the Resource and Environmental Price of Economic Growth in Hunan Province*, Resource and Environment of Yangtze River Basin, 2nd Issue in 2007; the prices in other years were calculated based on the comparable prices in previous years.

Table 29.7 (continued)

年份	Year
固体废物排放量（万吨）	Discharge volume of solid wastes (10 thousand tons)
固体废弃物治理价格（万元/万吨）	Treatment price of solid wastes (RMB 10 thousand/10 thousand tons)
1980-2008 年可比价格	Comparable price during 1980-2008
固体废弃物治理成本	Treatment cost of solid wastes

Chapter 30
Cost of Disasters in China

30.1 Overview of the Cost of Disasters in China

From the existing scientific researches, people's different understanding of natural disasters could be generally classified into two categories, one is from the perspective of scientific classification, and the other one is to understand natural disasters from the perspective of cross-disciplines. The former considers natural disaster as a kind of purely natural phenomenon; while the latter uses the method of system theory and understands disasters from the causes of disasters and their influence on the overall system as a sub-system of the overall society (for more details, please refer to Table 30.1).

When we study natural disaster, we would not only study natural disaster itself, but also study the important influence generated from the interest relationship between human and nature, between different people, and between exchange and distribution, etc. From the viewpoint of economics, natural disaster essentially could be interpreted as a kind of economic issue, because the consequence eventually caused by natural disaster is loss of the human's material fortune. Cost is defined by Economics as the entire expenses needed for producing a product, including the expenses invested for equipments, raw materials, labors, etc.[1] Therefore, this book defines the cost of natural disaster as the entire expenses paid to prevent disaster, and for reconstruction during and after disaster as well as all the losses incurred hereof.

China is one of the countries with the severest natural disasters in the world, the average annual economic loss caused from natural disasters is RMB 174.7 billion, accounting for about 3% of GDP, the ratios of disaster-striking agricultural area in total farming area and disaster-striking area exceed 20 and 50% respectively.

---

[1] Xu Juan: *Disaster Relief Input and Cost Issue in the Economics of Disasters,* Science of Disaster, 2nd Issue, Volume 21, 2006.

Table 30.1 Definition of natural disaster from different perspectives

No.	Perspective of pure natural phenomenon	Interdisciplinary perspective
1	Disaster is a kind of natural phenomenon, having close correlation with human being, and would usually bring harm to human survival or damage human's living environment[a]	Disasters occur in human's living environment, and often correlated with change of social environment, they would be intensified with the complication of social environment, and may also have new types of disasters[b]
2	Disasters are some physical geographic events that are far beyond human's expectation, no matter what kind of scale and frequency they have, they would bring obvious material destruction and life loss to human, so as to place human in painful state[c]	Disaster is normally interpreted as a kind of natural event that would cause life or property loss to people, and mostly is a emergency process, from a broad sense of it, any and all natural and social events that could cause or bring (even destructive) harm to the ecological system that human rely on for lives and breeds, construction and development of material and spiritual civilization can be referred to as disaster[d]
3	Disaster is a type or a series of phenomenon that causes personal injury or death, property loss and social instability	Disaster is the negative consequence brought to human survival and social development due to natural reasons, artificial factors or both. Disaster is not a purely natural phenomenon or social phenomenon, but a natural-social phenomenon, and a product of interactive effect between natural system and human material cultural system[e]
4	Disaster system is an earth surface changing system that is complex in nature and composed of hazardous environment, hazard-affected body, disaster-inducing factors and disaster situation, it is an important composing part of the earth surface system[f]	The phenomenon or process that would cause damage or loss to human life, property and the environment for human survival and development due to natural variation, artificial factor or combination of the two[g]

[a]Jinzi Shilang: *Major Disasters of the World,* Shandong Science and Technology Press, 1981 Edition, p. 2
[b]Ren Luchuan: *Several Philosophical Thoughts about the Nature of Disasters,* Oriental Forum, 4th Issue in 1995
[c]Zeng Weihua and Cheng Shengtong: *Guiding Theory of Environmental Disaster Science,* China Environment Science Press, 2000 Edition, p. 15
[d]Li Yongshan: *Discussion of Disaster System and Science of Disaster,* First Issue in 1986
[e]Luo Zude and Xu Changle: *Science of Disasters,* Zhejiang Education Press, 1998 Edition, p. 24
[f]Shi Peijun: *Disaster and Science of Disaster,* Geographic Knowledge, 1st Issue in 1991
[g]Ma Zongjin: *Guide of the Science of Disasters,* Hunan People's Publishing House, 1998 Edition

According to statistics, rural economic loss caused from natural disasters accounts for about 7% of total agricultural output. Compared with world average level, the occurring frequency of disasters in China is 17% higher than the world average level, China's nature protection cost is 27% higher than the world average level, and China's occurring frequency of disasters is 36% higher than the world average

30.1 Overview of the Cost of Disasters in China

Table 30.2 China's common natural disasters and statistics of disasters

Department	Type of disasters in statistics	Contents of statistics
Ministry of civil affairs	Drought, flood, hailstorm (including tornado, hurricane, dust storm, etc.) typhoon (including tropical storm), earthquake, low-temperature freezing, snow storm, landslide and debris flow, plague of disease and insects	Basic situation, disaster-affected population situation, disaster-affected agricultural crops situation, loss situation, disaster relief work
National disaster defense administration	Flood (including typhoon, tropical storm, storm surge), drought	Basic situation at occurrence of flood, disaster-affected population situation, loss situation; occurring time, place and degree of drought, influence situation, drought relief situation, drought relief effect, etc.
Ministry of land and resources	Geological disasters such as landslide, collapse, surface collapse, debris flow, etc.	Basic disaster situation, disaster-caused loss situation, prevention and control work
Ministry of agriculture	Drought, flood, hailstorm and low-temperature freezing	Disaster-affected agricultural crops situation, natural-disaster-affected loss situation
China seismological Bureau	Earthquake	Earthquake situation, disaster-affected population situation, disaster-affected building and other engineering structure, economic loss situation, ground and other destruction situation
China meteorological administration	Meteorological disaster	Disaster weather information
State oceanic administration	Oceanic disaster (storm surge, sea wave disaster, red tide disaster, oil spill disaster, coastal erosion)	Basic disaster information, disaster situation
State forestry Bureau	Forest fire	Affected area, economic loss from forest fire

Source Yuan Yi and Zhang Lei: *Current Situation and Prospect of the Statistics of China's Natural Disasters*, Vol. 21, 4th Issue in 2006

level.[2] So this increases the necessity of us studying natural disasters and their forming cost. Besides, natural disaster situation is an important part of China's situation, is a basis to be referred to in the formation of disaster resistance, rescue and reduction countermeasures, and also the most important basic materials in scientific disaster research. Due to the importance of natural disaster, some departments in China have also made classification and sorting-out of the types of natural disasters (for more details, please refer to Table 30.2)

---

[2] Xu Songling and others: *Concept of Disaster's Economic Loss and Measurement of Industry-related Indirect Economic Loss*, Journal of Natural Disasters, 7th Issue in 1998.

Table 30.3 Statistics by the Ministry of civil affairs about direct economic loss from natural disasters

Type of disaster	Direct economic loss
Drought	Agricultural loss: direct economic loss to planting industry, forestry, husbandry and fishery caused from disaster
Flood	
Hailstorm	Loss of industrial and mining enterprises: direct economic loss of mining, manufacturing, architectural and commercial enterprises caused from disasters
Typhoon	
Low-temperature freezing and snow storm	Infrastructure loss, direct economic loss of transport, electricity, water conservancy, telecommunication, municipal and other public facilities caused from disasters
High temperature and heat wave	
Earthquake	Public facility loss, direct economic loss of education, health, scientific research, cultural, sports, social security, social welfare and other public facilities caused from disaster
Landslide and debris flow	
Plant diseases and insect pests	Household property loss, direct economic loss of resident houses and their indoor auxiliary equipments, indoor properties, agricultural equipments, transport vehicles, live stocks caused from disaster
Others	

Source Ministry of Civil Affairs: *Statistic System of Natural Disasters,* MH (2008) No. 119

30.2 Forms of the Cost of Disasters in China

There are many forms of disaster cost, the Chinese Ministry of Civil Affairs has preliminary statistics about the direct economic loss caused from natural disasters (for more details, please refer to Table 30.3), there are more than ten types of common natural disasters, the direct economic loss caused and affected range are huge.

While this book mainly implements quantitative study on the cost of geological disasters, the cost of forest fire and the cost of oceanic disaster among natural disasters.

30.2.1 Cost of Geological Disaster

Geological disaster means the geological effect (phenomenon)[3] that is formed under the effect of nature or artificial factor and causes damage or loss on the lives, assets and environment of human being. China is a country with many geological disasters. According to statistics by the State Statistics Bureau, the economic losses caused from geological disasters in China are very severe, for more details, please refer to Chart 30.1.

---

[3]Such as collapse, landslide, debris flow, ground fracture, loss of water and soil, land desertification and swampiness, salinization of soil, as well as earthquake, volcano, terrestrial heat disaster, etc.

30.2 Forms of the Cost of Disasters in China

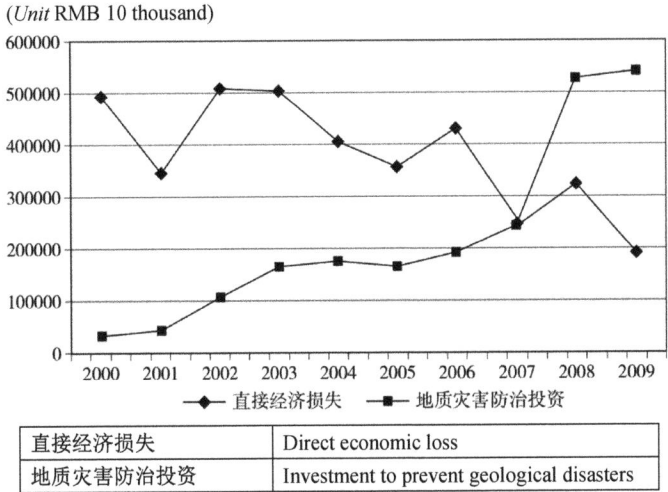

| 直接经济损失 | Direct economic loss |
| 地质灾害防治投资 | Investment to prevent geological disasters |

Chart 30.1 China's cost of geological disasters during 2000–2009. *Source* Sorted out and calculated based on *China Statistic Yearbooks* in previous years

We can see from Chart 30.1 that China's investments since 2000 in the aspect of prevention of geological disasters have always been increasing, it increased from RMB 331.97 million in 2000 to RMB 5.42368 billion in 2009, in which the increase from 2007 to 2008 was particularly prominent, it increased from RMB 2.44885 billion in 2007 to RMB 5.29939 billion in 2008, accounting for 56% of the increment from 2000 to 2009. While during the period from 2000 to 2009, the direct economic losses brought by China's geological disasters reflected the trend of fluctuated-style decline, which eventually reached the lowest point of RMB 1.90109 billion in 2009. From the relationship between the two variables, namely the direct economic losses and preventive investment of geological disasters during 2000–2009, it can be concluded that the investment in prevention of geological disasters and the direct economic losses brought by geological disasters reflect inverse relationship. The more the investment in prevention of geological disasters, the lower the direct economic losses brought by geological disasters

30.2.2 Cost of Forest Fire

In broad sense, forest fire means any and all forest fires that are out of artificial control, freely spreading and expanding in forest land, bring certain damage and loss to forest, forest ecological system and human being. In narrow sense, forest fire is a natural disaster that has high abruptness, big destruction and relatively difficult handling and evacuation, for detailed statistics, please refer to Chart 30.2.

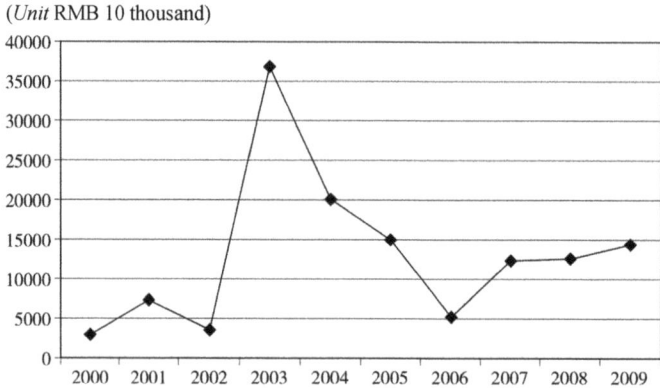

Chart 30.2 China's direct economic losses from forest fires during 2000–2009. *Source* Sorted out and calculated based on *China Statistic Yearbooks* in previous years

We can see from Chart 30.2 that the direct economic losses of China caused from forest fires since 2000 had relatively high fluctuation, relatively strong abruptness and very high difficulty in advance prevention. In 2003 alone, there were as high as 10,463 times of forest fires, the direct economic losses caused reached RMB 369.99 million, accounting for 28% of the total direct economic losses caused from forest fires during the 10 years from 2000 to 2009. Since 2003, the times of forest fires occurred in China and the direct economic losses caused from forest fires gradually declined, in 2006, it reached the lowest level of RMB 53.74 million.

30.2.3 Cost of Oceanic Disaster

Oceanic disaster means the vents that are due to abnormal or severe changes of oceanic natural environment, occurred on sea or at shore, and causing severe damage to economy as well as people's lives and properties. Main oceanic disaster includes disastrous sea wave, sea ice,[4] red tide,[5] tsunami and storm tide,[6] the

---

[4] Sea ice means all ices on sea, including salt water ice, river ice, iceberg, etc.

[5] The abnormal phenomenon of water color caused from explosive breeding of plankton in water is referred to as red tide, it is a eutrophication that mainly occurs in offshore sea. Under the influence of human activities, nitrogen, phosphorus and other nutritious substances needed by organisms entered seas in large scale, which causes rapid breeding of alga and other planktons, which would greatly consume the dissolved oxygen level in water, causing water quality deterioration, large-scale death of fish and other organisms, this is the root cause of red tide. Due to the increasing severity of environmental pollution of sea, the frequency of red tides also increases year by year.

[6] Storm tide is the phenomenon of abnormal sea surface rise or decline caused from the strong wind effect of typhoon, extra-tropical cyclone, cold front, and sudden change of air pressure and other strong weather systems.

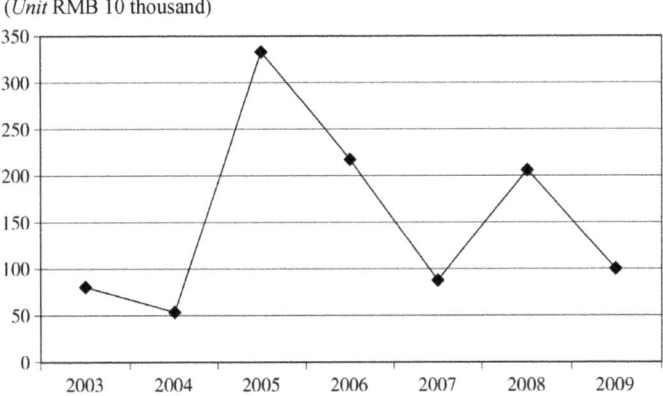

Chart 30.3 Direct economic losses from China's oceanic disasters during 2003–2009. *Source* Sorted out and calculated based on *China Statistic Yearbooks* in previous years

disastrous phenomenon relating to ocean also includes El Nino phenomenon,[7] La Nina phenomenon,[8] typhoon, etc. For the detailed economic losses caused by oceanic disasters occurred in China in previous years, please refer to Chart 30.3.

Similar to the direct economic losses caused from forest fire, the direct economic losses from oceanic disasters in China have high fluctuation range and unpredictability. We can see from Chart 30.3 that the direct economic losses of China generated from oceanic disasters reached the highest amount in 2005 at RMB 33.24 billion, it quickly declined to RMB 8.837 billion in 2007 with decrease of 73.4%.

30.3 Forming Causes of the Cost of Disaster in China

Due to the diversification of China's natural environment, geographic environment and climate condition, the factors causing natural disasters are also in great variety. Basically, the formation of disasters is closely connected in terms of space, time and generating mechanism, so occurrence of one disaster could cause the formation of other disasters, therefore, China widely classifies natural disasters into three types

---

[7]El Nino phenomenon is a term used by the fishermen in Peru, Ecuador to refer to abnormal climate phenomenon. It mainly means the abnormal continuous warming of the sea water temperature of the tropical ocean in east and central parts of Pacific, which make the climate model of entire world change and cause certain regions to have drought while other regions to have excessive precipitation.

[8]La Nina phenomenon means that within the scope of central and east parts of Pacific around the equator of over 10 thousand km from east to west and over 1 thousand km from south to north, the temperature of ocean is 0.2 °C lower than the normal sea surface temperature at east and central part of Pacific, and would continue for six months (opposite of El Nino phenomenon).

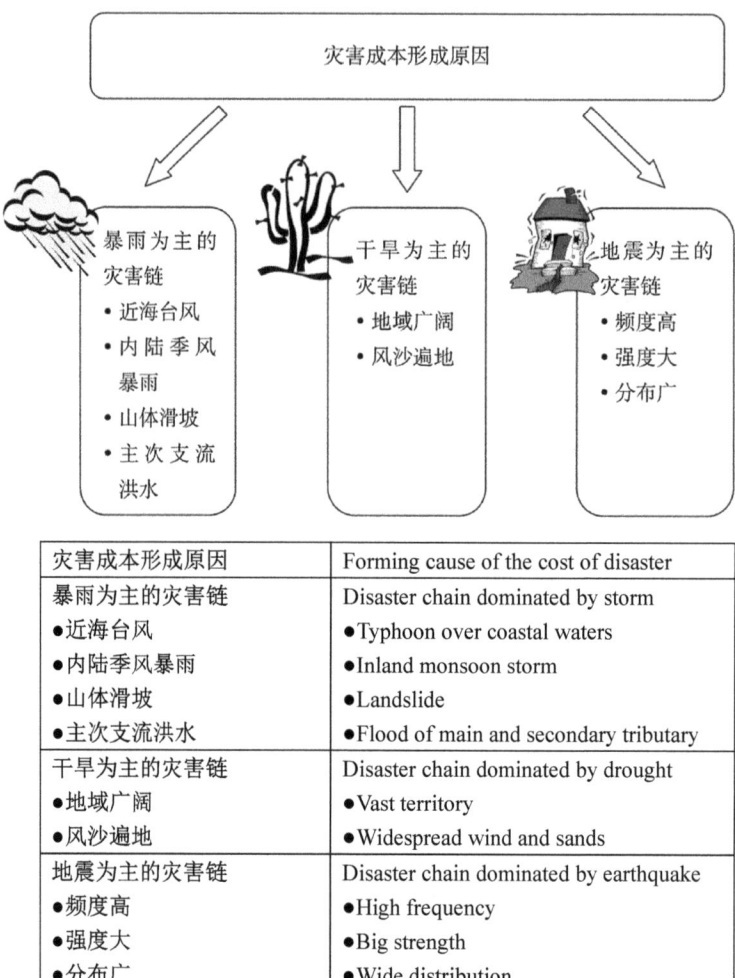

Chart 30.4 Forming causes of the cost of disaster

of natural disasters with three natural disaster chains as the main line, including the disaster chain of storm, disaster chain of drought and disaster chain of earthquake (for more details, please refer to Chart 30.4). The natural disasters dominated by these three disaster chains could be considered as the main influencing factors of the generation of the cost of China's natural disasters.

The first is the disaster chain of storm, from the average rainfall of Chinese inland regions during the past several decades, using the average annual precipitation of 500 mm as the boundary, China could be generally divided into two

30.3 Forming Causes of the Cost of Disaster in China

regions, the Southeast and the Northwest.[9] Disasters in this region are mainly storms, landslides brought by coastwise typhoon and inland monsoon and the flood in mid and downstream of primary and secondary tributary. The occurrences of these four disasters are widely connected. The three deciding factors of disaster in this region are population, economy and disaster degree. According to statistics, the disasters in this region are the most, strongest and heaviest in Mainland China, and have the biggest destruction on the industry, agriculture and commerce in this region.[10]

Secondly, the Northwest Region has vast territory and widespread wind and sand, its disasters are dominated by drought. This is also one of the primary factors affecting economic development in this region. Long-term drought, wind, sand, and glacier change would directly and indirectly affect the agriculture in the Northwest Region, even the development of population and mining industry. The loess plateau region in the Northwest has always been experienced the problem of rainfall gully and soil loss. Besides, China's Northwest Region is adjacent to the Central Asian Continent and Africa, which are the most prominent drought belts in the world. The special research team in China found out through research that the part of Northwest China adjacent to Central Asia and Africa is the most prominent drought belt in the world. Central Asia's high mountains have many layers, and their ice peaks have many edges, but the current global warming has started obvious melting, water loss of ice mountain, dry soil in farmland and forest land, water and soil loss, expansion brought by vulnerable ecology, it is also a disaster chain that is developing in long term, in China's Qinghai-Tibet Plateau and its extended regions, the livelihood issues of drought, wind, sand and long-term water shortage are also worth to be deeply studied.[11]

Finally, it's the third disaster chain of earthquake. China is a country with severe earthquake disasters, which have high frequency, big strength and wide distribution. China's earthquake activities are mainly distributed in 23 earthquake belts in five regions (for more details, please refer to Table 30.4). Because the human casualty incurred from earthquakes and economic losses caused from earthquakes are very severe, the severity of earthquakes and other natural disasters constitute one of the basic national situations of China.

---

[9]In which the precipitation of the Southeast region is relatively more, the precipitation of Northwest region is commonly lower than 500 mm. In which the Southeast region includes the coastal belt, South China, North China and part of Northeast Plain. This region has dense population, the number of economically developed medium and small cities accounts for 5/6 of the whole nation.

[10]Ma Zongjin: *Countermeasures for China's Natural Disasters and Disaster Relief—Enhance Disaster Efficiency, Using Scientific Ideas to Guide Various Aspects of Disaster Relief Work*, Journal of Disaster Prevention Science and Technology College, 1st Issue, Volume 10, 2008.

[11]National Major Natural Disaster Comprehensive Research Group of State Science and Technology Commission: *China's Major Natural Disasters and Disaster-relief Countermeasures*, Science Press, 1994 Edition, p. 56.

Table 30.4 Five main Earthquake regions of China

No.	Earthquake regions
1	Chinese Taiwan province and its nearby waters
2	Southwest region, mainly Tibet, West Sichuan, Central and West Yunnan
3	Northwest region, mainly at Gansu's Hexi Corridor, Qinghai, Ningxia, and north side of Tianshan Mountain
4	North China region, mainly at both sides of Taihang Mountain, Fengwei River Valley, Yingshan Mountain—Yanshan Mountain belt, Central Shandong and Bohai Bay
5	Southeast coastal regions, including Guangdong and Fujian. China's Taiwan Province is located at Pacific-ring earthquake belt, the provinces such as Tibet, Xinjiang, Yunnan, Sichuan and Qinghai are located on the Pacific-ring earthquake belt, the provinces such as Tibet, Xinjiang, Yunnan, Sichuan, Qinghai are located on the Himalaya—Mediterranean Earthquake Belt, the other provinces are located on the related earthquake belt

Source Li Jiafa, Wu Yizhu, An Zhenchang, Wang Yuehua: *Relationship between Regional Magnetic Anomaly and Earthquake Distribution in China*, South China Earthquake, 1st Issue in 1992

30.4 Countermeasures for the Control of the Cost of Disaster in China

The main countermeasure for the control of natural disasters is to fundamentally control the occurrence of disasters. China is a country with frequent occurrence of disasters, therefore, in order to effective control the cost of disasters, we should control the occurrence of disasters from many aspects, for the detailed countermeasures, please refer to Chart 30.5.

First of all, establishment of a comprehensive disaster prevention and relief system is one of the effective means to control the cost of disasters. Considering China's national situation, China should integrate the aspects of biological resource development, national land resource, weather, hydrology, earthquake, geology, agriculture, forestry, etc. to establish a testing, researching and loss reduction mechanism that could effectively handle natural disasters in China. The problem that has long been existing in China is that the departments of earthquake, weather, hydrology, geology, agriculture and forestry respectively do their own business, there will be certain difficulty in the implementation of a unified analysis, evaluation and temporary alarming. Therefore, China should integrate dispersed resources, establish a response mechanism targeting different disaster prevention and disaster reduction, realize unified predication before disaster, resource sharing during disaster and effective emergency after disaster, so as to control and reduce the losses of natural disaster to minimum, and reduce the cost of disaster.

Secondly, China should establish and improve the disaster risk dispersing mechanism, expand the channel of reconstruction fund. Similarly, using financial means could control losses after disaster to minimum and enable the infrastructure at disaster-striking areas to be recovered, and enable people's lives to relatively

30.4 Countermeasures for the Control of the Cost of Disaster in China

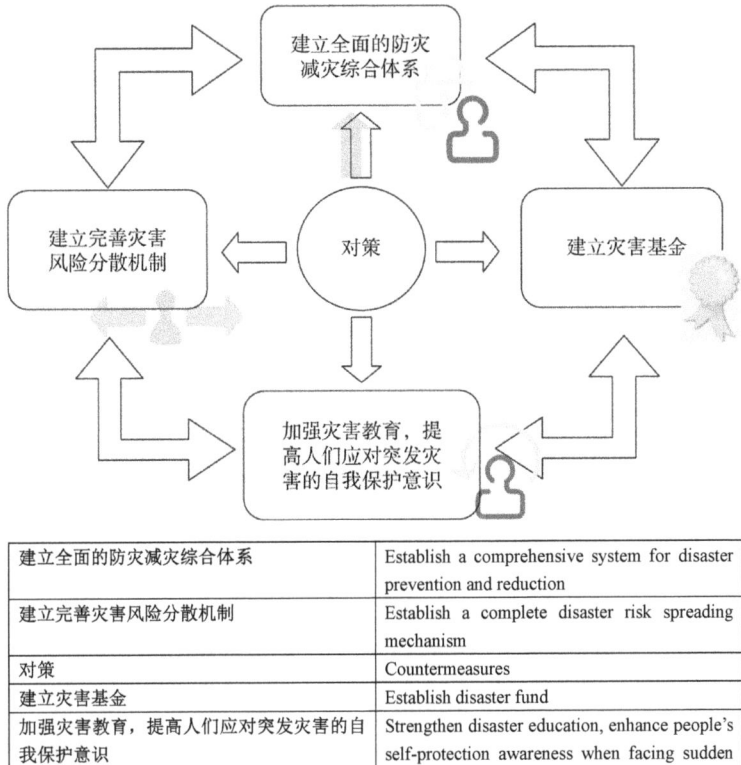

建立全面的防灾减灾综合体系	Establish a comprehensive system for disaster prevention and reduction
建立完善灾害风险分散机制	Establish a complete disaster risk spreading mechanism
对策	Countermeasures
建立灾害基金	Establish disaster fund
加强灾害教育，提高人们应对突发灾害的自我保护意识	Strengthen disaster education, enhance people's self-protection awareness when facing sudden disasters

Chart 30.5 Countermeasures for the control of the cost of disaster

quickly recover to stable state. Some western scholars believed that for the compensation and risk prevention of the losses of natural disasters, a comprehensive insurance system has more benefit to economic development than direct compensation by government.[12] Besides, Qiang and Wan (2007) believed that "In a poor and backward economy, individual disaster insurance is more difficult to form, so the government must play important role in establishing a sustainable social insurance system. Such insurance system could not only include temporary starvation relief when disasters approach, but also include emergent public employment and unemployment insurance, income subsidy and other social welfare programs. Capital market could use disaster bonds, futures and options to provide protection for disaster risks".[13] Compared with some developed countries in the world,

---

[12] Dacy, D.C., Xunreuther, H., *Economics of Natural Disasters: Implications for Federal Policy*, Free Press, New York, 1969.

[13] Qiao Haishu and Wan Ying: *Research and Review of Natural Disaster Economic Theories*, Dynamic of Economics, 9th Issue in 2007.

China's post-disaster economic compensation mechanism is still relatively backward, so some scholars suggested the establishment of national big disaster risk management and coordination institution and establish China Big Disasters Reinsurance Company, design and issue the Chinese big disaster insurance and risk transfer mechanism that is suitable for China's national situation, as well as establish China's big disaster risk reserve fund system,[14] so as to control China's cost of natural disasters within acceptance range.

Thirdly, establishment of disaster fund is also one of the effective means to control disaster cost. From the experiences of some developed countries, the US, Spain, France, etc. all have special insurance and disaster funds, which have achieved excellent performances in practice.

Special Column 30-1 Source of Reconstruction Funds after the Osaka Kobe Earthquake

After the Osaka Kobe Earthquake, according to the *Special Fiscal Aid Law after Severe Disasters,* the *Law on Treasury Bearing of Disaster Reconstruction Project Expense of Public Facilities*, the *Promotion Law of Earthquake Resistance and Repairs* as well as other laws and regulations, Japanese government actively bore government responsibilities and obligations:

(1) Enlarge government investments. As estimated, during the recovery process of Osaka Kobe Earthquake, different departments of Japanese government input JPY 10 trillion of reconstruction expenditure in total, the per capita reconstruction expense was about JPY 30 million, or RMB 1.9 million on average.
(2) Establish reconstruction fund for disaster recovery. The first was to establish the basic fund dominated by government input, the second was to establish the investment fund dominated with social capital, ratio of the two funds was 1:50, the former was mainly used in the construction of infrastructure and basic public facilities, while the latter was mainly used to invest in the industrial projects with relatively high yields.
(3) Aid and supporting policies of financial institutions.
(4) Private donation. Hyogo Prefecture, a severely affected area of Osaka Kobe Earthquake alone received JPY 179.1 billion of donations from all over Japan, as well as 420 thousand bags of aid materials.

Source Zheng: *Improve the Natural Disaster Economic Management Mechanism with Chinese Characteristics*, China Management Science, 10th Issue in 2009.

After referring to some experiences in developed countries, in view of China's national situation, the source of funds in China's post-disaster infrastructure reconstruction process is mainly dominated by government input and social donation. For the detailed obtaining channels, please refer to Table 30.5.

---

[14]Shi Peijun, Li Chang'an, Zou Minsheng and Le Jiachun: *Build Comprehensive Preventive and Aiding System to Handle the Risk of Major Disasters,* Financial and Accounting Research, 10th Issue in 2008.

Table 30.5 Obtaining channels of post-disaster funds in China

No.	Specific obtaining channels
1	Insurance compensation. Establish government-invested reinsurance fund, encourage financial institutions to develop major natural disaster legal person property loss insurance, household loss insurance, personal injury insurance products and the related financial derivative products
2	Fund sponsoring. Learn from the lesson from the late response and weak relief by the US government since the "Hurricane Katrina", take full advantage of the socialism that could concentrate power to do big things, establish the major natural disaster fund with government as the majority and participated by social capitals
3	Bank loans. The government provides interest subsidy and provide guarantee, provide low-interest or interest-free small-amount reconstruction loans to the small and medium-sized enterprises in disaster areas and the houses of disaster-affected people

Source Zheng Yaping: *Improve the Natural Disaster Economic Management Mechanism with Chinese Characteristics*, China Management Science, 10th Issue in 2009

Finally, strengthening disaster education to enhance people's self-protection awareness[15] in response to disasters could also to certain extent reduce the losses of disasters on people. If people have the awareness of disaster prevention and reduction, when large-scale or even global environmental disasters approach, it could very effectively enhance people's self-protection awareness, and reduce casualty. Besides, the occurrences of most environmental disasters are caused from the irrational production and living activities of human being themselves, therefore, people should fully recognize the harm of such activities, actively restrain the behaviors that could cause the occurrence of natural disasters, gradually recover and improve ecological environment. Meanwhile, targeting such circumstance, some scholars pointed out that the cause of such circumstance is due to people's grossness, blindness and exploitation of people in their transformation of nature, which stimulated and intensified the abnormal power in technical circle that destructs the ecological balance of natural circle on the surface of earth, and threatens the existence of human being, so a series of environmental disasters integrated between natural factors and artificial factors are increasingly severe.[16] So people are very necessary to rethink their own behaviors, establish correct environmental protection idea, and implement the resource-intensive economic-social development model, so as to fundamentally realize the harmonious coexistence between human and nature.

---

[15]People's disaster awareness means the alertness of disaster prevention and reduction before occurrence of disaster.
[16]Yang Jidong: *Characteristics, Cause Types of Environmental Disasters and Disaster-relief Countermeasures,* Shandong Environment, 3rd, 1995.

Appendix

See Tables 30.6, 30.7 and 30.8.

Table 30.6 China's economic losses caused from geological disasters during 2000–2009

年份	发生地质灾害起数（次）	人员伤亡（人）	直接经济损失	地质灾害防治投资（万元）
2000	19653	27697	494201	33197
2001	5793	1675	348699	44639
2002	40246	2759	509740	110022
2003	15489	1333	504325	166514
2004	13555	1407	408828	175231
2005	17751	1223	357678	166860
年份	发生地质灾害起数（次）	人员伤亡（人）	直接经济损失	地质灾害防治投资（万元）
2006	102804	1227	431590	193570
2007	25364	1123	247528	244885
2008	26580	1598	326936	529939
2009	10580	845	190109	542368

Source the data were sorted out from the *China Statistics Yearbooks from 2000 to 2009*.

年份	Year
发生地质灾害起数（次）	Occurring times of geological disasters (times)
人员伤亡（人）	Personal injury and death (person)
直接经济损失	Direct economic loss
地质灾害防治投资（万元）	Prevention and control investment of geological disasters (RMB 10 thousand)

Table 30.7 China's economic losses caused from forest fire disasters during 2000–2009

年份	发生森林火灾灾害起数（次）	人员伤亡（人）	直接经济损失（万元）
2000	5934	178	3069
2001	4933	58	7408
2002	7527	98	3609
2003	10463	142	36999
2004	13466	252	20213
2005	11542	152	15028
2006	8170	102	5374
2007	9260	94	12415
2008	14144	174	12593
2009	8859	110	14511.4

Source the data were sorted out from the *China Statistics Yearbooks from 2000 to 2009*.

年份	Year
发生森林火灾灾害起数（次数）	Occurring times of forest fire disasters (times)
人员伤亡（人）	Personal injury and death (person)
直接经济损失（万元）	Direct economic loss (RMB 10 thousand)

Table 30.8 China's economic losses caused from oceanic disasters during 2003–2009

年份	发生海洋灾害起数（次）	人员伤亡（人）	直接经济损失（万元）
2003	172	128	80.52
2004	155	140	54.22
2005	176	371	332.4
2006	180	492	218.45
2007	163	161	88.37
2008	128	152	206.05
2009	132	95	100.23

Source the data were sorted out from the *China Statistics Yearbooks from 2003 to 2009*.

年份	Year
发生海洋灾害起数（次数）	Occurring times of oceanic disasters (times)
人员伤亡（人）	Personal injury and death (person)
直接经济损失（万元）	Direct economic loss (RMB 10 thousand)

Chapter 31
China's Cost of Handling Climate Change

31.1 Overview of China's Cost of Handling Climate Change

Climate change has become one of the main environmental issues faced by human being that is widely recognized by the international community. Climate change means the climate change that is concluded after considerable period of observation, caused by the change of global atmosphere that is directly or indirectly resulted from human activities beyond natural climate change.[1] Due to human's production and living activities as well as the development of industrialization, large volume of green house gases have been emitted to the atmosphere, and caused the change of the chemical composition of atmosphere.

Among others, carbon dioxide (CO_2), methane (CH_4) and nitrous oxide (N_2O), etc. are the most common green house gases. Because these gases have strong ability to absorb infrared radiation from ground surface, plus the characteristic that they could stay long period in atmosphere, which would change the radiation balance of earth, resulting in green house effect and accelerate the rate of global climate warming. As estimated by Intergovernmental Panel on Climate Change (IPCC): in order to maintain green house gas at certain concentration, we need to reduce the carbon dioxide emission volume generated from human society activities by 2/3.[2] Since the industrial revolution, the burning of large volume of fossil fuel in the process of human activities has to certain extent increased the contents of carbon dioxide, methane and nitrous oxide in atmosphere.[3] According to statistics,

---

[1] Pan Jiahua: *Disputes of Economics Caused from Climate Change*, Ecological Civilization Theories, 2nd Issue in 2007.
[2] IPCC, *Climate Change 2001: The Scientific Basis*, Published by IPCC, 2001, p. 92.
[3] During the past 100 years, the ground air temperature across the globe generally tended to rise, the average surface temperature increased by 0.3–0.6 °C, in which the temperature increased by 1 °C in the continental regions in Northern Hemisphere.

China's emission of green house gas ranks second in the world, only next to the US, the green house gas emitted by fossil fuel accounts for 17.9% of total volume in the world. As predicted by International Energy Agency, the total volume of green house gas emitted by China will surpass that of the US, so China will become the biggest green house gas emitting country in the world. The per capita emitting level of China has also approached the world average.[4] Therefore, the issue of China's green house gas emission should draw attention from people.

The influence of climate change on human being is severe, usually such influence takes effect through water. Based on the current temperature increase due to emission of green house gas, the air temperature will increase by 2–3 °C within 50 years. If left uncontrolled, its influence on human being will be huge. For more details about the specific influence of green house effect on the lives of human being, please refer to Table 31.1.

Therefore, the influence of climate change on human being is huge. Climate, as a kind of environmental resource, is closely related to the social economic activities and production activities of human being. As part of the general system of human society, climate change is bound to generate great influence on the social and economic activities of human being, so consideration of the costs brought by climate change is very necessary.

People have different understandings towards the cost of climate change, with the continuous development of human society, the influence of human production and lives on the nature becomes bigger and bigger, the cost of climate change has become one of the most active and most complicated in changing of the cost generated by the economic activities of human being. One of the understandings of the cost of climate change is to interpret it as the influence of the production and living activities of human being on the climate, and the cost generated from the loss of caused disasters.[5] Another kind of climate change could be interpreted as the cost generated from the process of removing the green house gas generated from human production activities. Because the influencing factors of former is very complex, and difficult to have specific measurement from economic perspective, therefore, the measurement of the cost of climate change in this book would tend to the latter.

---

[4]Pan Jiahua: *Disputes of Economics Caused from Climate Change,* Ecological Civilization Theories, 2nd Issue in 2007.

[5]Taking the coastal region for an example, when considering climate cost, we also need to pay attention to population, human residence environment, and ecological support ability. Besides considering the economic loss from typhoon, flood, rising sea level brought by climate change, we also need to consider a series of indirect loss brought by climate change, such as personal injury and death, damage and destruction of public facilities, epidemic disease after disasters, psychological influence, social stability, rising inflation, etc.

Table 31.1 Specific influence of greenhouse effect on the lives of human being

No.	Specific influence of greenhouse effect on human lives
1	Melting iceberg at the beginning would increase the risk of flood, then it will severely reduce the supply of water, and would eventually threaten 1/6 of world population. These people mainly live in Indian Subcontinent, some regions in China and the Andes Mountains in South America
2	Grain production gradually decline, especially in Africa, which may cause hundreds of millions of people to lose their abilities to produce or purchase sufficient grains. In the high latitude regions of China, although grain product would increase due to slight temperature rise (2–3 °C), but the continuous rising temperature would cause decline of output. When the temperature reaches and surpass 4 °C, the global grain production may be subject to great influence
3	In the regions with even higher latitude, death related to coldness will decline. But climate change would cause the number of deaths due to malnutrition to rise. If effective control measures are not in place, the spreading scope of vector-borne diseases such as malaria and dengue[a] would get bigger
4	When the globe is 3–4 °C warmer, the population that is subject to flood in each year will increase by over ten million or even over 100 million due to rising sea level. Southeast Asia, Caribbean and small islands in the Pacific and some coastal cities will face severe risks
5	Ecological system will particularly be subject to influence of climate change. Around 15–40% of species may face the danger of extinction after 2 °C of temperature increase. Besides, it may cause ocean acidification[b], which will have severe influence on oceanic ecological system[c]

Source Chen Ying, Pan Jiahua: *Comment and Interpretation of the New Report by Stern*, Progress of Climate Change Research, 5th Issue in 2008

[a]Dengue is an acute infectious disease caused from dengue virus and spread through mosquitoes. Its clinical symptoms include acute symptoms, high fever, pain in muscles, marrow and joints, extremely fatigue, some patients may have rash, bleeding tendency and lymphadenectasis

[b]Ocean acidification means that due to the ocean absorbing and release excessive carbon dioxide (CO_2) in the atmosphere, the sea water become more acid. Since the industrial revolution, the pH value has decreased by 0.1. The increasing acid of sea water will change the various chemical balance in sea water, and make various ocean organisms and even ecological system that depend on stable chemical environment to face great threat

[c]Oceanic ecological system is the natural system constituted by the biocenosis and its environment after interaction. In broad sense, the global ocean is a large ecological system, including many different grades of secondary ecological systems. Each secondary ecological system would occupy certain space, forming a unity with certain structure and function by the interactive biology and non-organisms through energy flow and material flow

31.2 Forms of China's Cost in Handling Climate Change

It is believed in this book that the forms of the cost of climate change mainly include: the cost of carbon dioxide emission, the cost of methane emission, the cost of nitrous oxide emission, etc. (for detailed contents, please refer to Table 31.2).

Table 31.2 Forms of the cost of climate change

Classification of cost	Forms	Specific contents
Cost of climate change	Emission cost of carbon dioxide (CO$_2$)	Cost from treating the carbon dioxide emitted from people's production and lives
	Emission cost of methane (CH$_4$)	Cost from treating the methane emitted from people's production and lives
	Emission cost of nitrous oxide (N$_2$O)	Cost from treating the nitrous oxide emitted from people's production and lives

31.2.1 Cost of Carbon Dioxide Emission

In previous section, we mentioned that this book will mainly study the cost of handling green house gas such as CO_2, CH_4 and N_2O. The emission cost is mainly generated from burning of energy, the specific calculation formula is as follows:

$$\text{Emission Cost of Carbon Dioxide} = \text{Emission volume of Carbon Dioxide} \times \text{Treatment price of Carbon Dioxide} \tag{31.1}$$

Through the formula above, substitute China's statistics data[6] from 1978 to 2009, we can calculate China's cost from handling of Carbon Dioxide from 1978 to 2008 (for more details, please refer to Chart 31.1), through the chart we can see that China's cost from handling Carbon Dioxide emission from 1978 to 1997 has reflected increasing trend in each year, during the six years from 1998 to 2003, it entered the "Flat Stage", which generally maintained at RMB 265 million per year, then we can see from the Chart that China's cost generated from handling Carbon Dioxide emission quickly increased during the period from 2004 to 2008, as of 2008, it reached the peak at RMB 65 million per year. It can be seen from this trend that at this stage, China's attentions paid on the issue of handling green house gas emission quickly increased, and China actively adopted handling measures, the degree of treating Carbon Dioxide emission has been continuously strengthening.

31.2.2 Cost of Methane Emission

Similarly, the cost of methane emission mainly focuses on the research of the cost from treating CH$_4$ that is generated from the production activities of human being, the specific calculation formula is as follows:

---

[6]Calculating China's cost of handling carbon dioxide emission during 1978–2009, for more details relating to each variable in the formula, please refer to the Appendix.

31.2 Forms of China's Cost in Handling Climate Change

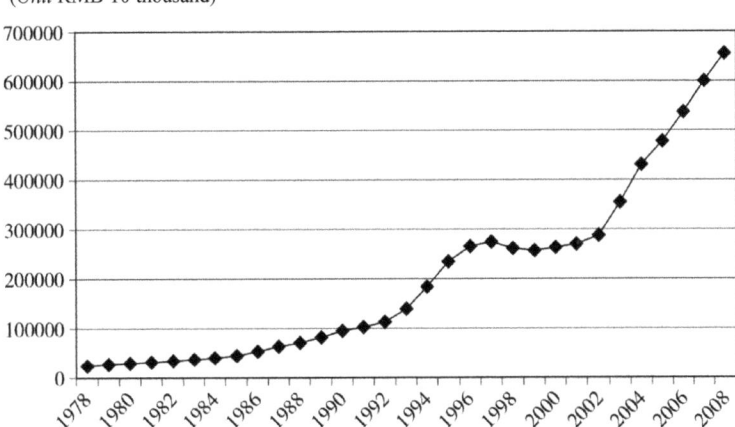

Chart 31.1 China's cost from handling carbon dioxide emission during 1978–2008. *Source* Calculated based on the date in Table 31.1

$$\text{Emission cost of methane} = \text{Emission volume of methane} \\ \times \text{Treatment price of methane} \quad (31.2)$$

Through the formula above, substitute China's statistic data[7] from 2000 to 2009, China's cost generated from handling of methane emission from 2000 to 2009 can be calculated (for more details, please refer to Chart 31.2), we can see from the chart that during the period from 2000 to 2009, China's cost input for handling methane emission had always been relatively high level, and it reached the peak value in 2009 at RMB 920 billion.

31.2.3 Emission Cost of Nitrous Oxide

Similarly, the cost of nitrous oxide emission mainly focuses on the research of the cost from treating nitrous oxide that is generated from the production activities of human being, the specific calculation formula is as follows:

$$\text{Emission cost of nitrous oxide} = \text{Emission volume of nitrous oxide} \\ \times \text{Treatment price of nitrous oxide} \quad (31.3)$$

---

[7]Calculating China's cost of handling methane emission during 2000–2009, for more details relating to each variable in the formula, please refer to the Appendix.

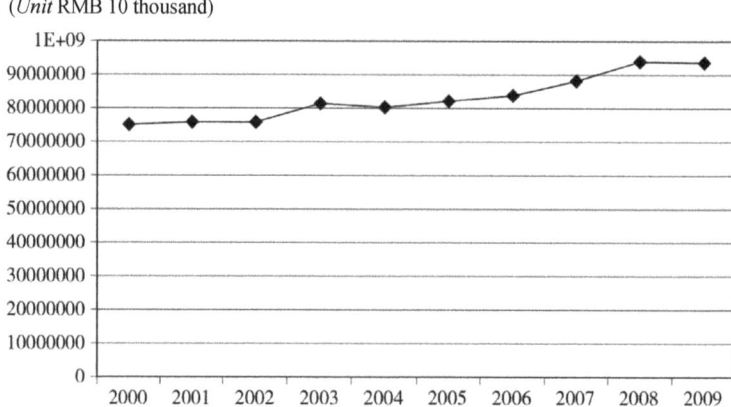

Chart 31.2 China's cost of methane emission during 1978–2009. *Source* Calculated based on the date in Table 31.2

Through the formula above, substitute China's statistic data[8] from 2000 to 2009, China's cost generated from handling of nitrous oxide emission from 2000 to 2009 can be calculated (for more details, please refer to Chart 31.3), we can see from the chart that during the period from 2000 to 2009, China's cost input for handling nitrous oxide emission had always been relatively high level, and it grew at a relatively even rate, as of 2008, it reached the peak value of RMB 5.3492039 trillion, in 2009, China's cost of nitrous oxide emission slightly declined.

31.3 Forming Causes of the Cost of Handling Climate Change

Due to human activities and continuous development of society, the stock of green house gas in atmosphere around the world gradually increases. For the specific sources of green house gas, please refer to Chart 31.4. Currently, the level of green house gas in atmosphere has reached 430 ppm,[9] while before the industrial revolution, the content of carbon dioxide in atmosphere was only 280 ppm. Such trend of continuously increasing green house gas concentration increased the global air temperature by about 0.5 °C. It is estimated that by 2050 the stock of green house gas in atmosphere will reach twice the level before the industrial revolution, and

---

[8]Calculating China's cost of handling nitrous oxide emission during 2000–2009, for more details relating to each variable in the formula, please refer to the Appendix.

[9]ppm is a concentration unit, for pollutants in atmosphere, using common volume concentration and quality—volume concentration to reflect its content in atmosphere. 1 ppm means 1 cm³ of pollutant in 1 m³ of atmosphere.

31.3 Forming Causes of the Cost of Handling Climate Change

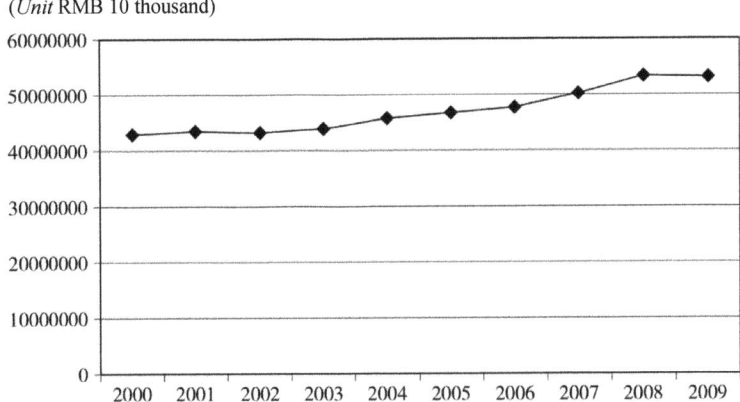

Chart 31.3 China's cost of nitrous oxide emission during 2000–2009. *Source* Calculated based on the date in Table 31.3

this trend will go on. Besides, with the rapid economic development, different countries will also have increasing demands for energy, according to statistics, the stock of carbon dioxide will reach 550 ppm by 2035,[10] and this trend is concluded from large quantity of observed factors and various researches from the oil industry (for more details, please refer to Table 31.3). Considering China's national situation, China's energy structure is mainly composed of coal, then oil and natural gas. So China's current energy consumption depends to most extent on fossil fuel that is dominated with coal, plus the widespread low utilization rate of energy in China, so China's carbon emission gradually increases. Particularly since the reform and opening-up, China's demand for energy has been continuously increasing, currently China has become the second biggest emitting country in the world, if left unrestricted, the consequence caused to human ecological environment will be beyond imagination.

The main cause of climate change resulted from the concentration increase of green house gas across the world is due to the production and living activities of human being. The influence of human being on climate change could be mainly reflected in two aspects, the first is direct influence, directly emitting green house gas into the atmosphere, such as the CO_2 and N_2O generated from the burning of fossil fuel, or the CO_2, CH_4, CFCs generated from industrial production process, etc., the second is the that human production activities have changed the sources absorbing green house gas, such as large-scale deforestation, which reduced the medium absorbing green house gas, so as to cause large volume of green house gas staying in atmosphere and unable to have good circulation. This book mainly explained the cost of climate change generated from CO_2, CH_4 and N_2O that are

---

[10]Yu Xinwen: *Scenario Selection of China's Participation in the Long-term* 2000–2050 *Carbon Dioxide Emission Reduction,* Progress of Climate Change Research, 1st Issue in 2010.

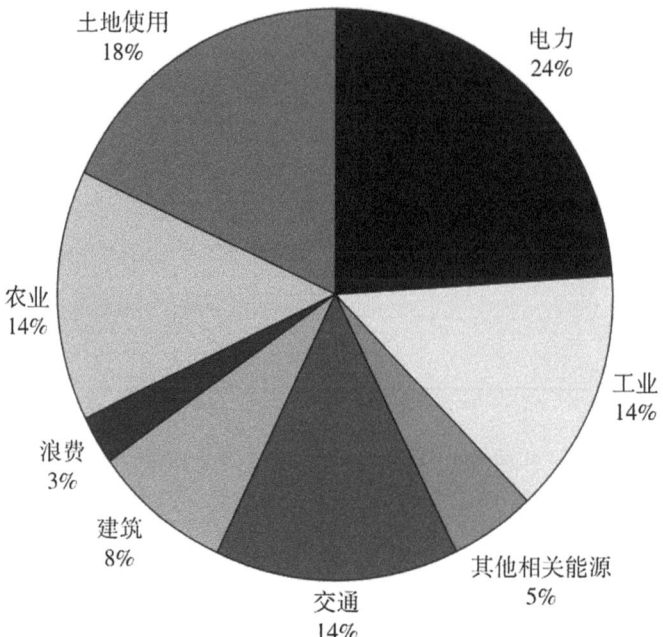

土地使用 18%	Land use 18%	农业 14%	Agriculture 14%
浪费 3%	Waste 3%	建筑 8%	Building construction 8%
交通 14%	Transport 14%	其他相关能源 5%	Other related energy 5%
工业 14%	Industry 14%	电力 24%	Electricity 24%

Chart 31.4 Ratios of green house gas emission by energies in 2000. *Source* Based on drawing by edition 3.0 of the climate analytical indicator instrument online database of the world resource institute

generated from the use of energy in human activities. The sources of these three green house gases are concluded in Tables 31.4, 31.5 and 31.6.

The main sources of green house gases can be seen from a series of tables above, in which the emission of CO_2 is mainly due to burning of fossil fuel, according to statistics, the CO_2 emitted from burning of fossils account for over 70% of the total emission volume. The Global Climate Change Observation Institute pointed out in its Annual Report in 1990 that the production and burning of fossil fuel across the globe in 1988 emitted about 5.66 ± 0.5 Gt-c, equivalent to 5 times of carbon released from deforestation in tropical regions and burning of forests in the same year.[11] CH_4 is mainly from swamps and rice fields, etc. only 16% of the CH_4

---

[11] Wang Mingxing and others: *Inspection and the Concentration and Emission of Greenhouse Gas and Related Process,* China Environment Science Press, 1996 Edition, p. 8.

31.3 Forming Causes of the Cost of Handling Climate Change

Table 31.3 Change of concentration of main green house gases in atmosphere (ppm, ppb and ppt respectively mean the volume fraction of certain component is 10-6, 10-9 and 10-12)

气体	CO_2	CH_4	N_2O	CFC—12	HCFC—22	CF_4
工业化前浓度	280ppm	0.7ppm	0.275ppm	0	0	0
1992年浓度	356ppm	1.714ppm	0.311ppm	503ppt	105	70ppt
当前年增加率	1.5ppm	0.013ppm	0.75ppb	18—20ppt	7—8ppt	1.1—1.3ppt
大气寿命（年）	50—200	12—17	120	102	13.3	5000

Source IPCC. *Climate Change*. UK: Cambridge University Press. 1994. P.339

气体	Gas
工业化前浓度	Concentration before industrialization
1992年浓度	Concentration in 1992
当前年增加率	Current annual growth rate
大气寿命（年）	Atmospheric lifetime (year)

Table 31.4 Main sources of CO_2 in atmosphere (GtC/a)

主要来源	排放量 （GtC/a）
化石燃料燃烧	6.3±0.4
土地利用变化	1.6±1.0
海洋吸收	1.4±0.7
中纬度森林吸收	0.5±0.5

Source jointly compiled by [US] IUCN, UNEP and WWF: *Protection of the Earth --- Sustainable Survival Strategy*, translated by the Foreign Affairs Office of the State Department of Environmental Conservation, China Environmental Science Press, 1992 Edition, Page 97.

主要来源	Main source
排放量(GtC/a)	Emission volume (GtC/a)
化石燃料燃烧	Burning of fossil fuel
土地利用变化	Change of land use
海洋吸收	Absorption by sea
中纬度森林吸收	Absorption by mid-latitude forests

emission was sourced from energy activities. N_2O in the atmosphere will stay for relatively long period of time in the atmosphere, its main emission into the atmosphere is mostly from soil. Table 31.7 takes the example of industrial boiler as the source of carbon emission.

Table 31.5 Main sources of CH$_4$ in atmosphere (Tg/a)

主要来源	变化范围
海洋	5—20
湖沼	100—200
苔原	1.3—13
森林	0—10
稻田	35—50
动物	65—100
白蚁	0—150

Source jointly compiled by [US] IUCN, UNEP and WWF: *Protection of the Earth --- Sustainable Survival Strategy*, translated by the Foreign Affairs Office of the State Department of Environmental Conservation, China Environmental Science Press, 1992 Edition, Page 99

主要来源	Main source
变化范围	Range of variation
海洋	Sea
湖沼	Lakes and marshes
苔原	Tundra
森林	Forest
稻田	Rice field
动物	Animal
白蚁	Termite
化石燃料生产和利用	Production and utilization of fossil fuel
大气中氧化	Oxidation in atmosphere
土地吸收	Absorption by soil

It can be seen from above that there are diversified forming causes of China's cost of climate change, on one hand, it is the cost generated from treatment of the green house gases that are generated from production and lives of human being and directly emitted into the atmosphere; on the other hand, it is the price paid by people for damaging these "Absorbing Sources" of green house gas. The forming factors of the cost of climate change could be concluded as Chart 31.5.

31.4 Countermeasures for the Cost of Handling Climate Change

Due to the continuous development and changes of the society, more and more people will face the crisis of climate change. The issue of global climate change should draw sufficient attentions from people. We should form a series of

31.4 Countermeasures for the Cost of Handling Climate Change

Table 31.6 Main sources of N_2O in atmosphere

主要来源	排放量（Gg）①
森林	89.5
草原	112.13
农田	143.33
化石燃料燃烧	38.4
生物质燃烧	19.7
己二酸生产	7.31

Source Jointly compiled by [US] IUCN, UNEP and WWF: *Protection of the Earth --- Sustainable Survival Strategy*, translated by the Foreign Affairs Office of the State Department of Environmental Conservation, China Environmental Science Press, 1992 Edition, p. 101.

主要来源	Main source
排放量（Gg²）	Emission volume (Gg)
森林	Forest
草原	Grassland
农田	Farmland
化石燃料燃烧	Burning of fossil fuel
生物质燃烧	Biomass burning
己二酸生产	Production of adipic acid

Table 31.7 Carbon emission source of industrial boiler

Type of loss	Specific reflection
Loss of ash	Unburned or not fully burned carbon particles in ash
Loss of flying ash	Carbon particles that are not fully burned are emitted out of furnace with smoke
Loss of leaked coal	Loss incurred from partial fuel fall into ashcan through fire grate

Source Ma ZHonghai, Pan Ziqiang, etc.: *Comparison between Green House Gas Emission from China's Coal Power China and That of Nuclear Power Chain*, Science and Project, 3rd Issue in 1999

countermeasures to reduce the consumption of fossil fuel, enhance the energy use efficiency, develop alternative energy, reduce the use of automobile, reduce the high-energy-consuming buildings, strengthen forest management, etc., so as to handle the continuously deteriorating ecological environment.[12]

---

[12]Since 2007, the Intergovernmental Panel on Climate Change (IPCC) released four times of climate change evaluation report, and indicated the fact that human activities have caused global warning. The report indicated that climate change will increase the times of natural disasters on global scale, as for involved number, over 100 million people will be affected and face the threat of starvation.

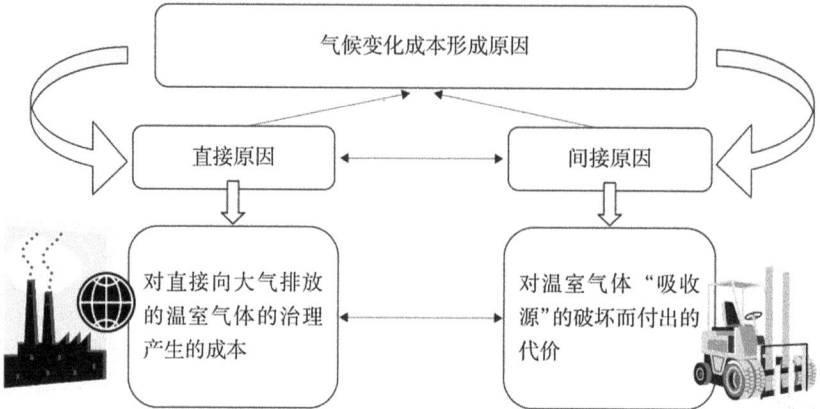

气候变化成本形成原因	Forming cause of the cost of climate change
直接原因	Direct cause
间接原因	Indirect cause
对直接向大气排放的温室气体的治理产生的成本	Cost generated from treatment of the greenhouse gas directly emitted into the atmosphere
对温室气体"吸收源"的破坏而付出的代价	Price paid for destruction of "Absorption Source" of greenhouse gas

Chart 31.5 Forming causes of the cost of climate change

31.4.1 Strengthen the Ecological Protection Countermeasure in Water Source Management

The temperature increase resulted from green house effect will to certain extent reduce the runoff volume of some river basins in China, so it will to certain extent accelerate the environmental deterioration and land desertification degree of China's drought regions. So targeting this issue, China started a series of measures, including large-scale forestation, returning farmlands to forests, and saving of water resources in irrigation, etc. A series of countermeasures in the aspect of water resource management includes driver dredging, wetland protection, purification of water pollution, etc. Besides, pricing mechanism of reasonable water fee, development of water-saving products, improvement of demand management are all ecological protection countermeasures to effectively strengthen water resource management. Besides, in nationwide scope, China's water resource management should integrate with regional development, implement comprehensive treatment, and further strengthen the emergency countermeasures in the aspect of water resource management.

31.4.2 Strengthen Environmental Construction in Coastal Regions

According to statistics, 70% of big cities and over 50% of population in China are distributed in the eastern coastal regions, due to the green house effect and the continuous increase of air temperature, the crisis faced by the coastal region becomes increasingly prominent. In the past 50 years, the sea level in China's coastal regions has been continuously rising, which is extremely easy to cause hazards such as seawater intrusion, farmland salinization or even collapse of coastal fending groyne.[13] According to the statistics by OECD, if ranking the global coastal cities that are exposed to flood risks in social total volume, Chinese cities such as Guangzhou, Shanghai, Tianjin, Hong Kong, Ningbo, Qingdao, etc. all ranked the top 20 cities with the highest risk.[14] Therefore, under such circumstance, China should strengthen the adapting measures of various aspects in coastal region, for more details, please refer to Table 31.8.

31.4.3 Strengthen International Cooperation to Handle Global Climate Change

Under the in-depth development of globalization and computerization, the connection among countries in the world has become increasingly closer, different countries in the world connect with each other and influence each other. Climate change, as a global issue, must be solved through close international cooperation.[15]

31.4.4 Reduce the Deforestation

Due to development and advancement of human society, large-scale deforestation has been made, at the preliminary stage of social development, there hasn't been a

---

[13]Pan and Zheng: *Analysis Framework and Policy Connotation of Adaptation to Climate Change*, China's Population Resource and Environment, 10th Issue in 2010.

[14]Nicholls. R.J., et al. "Ranking Port Cities with High Exposure and Vulnerability to Climate Extremes: Exposure Estimates". *OECD: OECD Environment Working Papers, 2008.*

[15]The issuing and signing of policies such as *United Nations Framework Convention on Climate Change* and *Kyoto Protocol* set foundation for the international cooperation in response to climate change. International cooperation should also consider each aspect of emission reduction, not only strengthening cooperation in the aspects of pricing, technology and elimination of behavior obstacles, but should also take actions on the emission generated from use of land. To reach continuous and wide cooperation, we must adhere to fair principle, therefore, some developed countries should take responsibilities, and make efforts to reduce the emission volume by 60–80% to 2050 compared with that in 1990.

Table 31.8 Adapting measures of chinese coastal regions in handling climate change and to be strengthened

Type of measure	Specific reflection
Engineering type	Build seawalls, adopt anti-flood measures, consolidate buildings, transferring personnel and properties, etc.
Technological type	Improve water resource management model, change the production method of agriculture and fishery in coastal regions, such as promoting the crops that could resist flood and salt and alkali, adopt new-type seepy floor materials, etc.
System type	Building standard, legislation, taxation subsidy, property insurance, social security system construction, etc.

Source Chen Xuanyu, etc.: *Evolution of Chinese Climate and Environment (Volume II)*, Science Press, 2005 Edition, p. 45

series of national policies and regulation to standardize deforesting behaviors. So it caused large-scale depletion of China's forest resource, so as to form China's huge cost of forest depletion. Therefore, in order to implement rational and highly-efficient control on the cost of forest depletion, we should on national level clearly define forest property rights, land owner and deforesting right,[16] so as to effectively control of quantity of deforestation. Therefore, adopting a series of actions to ban deforestation is very important for protecting the residual forest resource and effectively control the global temperature increase that is caused from green house effect.

31.4.5 *Innovate the Domestic Carbon Trade Mechanism*

Firstly, we should establish the carbon accounts for various regions, main industries, forest departments, and central state-owned enterprises, including the carbon sequestration capacity and oxygen emitting capacity, which shall be opened to public, so as to form strong social supervision mechanism; secondly, we should establish a national carbon trade market, promote different regions to have carbon trade, particularly the big carbon-emission provinces should directly purchase carbon emitting quota from provinces with large forests; thirdly, the over 100 state-owned industrial enterprises that are directly governed by the Central

---

[16] Because the emission generated from deforestation accounts for about 18% of total emission in the world, this has exceeded the emission load generated from traffic in the world.

31.4 Countermeasures for the Cost of Handling Climate Change

Government should directly purchase forest carbon sequestration quota from various regions; fourthly, promote the establishment of internal trading mechanism with province as unit.[17]

Appendix

See Tables 31.9, 31.10 and 31.11.

---

[17] Hu Angang: *Taking "Forest Chongqing" for An Example,* National Situation Report, 28th Issue in 2009.

Table 31.9 China's cost of carbon dioxide emission during 1978–2008

年份	CO_2排放量（千吨）	CO_2治理价格	可比价格	CO_2排放成本
1978	1443885	178.66	100	25795.82
1979	1560672	186.87	104.6	29164.88
1980	1480575	195.63	109.5	28964.19
1981	1501080	202.95	113.6	30464.85
1982	1626464	210.46	117.8	34229.97
1983	1648128	218.14	122.1	35952.03
1984	1774069	226.18	126.6	40125.55
1985	2011169	234.22	131.1	47105.11
1986	2042633	258.87	144.9	52878.06
1987	2186000	286.03	160.1	62525.66
1988	2331546	316.04	176.9	73686.62
1989	2366784	349.27	195.5	82665.10
1990	2460744	386.61	216.4	95135.04
1991	2584538	399.83	223.8	103337.94
1992	2695982	425.38	238.1	114681.45
1993	2878694	487.91	273.1	140453.97
1994	3058241	605.64	339	185220.14
1995	3320285	709.08	396.9	235436.22
1996	3463089	768.04	429.9	265979.34
1997	3469510	789.48	441.9	273910.67
1998	3324345	783.23	438.4	260371.50
1999	3318056	772.15	432.2	256203.63
2000	3405180	775.37	434	264025.95
2001	3487566	780.73	437	272283.10
2002	3694242	774.47	433.5	286108.82
2003	4525177	783.76	438.7	354666.37
2004	5288166	814.31	455.8	430622.00
2005	5790017	828.96	464	479970.59
2006	6414463	841.47	471	539756.67
2007	6791805	881.84	493.6	598931.52
2008	7031916	933.83	522.7	656663.66

Source The carbon dioxide emission volume during 1978-2008 was from UN data, A World of Information. http://data.un.org/Data.aspx? q=CO2+&d=MDG&f=seriesRowID%3a749, Millennium Development Goals Database, United Nations Statistics Division; the carbon dioxide price in 1991 was cited from Wang Kaifeng, etc.; *Evaluation of the Resource and Environmental Price of Economic Growth in Hunan Province*, Resource and Environment in Yangtze River Basin, 2nd Issue in 2007; the prices in other years were calculated based on the comparable prices in previous years.

年份	Year
CO2排放量（千吨）	CO2 emission load (thousand tons)
CO2治理价格	CO2 treatment price
可比价格	Comparable price
CO2排放成本	CO2 emission cost

Table 31.10 China's cost of methane emission during 2000–2009

年份	甲烷排放量 （千吨/二氧化碳当量）	治理价格	甲烷排放成本 （万元）
2000	973730	775.37	754996763.7
2001	978111.785	780.73	763636605.7
2002	982513.288	774.47	760929365.4
2003	986934.5978	783.76	818131594.2
2004	991375.8035	814.31	807289763.7
2005	995760	828.96	825447520
2006	1000240.92	841.47	841670943.9
2007	1004742.004	881.84	886026109.1
2008	1009263.343	933.83	942483617.4
2009	1013805.028	927.22	940023279

Source The emission volumes of methane in 2000 in 2005 were cited from the World Bank WDI database and the World Bank: *2009 World Development Index,* methane emitting volume in other years were estimated by the author. The treatment price in 1991 was cited from Wang Kaifeng, etc.; *Evaluation of the Resource and Environmental Price of Economic Growth in Hunan Province,* Resource and Environment in Yangtze River Basin, 2nd Issue in 2007; the prices in other years were calculated based on the comparable prices in previous years.

年份	Year
甲烷排放量（千吨/二氧化碳当量）	Methane emission load (thousand tons/ carbon dioxide equivalence)
治理价格	Treatment price
甲烷排放成本（万元）	Methane emission cost (RMB 10 thousand)

Table 31.11 China's cost of nitrous oxide emission during 2000–2009

年份	氧化亚氮排放量 （千吨/二氧化碳当量）	治理价格	氧化亚氮排放成本 （万元）
2000	556620	775.37	431584010.6
2001	558623.832	780.73	436131752.5
2002	560634.8778	774.47	434196205.8
2003	562653.1634	783.76	440986413.4
2004	564678.7147	814.31	459824967
2005	566680	828.96	469756367.6
2006	568720.048	841.47	478559845
2007	570767.4402	881.84	503328070.4
2008	572822.203	933.83	534920390.8
2009	574884.3629	927.22	533045969.2

Source The emission volumes of nitrous oxide in 2000 in 2005 were cited from the World Bank WDI database and the World Bank: *2009 World Development Index,* nitrous oxide emitting volume in other years were estimated by the author. The treatment price of nitrous oxide in 1991 was cited from Wang Kaifeng, etc.; *Evaluation of the Resource and Environmental Price of Economic Growth in Hunan Province*, Resource and Environment in Yangtze River Basin, 2nd Issue in 2007; the prices in other years were calculated based on the comparable prices in previous years.

年份	Year
氧化亚氮排放量（千吨/二氧化碳当量）	Nitrous oxide emission load (thousand tons/carbon dioxide equivalence)
治理价格	Treatment price
甲烷排放成本（万元）	Nitrous oxide emission cost (RMB 10 thousand)

Chapter 32
Cost of Harmony Between Human and Nature

32.1 Overview of the Cost of Harmony Between Human and Nature

Harmony means that different factors and various elements of a system live in harmony and properly coordinate with each other. Taking the nature for an example, harmony of an ecological system means the state that different elements in this system, such as human being, water, soil, atmosphere, rock, forest, etc., depend on each other, restrain each other, balance each other and maintain relative stability.

Some experts and scholars conclude harmony between human and nature into the following four aspects (refer to Table 32.1).

There are two important standards to measure if human and nature are in harmony: the first is the carrying capacity of resources, the second is the carrying capacity of the environment.[1] The advancement and development of human society that are not beyond the two carrying capacities above can be called sustainable development, for the specific carrying capacity of resource and environment, please refer to Chart 32.1.

Why do we say that taking care of the relationship between human being and these two aspects can be called as harmony between human and nature? From the perspective of the carrying capacity of resources, of which fresh water resource, energy resource and farmland resource are limited, most energies and resources are non-renewable, in the continuous development process of human society, large quantity of construction of plants, residences and public facilities resulted in use of lands, while land resource is limited, if the development degree of human being exceeds the carrying capacity of limited resource, we would never achieve sustainable development. The carrying capacity of environment mainly means destruction of ecological system, such as river interruption, desertification, decrease

---
[1]Wang Shucheng: *Model C: Self-disciplined Development,* National Situation Report, 31st Issue in 2005.

Table 32.1 Reflection of harmony between humana and nature in four aspects

No.	Aspects of reflection	Specific contents
1	Optimal state of living environment for human	When obtaining, distributing and using various materials, energies and information, human should make the most efforts to reduce destruction of ecological environment, so as to avoid deterioration or degradation of ecological environment
2	Maintain ecological balance	Prevent abuse of various resources or hunting of various animals to maintain ecological diversity
3	Enhance the using efficiency of materials and energy to most extent	Enhance the using efficiency of resources and energy to reduce the waste and unreasonable loss of materials and energy[a]
4	Limit the exploitation of resources and energy	Develop and exploit various resources in the nature with plan, so as to realize sustainable use of natural resources

Source Concluded by the author

[a]The price of continuous human development is the large-scale consumption of resources and energy and excessive seeking from the nature. According to researches, the mineral fuel that human consumes in 1 year is equivalent to the quantity that the nature has to accumulate for 1 million year

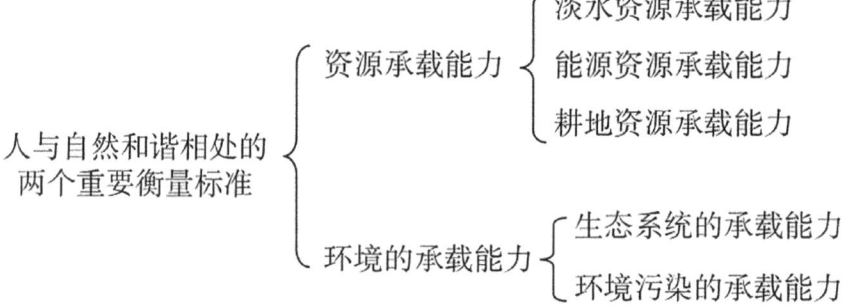

人与自然和谐相处的两个重要衡量标准	Two important measuring standards of harmonious coexistence between human and nature
资源承载能力	Carrying capabilities of resources
环境的承载能力	Carrying capabilities of environment
淡水资源承载能力	Carrying capabilities of fresh water resource
能源资源承载能力	Carrying capabilities of energy resource
耕地资源承载能力	Carrying capabilities of farmland resource
生态系统的承载能力	Carrying capabilities of ecological system
环境污染的承载能力	Carrying capabilities of environmental pollution

Chart 32.1 Specific reflection of the carrying capacity of resource and environment

of biological diversity, etc. Besides, the consideration of environmental carrying capacity also includes the emission of polluted water, emission of toxic or harmful gases, and influence causes from discharge of solid garbage. In short, in the continuous process of industrialization, it is very unwise to trade the most precious water resource, most precious energy and most precious land for a period of GDP growth, and causes irreversible pollution. The materialistic dialectics told us that anything has two sides, the quickly developing society has increased people's incomes and enhanced people's living quality, but if we don't pay attention to the negative influences caused to the nature, then the loss of human being will be very severe.

The nature is the foundation of the living and development of human being, in the process of human being realizing better living and development, we obtains large quantity of resources from the nature, all the material resources needed for the living and development of human being are from the nature. Over one hundred years ago, Engels pointed out that the human being could make the nature serve us by changing the nature, and could dominate the nature, but you must remember that the human being can't dominate the nature by standing out of the nature.[2] While explaining this important thinking, Engels also indicated that "we shouldn't be over excited about our victory over nature. For every such victory, the nature will get back on us. For every victory, in the beginning, we indeed obtains the results we expected, but it would have completely different and unexpected influences later and after that, which would often erase the preliminary result".[3] Therefore, Engels told people that "we are, step-by-step, learning to more correctly understand the rules of nature, learning to recognize the soon or later consequences caused from our interventions on the nature on day-to-day basis. Especially since the big-step advancement of the natural science since this century, we are more likely to learn to recognize and therefore control those soon or later natural consequences that are caused from our most common production behaviors".[4]

In the process of continuous coexistence between human and nature, people's dominating ability and using ability of the materials, power and information in the nature have been continuously strengthening. While in this development process, the human being has also been continuously breaking the bottom line of harmony between human and nature. The achievements obtained in human's development and social development are obvious, according to the statistics of World Development Report, the number of developing countries in the world that are in poverty has decreased from half to 1/4, and technological innovation also promoted the continuous advancement of the society. In middle-income countries, the average income has doubled. Meanwhile, the environmental problems brought from human development also become increasingly prominent. The frequent droughts, floods, storms and other natural disasters are constantly threatening the ecological environment that human being rely on. By the end of 21st Century, the air temperature

---

[2]*Selected Works of Jiang Zemin*, Volume 2, People's Publishing House, 2006 Edition, p. 233.
[3]*Karl Marx and Frederick Engels*, Vol. 9, People's Publishing House, 2009 Edition, p. 599.
[4]*Karl Marx and Frederick Engels*, Vol. 4, People's Publishing House, 1995 Edition, p. 384.

cold be 5 °C higher than that before the industrialization.[5] The problems in developed countries that occurred in nearly one hundred years of industrialization process have been concentrated in the short 30 years of China's reform and opening-up. According to statistics, the environmental problems faced by China mainly cover three aspects; the first is environmental pollution problem, many rivers have been polluted since the upstream, a large part of rivers and lakes are having the phenomenon of eutrophication. The degree of air pollution in cities is relatively severe, the area of lands subject to damage from acid rain is relatively large. The second is the problem of ecological environment destruction, in recent years, China's area of water and land loss and the area of land desertification have been increasing, the forest coverage rate is insufficient, the grassland degradation is also very obvious. The third is that aspect of pressure from international environmental issue is under great challenge, such as fulfillment of the related international treaties, etc.

Therefore, we could have multiple interpretations about the cost of harmony between human and nature. The first is the price paid by human after the destruction of nature by human production and living activities during the continuous development of human society[6]; the second is to interpret the cost of harmony between human and nature as the efforts made by human being in order to reach the state of harmony between human and nature, in other words, it means the measurement of the cost paid by human being in order to stop further deterioration of natural environment or recover to ideal state from the perspective of economics. The cost of harmony between human and nature in this book will focus on the former.

32.2 Forms of the Cost of Harmony Between Human and Nature

Cost of resource and energy, cost of ecological environment, cost of disaster and cost of handling climate change are all passive cost and need solving and handling after actual occurrence of facts. While the cost of harmony between human and nature includes all active handling of the relationship between human and nature, including making up the loss of ecological environment, the most important thing is to make up the loss of human health brought by environmental pollution, as well as including turning forest deficit into forest surplus, actively handling various disaster

---

[5]World Bank: *World Development Report—Development and Climate Change*, Tsinghua University Press, 2010 Edition, p. 187.

[6]For example, human disease, decrease of biological diversity brought by the disasters, rising sea level, destruction of farmland, environmental pollution and destruction of ecological environment that are the results of climate change, while the series of price paid by human being in the process of coexistence between human and nature, if measured from economic perspective, is the cost of harmonious coexistence between human and nature.

prevention and relief projects, and the most important thing is to constantly implement and improve the design of system.

32.2.1 Cost of Health Loss Resulted from Environmental Pollution

The most direct victim of climate pollution brought by energy burning is human being. The over one billion Chinese people are severe victims of atmospheric pollution. The continuous growth of carbon dioxide emission in China has accelerated the global crisis of climate change.[7] In 2005, the number of people affected by artificial particulate matter pollution was as high as 1.163 billion per year, which was far higher than that of India (432 million people per year) at second place, and the total quantity was equivalent to 2.69 times of that of India. We can see from the data that the influence of atmospheric pollution on Chinese people is equivalent to 1 year less of average life expectancy of the 1.3 billion of Chinese people. Under the basic-scenario emission prediction, this data of China in 2035 will be far higher than the 1.215 billion people per year under the emission reduction scenario at 450, and reach 1.573 billion people per year[8] (refer to Table 32.2).

From the perspective of trend in the future, the loss of total life expectancy of China due to artificial particular matter will continue to rise. By 2020, the total life expectancy loss under the scenario 450 will increase by 28% compared with that in 2005; and the basic scenario will increase by 35% compared with that in 2005. By 2035, the basic solution will still increase by 35% compared with that in 2005, but the 450 solution only increases by 4% compared with that in 2005. If calculating based on 1.462 billion of Chinese population in 2035, the average life expectancy of China under the basic scenario will decrease by 0.245 year compared that under scenario 450, the loss of people's life welfare caused hereof and the loss of total national output would be huge, and we need to come up solutions to make up these losses.

32.2.2 Transformation from "Forest Deficit" to "Forest Surplus"

"Forest Deficit", the loss brought to people from continuous degradation of forests could be considered as important part of the cost of ecological environment. Since

---

[7]As indicated by the data by International Institute of Applied System Analysis, IIASA, China ranks No. 1 in the total loss of citizen life expectancy caused from artificial particles in the world.
[8]Hu Angang and Liang Jiaocheng: *China and the World in Post-Copenhagen Era: Trend of Energy Conservation and Emission Reduction,* National Situation Report, 41st Issue in 2010.

Table 32.2 Loss of total life expectancy due to artificial particulate matter pollution (2005–2035)

(Unit: Million People / year)

		基准情景		450 情景	
	2005	2020	2035	2020	2035
中国	1163	1565	1573	1491	1215
印度	432	854	1466	792	1085
俄罗斯	53	49	49	47	46
欧盟	234	146	119	138	108

Source: IEA: *WEO2010*, cited from HASA (2010).

基准情景	Benchmark scenario
450 情景	450 scenario
中国	China
印度	India
俄罗斯	Russia
欧盟	EU

the establishment of New China, China's social development and people's economic production activities have never stopped consumption of China's forest resources. Basically, China's state of forest has always been reflecting mixed situation. On one hand, in order to alleviate the continuous degrading trend of forest, China has made huge investments in forest planting and the construction of protective forest. This is a very big achievement for a big nation of agricultural population. On the other hand, due to the demand of social development, China's natural forests have always been in sliding trend for the 50 years after the establishment of New China, and will continue this way based on the trend.

Such trend also caused the generation of China's "Forest Deficit[9]". China's forest deficit is very considerable. According to statistics, the annual consumption of China's ripe woods in the "Commercial Forest" exceeds the growing volume of ripe woods and growth volume of nearly ripe woods entering ripe woods combined, the deficit is about 170 million m^3, if calculating for 50 years, the accumulated deficit has reached 8.5 billion m^3.[10]

---

[9]The so-called "Forest Deficit" means that the exploitation volume exceeds the growing volume of forest. On the contrary, "Forest Surplus" means that the growing volume has far exceeded the exploitation volume of forest.

[10]Hu Angang: *Let the Nature Rest for 50 Years: Major Forestry Strategic Change from Forest Deficit to Forest Surplus,* National Situation Report, 93rd Issue in 2002.

From the perspective of economics, "Forest Deficit" could be represented by the ratio of net forest depletion[11] in Gross National Income (GNI). The ratio of China's net forest depletion in gross national income from 1976 to 2000, based on this economic indicator, we can calculate and make objective evaluation of the benefit of economic policies and forestry policies at different stages, especially their ecological economic loss. Since 1970s, China's forest deficit kept growing and reached a relatively high level in 1980s, and began to quickly decease from 0.45% in 1995 to 0.07% in 2000, but it still had certain gap with the word average level of 0.037%. Besides, since 1993, China began to extensively reduce the deforestation of natural woods, the ratio of China's forest depletion in national gross income had sharp decline and directly reduced the cost of forest deficit.[12]

As studied by Hu and Liu (2011): China's forest resources generally reflected "U" shaped change, firstly decrease, then increase. The path of forestry had a winding development, the forestry development in China could be divided into pre-new policy period and new policy period, and could be sub-divided into six stages. During the period of first new forestry policy, China changed the situation of "Forest Deficit" and had the golden period of "Forest Surplus".[13] The second period of green forestry new policy, the key was the transformation of forestry development, meaning that the forestry fully takes its natural advantages and comprehensively transform into green development.

32.3 Cost of Maintaining Harmony Between Human and Nature

With the development of human being and continuous advancement of society, the relationship between human and nature also had gradual transformation, human being gradually changed from the state of "Respecting the Nature" to the state of "Conquering the Nature", and the nature has gradually become the object of transformation by human. Since the agricultural society, the transformation of nature by human being has formed large scale, the farmland areas continuously expanded, and the demands for grains also continuously increased, but such advancement has resulted in a series of environmental problems. Especially after

---

[11]Calculating based on the green total national income account method used by World Bank, net forest depletion equals the multiplying between the rent of unit resource and the collection volume of raw forest from natural growth.

[12]Hu Angang: *Let the Nature Rest for 50 Years: Major Forestry Strategic Change from Forest Deficit to Forest Surplus,* National Situation Report, 93rd Issue in 2002.

[13]Hu Angang and Liu Min: *From Forest Deficit to Forest Surplus: Forestry Development Transformation and New Green Policy,* National Situation Report, 3rd Issue in 2011.

the occurrence of industrial civilization, the relationship between human and nature had fundamental transformation. The power of nature was no longer mysterious as before, the human's reliance and respect to nature had gradually transformed into large-scale development and utilization of nature, so as to cause a series of losses generated from constant deterioration of nature and ecological environment. After entering the period of modern ecological civilization construction, human being has actively made thinking and actions about the sustainable development of the resources, energies, environment and ecology, and actively implemented disaster prevention and relief as well as design of systems, input of such cost has not only realistic benefit, but also future benefit, we should actively maintain the "Beautiful China" and ecological civilization construction as stressed by the party's 18th National Congress, this is also the key and focus of reflecting the system design and strategic guarantee in the planning.

32.4 Countermeasures of the Cost of Harmony Between Human and Nature

Regarding the cost of environmental pollution and the cost of ecological environment generated from inharmonious relationship between human and nature as mentioned in previous section, this section will integrate China's national situation at current stage and make further explanation of the countermeasure for the formation of the costs in these two aspects, so as to help us correctly handle the relationship between human and nature as well as promote the sustainable development of the society in the future.

Environmental pollution is mainly generated from the industrialization process of human being, the early-stage industrialization was realized through the "Black Mode", featuring high input, high consumption and high pollution emission. From the development process of industrialization across the world, each industrial revolution had its different driving force. Since the end of 18th Century, there were three industrial revolutions in the world, each industrial revolution had its typical model and representation. After entering the 21st Century, due to the influence of scientific and technological advancement, globalization and computerization, the human being is facing another round of new industrial revolution.

Under the grand background of the fourth industrial revolution, targeting China's problem in the aspect of environmental pollution, in 2006, China integrated its own national situation and wrote down the "Construction of a Resource-saving and Environmentally Friendly Society" into the nation's "Eleventh Five Year Plan" outline for the first time, which proposed enforcement of the basic national policy of saving resources and protecting the environment, building the national economic system of low input, high output, low consumption, less emission, circulation and sustainability and a resource-saving and environmentally friendly society.[14] Therefore, green development is the necessary output of green industrial revolution, and could also be deemed as China's efforts to rationally control the continuously deteriorating environmental pollution of China at current stage.

---

[14]Hu Angang and Liao Jiaocheng: *China's Green Development Strategy and the "Twelfth Five Year" Plan,* 16th Issue in 2011.

Summary of China's Cost of Nature Development

Through quantitative studies of various costs, generally speaking, China's nature development cost has always been reflecting growth since the reform and opening-up. During the period from the reform and opening-up to 1990s, China's cost of nature development has been growing slowly. With the high-speed growth of China's economy, the demands for natural resources by human development have been continuously increasing during social development process, most nature development cost except disaster cost began to enter rapid growth period since 1990s. Such characteristic was the most significant on the cost of natural gas in China. From 1978 to 1993, the cost of natural gas in China had maintained the annual average growth rate of 8.4%, the cost of natural gas in 1993 was about 3 times of that in 1978. However, during the period from 1993 to 2009, the cost of natural gas in China had the annual average growth rate of 15.2%, in 2009, the cost of natural gas was as high as 10 times of that in 1993. Due to its own suddenness and unpredictability, China's cost of disasters generally reflected fluctuating state. However, from the perspective of the input to prevent disaster cost in China, China's input in the aspect of disaster reflected the trend of growth year by year. Therefore, generally speaking, with the acceleration of China's economic development and increase of demands for resources, the cost of nature development will also grow accordingly.

Under the influence of the model of excessive industrialization and excessively gross growth since 1990s, there is no doubt about the influence of the obvious natural environment cost on human environment, however, the influence of hidden cost of natural environment (such as the cost of health loss from environmental pollution as mentioned in this book) on the living environment of human being is also continuously strengthening. Excessive utilization of natural resources without restraint will inevitably cause inharmonious relationship between human and nature, and will bring negative influence on the sustainable development of human being and society. Therefore, we need to integrate China's national situation and long-term development goals to well handle the relationship between human being and nature, play the role of a responsible power and become the leader in promotion of the establishment of a harmonious world.

Part VIII
Inspiration of China's Development Cost Theory for China's Development

Chapter 33
Indicator System Foundation of Development Cost

Through the studies and thinking from Part II to Part VIII, it is believed in this book that development cost has already been able to form a cost system, which generally includes five parts, the traditional economy, politics, society, culture and ecology, the indicator system of development cost of which the foundation could be listed are as follows in Table 33.1.

The development cost system includes 32 Class I indicators, 12 of which are passive, meaning those mainly reflecting the price of development, or the already occurred and caused losses, some have drawn attentions, while some haven't drawn full attentions. These 12 indicators shall be solved in priority.

Table 33.1 Indicator system foundation of development cost

Cost of economic development (internal)	
Class I indicator	Refining
A1 Cost of economic growth (active)	A1.1 Expansion of investment A1.2 Expansion of domestic demands A1.3 Expansion of exports
A2 Cost of economic transformation (active)	A2.1 System transformation A2.2 Structural transformation A2.3 Development method transformation A2.4 Globalization transformation
A3 Cost of economic disturbance (passive)	A3.1 Inflation A3.2 Change of exchange rate A3.3 Risk of economic crisis
A4 Cost of economic regulation (active)	A4.1 Regulation by economic means A4.2 Regulation by legal means A4.3 Regulation by administrative means
Cost of political development	
Class I indicator	Class II indicator
B1 Cost of political reform (passive)	B1.1 Political struggle B1.2 Political risk
B2 Cost of political system construction (active)	B2.1 Governance of corruption B2.2 Political system reform B2.3 Political system design
B3 Cost of political decision making (active)	B3.1 Establishment of decision-making system B3.2 Response to decision-making mistakes B3.3 Decision-making completing mechanism
B4 Cost of the construction of ruling party itself (active)	B4.1 Organization construction B4.2 Working style construction B4.3 Ideological and cultural construction B4.4 Construction of democracy within the party B4.5 Construction of political civilization
B5 Cost of political consultation and cost the crossing information gap (active)	B5.1 Construction of political consultation system B5.2 Crossing of information gap
B6 Cost of democracy construction (active)	B6.1 Integration of democracy B6.2 Construction of democracy B6.3 Political disturbance
Cost of social development	
Class I indicator	Class II indicator
C1 Cost of solving livelihood issues	C1.1 Balanced development of education C1.2 Employment promotion C1.3 Social security C1.4 Unbalance of income distribution C1.5 Development of medical and health
C2 Cost of population change (passive)	C2.1 Pressure from increasing population C2.2 Response to aging and low birth rate

(continued)

Table 33.1 (continued)

Cost of economic development (internal)	
Class I indicator	Refining
	C2.3 Scientific management of population
C3 Cost of social management (active)	C3.1 Administrative management expense
C4 Cost of social stability (passive)	C4.1 National defense expenditure C4.2 Expense of stabilization C4.3 Economic and social price for destruction of social stability events
C5 Cost of social advancement (active)	C5.1 Encourage innovation and vitality C5.2 Full information symmetry C5.3 Comprehensive development
Cost of cultural development	
Class I indicator	Class II indicator
D1 Cost of cultural transfer (active)	D1.1 Maintain peace in civilization conflict D1.2 Elimination and evolution of civilization D1.3 Cost of civilization protection and transfer
D2 Cost of protecting ethnic custom	D2.1 Avoid ethnic conflict D2.2 Promote ethnic integration D2.3 Protect religious faith
D3 Cost of ideological evolution (passive)	D3.1 Cost of ideological evolution D3.2 Cost of regular ideological construction D3.3 Risk of ideological construction D3.4 Cost of international ideological confrontation and conflict
D4 Cost of constructing modern media system (active)	D4.1 General management cost of modern media construction D4.2 Reform cost of modern media system
D5 Cost of soft power construction (active)	D5.1 Cost of competition for international speaking right D5.2 Cost of regional soft power construction D5.3 Cost of modern organization system construction D5.4 Cost of scientific quality and cultural quality enhancement
D6 Cost of change in social civilization (active)	D6.1 Cost of sliding morality of social groups D6.2 Cost of change in social relations D6.3 Cost of conflicts and contradictions caused from formation of new social groups D6.4 Opening and innovation of social advanced thinking
Cost of economic development (Cost of opening-up and development)	
Class I indicator	Class II indicator
E1 Cost of opening and developing thinking an ideas	E1.1 Cost of no opening and thinking competition E1.2 From gradual opening-up to comprehensive opening-up

(continued)

Table 33.1 (continued)

Cost of economic development (internal)	
Class I indicator	Refining
E2 Cost of international exchange and consensus formation (active)	E2.1 Participate in international exchange E2.2 Join international organization E2.3 Fulfill international consensus
E3 Cost of investment promotion and investment environment construction (active)	E3.1 Foreign investment E3.2 Internal construction of investment environment
E4 Cost of international trade (passive)	E4.1 Tax rebate burden E4.2 Opportunity cost and risk of high foreign exchange reserve E4.3 Response to trade friction
E5 Cost of construction and development of three-dimensional transport (active)	E5.1 Infrastructure investment E5.2 Opportunity of infrastructure investment E5.3 Loss of management efficiency
E6 Cost of creditability construction and response to threats (passive)	E6.1 Maintenance of territory and sovereign integrity E6.2 National unity E6.3 Foreign aids E6.4 Bearing of international responsibilities E6.5 Communication of big power image
Cost of nature development	
Class I indicator	Class II indicator
F1 Cost of resources and energy (passive)	F1.1 Decrease of farmland F1.2 Decrease of forest land F1.3 Depletion of mineral resources F1.4 Depletion of coal F1.5 Depletion of oil F1.6 Depletion of natural gas
F2 Cost of ecological environment (passive)	F2.1 Water pollution F2.2 Emission of sulfur dioxide and smoke & dust F2.3 Emission of industrial dust F2.4 Emission of solid wastes
F3 Cost of disaster (passive)	F3.1 Geological disaster F3.2 Forest fire disaster F3.3 Oceanic disaster
F4 Cost of response to climate change (passive)	F4.1 Emission of carbon dioxide F4.2 Emission of methane F4.3 Emission of nitrous oxide
F5 Cost of harmonious coexistence between human and nature (active)	F5.1 Promote "forest surplus" F5.2 Cover the health loss caused from environmental pollution F5.3 Disaster prevention and reduction F5.4 Design of ecological civilization system

Chapter 34
Promote the Enhancement of Total Factor Productivity and Reduce Development Cost

The welfare of human development comes from creation of fortune; so economic development was for a period the main line of China's development. The main driving force of economic development is economic growth, but simply using GDP that reflects economic growth to measure economic growth is not sufficient in current theoretical development and realistic development background. The author believes through the analysis above that if understanding from the perspective of input, the development of economy, politics, society, culture and ecology all needs input of resources (labor, material and fund), while the general meaning of total factor productivity (TFP) represents the development and utilization efficiency of resources, the sources of total factor productivity include technical advancement, system reform, organizational innovation, factor optimization and allocation, enhancement of labor quality, etc.

Abramovitz (1993) indicated that at early stage of industrialization, the degree of technical advancement highly focused on the using direction of material capital. The rapid expansion of market encouraged large-scale production, and encouraged large-scale investment in plants and urban infrastructure. Therefore, by dissolving the factors of economic growth at the stage of early industrialization, it can be found that there were huge contribution of capital accumulation and relatively low TFP growth. However, once the first stage of industrialization is completed, the technical advancement focused on material capital would decrease, and technical advancement would focus on the use of intangible capital, so as to generate relatively high TFP growth.[1]

For the estimation of total factor productivity of China since the reform and opening-up, it is already a relatively mature research, for instance, Liu and Li (2011) believed that the main factors of the rapid growth of China's total factor productivity include technical advancement, system reform, factor optimization and allocation, opening-up, foreign direct investment, labor quality enhancement,

---

[1] Liu and Li: *Change of China's Total Factor Productivity since Reform and Opening-up and Growth Trend in the Future,* Economic Research Reference, 33rd Issue in 2011.

Table 34.1 Factor decomposition of China's economic growth

年份		1978—1990	1991—2000	2001—2008	1978—2008
GDP 增长率（%）		9.02	10.56	10.46	9.80
各要素的贡献	资本	4.29	5.25	6.02	5.27
	劳动	2.13	0.54	0.36	1.02
	TFP	2.60	4.77	4.08	3.51

Source: Liu Ming and Li Shantong: *Change of China's Total Factor Productivity since Reform and Opening-up and Growth Trend in the Future*, Economic Research Reference, 33rd Issue in 2011.

年份	Year
GDP 增长率（%）	GDP growth rate (%)
各要素的贡献	Contribution of each factor
资本	Capital
劳动	Labor

development of infrastructure and service industry, etc., since the reform and opening-up, especially after 1990, the increase of TFP contribution was very significant (as shown in Table 34.1). Zhao and Yang (2011) believed that technical advancement was the main cause of the change of total factor productivity since the reform and opening-up, especially the role of system change since 1994 on economic growth promoted the growth of capital and labor to the growth track of total factor productivity, but there are still certain problems, for example, although the research and development expenditures have been growing in each year, the direct effect was only to greatly increase the stock of technical knowledge, rather than effectively transform into the enhancement of total factor productivity.[2]

In China, the enhancement of total factor productivity is correlated with economic transformation, correlated with system construction and correlated with human development. The author believes that it can be assumed in this way: the scope and influencing factors of development cost are also correlated with these aspects, the decrease or increase of development cost may influence the change of total factor productivity. But there is always one fact, in order to examine the effect and result of economic growth, on one hand, we need to use total factor productivity to research internal structural change and optimization, and on the other hand, we also need to use development cost to evaluate from external influence.

---

[2]Zhao and Yang: *Calculation and Interpretation of China's Full Factor Productivity: 1970–2009*, Economic and Financial Issue Research, 9th Issue in 2011.

Chapter 35
Benchmarking of Scientific Achievement View

The so-called benchmarking means finding gap by targeting benchmark, scientific achievement view stresses taking maximization of net welfare as priority, plus stable increase of total factor productivity and reasonable decrease of development cost. The calculation of The Cost of Development in China should be compared with the development cost in other countries, find the countries with relatively high net welfare level, stable growth of total factor productivity and rational development cost to redesign scientific achievement view.

We should consider development as a means of objective-based regulation, integrate "Development Cost" into the full process of planning design, policy formation and supervision management, establish the coordinated mechanism between development cost and economic growth and development, establish its organic connection with political development, social development, cultural development and ecological development, actually only input of cost could create fortune, but the structure of cost input is different and the structure of fortune and welfare are also different. Similarly, this book proposed another perspective, which is also called price, price means loss of opportunity cost, and also a negative externality, these price-type costs are worth to be noted by system builders and reformers. Because they are usually hidden, or have always been ignored, hidden, amortized or accumulated (Table 35.1).

The government is the leader and promoter of system construction and reform, as well as the main bearer of the calculation of "Development Cost". Governments on different levels should actively bear the accounting work of "Development Cost" of the respective regions, provide scientific reference for establishing low-cost development strategy, so as to provide inseparable data basis for the establishment of a systematic and sustainably developing policy system, we should devote ourselves to designing low-cost strategy system in long term, include comprehensive view of development cost into scientific development practice, which is exactly the thing that has been greatly promoted in this book.

Table 35.1 Comparison of current attention and importance of development costs

Type of development cost	Degree of attention	Importance	Type of development cost	Degree of attention	Importance
A1 Cost of economic growth	++++	++++	D1 Cost of cultural transfer	+++	++++
A2 Cost of economic transformation	+++	++++	D2 Cost of protecting ethnic custom	++	++++
A3 Cost of economic disturbance	+++	++++	D3 Cost of ideological evolution	++	+++
A4 Cost of economic regulation	++	++++	D4 Cost of constructing modern media system	++	++++
B1 Cost of political reform	+++	++++	D5 Cost of soft power construction	+++	++++
B2 Cost of political system construction	++++	++++	D6 Cost of change in social civilization	+	++
B3 Cost of political decision making	+++	++++	E1 Cost of opening and developing thinking an ideas	++	+++
B4 Cost of the construction of ruling party itself	++++	++++	E2 Cost of international exchange and consensus formation	++++	++++
B5 Cost of political consultation and cost the crossing information gap	++	+++	E3 Cost of investment promotion and investment environment construction	++++	++++
B6 Cost of democracy construction	++	+++	E4 Cost of international trade	++++	++++
C1 Cost of solving livelihood issues	++++	++++	E5 Cost of construction and development of three-dimensional transport	+++	++++
C2 Cost of population change	+++	++++	E6 Cost of creditability construction and response to threats	++	++++
C3 Cost of social management	+++	++++	F1 Cost of resources and energy	+++	++++
C4 Cost of social stability	+++	++++	F2 Cost of ecological environment	+++	++++
C5 Cost of social advancement	+	++++	F3 Cost of disaster	+++	++++
			F4 Cost of response to climate change	+++	++++
			F5 Cost of harmonious coexistence between human and nature	+++	++++

References

English Part

Chan G (2006) China's compliance in global affairs. World Scientific Publishing Company, New Jersey
Dacy DC, Xunreuther H (1969) Economics of natural disasters: implications for federal policy. Free Press, New York
HIS Global Insight (2010) World industry service database; national science foundation (NSF). Sci Eng Indic
IPCC. Climate Change (2001) The scientific basis. published by IPCC
Kwan Y, Chow C (1996) Estimating economic effects of political movements in China. J Comp Econ 23
Nye JS (1990) The Boston Glob. Foreign Policy Fall (2)
Nicholls RJ et al (2008) Ranking port cities with high exposure and vulnerability to climate extremes: exposure estimates. OECD: OECD environment working papers
World Bank (2006) World development report 2006. Oxford University Press

Chinese Part

Bai X (2005) Traditional common law standardization and ecological protection of minorities. J Qinghai Nationalities Coll (1)
Bai X, Wang Y (2009) Reform and development of Chinese traffic & transportation industry. Economic Management Press, 2009 Edition
Bao J (2006) Natural process and political control of economic transformation—theoretical assumption and a new economic explanation on its economic transformation method and performance difference. Res Syst Econ (1)
Bo Y (1993) Review of several major decision-making and events, Vol B. CCP Central School Press, 1993 Edition
Cai F, Duyang (2003) Destruction of physical capital and human capital by "cultural revolution". Economics 2(4)
Cai G (2002) Discussion on China's legal protection of the freedom of religious belief—observation of Chinese religious legislation from the perspective of international convention of human rights. Soc Sci J Xiangtan Univ (S1)
Cai X (2006) Flaw and modern transformation of Chinese traditional administrative culture. J Tianshui Adm Coll (1)
Cao M (2011) What is the "Troika" driving economic growth. Stat Sci Pract (4)
Capital (1975) Volume 3. People's Publishing House, Edition

CCP Central Committee Archives (1987) Selected Documents of CCP Central Committee, Volume 14. CCP Central Committee Party School Press, 1987 Edition

CCP Central Committee Literature Research Office (1991) Compilation of selected important literature since the party's 13th national congress. People's Publishing House, 1991 Edition

Compilation Group of China's Military History (1986) Chronology of wars in history. People's Liberation Army Press, 1986 Edition

Chang G (2009) China shouldn't have self-inflicted setbacks: China's attitude, global role and no self-inflicted setbacks. Jiuzhou Press, 2009 Edition

Chang J (1991) Discussion on the generating causes of China's existing secondary environmental resource problems. Environ Dev 14(2)

Chang R, Xie W (2011) Empirical analysis of the changes of china's administrative fees and expenditures. Orient Enterp Cult (2)

Chen B (2002) Strategy of economic globalization and Chinese social development. J Anhui Educ Coll

Chen S (2004) Law breaking and crime problems of floating population and thinking of countermeasures. Acad J Popul (5)

Chen J (2009a) China's issue of population scale and structure and related policy adjustment. Popul Dev (2)

Chen L (2009b) Needham problem and the missing of Chinese development opportunities since modern times. J Univ S China (4)

Chen Z (2009c) Analysis of the construction strategy of Chinese media and Chinese soft power. Sci Technol Commun (4)

Chen B (2010a) Influence of new round of information technology revolution wave on China. Sci Decis Mak (11)

Chen D (2010b) Measurement of the cost of Chinese economic transformation: 1978–2008. Res Quant Econ Tech Econ (2)

Chen L (2010c) Observation of Chinese ideological construction from the connotation of ideology. Zhejiang Acad J (1)

Chen Y, Chen L (2007) Causes of government administrative decision-making mistakes and countermeasure suggestions. J Chongqing Univ Ind Commer (Soc Sci Ed) (4)

Chief Editor of "Party Building Research" Magazine: Chinese Communist Party has 80.269 million members and 3.892 million basic-level organizations, Party Building Research, 7th Issue in 2011

Comprehensive Research Group of National Major Natural Disasters of State Science Commission (1994) Chinese major natural disasters and relief countermeasures. Science Press, 1994 Edition

Ci H (1999) Editing Committee: Ci Hai (Middle Volume). Shanghai Cishu Press

Cui B (2007) Chinese media industry development report. Social Science Literature Press, 2007 Edition

Cui J (2008) Historical evolution of the effective room of fiscal and monetary policies and its inspiration—based on the practices of Chinese fiscal and monetary policies, 3rd Issue in 2008

Dai L (2001) Beyond growth—economics of sustainable development. Shanghai Translation Press, 2001 Edition

Dai C (2008) Discussion of the role of foreign Aids in the enhancement of Chinese soft power. Rule Law Soc (9)

Dallmayr (1992) Twilight of subjectivity. Shanghai People's Publishing House, Edition

Deng X (1994) Volume 2. People's Publishing House, 1994 Edition

Deng X (1993) Volume 3. People's Publishing House, 1993 Edition

Deng X (1994a) Volume 1. People's Publishing House, 1994 Edition

Deng X (1994b) Report regarding amendment of party constitution, selected works of Deng Xiaoping, Vol 1. People's Publishing House, 1994 Edition

Ding J (2002) Analysis and countermeasure of China's aging population in the 21st century on sustainable development. Theor Mon (10)

Dong Z (2008) Current status of China's administrative management expenses and its control. Adm Forum (1)

Edited by CCP Central Committee Literature Research Office (1982) Compilation of important literatures since the third plenary meeting (B). People's Publishing House, 1982 Edition

Edited by CCP Central Committee Party History Research Office (1991) Seventy years of Chinese communist party. CCP Party History Press, 1991 Edition

Energy Statistics Department of State Statistics Bureau (2006) China energy statistics yearbook 2006. China Statistics Press, 2006 Edition

Edited by CCP Central Committee Literature Research Office (2004) Chronicle of Deng Xiaoping (1975–1997). CCP Central Committee Literature Press, 2004 Edition

Edited by CCP Central Committee Literature Research Office (2006) Compilation of selected important literature since the party's 16th national congress. CCP Central Committee Literature Press, 2006 Edition

Edited by Zhou T, Wang C, Wang A (2008) Problem solving: research report of China's political system reform. Xinjiang Production and Construction Corps Press, 2008 Edition

Fan Y (2001) Historical observation and realistic analysis of social development cost issue. J Wuhan Univ (4)

Gao F (2005a) Respond to new wave of globalization, develop socialist democracy. J Ningxia Party Sch 3(3)

Gao F (2005b) Investment of transportation infrastructure and economic growth. Chinese Fiscal Economy Press, 2005 Edition

Gao L (2002) Social security level of Shandong Province and its proper selection. Popul Econ (5)

Ge W (2011) Media system with Chinese characteristics: historical evolution and its development and improvement. Chinese Adm Manag (6)

Gong Y (1994) From Mao Zedong to Deng Xiaoping. CPC Party History Press, 1994 Edition

Gu D (2007) In-depth historical significance of Deng Xiaoping's thinking of strengthening comprehensive national power. J Nantong Text Vocat Technol Coll (Compr Ed) (3)

Guan L, Wang X (2008) Discussion on the promotion and popularization of Chinese traditional classic culture. J Soc Sci Jiamusi Univ (4)

Guan T (2011) Chinese countermeasure in currency war: actively respond to the impact of international capital flow. Int Econ Rev (2)

Gui X (2008) Problems and countermeasures faced by China's sustainable use of farmland resource. Theoret Guide (7)

Guo D (1997) Preliminary discussion on the control of green cost. Mon J Account (5)

Ha Z (2009) Brief Review of China's Minority and Religion Policies in Modern Times. J Party Sch CPC Jinan Municipal Committee (2)

Han W, Ying Y (2003) General research statement of the income gap of Chinese residents. Econ Res Ref (83)

He X (2004) Chinese minorities' traditional culture and ecological protection. J Yunnan Nationalities Univ (Philos Soc Sci Ed) (1)

Hou Q (2010) Whether China has optimal span of inflation—empirical research of the relationship between inflation and economic growth. Value Project (2)

Hu A, Liang J (2010) China and the world in post-copenhagen era: trend of energy conservation and emission reduction. Nat Situation Rep (41)

Hu A, Liao J (2011) China's green development strategy and the "Twelfth Five Year" plan, 16th Issue in 2011

Hu A (2002a) Let the nature rest for 50 years: major forestry strategic change from forest deficit to forest surplus. Nat Situation Rep (93)

Hu A (2002b) How China Respond the Increasing Digital Gap. Nat Situation Rep (3)

Hu A (2006) China's path of modernization: review and prospect (1950–2050). National Situation Report, 7th Special Issue in 2006

Hu J (2007) Report on the 17th National Congress of Chinese Communist Party. People's Daily, 25 Oct 2007

Hu A (2008a) China: Livelihood and development. China Economics Press, 2008 Edition

Hu A (2008b) Historical review of China's politics and economy (1949–1976). Tsinghua University Press, 2008 Edition

Hu A (2008c) China's mid-and-long-term population and comprehensive development strategy. Tsinghua University Press, 2008 Edition

Hu A (2009) Practice of green development—taking "forest Chongqing" for an example. Nat Situation Rep (28)

Hu A, et al (2003) Comparison of national defense strength of China, the US, Japan and India. Strategy Manage (6)

Huang G (1993) Nationalities work of modern China, Volume A. Modern China Press, 1993 Edition

Huang T, Cao Z (2011) Wastes caused from decision-making mistakes are the worst. Nanfang Daily, 4 Nov 2011

International Public Relations Research Center of Fudan University (Investigation and Research Group of the Soft Power of Chinese Cities) (2009) Investigation of the soft power of Chinese Cities in 2009. Int Pub Relat (4)

Huang X (2010) Origin and development of "China Threat Theory". Jiangxi Higher Education Press, 2010 Edition

Jiang Z (2006) Volume 2. People's Publishing House, 2006 Edition

Jiang Z (2006) Volume 3. People's Publishing House, 2006 Edition

Jiang Y (2010) Concept, factor and evaluation indicator system of regional soft power. Guihai Rev (3)

Jiang Y, Ye J (2009) Research and review of national soft power. J Wuhan Univ (Philos Soc Sci Ed) (2)

Jin M (2009) Characteristics and handling strategy of China's mass disturbance at current stage. Theore Front (24)

Jing Y (2011) Research on the influence and response strategy of high foreign trade dependence on China's economy. J Anhui Electron Inform Occup Coll (2)

Ju W (2011) Historical observation and inspiration of socialist common prosperity. J Ningxia Party Sch (1)

Kaneko Shiro (1981) Calamity of the world. Shandong Science & Technology Press, Edition

Kong Q (2010) How to differentiate the standard of "Elegance" and "Vulgar". People's Forum, 24th Issue in 2010

Law Research Institute of China Academy of Social Science (2010) China rule of law development report no. 8 (2010). Social Science Literature Press, 2010 Edition

Li C (2006) Why petitioners increasingly concentrate in Beijing. Chinese agriculture, Rural and Farmer Issue, 4th Issue in 2006

Li C (2004) New trend of China's social classification and lifestyle. Sci Soc (1)

Li G (2008) Growth of big nation and building of China's national image. Mod Int Relat (10)

Li H, Li Z (2011) Influence of "hot money" on Chinese economy and response measures. Fujian Finance (4)

Li H (2011) Preliminary discussion of the current situation and countermeasures of China's protection of intangible cultural heritage. Economist (7)

Li H (2008) Responsible power in international society—identity resort and practical construction of modern China. J Univ Int Relat (1)

Li J (1992) Several issues regarding natural resources. J Nat Resour (7)

Li J (2011) Correct the concept of political achievements to accelerate economic development ways, 12th Issue of China's Foreign Trade in 2011

Li J (2005) Discussion on the achievements, problems and countermeasures of the social management by Chinese government. J Hubei Adm Coll (1)

Li S, Zhou L, Qiu H (1997) Legal cost and the construction of china's economic rule of law. Chinese Social Science (4)

Li S, et al (1997) Legal cost and the construction of China's economic rule of law. Chinese Social Science (4)
Li X (2004) Autocracy by first leaders must be renovated. Party Constr (4)
Li Y, Zhang Q, Zhang C (2010) Analysis of the value loss of Chinese oil and gas resources based on user cost method. Fin Econ (6)
Li Y (1986) Dicussion of disaster system and science of disaster, First Issue in 1986
Li Y (2000) Several thoughts regarding expanding investment to drive economic growth. Shandong Soc Sci (1)
Liang Y (2004) A harmonious society towards modernization—interview with professor Wu Zhongmin from central party school. Forum Chin Party Pol Officials (11)
Liang L (2007) Environment—economic-oriented analysis of enterprise environment income and cost. Financ Account Commun (6)
Liang M (2005) Will Chinese people's income gap be further widened? Chin Bus J
Liang W (2011) Dilemma of stabilization and lack of citizen society. Lingnan J (3)
Liao X (2003) Discussion of reform and improvement of government decision-making mechanism. J Chengdu Adm Coll (4)
Li B, He X (2008) Cost of depletion of Chinese oil and gas resources and the macroeconomic influence of policy selection. Econ Res (5)
Lin B, Wei W, Li P (2007) China's long-term coal demand: influence and policy selection. Econ Res (2)
Lin J (1993) Literati and officialdom during Ming and Qing dynasty and dispute of etiquette between China and the West. Hist Res (2)
Lin Z, et al (2003) Model B—saving the earth and sustaining the civilization. Oriental Press, 2003 Edition
Liu M (2004) Century of China: loss and reconstruction of cultural tradition. J Sch Lit Nanjing Norm Univ (1)
Liu M, Li S (2011) Change of China's total factor productivity since reform and opening-up and growth trend in the future. Econ Res Ref (33)
Liu S (1981) Report regarding amendment of party constitution, May 1945. Volume A of the Selected Works of Liu Shaoqi, People's Publishing House, 1981 Edition
Liu S (2004) Several issues regarding the outlook of scientific development. J Inner Mongolian Coll Fin Econ (6)
Liu X (2006) Operating mechanism theory of interest rates. Fujian People's Publishing House, 2006 Edition
Liu Z (2007) Discussion regarding resource cost awareness theory. Chin Econ Land Resour (5)
Liu Z (2006) China's grain security and farmland protection. Sci Fin Econ (7)
Long X, Zhang Y (2010) Observation of the construction of China's image as responsible major power from Aids to Haiti. Foreign Commun
Lu F (2003) After enlightenment. Hunan University Press, 2003 Edition
Lu X, Zhang N (2005) Analysis of the current situation and problems of Chinese government's social management. Acad Southeast (4)
Lu X (2010) Construction of cultural soft power and protection of China's ideological safety. J Shandong Univ (Philos Soc Sci Ed)
Lu Z (2001) Brief discussion of the influence of aging population on the adjustment of China's industrial structure. J Shenzhen Univ (2)
Lv W (2009) Chinese-style transition: inherent characteristics, evolving logics and prospect—in memorial of the 30th anniversary of China's reform and opening-up. Res Fin Econ Issues (3)
Luan Y (2011) New media and the construction and communication of China's soft power. Res Cult Art (2)
Luo G, et al (1997) Green marketing. Economic Science Press, 1997 Edition
Luo N, Guo G, Xie L (2010) Evaluation and research of China's regional cultural soft power. Econ Geogr (9)

Luo S, Xu J (1999) Thinking of adjustment of China's nationalities policies at transition of centuries. J Guangxi National Coll (Philos Soc Sci Ed) (2)

Luo Z, Xu C (1998) Science of disasters. Zhejiang Education Press, 1998 Edition

Ma H (1998) Critical biography of Kang Youwei. Nanjing University Press, 1998 Edition

Ma J (2007) Theoretical analysis of the internalization of the external cost of environment—theoretical framework of pollution emission reduction and policy suggestions. Qinghai Environ 17(3)

Ma Zo (1998) Guide of the science of disasters. Hunan People's Publishing House, 1998 Edition

Ma Z (2008) Countermeasures for China's natural disasters and disaster relief—enhance disaster efficiency, using scientific ideas to guide various aspects of disaster relief work. J Disaster Prev Sci Technol Coll 10(1)

Mao Z (1955) Speech at the national congress of Chinese communist party, March, 1955. Volume 6 of the Selected Works of Mao Zedong, People's Publishing House, 1999 Edition

Mao Z (1999) Volume 6. People's Publishing House, 1999 Edition

Marx K, Engels F (1964) Volume 16. People's Publishing House, Edition

Marx K, Engels F (1995) Volume 1. People's Publishing House, Edition

Marx K, Engels F (2009) Volume 9. People's Publishing House, Edition

Meng L, Shang K (2007) Comparison analysis of oil tax systems in China and foreign countries and suggestions for policy reform. Taxation Econ (6)

Mi J (2011) Causes of China's slowing economic growth rate and discussion of countermeasures. Bus Era (20)

National Science Quality Guideline Implementing Work Office (2010) Annual report of national scientific quality action plan guideline. Science Popularization Press, 2010 Edition

Nye J (2005) Hard power and soft power, translated by Men Honghua. Peking University Press, 2005 Edition

Pan J, Zheng Y (2010) Analysis framework and policy connotation of adaptation to climate change. Chin Popul Resour Environ (10)

Pan J (2007) Disputes of economics caused from climate change. Ecol Civilization Theor (2)

Pan M (2006) Issues to be solved in implementation of strict farmland protection system. Theore Vis (4)

Pang S (2003) China in the Era of Comrade Mao Zedong (1949–1976) (I). CPC Party History Press, 2003 Edition

Pei C, Gao P (2008) Export tax rebates and China's foreign trade. Social Science Literature Press, 2008 Edition

Pei S (2005) Schumpeter's innovation theory and new development outlook. J Yancheng Ind Coll (Soc Sci Ed) (3)

Peng J (2009) Research on the development of cultural industry and ideological changes during 30 years of reform and opening-up. Soc Sci J Hunan Norm Univ (1)

Pu S (2005) Treating public security as basic national policy. Guid Chin Soc (4)

Qiao H, Wan Y (2007) Research and review of natural disaster economic theories. Dyn Econ (9)

Qing N (2011) Several thoughts about the causes of China's inflation. Sci Manage (3)

Ren L (1995) Several philosophical thoughts about the nature of disasters. Orient Forum (4)

Shen P (2011) Observation of traffic resource optimization from Beijing-Shanghai high speed rail. Compr Transp (7)

Shen R (2003) Government mechanism. National Administrative College Press, 2003 Edition

Shi X, Yang D (2009) Building of harmonious society needs social capital construction. J Beijing Adm Coll (5)

Shi Z (2000) Research of foreign comprehensive national strength theory. Foreign Econ Manage (1)

Shi P, Li C, Zou M, Le J (2008) Build comprehensive preventive and aiding system to handle the risk of major disasters. Financ Account Res (10)

Shi P (1991) Disaster and science of disaster. Geogr Knowl (1)

Shi H (1996) Strictly controlling population growth is an important guarantee to realize the goal of reducing the gap with the eastern region. Econ Reform (6)

Song J (2008) Discussion of the connotation and basic forms of political negotiation. J Shayang Norm Junior Coll (1)

Spence M (2008) Successful experiences and new challenges. Overseas Chinese Research, 8th Issue in 2008

State Statistics Bureau (1993) China statistics yearbook 1993. China Statistics Press, 1993 Edition

State Statistics Bureau (1999) Compilation of statistic materials of new China for fifty years. China Statistics Press, 1999 Edition

State Statistics Bureau (2011) China development report 2011. China Statistics Press, 2011 Edition

Sun F, Wei X (2006) Geographic research of Chinese crimes. J Liaoning Norm Univ (Nat Sci Ed)

Sun X (2005) Preliminary analysis of the origin and challenges of strengthening social construction and management. J CPC Fujian Provincial Committee Party Sch (12)

Sun Y (2003) Brief analysis of risks in Urban operation. Jiangsu Constr (1)

Sun Z (1981) People's Publishing House, 1981 Edition

Tang J (2004) Causes of China's current inflation pressure and countermeasures of relief. Young Thinkers (3)

Theory Bureau of CCP Central Committee Propaganda Ministry (2010) Seven how to look at it. Studying Press and People's Publishing House, 2010 Edition

UNCTAD (2011) World investment report. Economic Management Press, Edition

Wang S (1998) China: facing the challenge of inequality and response. Nat Situation Rep (34)

Wang Y (1999) Discussion on the defense countermeasures of developing countries facing external impacts. J Qinghai Norm Univ (Philos Soc Sci Ed) (4)

Wang L (2000) Historical evolution of the party's religious policy development since the third plenary session of the 11th central committee of the Chinese communist party. J Guangzhou Norm Coll (Soc Sci Ed) (12)

Wang S (2005) Model C: self-disciplined development. Nat Situation Rep (31)

Wang H (2006) Preliminary discussion of 'Chinese Threat Theories'. Western Asia and Africa, Aug 2006

Wang X (2007) Mass media and ideological communication—taking the example of the influence of neo-liberalism on China. Soc Sci (1)

Wang X (2008) Discussion on the current situation of China's foreign exchange reserve. Chin Bus Circle (4)

Wang B (2010a) Tell "Chinese Story" to the world. World Knowl (1)

Wang M (2010b) Preliminary discussion of the transformation from rigid stabilization to flexible stabilization. J Party Political Officials (4)

Wang X (2010c) Grey income and distribution of national income. Comparison, 48th Issue, CITIC Press, 2010 Edition

Wang Z (2010d) Preliminary discussion of environmental cost and its measurement. Financ Account Res (24)

Wang J (2011a) Discussion on the meaning and characteristics of natural resources. J Dezhou Coll 7(2)

Wang X (2011b) Discussion on the exploration by Mao Zedong on the development model of the world of Universal Harmony. Res Marxism (2)

Wang M, et al (1996) Inspection and the concentration and emission of greenhouse gas and related process. China Environment Science Press, 1996 Edition

Wang Y, Zhang Y (2010) Soft power construction under the vision of China's peaceful development strategy. J Shaanxi Adm Coll (2)

Wang R, Sun C (1996) Fork of black and green. Shanxi Economic Press, 1996 Edition

World Bank (2008) World development indicator (Chinese Edition). China Financing and Economic Press, Edition

World Bank (2010) World development report—development and climate change. Tsinghua University Press, 2010 Edition

Wu S (2007) How enterprises enhance the financial fine management capability. Econ Trade Time (11)
Wu Z (2008) Basic definition and characteristics of livelihood. Forum Chin Party Polit Officials (5)
Xiang S (1998) Calculation and regression analysis of the GINI coefficient of national resident income distribution. Financ Econ Theor Pract (1)
Xiao Z, Zhou S (1998) Chinese population and sustainable development. Chinese Population Press, 1998 Edition
Xie C (2002) 50 years of cultural relics protection works in New China. Res Mod Chin Hist (3)
Xie X, Mo H (2008) Flaws and inspirations of ancient Chinese thinking and cultural policies. J Fujian Provincial Socialist Coll (1)
Xiong Y (2008) Defining of intangible cultural heritage. J Chin Univ Geo-sci (Soc Sci Ed) (5)
Xu B (2011) Comparison research of the economic transformation and its cost in China and Russia. Siberia Res 38(2)
Xu G (2007) Preliminary discussion of the input of China's national defense budget. Econ Horiz (6)
Xu J (2006) Disaster relief input and cost issue in the economics of disasters. Sci Disaster 21(2)
Xu Q (2005) Export expansion and economic growth. Northern Econ Trades (7)
Xu S, et al (1998) Concept of disaster's economic loss and measurement of industry-related indirect economic loss. J Nat Disasters (7)
Xue J (2008) Theoretical guidelines of ruling party construction of the three generation of central leadership groups. CPC Central Party School Press, 2008 Edition
Yan X, Wu B (2010) Economic interpretation of the characteristics of China's economic growth since reform and opening-up. J Xiaogan Coll 30(2)
Yang X (1995) Characteristics and cause types of environmental disasters and disaster relief countermeasures. Shandong Environ 1995
Yang L (2004) From traditional development outlook to scientific development outlook: modern transition of development outlook. J Yan'an Univ (Soc Sci Ed) (5)
Yang M (1995) Economic benefit and resource price. Res Quant Econ Tech Econ (7)
Yang Z (2000) Aging and industrial structure adjustment. Guangxi People's Publishing House, 2000 Edition
Ying H (2001) Changes of China's economic growth and labor market since reform and opening-up and prospect of the future. Popul J (5)
Ying W, Zhang H (2009) Scale calculation of escaped capital from China (1985–2008). Financ Dev Res (8)
Yu H, et al (2010) Empirical analysis of the regional economic difference of shandong province based on GINI coefficient decomposition. J Ludong Univ
Yu J (2008) Causes and countermeasures of the gender unbalance of newborn population in China. Acad Exch (1)
Yu X (2010) Scenario selection of China's participation in the long-term 2000–2050 carbon dioxide emission reduction. Progr Clim Chang Res (1)
Yu L (2010) Causes and controlling measures for the growth of China's administrative management expenses. Mod Econ Inform (9)
Yu P (2006) Chinese model and "Beijing Consensus"; beyond "Washington Consensus". Social Science Literature Press, 2006 Edition
Zeng G (2001) Discussion on the characteristics, causes and countermeasures of Chinese residents' income gap. J Chin Univ Geo-sci (Soc Sci Ed) (4)
Zeng W, Cheng S (2000) Guiding theory of environmental disaster science. China Environment Science Press, 2000 Edition
Zhai Z (2007) Chinese population development: new challenges and choices. Theore Vis (9)
Zhai Z (2001) Analysis of Chinese population scale and age structure conflict. Popul Res (3)
Zhang C (2010) Research on the latest development trend, causes and countermeasures of China's mass disturbance events. Shandong Soc Sci (5)

Zhang D (2004) Establishment of price awareness in foreign economic relations. Econ Issue Explor (11)

Zhang J (2009) Causes and prospect of China's high economic growth rate since the reform and opening-up. Econ Horiz (3)

Zhang J (2011) High cost dilemma of "rights protection" and "stabilization". Theory Reform (3)

Zhang L (2004) Discussion on "economic net welfare growth": redefining of "fourth industry" and economic growth. Qinghai Soc Sci (5)

Zhang M (1994) China's "political person". China Social Science Press, 1994 Edition

Zhang P (2008) China's economic growth and structural reform for 30 years of reform and opening-up. Mod Econ Discuss (7)

Zhang Q, Yang M (2004) Farmland protection and sustainable development. J Zhoukou Norm Coll (5)

Zhang X, Hu H, Zhang J (2009) China's cultural industry development report in 2009. Social Science Literature Press, 2009 Edition

Zhang X (2004) Moral issue of new idea of moral and profit and market economy—analysis of the moral construction of China in social transformation period. J East Chin Jiaotong Univ (3)

Zhao L (2007) Three dimensions to understand China's soft power: cultural diplomacy, multilateral diplomacy and foreign aids policy. Soc Sci Forum (Vol Acad Rev) (5)

Zhang X (2008) Comparison research of Kang Youwei's World of Universal Harmony and socialist harmonious society. Knowl Econ (2)

Zhao X (2009) General establishment of the development path of regional cultural soft power. J Henan Norm Univ (Philos Soc Sci Ed) (2)

Zhao Q, et al (2006) Change of China's farmland resources and its sustainable use and protection countermeasures. J Soil (4)

Zhao Z, Yang Z (2011) Calculation and interpretation of China's full factor productivity: 1970–2009. Econ Financ Issue Res (9)

Zheng H, Hong D (2004) Potential social security hazards and countermeasures in China's transformation period. J Chin Renmin Univ (2)

Zheng J (2002) Impact of "Silver Wave" on China's endowment insurance system and countermeasures. Stat Res (1)

Zheng R (2003) Preliminary discussion on media's function of supervision of public opinions. J Fujian Party Sch (4)

Zheng W (2004) Adhere to and improve multiple party cooperation and political consultation system under the leadership of Chinese communist party, consolidate and develop broad patriotic united front, refer to the editing group of CCP central committee's decision regarding strengthening the party's ruling capacity, guiding book of CCP central committee's decision regarding strengthening the party's ruling capacity. People's Publishing House, 2004 Edition

Zhou X (1999) Thoughts regarding reducing economic legislation cost and enhancing economic legislation quality. J Jinan (Philos Soc Sci) (5)

Zou H (2011) Development is the absolute principle and stability is the absolute task. Mass Mag (9)

Zheng K, Li Z (2002) Research of the important thought of "three represents". Sichuan People's Publishing House, 2002 Edition

Lightning Source UK Ltd.
Milton Keynes UK
UKHW02n2145240418
321576UK00003B/215/P